Vascular Mechanics and Pathology

Mano J. Thubrikar

Vascular Mechanics and Pathology

Mano J. Thubrikar, Ph.D., FAHA
Formerly, Research Associate Professor of Surgery,
University of Virginia Health Sciences Center, Charlottesville, VA

Associate Director, Heineman Medical Research Center,
Carolinas Medical Center, Charlotte, NC

Adjunct Professor, Dept. of Mechanical Engineering,
University of North Carolina at Charlotte, NC

Currently, Edwards Distinguished Research Scientist,
Edwards Lifesciences, Irvine, CA

Additional material to this book can be downloaded from http://extras.springer.com

ISBN-10: 1-489-98924-6
ISBN-13: 978-1-4899-8924-6
ISBN-10: 0-387-68234-1 (eBook)
ISBN-13: 978-0-387-68234-1 (eBook)

Printed on acid-free paper.

9 8 7 6 5 4 3 2 1

springer.com

Dedicated to my family—
*wife **Sudha**, daughter **Vaishalee**, son **Vishal**,*
*father **Jumdeoji**, and mother **Varanasi***

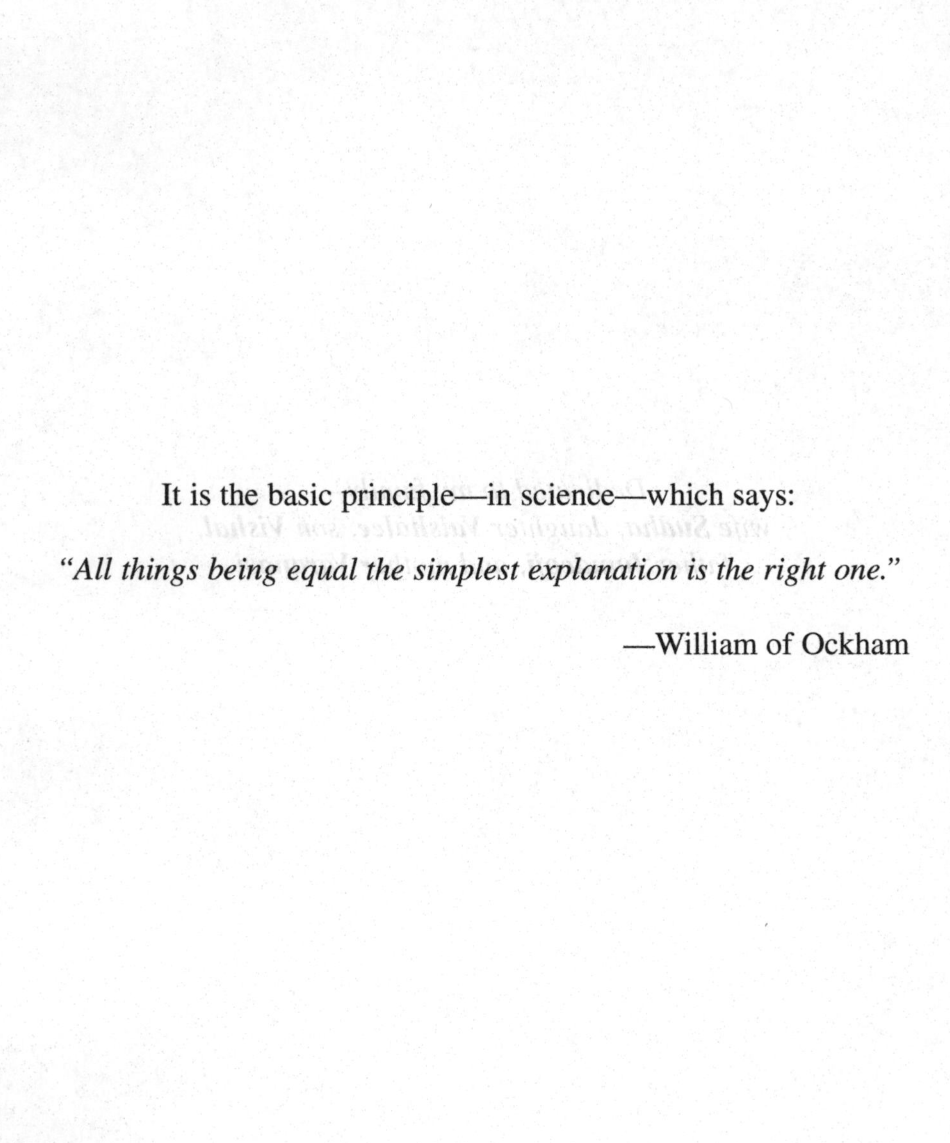

It is the basic principle—in science—which says:

"All things being equal the simplest explanation is the right one."

—William of Ockham

Preface

Nowhere in the scientific progress has the schism in the knowledge been as striking as in the case of vascular mechanics and pathology. This joint subject would serve as a classic example of science developed in two different directions. It provided the motivation to put forth this book and establish a correlation between vascular mechanics and pathology. The book focuses on the artery and arterial diseases. The most fundamental functions of the artery are (1) to serve as a conduit of blood flow and (2) to serve as a container of blood pressure. The artery carries the blood to all organs of the body and it uses pressure to drive the blood through the tissue to provide nourishment. Hence, the artery is both a pipe and a pressure vessel. The artery pulsates about 103,000 times a day along with the beating heart. In a lifetime, the artery sustains cyclic pressure for about 3.8 billion cycles. This obviously poses a significant challenge to the artery and therefore the artery must be endowed with special structure and properties to meet this challenge. In the event that additional challenges are imposed, such as high blood pressure, it would not be surprising that the artery could "break down" or become diseased. In the book, we examine the structure and properties of the artery and study the challenges imposed on it with a view to understand the survival of and the development of the diseases in the artery.

Two separate bodies of knowledge have developed in great detail, which are of interest to us. They are (1) arterial diseases, particularly atherosclerosis and aneurysm and (2) engineering analysis of pressure vessels. In arterial diseases, the most common is atherosclerotic plaque, which predominantly forms at branches, bifurcations, and curvatures. The second most common is the aneurysm, which, among other locations, also forms at the branches of cerebral vessels. Other situations concerning the artery are hypertension, use of beta-blockers to reduce the heart rate, exercise which increases and then reduces heart rate, aortic dissection, orientation of cells, and balloon angioplasty. Development of hyperplasia or aneurysm at the arterial anastomosis is also an interesting example. Why do we use veins from the lower leg for coronary artery bypass grafts is not obvious to many. These are some of the topics that need explanations and the book addresses them in terms of vascular mechanics.

The other body of information exists in the field of pressure vessel engineering. For mechanical engineers pressure vessels are the means by which outer space and ocean depths are reached, nuclear power harnessed, energy systems controlled, and petroleum processes operated. The demand for high pressure and large diameter vessels has been accompanied also by a demand for weight reduction and structure fabrication, posing a challenge for design engineers. In the body, nature has taken on a challenge for making the most efficient structures that will achieve storage, compliance, and flexibility all at once. In pressure vessels, it is recognized that the vessels fail from fatigue in locations where stress concentration occurs. The stress concentration occurs in localized regions of vessel intersections, bifurcations, curvatures, and holes. The pressure vessels are reinforced by such means as increased thickness and use of stronger materials.

For the mechanical pressure vessels, the design engineers can consider pressure as the direct cause of rupture or failure, but for the artery, a living biological material, pressure alone can seldom be considered a cause of the disease. Cellular, biochemical, genetic, and other biological processes will be intimately involved, and a mechanical entity, such as pressure, may have its consequences either directly or through these pathways. For physicians, high blood pressure in patients sends alarming signals that put all on alert, from blowout of an aneurysm in the aorta, from blowout of an aorta due to dissecting aneurysm, from blowout of a berry aneurysm in the cranium, from enhancement of atherosclerotic disease, exhaustion of the heart working harder against the pressure, and alike.

When one acquires the knowledge of these two fields, the similarity between the failure of pressure vessels and the occurrence of arterial diseases is inescapable and this book serves to point that out. It is known that the heart muscle becomes thicker in aortic stenosis and the heart chamber volume becomes larger in aortic insufficiency; both conditions are indicative of a cause and an adaptive response. It is also known that arteries become thicker in hypertension. These correlative relationships, however, have not been extended previously to the most important and pervasive diseases of the artery. The book explains that the parameters that determine the mechanical failure of the pressure vessels are also the ones that matter the most in the "failure" of the blood vessels and that the failure appears in the form of atherosclerosis or aneurysm.

The book begins with the description of previously reported mechanisms of atherosclerosis and aneurysm, and points out why these mechanisms appear incomplete. It then describes the structure of the artery and how it is suited to the dual function of the artery. For the first time, it describes the structure of the arterial branch. The principles of pressure vessels are then described and applied to the coronary arteries, carotid bifurcations, aortic bifurcations, aortic arch, etc. Both the general principles and the occurrence of stress concentration at the pressure vessel junctions are described. The orientation of endothelial cells and smooth muscle cells, and higher permeability of specific regions of the artery to low density lipoprotein are described. The proliferation of cells in balloon angioplasty as a response to strain is described. The reduction of stress and the role of beta-blockers in the reduction of atherosclerosis and/or related complications are described. The principles of reinforcements and reduction of diseases are applied to intramyocardial coronary arteries and vertebral arteries. The book describes the use of veins as the arterial grafts and the role of vein valve in graft stenosis.

The principles of stress concentration are further illustrated in case of anastomosis. The development of intimal hyperplasia at the anastomosis and the development of anastomotic aneurysms are covered once again from the point of view of vascular mechanics. The intracranial aneurysms and the aortic aneurysms are described with emphasis on the stress in the aortic wall. Aneurysm formation, growth, and rupture are described using pressure vessel principles. Finally, the aortic dissection is described and, for the first time, it is shown that the aortic root mechanics plays a very important role in the development of this pathology. Both pulling and twisting motions of the aortic root are analyzed for their role in the aortic dissection.

Overall, the two fields of science are brought together to enrich our understanding of the role of vascular mechanics in pathology. A broad range of subjects covered in the book provides one of the most comprehensive treatments of vascular pathology in a single document and makes this document four books in one. It enhances our knowledge of both engineering and medicine by pointing out the important link between them in the area of vascular pathology. It also promotes understanding of a common phenomenon in such varied subjects as atherosclerosis, aneurysms, pressure vessel, stress concentration, vein grafts, anastomosis, cell proliferation, and beta-blocker treatment, all associated with arterial diseases. It points to newer applications of engineering principles in medicine thereby opening new avenues for experimental research in both the fields. Usually, it is the philosophers' forte to try to understand "all things in terms of a single principle" but serendipitously, in this book, it is "the single principle" that has been brought to the forefront of multiple vascular diseases.

The book will be useful to cardiovascular surgeons, cardiologists, pathologists, radiologists, neurosurgeons, anatomists, and manufacturers of medical devices. It will also be useful to students in medicine and in biomedical engineering and to the researchers in various disciplines. It can serve as "the book" for a course on "vascular mechanics and pathology" in bioengineering. It is likely to open new doors for interdisciplinary research, which could lead to the reduction of vascular diseases and to the development of new treatments, thereby benefiting large number of patients.

Irvine, California, USA Mano J. Thubrikar, Ph.D., FAHA

Acknowledgments

Writing this book was a big task and a significant challenge because it is based on almost twenty years of laboratory research while I was at the Department of Surgery, University of Virginia, Charlottesville, Virginia, and at the Heineman Medical Research Center, at Carolinas Medical Center, Charlotte, North Carolina. The work included in the book came from the following collaborations:

At the University of Virginia, in the Health Sciences Center, collaborations with the Departments of Anatomy and Cell Biology, Biochemistry, Physiology, and Pharmacology; in the School of Engineering, collaborations with the Departments of Civil Engineering and Applied Mechanics, and Biomedical Engineering; collaborations with AB Hassle Pharmaceutical Company, Gothenberg, Sweden; at the University of North Carolina at Charlotte, collaborations with the Department of Mechanical Engineering; and at the Carolinas Medical Center, Charlotte, collaborations with Sanger Surgery. This comprehensive book would not have been possible without such broad collaborations and indeed because of that it has far-reaching implications.

There are special reasons to acknowledge certain people without whom the work, and therefore the book, would not have been possible. They are Dr. Stanton P. Nolan, Dr. J. David Deck, Dr. Richard T. Eppink, and Dr. Bengt Ablad. The work and the book are primarily based on the principles of mechanics applied to the arterial pathology, particularly atherosclerosis and aneurysm. The work was begun at a time when there was no support for such an approach and, in fact, there was resistance to pursuing these studies. At such a time one person, **Dr. Stanton P. Nolan** from the Department of Surgery, University of Virginia, was instrumental in supporting, encouraging, and participating in this approach. My sincere and special thanks go to him for being the enabler. Drs. Deck and Eppink soon joined, and the team was formed. **Dr. Bengt Ablad** (AB Hassle Pharmaceutical Co.) was also a key person for financially supporting the projects for several years as he saw the value of this approach in understanding the actions of beta-blockers.

Subsequently, **Dr. Francis Robicsek** of the Carolinas Medical Center supported and participated in this work, thereby enhancing it significantly. My very special thanks go to him. There are others who were quite instrumental in encouraging this work. They are Dr. Peter Holloway (UVA), Dr. D. Kenyon, Dr. Russell Ross (editor of Arteriosclerosis), Dr. Seymore Glagov (University of Chicago), Dr. Gardner McMillan (NIH), Dr. Y.C. Fung (UC San Diego) and Dr. Thomas Clarkson (Wake Forest University). These individuals played a determining role in my pursuit of this approach and I am truly indebted to them.

There were several individuals who participated in the experiments described in the book. I thank all of them. Among them Dr. Radha Moorthy, Dr. Michel Labrosse, and Brett Fowler contributed significantly.

Typing of the entire manuscript was done by Brigitte Dorn, and I cannot thank her enough for her tremendous efforts. My thanks also go to Dr. Chun-An Lin and Harold Rice for their invaluable help with the figures.

Lastly, and most importantly, my thanks and gratitude go to my wife, Sudha Thubrikar, whose patience, understanding, and support for five years, during the writing of this book, was exemplary and humbling. The book would not have been possible without her help.

Yorba Linda, California, USA Mano J. Thubrikar

Contents

About the Author

Mano J. Thubrikar, Ph.D., FAHA

Dr. Thubrikar obtained his Bachelor of Engineering degree (First in the Order of Merit) in Metallurgy and Materials Science from Visvesvaraya Regional College of Engineering, Nagpur University, India. From New York University, New York, he obtained his M.S. in the same field and his Ph.D. in Biomedical Engineering.

He served as a Research Associate Professor and a Director of Surgical Research in the Department of Surgery at the University of Virginia Health Sciences Center, Charlottesville, VA. He was the Associate Director at Heineman Medical Research Center and also the Director of the Biomedical Engineering Program at Carolinas Medical Center, Charlotte, NC. He was an Adjunct Professor in the Department of Mechanical Engineering, University of North Carolina at Charlotte, NC. Currently, he is an Edwards Distinguished Research Scientist at Edwards Lifesciences, Irvine, CA.

He is a recipient of the Research Career Development Award from the National Institutes of Health, the Certificate of Merit from the New York Academy of Medicine, and the Minna-James Heineman Stiftung Research Award for Outstanding Research Achievements in the Area of Lifesciences. He has been the recipient of research grants from the National Institutes of Health, American Heart Association, and various private industries. He is an elected Fellow (FAHA) of the American Heart Association, the Council on Cardio-Thoracic and Vascular Surgery, and (FAIME) of the American Institute for Medical and Biological Engineering.

He is also the author of the book *The Aortic Valve*, CRC Press, Boca Raton, Florida, 1990. He has published over 80 full-length papers, over 150 abstracts, and written chapters for eight books. He has been an invited speaker at several international symposiums. He is the North American Editor of the Journal of Medical Engineering & Technology, Taylor and Francis, UK. He has served on a number of Scientific Review Boards at NIH, and has been a reviewer for several journals including *J. of Thoracic and Cardiovascular Surgery, The Annals of Thoracic Surgery, American J. of Cardiology, J. of Heart Valve Diseases, and J. of Biomechanics*.

Dr. Thubrikar has also been the thesis advisor to several graduate students in their M.S. and Ph.D. degree programs in both the School of Engineering and the School of Medicine. He is a career scientist, and his research has contributed immensely to our understanding of the Aortic Valve, Bioprostheses, Atherosclerosis, Aneurysm, Vascular Grafts, and Biomechanics.

1
Atherosclerosis I

1. Introduction

Atherosclerosis is a disease in which deposits of yellowish plaques containing fatty substances are formed within the intima and inner media of large- and medium-sized arteries (Figs. 1.1 and 1.2). The plaque causes hardening of the arteries, reduces flow through the arteries, and it can also completely block the

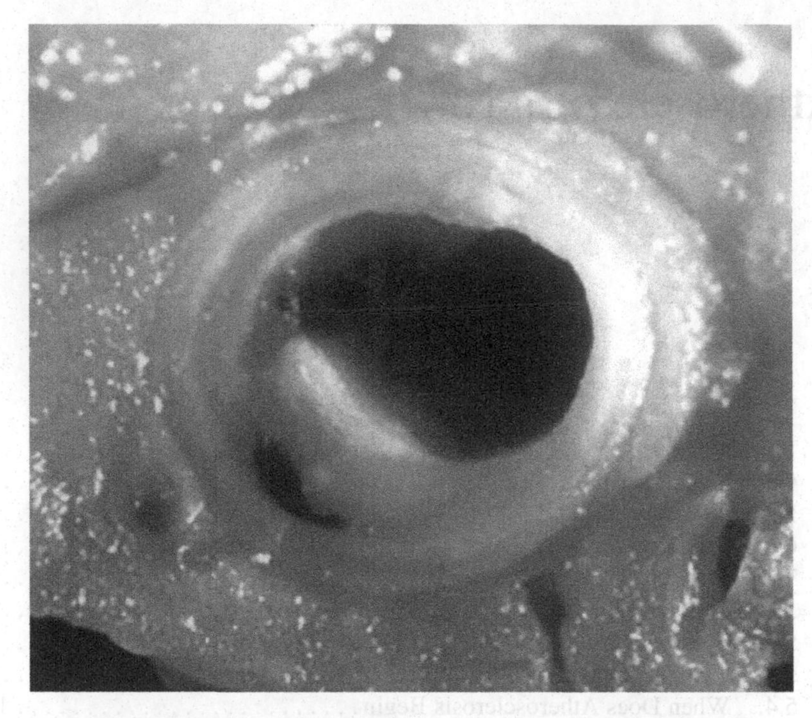

FIGURE 1.1. Atherosclerotic plaque in the artery. (*Please see color version on CD-ROM.*)

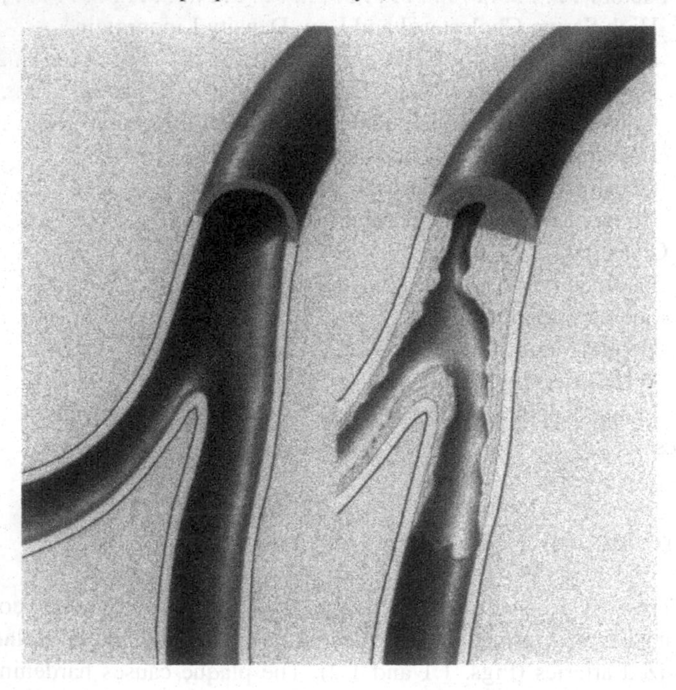

FIGURE 1.2. An illustration of a normal artery and the same artery narrowed by atherosclerotic plaque.

blood flow. It is a major cause of ischemic heart disease, and its complications include stroke and gangrene of extremities. It accounts for more than half of the annual mortality in the United States. As we expect, not only has a vast amount of material been written on the subject, but also a large number of articles continues to be written at the present time. Therefore, the description of the subject in the chapters here is by no means comprehensive. Nonetheless, the description is provided to the fullest extent as deemed necessary from the point of view of vascular mechanics.

2. Geographic Variations

The number one cause of death in America is *cardiovascular disease* (CVD). About 41.2% of all deaths are caused by CVD. Cancer is the number two cause of death. More than 70 million (M) Americans have one or more forms of CVD. More than 63 M have high blood pressure, more than 6 M coronary heart disease, more than 3 M stroke, and more than 1 M rheumatic heart disease. The probability of *heart attack* within 8 years, according to risk factors as determined in the Framingham Heart Study, is shown in Table 1.1 (1). For the purpose of illustration, the study used an abnormal blood pressure level of 150 mmHg systolic and a cholesterol level of 260 mg/dL in both a 55-year-old male and female. The average probability of heart attack in 8 years is 47/1000 in men and 8/1000 in women. As shown in Table 1.1, and it is common knowledge now, the risk of heart attack increases with the number of risk factors. Most scientists agree that high cholesterol is a risk factor, however, of 1.1 M heart attacks that occur each year in the United States, half of them have normal or even low cholesterol. Table 1.2 shows the distribution of the older population according to cholesterol levels in the United States (2). It is important to note that 35% of the males and 52% of the females above the age of 55 have a serum cholesterol level of more than 240 mg/dL as per a 1993 report (2). This is a higher percentage of the population than

TABLE 1.1. Probability of heart attack within 8 years by risk factors (U.S.) for a 55-year-old person with 150 mmHg systolic pressure and 260 mg/dL cholesterol level.

Number of people at risk per 1000		
Male	Female	Risk factors
31	5	None
46	9	1 risk factor: cigarettes
64	14	2 risk factors: cigarettes + cholesterol
95	23	3 risk factors: cigarettes + cholesterol + high blood pressure

TABLE 1.2. Total serum cholesterol levels (mg/dL) in percentage of people in age group 55–74 years in the United States (1993 report).

Cholesterol (mg/dL)	Male (%)	Female (%)
<200	29	15
200–239	36	33
≥240	35	52

TABLE 1.3. Distribution of risk factors in patients with known ischemic heart disease.

	India (%)	Singapore (%)
Hypertension	37	36
Diabetes mellitus	28	18.5
High lipids	14	6.3
Obesity	14	–
Unknown	7	–
Smoker	–	34
Familial history	–	4.8

that in Eastern countries. Therefore, in the United States, emphasis has been on the role of cholesterol in atherosclerosis.

In India, the distribution of risk factors in patients with known ischemic heart disease is shown in Table 1.3 (3). In that population, hypertension and diabetes are the most important risk factors for coronary artery disease, while serum cholesterol of about 250 mg/dL was present only in 14% of patients. In Singapore, the distribution of risk factors in patients who had coronary artery surgery is also shown in Table 1.3 (4). According to their data, the main risk factors for coronary artery disease are hypertension, smoking, and diabetes. A high cholesterol level was present in only 6.3% of the patients. Naturally, the research in these Eastern countries will focus on hypertension, diabetes, and smoking rather than on cholesterol.

These statistics have a profound effect on how we think about pathogenesis of atherosclerosis. The processes like hypertension, diabetes, high cholesterol, smoking, obesity, and so forth, are involved in aggravating the disease; however, the fundamental processes in pathogenesis of atherosclerosis should be independent of these factors. In other words, these factors are aggravating (risk) factors and may not be initiating factors in atherosclerosis.

3. Natural History

Atherosclerosis begins as intimal lipid deposits (fatty streaks) in childhood and adolescence (Fig. 1.3) (5). Fatty streaks in some arterial sites are converted into fibrous plaques by continued accumulation of lipid, smooth muscle, and connective tissue. In middle age, fibrous plaques undergo a variety of changes, some of which produce occlusion, ischemia, and clinical disease. Typically, clinical disease occurs 30 or more years after the process begins as fatty streaks, but it may be accelerated in persons with hypercholesterolemia. Only a few of the fatty streaks ever convert into atherosclerotic lesions, whereas others remain harmless. Also, atherosclerosis is considered by and large an old-age disease because a vast majority of clinical events, originating from atherosclerosis, occur after the age of 50. Cardiovascular disease causes 70% of all deaths beyond age 75. Beyond age 65, women become as vulnerable to cardiovascular mortality as men. The risk factors for coronary heart disease are similar in young and old and in men and

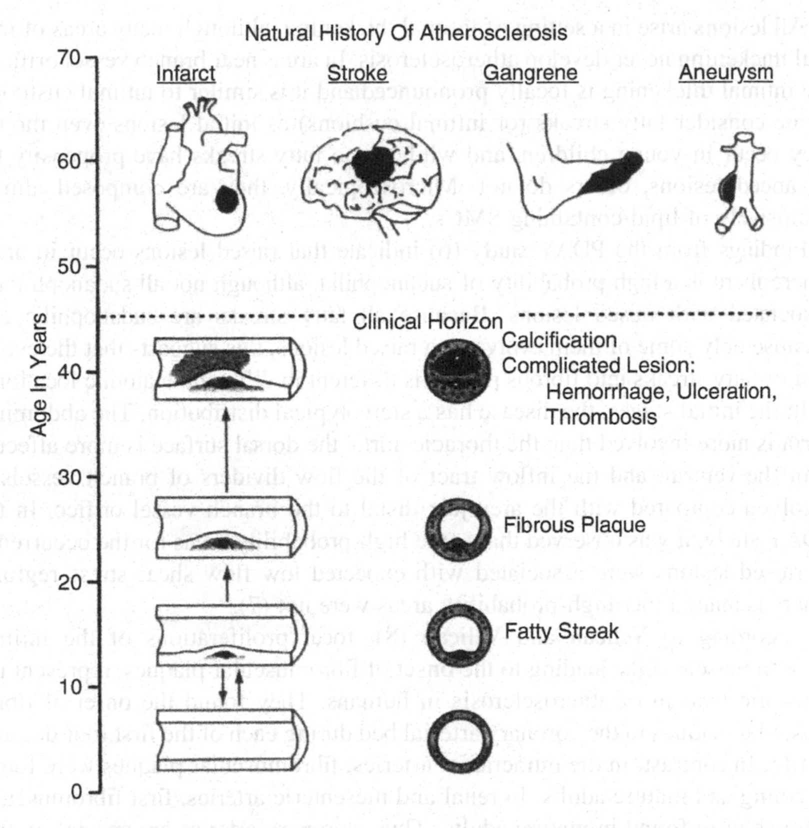

FIGURE 1.3. Natural history of human atherosclerosis. The earliest lesion is a deposit of lipid in the intima and inner media of large muscular and elastic arteries. Continued accumulation of lipid and proliferation of smooth muscle and connective tissue form fibrous plaques. The terminal occlusive episode usually results from rupture of plaque and thrombosis. Clinical manifestations vary with the artery involved.

women. The relevant risk factors include hypertension, dyslipidemia, diabetes, and cigarette smoking. Family history is also considered important, particularly for coronary heart disease (CHD) in younger patients.

4. Other Findings

Atherosclerosis is a focal disease. The major components of atherosclerotic plaques are smooth muscle cells (SMCs) that are set in a connective tissue matrix, with variable amounts of lipid in both intracellular and extracellular locations. Most of the intracellular lipid occurs in smooth muscle cells. A lesser amount of intracellular lipid is present in macrophages, usually most abundantly surrounding a central pool of lipid at the base of the plaque (6).

All lesions arise in a setting of intimal thickening, although many areas of intimal thickening never develop atherosclerosis. In areas near branch vessel orifices, the intimal thickening is focally pronounced and it is similar to intimal cushions. Some consider fatty streaks (or intimal cushions) as initial lesions even though they occur in young children, and while some fatty streaks have propensity for advanced lesions, others do not. Microscopically, they are composed almost exclusively of lipid-containing SMCs.

Findings from the PDAY study (6) indicate that raised lesions occur in areas where there is a high probability of sudanophilia, although not all sudanophilia is associated with raised lesions. Because all fatty streaks are sudanophilic and because only some of them evolve into raised lesions, this suggests that the evolution of fatty streaks into fibrous plaque is different in different anatomic locations.

In the initial stages, the disease has a stereotypical distribution. The abdominal aorta is more involved than the thoracic aorta, the dorsal surface is more affected than the ventral, and the inflow tract of the flow dividers of branch vessels is involved compared with the area just distal to the branch vessel orifice. In the PDAY study, it was observed that some high-probability areas for the occurrence of raised lesions were associated with expected low flow shear stress regions, whereas many other high-probability areas were not (7).

According to Velican and Velican (8), focal proliferations of the intimal smooth muscle cells, leading to the onset of fibromuscular plaques, represent the main mechanism of atherosclerosis in humans. They found the onset of fibromuscular plaques in the coronary arterial bed during each of the first four decades of life. In contrast, in the intracranial arteries, fibromuscular plaques were found in young and mature adults. In renal and mesenteric arteries, first fibromuscular plaques were found in mature adults. Thus, coronary arteries are special, in that the fibromuscular plaques develop there during childhood, adolescence, and adulthood. These observations suggest that the atherogenic agent(s), which determine the onset of lesions, act continuously on various arterial beds and give rise to fibromuscular lesions at susceptible sites.

Velican (9) also suggest the following interaction. The coronary artery wall can be regarded as a structure in which an abnormal acceleration of tension triggers a countercontraction and/or multiplication of SMCs, an abnormal rate of change of tension is absorbed by the elastic tissue, and an abnormal tensile force is met by the stretching of collagen fibers. When all conditions in which the preexisting contractile cells, elastin, and collagen fibers become inadequate to counteract hemodynamic stresses, then SMCs may be directly affected, and their response to injury is proliferation. This hypothesis is consistent with the new concepts proposed in Chapters 1 and 2.

5. The Pathogenesis of Atherosclerosis

Atherosclerosis is a multifactorial disease. Ross and Fuster (10) have described the process of atherosclerosis as follows. During the development of atherosclerotic plaque, the lesion contains all of the cellular responses that define an *inflammatory-fibroproliferative response to "injury."* Various stages in the initiation and progression of atherosclerotic lesions may be described as follows: i) The initial lesion consists of macrophage-derived foam cells that contain lipid droplets. ii) In the next stage, the lesion consists of both macrophages and smooth muscle cells with intracellular lipid droplets. iii) In the next stage, the lesion consists largely of smooth muscle cells, surrounded by extracellular lipid. iv) The next stage consists of a confluent cellular lesion with a great deal of extracellular lipids. v) The following stage in the lesion development has more than one possibility. One possibility is that the extracellular lipid is found in the core of the lesion, and the core is covered by a thin fibrous cap. Another possibility is that it evolves into a more stenotic and fibrotic type of lesion, which contains an increased number of SMCs and connective tissue, as well as lipid. Another variation is that the lesion may contain larger amounts of collagen and continue to become stenotic. vi) The last stage is "complicated" stenotic lesion. This lesion can rupture or fissure, thrombus can form there, and it can cause sudden occlusion of the artery. The advanced lesion often shows calcification in the plaque. Thus, the advanced lesion contains four essential components and they are excessive cells, connective tissue, lipid, and calcium. Different lesions contain different proportions of these elements. For instance, *soft plaque* may contain more lipid and no calcium, *fibrotic plaque* may contain mostly fibrotic and cellular material, and *hard plaque* may contain more calcium and less lipid.

5.1. The Fatty Streak

The earliest visibly detectable lesion of atherosclerosis is the fatty streak, which can be found in infants (10). The fatty streak consists of an accumulation of lipid within macrophages (foam cells) in the intima. The lipid appears to accumulate as a result of transport of lipoprotein particles through the endothelial cells into the artery wall. The fatty streak appears as an irregular yellow discoloration on the luminal surface, due to the accumulation of the lipid-filled macrophages (foam cells) that cluster beneath the endothelium. Fatty streaks primarily form at branches and bifurcations. With prolonged hyperlipidemia, the fatty streak can turn into fibrofatty or intermediate lesion. This progressed lesion may consist of multiple, alternating layers of foam cells and SMCs, which may be surrounded by some connective tissue. If the lesion increases in size, the intima may become thicker and somewhat raised. Further progression of the lesion shows a fibrous cap of connective tissue with SMCs intermixed with monocyte-derived macrophages, surrounded by a matrix containing collagen, proteoglycans, and elastic fibers. Further progression makes the lesion stenotic and fibrotic.

5.2. The Response-to-Injury Hypothesis of Atherosclerosis

In 1856, Rudolph Virchow suggested that the lesions of atherosclerosis result from some form of injury to the artery wall (11). In recent years, it became apparent that smooth muscle accumulation or proliferation was a key element in the development of the lesion. Ross et al. (11) suggested that some form of "injury" to the lining of endothelium, and possibly to the underlying SMCs, results in what would be considered a protective response. The injured cells somehow induce a chronic inflammatory response followed by a healing or fibroproliferative response. The most recent theory is that dysfunctional changes occur in the endothelium from various risk factors, resulting in formation of several adhesive molecules on the surface of the endothelium. These molecules bind to receptors on monocytes and lymphocytes. The leukocytes, after binding to the endothelial surface, are attracted to migrate into the intima. Once in the intima, they become activated and ingest lipids that have entered the same space before them. Activated macrophages can express growth factors and cytokines, which can have a profound effect on several cell types including SMC.

Hence, three critical processes involved in the formation of lesions of atherosclerosis are (Fig. 1.4) 1) focal intimal migration, proliferation, and accumulation of macrophages, T-lymphocytes, and SMCs; 2) formation of a connective tissue matrix by the accumulated SMCs; and 3) accumulation of lipid within macrophages and SMCs, as well as in the surrounding extracellular matrix.

If the stimuli that result in accumulation of SMCs and macrophages continue to exist, then the response to injury may become excessive. In other words, what begins as a protective, inflammatory-fibroproliferative response can become sufficiently excessive and become the disease entity itself that leads to atherosclerosis.

5.3. *Cellular Interactions in Atherosclerosis*

5.3.1. Endothelium

The response-to-injury hypothesis states that the injurious response initially occurs on the lining endothelium of the artery, leading to endothelial dysfunction. The principal functions of the endothelium may be listed as follows: 1) it serves as a permeability barrier to various substances in the plasma; 2) it provides a nonthrombogenic surface; 3) it maintains a vascular tone by formation of molecules such as nitric oxide (NO); 4) it forms growth regulatory molecules and cytokines; 5) it forms and maintains a connective tissue matrix; and 6) it modifies lipoproteins as they enter the artery wall. If some of these functions are altered as a result of hyperlipidemia, hypertension, diabetes, smoking, or other risk factors, then change in the properties of the endothelium could become important in the genesis of atherosclerosis (11).

5.3.2. Monocytes/Macrophages

The usual roles of the macrophages are to present antigen, act as a scavenger cell, and produce bioactive molecules. Macrophages are the main cells to mediate inflammatory response in atherosclerosis. They are able to proliferate and to provide stimulus for SMCs to proliferate. Proliferating macrophages together with SMC proliferation represents a key element of cellular accumulation in atherosclerosis.

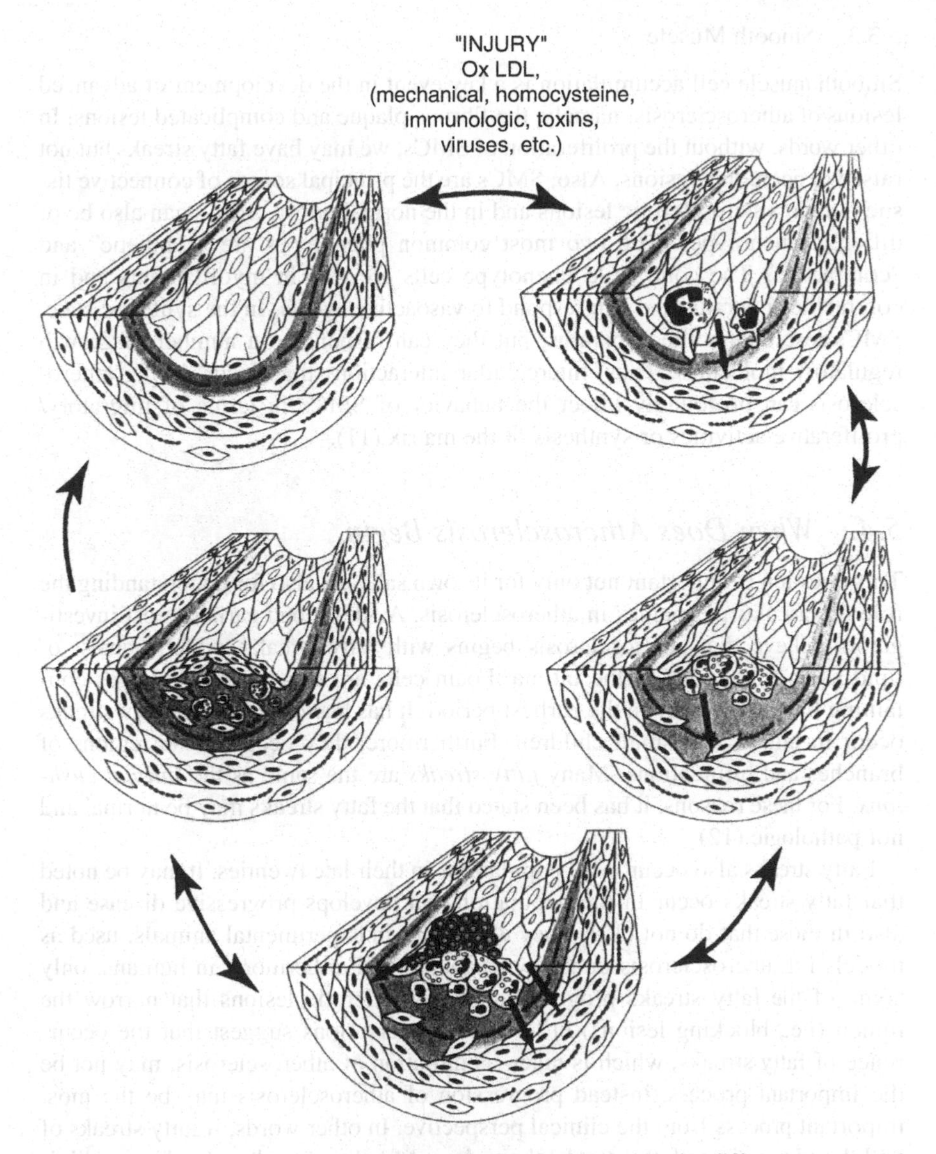

FIGURE 1.4. In the response-to-injury hypothesis of atherosclerosis, different sources of injury to the endothelium can lead to endothelial dysfunction. This can lead to adherence of monocytes/macrophages to the intima. These cells then migrate and localize subendothelially. Macrophages accumulate lipid and become large foam cells and together with smooth muscle form a fatty streak. The fatty streak can progress to an intermediate, fibrofatty lesion and ultimately to a fibrous plaque.

5.3.3. Smooth Muscle

Smooth muscle cell accumulation is a key event in the development of advanced lesions of atherosclerosis, namely, the fibrous plaque and complicated lesions. In other words, without the proliferation of SMCs, we may have fatty streaks but not raised or occluding lesions. Also, SMCs are the principal source of connective tissue both in atherosclerotic lesions and in the normal artery. SMCs can also be of different phenotypes. The two most common phenotypes are "synthetic" and "contractile." The contractile phenotype cells are rich in myofilaments and in contractile apparatus and can respond to vasoactive agents. In the synthetic state, SMCs lose contractile apparatus, but they can respond to a number of growth regulatory molecules. Thus, intercellular interactions that occur during atherosclerosis can profoundly affect the behavior of SMCs in terms of migratory/proliferative activities or synthesis of the matrix (11).

5.4. *When Does Atherosclerosis Begin*

This question is important not only for its own sake but also for understanding the role of vascular mechanics in atherosclerosis. As described earlier, most investigators believe that atherosclerosis begins with fatty streaks, which consist of lipid-laden foam cells in the intima. Foam cells are usually macrophages containing lipid, especially in the earliest period. It has been known that fatty streaks occur in infants and small children. Furthermore, they occur in the regions of branches and bifurcations. Many *fatty streaks* are the same as the *intimal cushions*. For these reasons, it has been stated that the fatty streaks may be normal and not pathologic (12).

Fatty streaks also occur in all populations in their late twenties. It may be noted that fatty streaks occur in the population that develops progressive disease and also in those that do not (12). They also occur in experimental animals, used as models for atherosclerosis under hyperlipidemia. Furthermore, in humans, only some of the fatty streaks progress into atherosclerotic lesions that narrow the lumen (i.e., blocking lesions) (12). These observations suggest that the occurrence of fatty streaks, which is taken as initiation of atherosclerosis, may not be the important process, instead progression of atherosclerosis may be the most important process from the clinical perspective. In other words, if fatty streaks of childhood remain as fatty streaks throughout life, then one does not have a clinical problem. So, the most important question may be why do some fatty streaks progress into blocking lesions (12)?

The progression of fatty streaks into fibroproliferative lesions requires proliferation of SMCs and of monocytes, but primarily SMCs. The SMC proliferation provides a link between vascular mechanics and atherosclerosis. Also, if one was to consider that the fatty streaks are "normal" and do not represent a pathologic process, as they occur in childhood and everyone has them, then it may be possible to state that atherosclerosis begins with the cellular proliferation (i.e., when the fatty streaks change into fibroproliferative lesions). This concept allows us to propose that "vascular mechanics is involved in initiating atherosclerosis by initiating a proliferative response of SMCs." Because some of the fatty streaks undergo progression into fibrous plaques whereas others do not, it is of great interest to know which ones turn into lesions and why.

6. Risk Factors

6.1. High Serum Cholesterol and Low-Density Lipoprotein as a Risk Factor for CHD

A positive relationship exists between serum total cholesterol levels and the risk of coronary heart disease (CHD; Fig. 1.5). Many studies in experimental animals, including primates, showed that diet-induced hypercholesterolemia will produce arterial lesions resembling human atherosclerosis. Also, in recent years several controlled clinical trials found that lowering cholesterol levels reduces risk of CHD and can retard the progression of atherosclerosis. Total cholesterol consists of various lipoproteins, which include LDL (low-density lipoprotein), HDL (high-density lipoprotein), VLDL (very-low-density lipoprotein), and IDL (intermediate-density lipoprotein). In a normal person, about two-thirds of the total cholesterol is carried in LDL. The data indicates that it is the LDL-cholesterol that is largely responsible for the positive relationship between the total cholesterol and CHD. Also, clinical trials in which LDL is reduced show that both the risk of CHD is reduced and progression of coronary atherosclerosis is retarded.

The serum total cholesterol is defined as (13) desirable (<200 mg/dL), borderline-high (200–239 mg/dL), and high (≥240 mg/dL). Corresponding definitions for LDL-cholesterol levels are desirable (<130 mg/dL), borderline-high risk

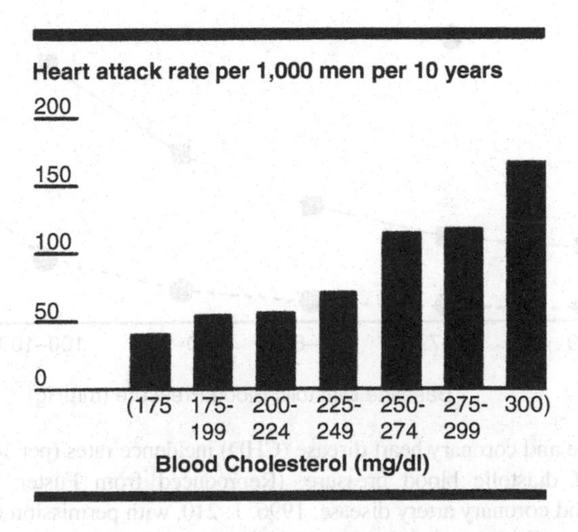

FIGURE 1.5. Risk of heart attack increases as blood cholesterol increases.

(130–159 mg/dL), and high risk (≥160 mg/dL). Levels of HDL-cholesterol are inversely correlated with risk of CHD. The mechanism by which HDL affects the atherosclerotic process has not been well understood.

6.2. Blood Pressure as a Major Cardiovascular Risk Factor

Both coronary artery disease (CAD) and stroke are strongly and positively associated with blood pressure in a graded, independent, and consistent fashion (Fig. 1.6) (14). Furthermore, in several clinical trials, treatment of hypertension markedly decreased the risk of stroke and modestly reduced CAD. Overall, the CAD end point was reduced by 17% while the stroke incidence declined by 38% with treatment that reduced blood pressure. These results establish hypertension as a major causal risk factor for cardiovascular disease. Furthermore, systolic blood pressure correlates with CAD risk just as well as does diastolic pressure. In fact, CAD risk increases progressively with systolic pressure independently of diastolic pressure (14).

One possible mechanism suggested for the role of hypertension in atherosclerosis is as follows (14). In LDL transport experiments at higher pressure, LDL concentration in the intima and inner media was several times greater than that at lower pressure. The investigators suggested that at higher pressure, the subintimal tissues are compacted and thereby retard the movement of the relatively large LDL parti-

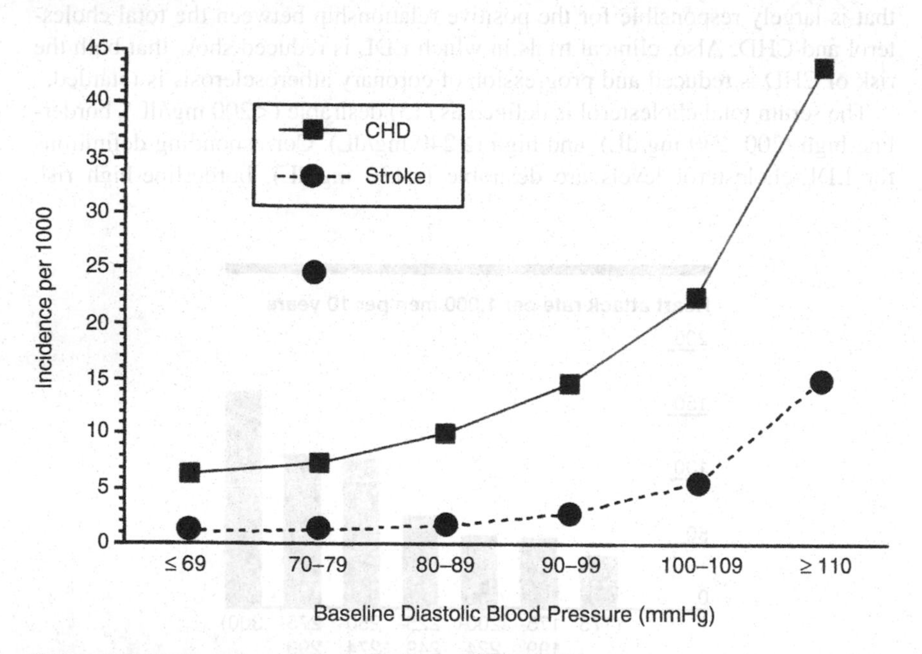

FIGURE 1.6. Stroke and coronary heart disease (CHD) incidence rates (per 1000 individuals) as a function of diastolic blood pressure. (Reproduced from Fuster V. et al. (eds): Atherosclerosis and coronary artery disease. 1996, 1: 210, with permission from Lippincott, Williams & Wilkins, Baltimore)

cles through the wall. Thus, trapping of LDL at internal elastic lamina combined with pressure-driven convection of LDL into the artery wall appears to be the major mechanism underlying the relationship between blood pressure and atherosclerosis.

From the vascular mechanics point of view, high pressure produces high wall tension and a much higher stress concentration (see Chapter 5), which will have a profound effect on the media and SMC. These effects are in addition to those on LDL transport described above.

6.3. *Cigarette Smoke Constituents and SMC Proliferation*

Cigarette smoking is a major independent risk factor for atherosclerosis (15). A positive correlation has been established between smoking and the severity of atherosclerosis in coronary and cerebral arteries and in the aorta. The increase in CVD in the United States parallels the trends observed in the prevalence of smoking. Also, the decrease in deaths from CVD has been correlated in part with reduction in smoking. Cigarette smoke can have effects on several parameters such as neurotransmitters, endothelial cells, lipoproteins, leukocytes, thrombosis, and so forth. It also seems to have a major effect on SMC proliferation, which directly influences atherosclerosis.

One possible mechanism, which addresses the effect of smoking on atherosclerosis, is the *monoclonal hypothesis* for plaque development. This hypothesis states that atherosclerotic plaques arise as a *monoclonal proliferation of SMCs* after a mutational event. There are a number of different mutagens and carcinogens in cigarette smoke, which can promote neoplasia. In smokers, there is an overexpression of p53, tumor suppressor protein, in a bladder with early stages of cancer and in cancers of the lung (15). It appears that smoking leads to changes whereby p53 is overexpressed. It is proposed that smoking may have an effect on oncoprotein, resulting in rapid growth of atherosclerotic plaque.

The mechanism of atherosclerosis, proposed from the vascular mechanics point of view (see Chapter 2), also considers SMC proliferation as a paramount event. However, an external entity like a carcinogen is not necessary for the proposed proliferation; instead, stress-concentration serves as the stimulus for SMC proliferation.

6.4. *Diabetes and Atherosclerosis*

Both insulin-dependent diabetes mellitus (IDDM) and non-insulin-dependent diabetes mellitus (NIDDM) are powerful and independent risk factors of atherosclerosis, CAD, stroke, and peripheral arterial disease (16). Atherosclerosis is a major problem in patients with diabetes, accounting for almost 80% of all deaths among diabetics in the United States. Three-quarters of these deaths are from CHD while the remaining include cerebral and peripheral vascular disease. Diabetes is also the most common cause of heart disease in young persons. Several autopsy studies have shown that diabetic subjects have more extensive atherosclerosis of both coronary and cerebral vessels than the matched controls. Although diabetes does not appear to be the primary cause of atherosclerosis, it accelerates the natural progression of atherosclerosis in all populations. Diabetics have a larger number of coronary vessels involved, a more diffuse distribution of atherosclerotic lesions, and a more severe narrowing of the left main coronary artery.

Diabetes affects atherogenesis through multiple mechanisms. It changes lipoprotein concentration as well as composition toward proatherogenic. These include increased VLDL, small and dense LDL, and decreased HDL. Hypertension is more prevalent among individuals with NIDDM and those with IDDM who develop renal dysfunction. High circulating insulin levels in patients with NIDDM may also be atherogenic. A procoagulant state exists in diabetes as a result of an increased level of clotting factors, increased platelet aggregability, and so forth. The effect of hyperglycemia, mediated through protein glycation and glycoxication, increases atherogenicity by affecting vessel wall structural protein. Glycation of LDL particles impairs its recognition by LDL receptors and renders it more susceptible to oxidative modification.

CHD is a major health problem in NIDDM. It is a leading cause of death among patients with NIDDM. Relative risk of CVD in NIDDM compared with the general population is increased two- to fourfold. In IDDM patients also an excess of cardiovascular mortality, compared with the general population, has been observed.

Clinical expression of CAD in diabetic patients differs from that in nondiabetic patients. Diabetics have a higher frequency of silent myocardial infarction and ischemia, greater morbidity and mortality after acute myocardial infarction, decreased reperfusion after thrombolytic therapy, and increased restenosis rate after coronary angioplasty.

6.5. *Obesity and Atherosclerosis*

A person is considered obese if the weight of that person is 20% or more above the desirable level. Obesity is clearly associated with increased prevalence of cardiovascular risk factors such as hypertension, diabetes, and dyslipidemia. However, a direct relationship between obesity and coronary artery disease is debatable. In the Framingham study, obesity was related to cardiovascular mortality and morbidity even after controlling for other risk factors (17). This would make obesity an independent risk factor. However, autopsy studies have shown no consistent relationship between weight and coronary atherosclerosis.

6.6. *Effect of β-Blockade in Atherosclerosis*

Hypertension has been recognized as probably the single most influential factor in cardiovascular diseases, including coronary heart disease, and stroke, originating usually from atherosclerosis and aneurysms. The overall effect of hypertension is considered so potent that in itself it has been referred to as a "hypertensive disease." In this context we may recall that localized regions of stress-concentration in the artery, under normal pressure, could be considered equivalent to those regions being under severe hypertension, while the rest of the artery is under normotension. This analogy may be helpful to those in the medical community who may not be familiar with the concept of stress-concentration.

The treatment of hypertension also has a long clinical history. The two types of treatment of hypertension (among many) we want to consider here relate to diuretics and β-blocker categories of drugs. Treatments with drugs in either category have been known not only to reduce the blood pressure but also to reduce the incidence of CHD. Furthermore, the reduction in incidence of CHD has been greater with β-blocker treatment than with diuretics, even if the reduction of pressure by both is similar. For example, in patients, three different multicenter trials have found that treatment with propranolol and metoprolol reduced mortality and complications of coronary atherosclerosis (18–22). A treatment with metoprolol (a potent beta-adrenoreceptor antagonist) was also associated with a significant reduction in the incidence of cerebrovascular events, as well as a tendency toward decreased atherosclerotic complication in the lower extremities (22). Because β-blockers are known to reduce both heart rate and blood pressure (21, 23), we may state, from the vascular mechanics point of view, that the added advantage of β-blockers is likely to be due to reduction of the heart rate. How the reduction of the heart rate, irrespective of the means with which it is achieved, can reduce the development of atherosclerotic lesions is described at length in Chapter 10.

7. New Concepts

One of the most important diseases in recent medical history is cardiovascular disease because it is the number one killer of people today. While exploring "new concepts" for the disease atherosclerosis, it may be useful to recall the statement by Dr. Julius Comroe: "An entirely *different type of knowledge* than we now have will be necessary to solve problems such as atherosclerosis" (24). Our understanding of this disease has been compromised by some very basic problems that are partly related to a crossover between the two different disciplines—engineering and medicine. Some examples below will serve to illustrate this point as well as highlight the role of vascular mechanics in cardiovascular disease.

7.1. *Parameters We Cannot Measure*

Some of the parameters in this category are so basic that at times we forget that we cannot directly measure them. The "point" cannot be measured but we can measure the distance between two points. The "resistance," electrical or flow, cannot be measured. Instead, the resistance (R) is *derived* from equations $R = \frac{V}{I}$ *or* $\frac{\Delta P}{F}$ where V is the voltage, I is the current, ΔP is the pressure difference, and F is the flow, and these are measurable. Still, we often say that we can "measure" the resistance, when in fact we measure the ΔP and F and *calculate* the resistance (R). The "stress" cannot be measured, but we can calculate it from other parameters. The stress (σ) is given by $\sigma = \frac{F}{A}$ where F is the force and A is the area over which the force is acting. The stress (σ) in the artery is the same way; we cannot measure it but we can calculate it on the basis of pressure-force and the cross-sectional area of the tissue. This sounds basic, however, when one attempts to correlate the artery-stress to the disease atherosclerosis, the medical scientists are likely to ask, "Can you measure the stress?" (25). The stress itself is not measurable. Often, the strain is measured to reflect the stress. It would help the reader to know that when the principles of vascular mechanics are applied to pathology, one may have additional barriers to cross because the engineering discipline is being connected with medicine.

7.2. Seeing Is Believing

The blood flow you can *see,* but the pressure may be the *real invisible enemy* when it comes to cardiovascular disease. It so happens that in recent decades, blood flow in the arteries and in the heart can be "visualized" with techniques of Doppler ultrasound and MRI and, therefore, it becomes possible to correlate flow with a variety of conditions such as arterial stenosis or valvular regurgitation. Unfortunately, diagnostic techniques do not exist that will allow us to "see" pressure or to see the result of pressure such as the wall stress. For instance, we cannot "see" pressure like we can "see" flow. We cannot "see" wall tension in the artery or in the heart. We know that the artery wall becomes thicker under hypertension and the left ventricle wall becomes thicker when it has to generate more pressure, but we cannot "see" pressure, and we cannot "see" wall tension. If we had a technique to "see" compression or tension in the bones, in the heart or in the artery, our thinking about many diseases—including cardiovascular disease—would change dramatically. For now though, we are likely to see a limited view, which is how blood flow correlates with a variety of conditions. Hopefully, the realization that we cannot "see" pressure will keep us from being completely swept away by the theories that singularly depend on blood flow.

7.3. Hemodynamics

The word *heme* refers to the blood and *dynamics* refers to the motion. So, *hemodynamics* refers to the blood motion or "blood flow," or a science related to blood flow. The very word *hemodynamics* makes one think of blood flow, and when it is used in the context of vascular disease (e.g., hemodynamics and atherosclerosis), it conveys to the reader "blood flow and atherosclerosis." By using such terminology, the reader is already mesmerized by the suggestion that the processes of atherosclerosis must be thought of in terms of blood flow. Obviously, by default, it tends to exclude the entire field of blood pressure and vascular mechanics. For the sake of science, therefore, we must use a terminology that will be inclusive of blood flow and blood pressure. Perhaps a phrase like "hemodynamics and vascular mechanics" could achieve that purpose.

7.4. Which Cell

When it comes to atherosclerosis, there are two key observations that have been known for a long time: 1) the disease is focal and occurs at arterial branches and bifurcations, and 2) high blood pressure promotes the disease. What could be so special about the branch and bifurcation regions of the artery to cause the disease there? It was simpler and obvious to think that the blood flow separates at the branches and bifurcations. This aspect of flow division was known to the medical community also. Therefore, it made sense to consider that flow separation, turbulence, high shear, low shear, and so forth, is somehow responsible for localization of the disease. The flow is in direct contact with the endothelial cells (ECs), and therefore it would be natural to explore the role of ECs in atherosclerosis. This approach is even more convincing when we consider that the atherosclerotic lesions always involve the intima. So, for the most part, the research path has been flow separation–EC–atherosclerosis. However, in spite of voluminous research over 40 years along this path, several basic observations such as the effect of hypertension or β-blocker treatment on atherosclerosis remain unexplained.

In contrast with the view above is the observation that the blood pressure has a profound influence on atherosclerosis. The problem in pursuing the *pressure theory* for atherosclerosis was that the pressure is the same at the branches and bifurcations. This is where the lack of knowledge and a breakdown of communications between engineering and medicine played a key role in turning scientists away from this path of exploration. For example, pressure produces wall stress and high blood pressure produces high wall stress, but what about the branch? In the areas of branches and bifurcations, there occurs a phenomenon called *stress-concentration* (see Chapters 4–7). Because the branch origin creates a "hole" in the artery, the wall stress due to pressure is very high in this localized region. This knowledge of stress-concentration has not been present in medical science and it is not intuitive to medical scientists or physicians. This singular aspect (i.e., *not knowing that the stress concentration occurs at the branch*), has been the turning point in the pursuit of research in atherosclerosis. Engineers in mechanical engineering, nuclear engineering, and pressure vessel technology have always known the stress-concentration phenomenon, however, there was a breakdown of communications between engineering and medicine on this subject. Consequently, much of the information along this path remains unknown, and the research in this area is unbelievably small. In spite of that, even the small amount of research in this direction has explained many basic observations, such as the effect of hypertension or β-blocker treatment on atherosclerosis.

The pressure in the artery and the resulting high wall stress at the branch suggests that the artery will undergo a breakdown or be diseased by the pressure load. The pressure in the artery is sustained by the media (connective tissue and SMCs) and generally not by the intima or ECs. This path would almost exclusively focus on the media. This direction of research would be pressure–wall stress–SMC and connective tissue of the media–atherosclerosis. In other words, the focus will be on SMC and how it participates in atherogenesis. So we have to ponder, which cell (EC or SMC) holds the key to understanding atherosclerosis. According to the pressure as a "cause" theory, it is the SMC that holds the key.

The reader may wonder, if the media is involved in sustaining the pressure, then why does the disease occur primarily in the intima? The answer is in the distribution of wall stress through the thickness of the artery. The wall stress from pressure is the highest on the inner surface of the artery and decreases toward the outer surface (see Chapter 4). Hence, the inner media is most affected by the stress, and therefore, the proliferative or apoptotic response will begin there. So, the atherosclerosis will primarily involve the luminal side of the media.

7.5. Plaque Rupture, Atherosclerosis, Aneurysm

Continuing along the lines that wall stress from pressure is the cause of arterial degenerative diseases, we can note a clear and direct link between atherosclerotic plaque rupture, atherosclerosis, and aneurysm. Fatigue injury to the artery will be "the one responsible" for artery degeneration in the region of stress concentration (see Chapters 2 and 10). This path is emphasized throughout the book, because it is a little-known phenomenon. Extending the fatigue theory to explain plaque rupture is then natural. The latter has been well recognized in the medical community. In other words, the mechanism that may form atherosclerosis is also the one that will rupture the plaque, and that mechanism is fatigue. While this mechanism is active, we obviously would like to know what might happen if the SMCs were unable to proliferate. The result will be aneurysm. We can speculate that the root cause of arterial diseases—atherosclerosis and aneurysm—is the injury to the media from pressure-generated wall stress. Furthermore, overwhelming proliferative response of SMCs could result in atherosclerosis, whereas inability to proliferate could result in aneurysm. These concepts are developed, supported, and described in detail in Chapter 2 and throughout the book.

References

1. American Heart Association. 1993 Heart and Stroke Facts Statistic. American Heart Association National Center, Dallas, 1992;9.
2. American Heart Association. 1993 Heart and Stroke Facts Statistic. American Heart Association National Center, Dallas, 1992;13-14.
3. Shetty KR, Parulkar GB (eds): Open Heart Surgery, Proceedings of the World Conference on Open Heart Surgery, 7-10 February 1985, Bombay, India. Tata McGraw-Hill Publishing Company Ltd, New Delhi, India, 1987:448.
4. Shetty KR, Parulkar GB (eds): Open Heart Surgery, Proceedings of the World Conference on Open Heart Surgery, 7-10 February 1985, Bombay, India. Tata McGraw-Hill Publishing Company Ltd, New Delhi, India, 1987:472.
5. Fuster V, Ross R, Topol EJ (eds): Atherosclerosis and Coronary Artery Disease. Volume 1. Lippincott-Raven Publishers, Philadelphia, 1996:31.
6. Damjanov I, Linder J (eds): Anderson's Pathology. Tenth Edition. Mosby, Chicago, 1996:1400-1402.

7. Cornhill JF, Herderick EE, Stary HC: Topography of human aortic sudanophilic lesions. In: Blood Flow in Large Arteries: Application to Atherogenesis and Clinical Medicine. Liepsch DW (ed), Monogr Atheroscler. Karger, Basel, 1990;15:13-19.
8. Velican C, Velican D: Natural History of Coronary Atherosclerosis. CRC Press, Boca Raton, FL, 1989:200.
9. Velican C, Velican D: Natural History of Coronary Atherosclerosis. CRC Press, Boca Raton, FL, 1989:204.
10. Fuster V, Ross R, Topol EJ (eds): Atherosclerosis and Coronary Artery Disease. Volume 1. Lippincott-Raven Publishers, Philadelphia, 1996:442.
11. Fuster V, Ross R, Topol EJ (eds): Atherosclerosis and Coronary Artery Disease. Volume 1. Lippincott-Raven Publishers, Philadelphia, 1996:447-449.
12. Glagov S, Newman III WP, Schaffer SA (eds): Pathobiology of the Human Atherosclerotic Plaque. Springer-Verlag, New York, 1990:7, 265.
13. Fuster V, Ross R, Topol EJ (eds): Atherosclerosis and Coronary Artery Disease. Volume 1. Lippincott-Raven Publishers, Philadelphia, 1996:47.
14. Fuster V, Ross R, Topol EJ (eds): Atherosclerosis and Coronary Artery Disease. Volume 1. Lippincott-Raven Publishers, Philadelphia, 1996:210-213.
15. Fuster V, Ross R, Topol EJ (eds): Atherosclerosis and Coronary Artery Disease. Volume 1. Lippincott-Raven Publishers, Philadelphia, 1996:304-310.
16. Fuster V, Ross R, Topol EJ (eds): Atherosclerosis and Coronary Artery Disease. Volume 1. Lippincott-Raven Publishers, Philadelphia, 1996:328.
17. Fuster V, Ross R, Topol EJ (eds): Atherosclerosis and Coronary Artery Disease. Volume 1. Lippincott-Raven Publishers, Philadelphia, 1996:349.
18. Thubrikar MJ, Moorthy RR, Holloway PW, Nolan SP: Effect of beta-blocker metoprolol on low density lipoprotein influx in isolated rabbit arteries. J Vasc Invest 1996;2:131-140.
19. Ablad B, Björkman JA, Gustafsson D, Hanson G, Östlund-Lindquist AM, Pettersson K: The role of sympathetic activity in atherogenesis: Effects of beta blockade. Am Heart J 1988;116:322-327.
20. Beta Blocker Heart Attack Trial Research Group. A randomized trial of propranolol in patients with acute myocardial infarction. I. Mortality results. JAMA 1982;247:1707-1714.
21. Hjalmarson Ä, Elmfeldt D, Herlitz J, et al. Effect on mortality of metoprolol in acute myocardial infarction: A double blind randomized trial. Lancet 1981;2:823-827.
22. Wikstrand J, Warnold I, Olsson G, Thomilehyo J, Elmfeldt D, Berglund G: Primary prevention with metoprolol in patients with hypertension. Mortality results from the MAPHY study. JAMA 1988;259:1976-1982.
23. The Norwegian Multicenter Study Group. Timolol induced reduction in mortality and reinfarction in patient surviving acute myocardial infarction. N Engl J Med 1981;304:801-807.
24. Comroe JH Jr: Exploring the heart. W.W. Norton and Company, New York, 1983:325.
25. Fuster V, Ross R, Topol EJ (eds): Atherosclerosis and Coronary Artery Disease. Volume 1. Lippincott-Raven Publishers, Philadelphia, 1996:597.

2
Atherosclerosis II

1. Anatomical Distribution of Atherosclerosis

Atherosclerosis commonly affects the artery only at certain well-defined locations rather than through its entire course. In humans, atherosclerotic lesions usually predominate at the origins of tributaries, bifurcations, and curvatures

(1, 2) (Fig. 2.1). Some authors (3–9) have thought this focal nature of the disease could be explained by local disturbances in the blood flow. Both the high-shear (3–5) and the low-shear (6–8) areas have been considered as primary sites of atheroma formation. In contrast with these views, we will consider here another view, which is that the *blood pressure–induced arterial wall stress* is the principal factor in the localization of the disease. With this view, we will explore how the arterial mechanics plays a role in this pathology. Pressure load produces mechanical stresses and strains in the entire thickness of the artery wall.

In this regard, we may note that atherosclerotic lesions do not develop in the veins in their normal environment of low pressure and high flow, but that the lesions do develop when the veins are used as arterial bypass grafts where they are subjected to high pressure. Similarly, atherosclerotic lesions develop in the

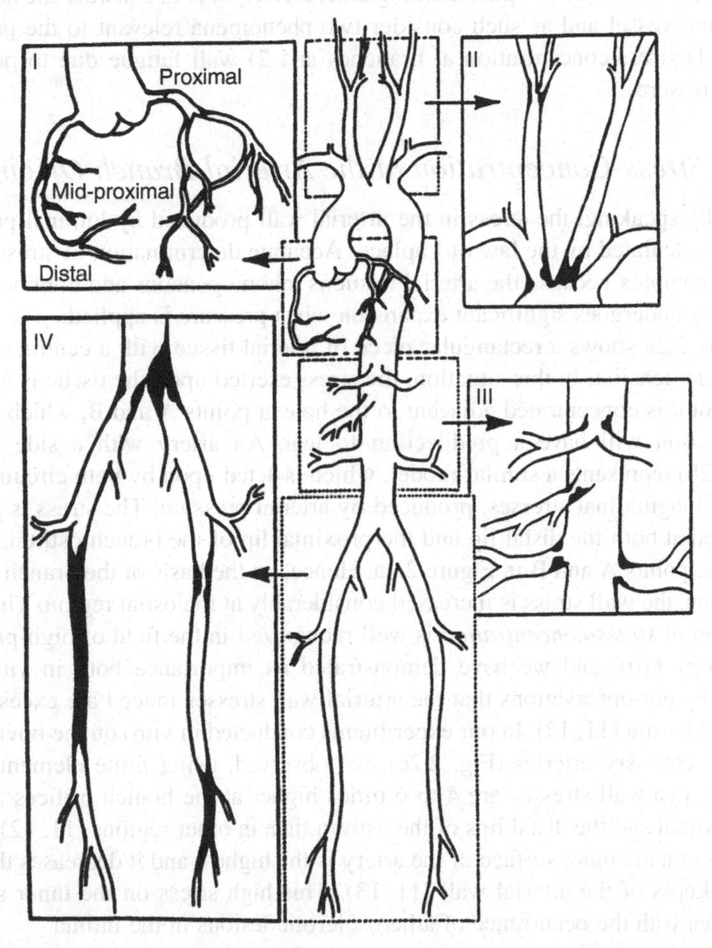

FIGURE 2.1. Distribution of atherosclerotic occlusive disease in humans. (Reproduced from the Annals of Surgery 1985;201:116 with permission from Lippincott, Williams & Wilkins, Baltimore, MD.)

pulmonary arteries only in pulmonary hypertension. This is not surprising because high blood pressure in general is a well-recognized risk factor in coronary heart disease, a phenomenon that fits well in the "arterial wall stress hypothesis," where the stress is produced by blood pressure and not by blood flow.

1.1. Atherosclerosis and Functions of the Artery

To serve as a conduit of blood flow is only one of the basic functions of the artery. The other basic function is to sustain blood pressure. The artery is therefore both, a conduit of blood flow and a container of pressure (pressure vessel). Although the first function has been studied in great detail, the artery as a pressure vessel scarcely has been the subject of investigation. In this chapter, we will attempt to establish that a key to understanding atherosclerosis is to consider the artery as a pressure vessel and as such consider two phenomena relevant to the pressure vessel: 1) stress-concentration at branches and 2) wall fatigue due to pulsatile blood pressure.

1.2. Stress-Concentration at the Arterial Branch Origins

Generally speaking, the stress in the arterial wall produced by luminal pressure may be calculated by the law of Laplace. Accurate determination of stress, however, is complex because the arterial tissue is inhomogeneous and nonlinear, and the artery undergoes significant expansion when pressure is applied.

Figure 2.2a shows a rectangular piece of arterial tissue with a central circular hole under tension. In this situation, the stress exerted upon the tissue is not uniform, but it is concentrated adjacent to the hole at points A and B, which for the same reason will have a predilection to tear. An artery with a side branch (Fig. 2.2b) represents a similar model, which is acted upon by both circumferential and longitudinal stresses, produced by arterial pressure. The stress is greatly increased at both the distal lip and the proximal lip of the branch ostium, which resemble points A and B in Figure 2.2a. Hence, on the basis of the branch geometry alone, the wall stress is increased considerably at the ostial region. This phenomenon of *stress-concentration* is well recognized in the field of high-pressure technology (10), and we have demonstrated its importance both in vitro and in vivo by our observations that the arterial wall stresses indeed are excessive at the branch ostia (11, 12). In our experiments conducted in vitro on the bovine circumflex coronary arteries (Fig. 2.2c), we observed, using finite element stress analysis, that wall stresses are 4 to 6 times higher at the branch orifices at both the proximal and the distal lips of the ostium than in other regions (11, 12). Also, the stress on the inner surface of the artery is the highest and it decreases through the thickness of the arterial wall (11–13). This high stress on the inner surface correlates with the occurrence of atherosclerotic lesions in the intima.

One of the most important consequences of stress concentration at the branch origins is that it produces a greater stretch at that location. In our experiments, we both measured and analytically calculated the stretch in the branch area (Fig. 2.3).

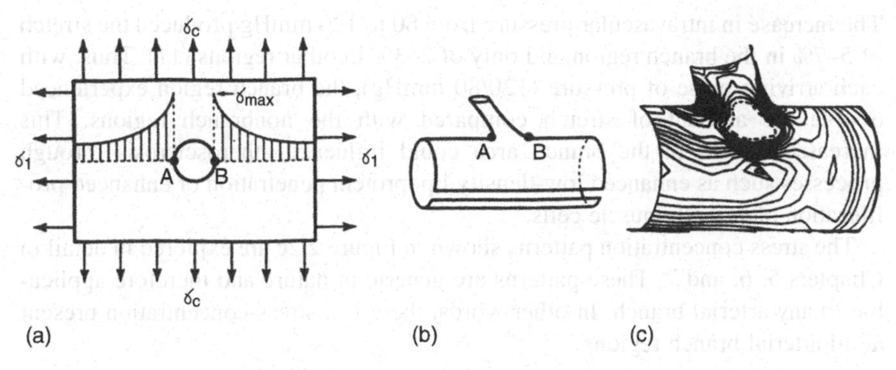

FIGURE 2.2. Stress-concentration at the arterial branch origins. (a) Stress distribution in a rectangular plate with a circular hole. σ_c and σ_1 represent, respectively, stresses in the circumferential and longitudinal directions in the artery. The stress in the plate increases toward the hold and reaches a maximum value (σ_{max}) at points A and B. (b) Schematic presentation of the artery with a branch. When the artery is opened in the longitudinal direction, the ostium of the branch appears similar to the hole in a plate. Points A and B at the ostium are similar to those in (a). (c) Maximum principal isostress contours on the inner surface of a bovine circumflex coronary arterial branch. The wall stress increases from one contour to the next toward the distal and proximal lips of the ostium. The stress was determined in vitro for a pressure increase from 80 to 120 mmHg using a finite element analysis method.

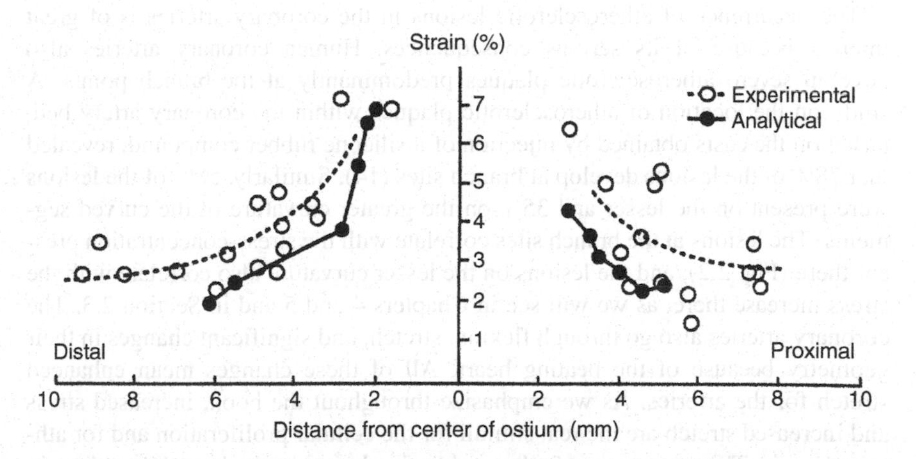

FIGURE 2.3. Distribution of strain around the ostium of a bovine circumflex coronary arterial branch. Experimentally measured strains near both the distal and the proximal lips of the ostium are as high as 5% to 7%, whereas those away from the ostium are 2% to 3%. The analytic strains were obtained from finite element analysis. The strains are for the pressure increase from 80 to 120 mmHg.

The increase in intravascular pressure from 80 to 120 mmHg produced the stretch of 5–7% in the branch region and only of 2–3% in other regions (11). Thus, with each arriving pulse of pressure (120/80 mmHg), the branch region experienced double the amount of stretch compared with the nonbranch regions. This increased stretch in the branch area could influence atherosclerosis through processes such as enhanced low-density lipoprotein penetration or enhanced proliferation of smooth muscle cells.

The stress concentration patterns shown in Figure 2.2c are explored in detail in Chapters 5, 6, and 7. These patterns are generic in nature and therefore applicable to any arterial branch. In other words, there is a stress-concentration present at all arterial branch regions.

2. Atherosclerosis and Stress-Concentration

2.1. *Ostial Lesions and Stress-Concentration*

Atherosclerotic lesions commonly occur at the ostia of *celiac, superior mesenteric, right renal, and left renal arteries* in the abdominal aorta, especially at the proximal and the distal lips of each ostia (Fig. 2.1). This phenomenon correlates well with the areas of high wall stress and high stretch at the branch as described earlier (Figs. 2.2 and 2.3). Because blood flow induces low shear at the proximal lip and high shear at the distal lip of the ostia, whereas blood pressure induces high wall stress at both the locations, it is logical that pressure-induced wall stress and not flow-induced shear stress correlates with the locations of atherosclerotic lesions.

The occurrence of atherosclerotic lesions in the coronary arteries is of great interest because of its serious consequences. Human coronary arteries also develop severe atherosclerotic plaques predominantly at the branch points. A study on the location of atherosclerotic plaques within the coronary artery bed, based on the casts obtained by injection of a silicone rubber compound, revealed that 78% of the lesions develop at branch sites (14). Similarly, 58% of the lesions were present on the lesser and 35% on the greater curvature of the curved segments. The lesions at the branch sites correlate with the stress-concentration present there (Fig. 2.2), and the lesions on the lesser curvature also correlate with the stress increase there, as we will see in Chapters 4 and 5 and in Section 2.3. The coronary arteries also go through flexion, stretch, and significant changes in their geometry because of the beating heart. All of these changes mean enhanced stretch for the arteries. As we emphasize throughout the book, increased stress and increased stretch are the real stimuli for the cellular proliferation and for atherosclerosis. The occurrence of atherosclerotic plaques at the branch locations in coronary arteries is generally less emphasized in the literature. It is, therefore, important to note that the dominant occurrence of atherosclerosis at the branch in coronaries is similar to the occurrence of atherosclerosis in other arteries.

2.2. *Aortic Arch Lesions and Stress/Strain Concentration*

Figure 2.4 shows the distribution of circumferential strain in the aortic arch. The details of calculations are presented in Chapter 7. Once again, in the aortic arch atherosclerotic lesions seem to be localized at the branch ostia (Fig. 2.1) where the circumferential strains are also localized. Thus, both the concentration of stress or strain can be correlated with the localization of the disease.

2.3. *Aortic Bifurcation Lesions and Stress*

Atherosclerotic lesions in the aortic bifurcation occur at both the "crotch" and the "hip" of the aortic bifurcation (Figs. 2.1 and 2.5). As shown on the bifurcation geometry (Fig. 2.5), the wall stress is very high at the crotch on the basis of the analysis presented earlier in Figure 2.2. Wall stress at the crotch is increased in a manner similar to that at the distal lip of the ostium where the branch angle is small.

To analyze wall stress occurring at the hip of the bifurcation, one must consider three parameters: 1) elliptical cross section, 2) wall thickness, and 3) surface curvatures (Fig. 2.5). Although the cross section of the main aorta and the two iliac arteries is circular, at the bifurcation the cross section becomes larger (15) and

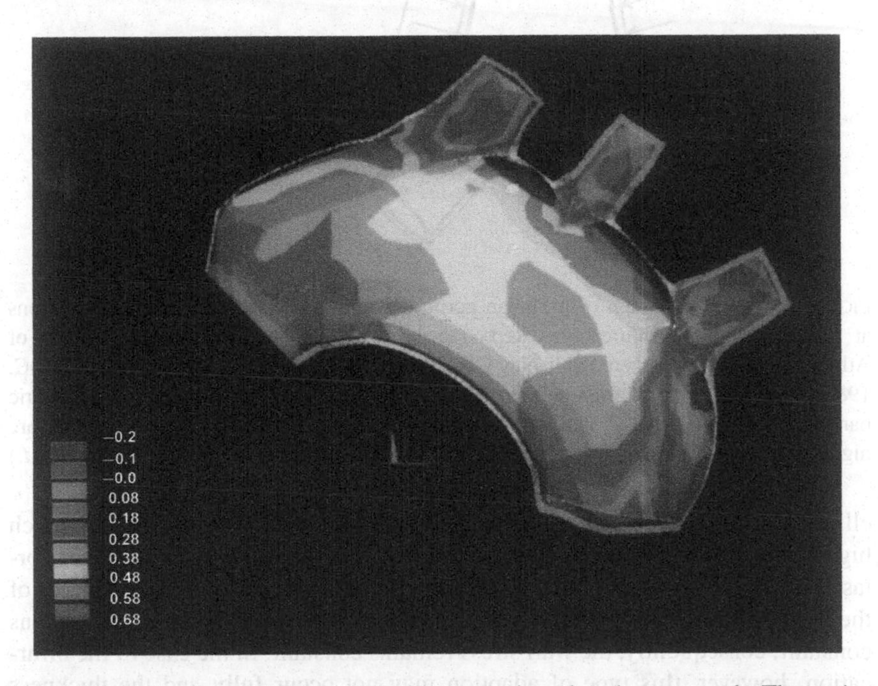

FIGURE 2.4. Circumferential strains in the human aortic arch and great vessels. The strains were determined for a pressure increase from 0 to 100 mmHg using a finite element analysis method. The strains are concentrated around the ostium of branches. See Chapter 7 for details. (*Please see color version on CD-ROM.*)

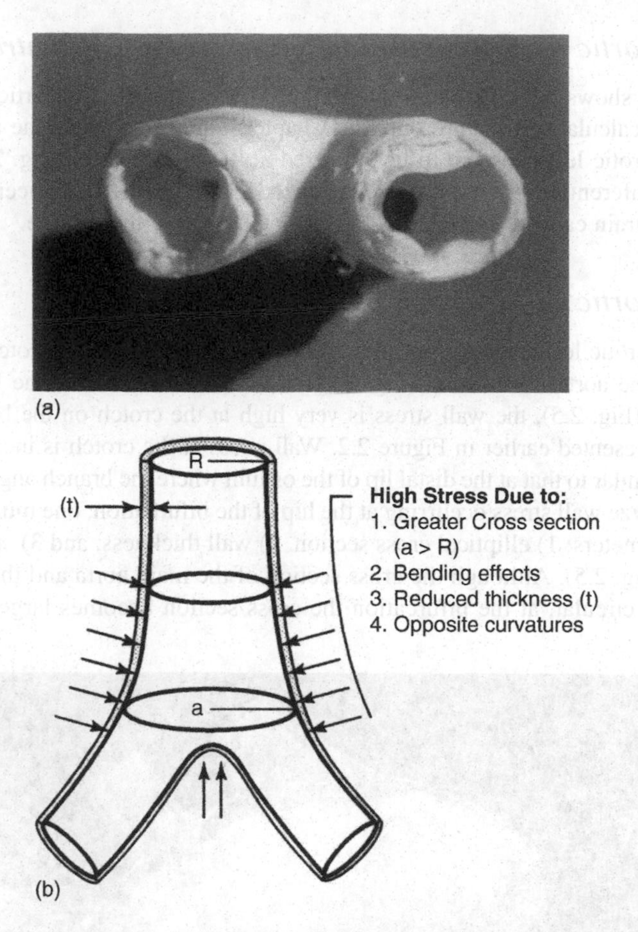

FIGURE 2.5. (a) Photograph of a human aortic bifurcation showing atherosclerotic lesions at the crotch of the bifurcation. (Reproduced from Texon M, Hemodynamic Basis of Atherosclerosis. Hemisphere Publishing Corporation/Taylor & Francis Inc., Washington, DC, 1980, Plate 10A, with permission.) (b) Schematic presentation showing various geometric parameters that may lead to uneven distribution of stress at the bifurcation. The stresses are high at both the crotch and the hip of the bifurcation. (*Please see color version on CD-ROM.*)

elliptical. Because the major axis of the ellipse is larger, the wall stress is much higher along the major axis compared with that in the main aorta. In larger aortas, which have higher wall tension, the wall is usually thicker. Thus, thickness of the aorta increases with its diameter and the diameter-to-thickness ratio remains constant; consequently, the wall stress remains constant. In the case of the bifurcation, however, this type of adaption may not occur fully and the thickness is reduced comparatively. As shown in Figure 2.5, the thickness gradually decreases from the main aorta to the iliac artery, and at the hip of the bifurcation the diameter-to-thickness ratio is the highest, leading to further increase in the wall stress. There is another phenomenon associated with the elliptical cross section. In an elastic artery, the pulse wave tends to change the shape of the cross section from elliptical to circular, which is the lesser energy configuration for the artery wall. This shape change produces additional bending stresses in the wall at the "hip" of the aortic bifurcation. These bending stresses add to the tensile stresses on the inner surface of the artery.

The last stress-enhancing factor we must consider is the arterial curvature (15). What we usually see is that arteriosclerotic plaques often develop on the inner curve of the aortic arch and the inner curve of the tortuous segments of any arteries (16) (Fig. 2.6a). The outer curve is a surface similar to that of a sphere with centers of the two principal radii of the curvatures on the same side of the surface (Fig. 2.6b). The inner surface, on the other hand, has the centers of the two principal radii of curvatures on opposite sides of the surface. It has been well established by Burton (17) and other texts of reference (18) that due to its geometry, the pressure-induced wall stress is much higher on the inner than on the outer curve. This correlates with the observations that atherosclerotic lesions usually favor the inner curve.

At the hip and the crotch of the aortic bifurcation, therefore, the wall stress is increased considerably. This correlates well with the occurrence of atherosclerotic lesions at those locations. The stress analysis of the aortic bifurcation is described in greater detail in Chapter 5.

2.4. Carotid Bifurcation Lesions and Stress

Studying the pathological anatomy of carotid bifurcation lesions, we found that most frequently the less advanced lesions occurred in the sinus bulb area, whereas the more advanced atheromas involved the crotch as well as the entire bifurcation (19) (Fig. 2.7). These observations agree with those reported in the literature (20, 21) and may be explained readily by the distribution of wall stress.

Figure 2.7c shows the geometric parameters that are responsible for producing uneven distribution of the wall stress just as it was the case in the aortic bifurcation. The elliptical cross section at the bifurcation produces high stresses in the "hip" region (areas A and C) as explained earlier. The thickness variation is another important parameter in the current case. The carotid sinus bulb happens to be unusually thin, even thinner than the wall opposite to it. This thin area of the sinus bulb is also the region that has baroreceptors. Obviously, this enhances its sensitivity to pressure because it could stretch more in response to pressure. The reduced wall thickness causes a significant increase in wall stress in the sinus bulb.

A word about baroreceptors may be in order. By far, the best known mechanisms for arterial pressure control is the *baroreceptor reflex*. Basically, this reflex is initiated by stretch receptors called baroreceptors. A rise in pressure stretches the baroreceptors and causes them to transmit signals into the central nervous system, and feedback signals are then sent back through the autonomic nervous system to the circulation to reduce arterial pressure downward toward normal level. Baroreceptors are spray-type nerve endings lying in the walls of the arteries and they are stimulated when stretched. Baroreceptors are extremely abundant in 1) the wall of each internal carotid artery in the area of the carotid

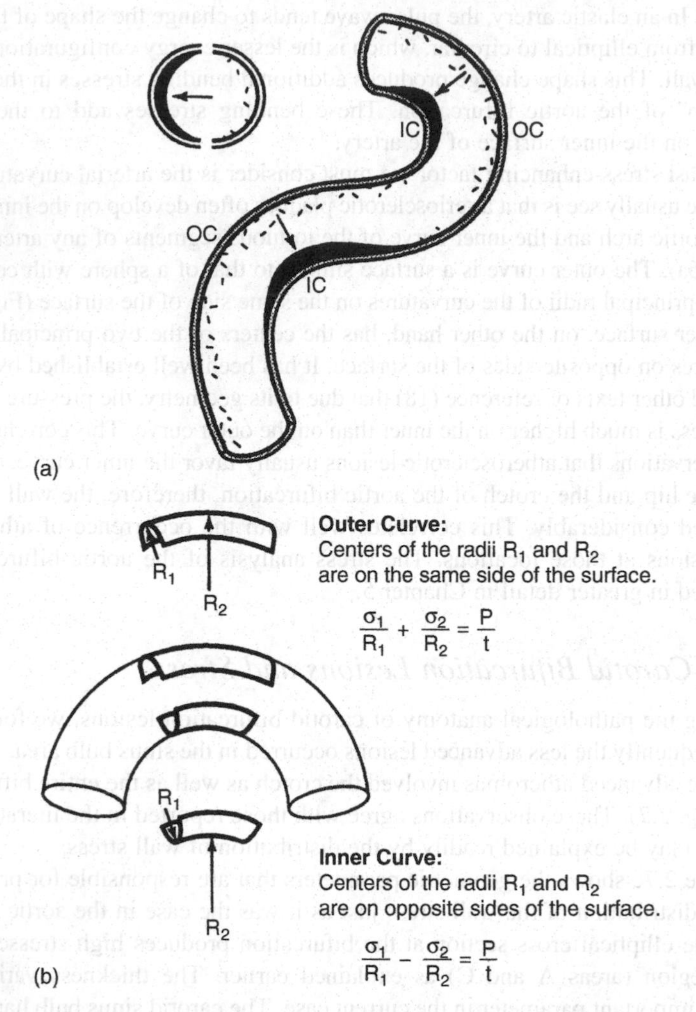
IC
OC
OC
IC
(a)
Outer Curve:
Centers of the radii R1 and R2 are on the same side of the surface.
R1
R2
σ1/R1 + σ2/R2 = P/t
R1
Inner Curve:
Centers of the radii R1 and R2 are on opposite sides of the surface.
R2
σ1/R1 − σ2/R2 = P/t
(b)

FIGURE 2.6. (a) Longitudinal section of the right carotid siphon, which illustrates the location of intimal cushion (IC). In childhood, the subendothelial elastic layer is well developed along the outer walls (OC) of the curvatures, whereas prominent intimal cushions (arrows) consisting mainly of loose connected tissue and sparse elastic elements are present along the inner walls of the curvatures. Inset shows a cross section of the vessel at the site of the arrows. (Reproduced from Wolf S, Werthessen NT, eds. Dynamics of Arterial Flow. Advances in Experimental Medicine and Biology, Vol. 115. Plenum, New York, 1976:357, with permission.) (b) Stress distribution in a curved artery: σ_1 and σ_2 represent, respectively, the circumferential and longitudinal stresses. The outer curve is part of a synclastic surface and the inner curve is part of an anticlastic surface. P represents luminal pressure and t represents wall thickness. Circumferential stresses are higher on the inner curve than on the outer curve.

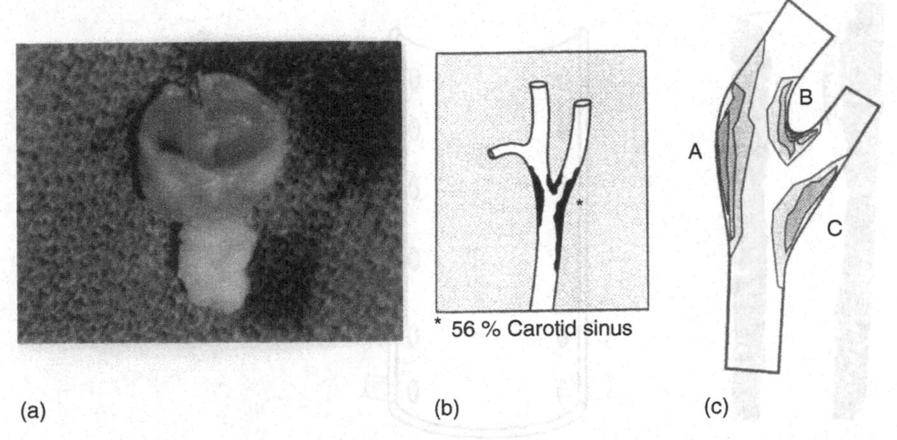

FIGURE 2.7. (a) Atherosclerotic plaque removed from a human carotic artery bifurcation. (b) Locations of atherosclerotic disease at the carotid bifurcation; 56% indicates the relative frequency of occlusive lesions in the brachiocephalic vascular system. Most often, the occlusive lesions occur involving the entire carotid bifurcation (a). (c) Maximum principal isostress contours at the human carotid artery bifurcation. The stresses are high at all three locations, B, A, and C, being the highest at B. The stress contours were obtained from the finite element analysis for a pressure (pulse) loading of 40 mmHg. See Chapter 6 for details. (*Please see color version on CD-ROM.*)

sinus bulb and 2) the wall of the aortic arch. As stated earlier, both the carotid sinus and the aortic arch are extra stretchable and that would make them prone to atherosclerosis.

We performed a stress analysis on the human carotid artery bifurcation using the method of finite element analysis (19). The results of this analysis are shown in Figure 2.7c. We found that the highest stress occurs at the crotch of the bifurcation, then in a decreasing order at the sinus bulb over a substantially large area and finally at the wall opposite the sinus bulb. Thus, the early lesions in the sinus bulb area as well as the advanced lesions involving the crotch and the entire bifurcation correlate well with the areas of high wall stress. The details of the stress analysis of the carotid artery bifurcation can be found in Chapter 6.

2.5. *The Descending Thoracic Aorta Lesions and Flexion Stress*

In the descending thoracic aorta also there is a well-defined pattern of atherosclerotic lesion development. Cornhill (2) and others (22) have described that lesions in this segment occur in the form of two parallel streaks that run along the sides of the intercostal arteries (Fig. 2.8). The flow pattern does not explain this localization but the wall stress hypothesis suggests that because the descending thoracic aorta is tethered to the spine by means of several pairs of intercostal

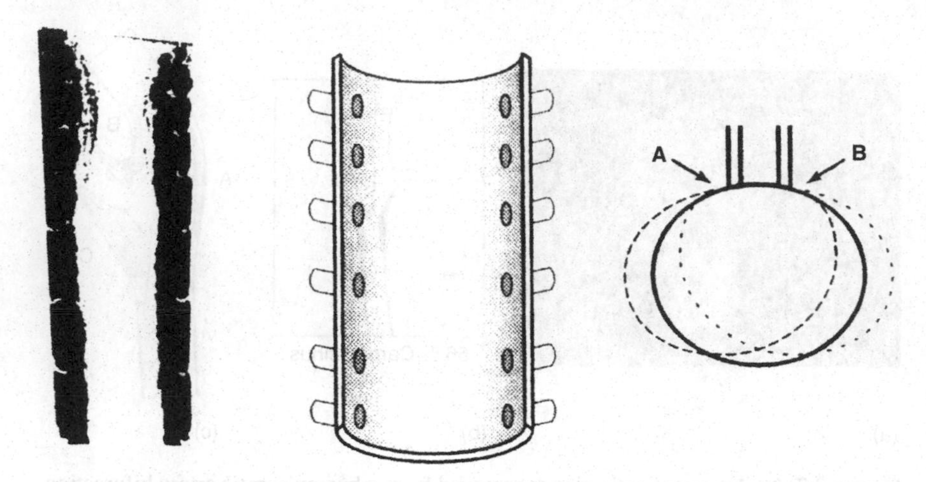

FIGURE 2.8. (Left) Distribution of atherosclerotic lesions in the thoracic aorta in humans. The lesions are present along two parallel lines just on the outside of the pairs of intercostal arteries. (Reproduced from Cornhill JF et al.: In Liepsch DW (ed): Monogram on Atherosclerosis, 1990; 15:15, published by S. Karger AG, Basel, Switzerland, with permission.) (Right) Thoracic aorta in the cross section where the intercostal arteries originate. Any lateral movement of the aorta will be achieved by deformation in regions A and B (i.e., flexion along the two parallel lines). These parallel lines correspond with those along which atherosclerotic lesions develop (left).

arteries, the wall stress may be responsible for this phenomenon. These pairs act as anchors for the aorta so that when the aorta bends during body movement, it changes shape along the two parallel lines by the sides of the intercostal arteries as shown in Figure 2.8. Thus, once again the parallel lines of mechanical flexion of the aorta correlate with the parallel streaks of atherosclerotic lesions.

2.6. *Effect of Blood Pressure and Lesions in the Lower Extremities*

It is well recognized that *elevated blood pressure*, which causes high wall stress, enhances atherosclerosis. As was mentioned earlier, atherosclerosis does not develop in veins under venous pressure but it does when the veins are subjected to arterial pressure. Similarly, atherosclerosis does not develop in *pulmonary arteries* under normal pressure, but it does in pulmonary hypertension. *Lower extremities* are more prone to atherosclerosis than upper extremities, and this also correlates with the pressure, because lower extremities have higher absolute pressure, due to the hydrostatic pressure head, than do the upper extremities (23).

The effect of blood pressure on atherosclerosis seems to be overwhelming in many respects. The pulmonary artery, for example, does not develop atherosclerotic plaques even if serum cholesterol levels are very high (400–500 mg/dL). On the other hand, the same vessels may develop severe atherosclerotic plaques, even in subjects with less than 200 mg/dL serum cholesterol, in the presence of *pulmonary hypertension*. Intimal thickening was consistently detected in the aorta immediately after birth, but it was only the pulmonary artery of the elderly individuals that presented similar thickening. Thus, aortic pressure of 120/80 mmHg can be considered atherogenic while pulmonary pressure of 25/10 mmHg does not have an atherogenic character (24).

In patients with *coarctation of the aorta*, there exists a severe hypertension in the brachial but not in other arterial beds. Postmortem examinations revealed extensive atherosclerosis in the areas subjected to hypertension making it possible to see a clear delineation between hypertensive arterial beds and nonhypertensive arterial beds in the same individual. In experimental animals also, a clear increase in atherosclerosis is seen in the artery proximal to coarctation and a decrease is seen distal to coarctation, when the stenosis is sufficient to cause an increase in pulse pressure of 30 mmHg or more in the proximal segment (24–26). Obviously, the most direct effect of pressure increase is the stress increase in the artery wall.

3. Atherosclerosis in the Aortic Valve and Stress

It needs to be stated that the diseases of the aortic valve, such as aortic stenosis and insufficiency, may not be related to atherosclerosis, however, the process of atherosclerosis has been implicated in the calcification of the valve cusps (27). Our interest in this section is to examine important observations made by Thubrikar et al. (28) on atherosclerotic lesions of the aortic valve in cholesterol-fed rabbits and how that might relate to the mechanical stress in the valve leaflets.

3.1. Methodology

Eight adult (3.2–4 kg weight, 6–8 months old) New Zealand white rabbits were fed a 2% cholesterol diet for 3–33 weeks. After that, the rabbits were sacrificed and the aortic valve, the aorta, and the major arteries were pressurized and stained with Sudan IV stain. The left ventricle was also flushed and stained with Sudan IV so as to stain the ventricular aspect of the aortic valve leaflets. The aortic valve and the arteries were fixed with 4% formaldehyde, in situ, under a diastolic pressure of 70 mmHg. The aortic valves were then dissected and photographed to record the stained areas. Finally, valves were sectioned and stained with hematoxylin and eosin for histologic examination.

In order to study the pressure-bearing area and the redundant area of the aortic valve leaflets, silicone rubber casts of the aortic valves were made under a diastolic pressure of 70 mmHg. To determine the stress in the aortic valve leaflet, they measured 1) the radius of the leaflet from silicone rubber casts, 2) the thickness of the leaflet, and 3) the pressure gradient across the leaflets. The details of the technique can be found in Ref. 28.

3.2. *Findings*

In rabbits fed with cholesterol diet, Thubrikar et al. found that fatty lesions, demonstrated by Sudan staining, developed in arteries and aortic valves after 3–6 weeks (28). In the major arteries, fatty lesions occurred primarily at arterial branching points. In the aortic valve, lesions appeared in load-bearing areas of the leaflets but not in redundant portions (Fig. 2.9). A histologic section shows the location of a 10-week fatty lesion (Fig. 2.10a). The fatty lesion was more cellular than deeper tissues of the leaflet and extended from the base of the leaflet to the fold between redundant and load-bearing areas of the leaflet. It was confined to the aortic surface of the leaflet. No such lesion was apparent on the ventricular surface of the leaflet at 10 weeks. The redundant part of the same leaflet was similarly free of cellular lesions on both faces. The lesion on the aortic face (Fig. 2.10b) consisted of a mass of so-called foam cells and less numerous cells of a second type in a dense feltwork of collagenous fibers. Foam cells in the fatty plaque were moderately large with relatively round or oval euchromatic nuclei and multiglobular, fatty or foamy cytoplasm. The second type of cells, in contrast, usually had darker nuclei and only one or two prominent cytoplasmic fat droplets, which might in turn impinge on and misshape the nucleus. In these features, the second cell type resembled fibroblasts of the deeper-lying leaflet tissue.

The initial atheromatous plaque appeared between 3 and 6 weeks in tissues of the aortic face at the base of the leaflet and consisted mainly of foam cells. After 7 weeks, the lesion had spread up the wall of the aortic sinus and farther out in the leaflet. Again, only the pressure-bearing portion of the aortic face of the leaflet was affected. By 10 weeks, all leaflet tissues were being affected but not in the same manner. The primary fatty plaque was still confined to the upper aortic face, but fibroblasts of the leaflet, both within its dense fibrous layer and in its

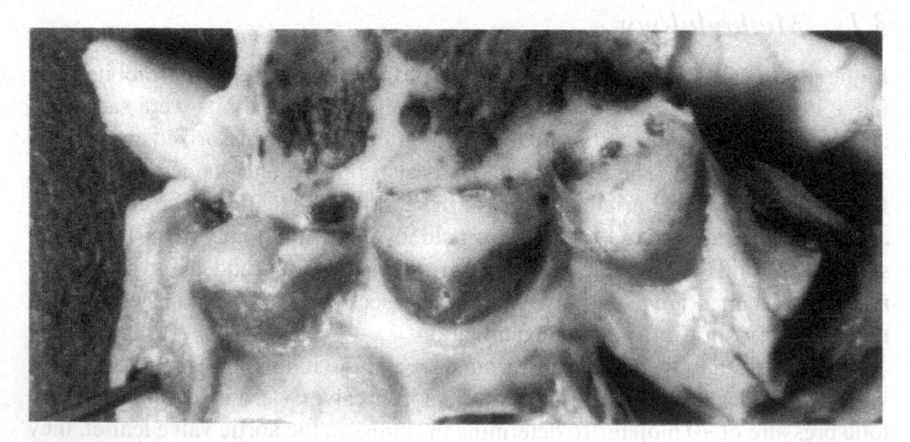

FIGURE 2.9. Photograph of the aortic valve from a rabbit on a 2% cholesterol diet for 11 weeks. Dark areas indicate fatty lesions on the aortic valve, in the coronary arteries, and in the ascending aorta. Fatty lesions occurred only on the load-bearing part of the leaflets.

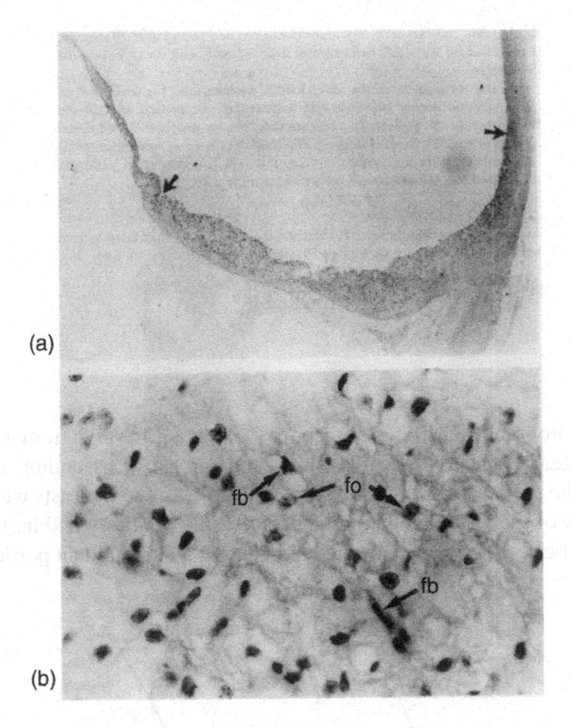

FIGURE 2.10. Histological details from the aortic valve of a rabbit on the cholesterol diet for 10 weeks. (a) Section cut in the radial plane of the leaflet, with free edge of the leaflet to the upper left and aortic sinus to the upper right. A fatty lesion occupies the aortic surface of the leaflet (between arrows). The lesion is characterized by large numbers of foam cells, the nuclei of which appear as dark dots. (b) Portion of the fatty lesion enlarged to show foam cells (fo) and fibroblasts (fb). Foam cells are characterized by a cytoplasm filled with small droplets of fatty material, providing the foamy appearance and have a large, lightly stained nucleus. Fibroblasts tend to be elongated cells with densely stained nuclei and may contain one or two large vacuoles from which lipid has been extracted by the preparation technique.

spongy loose tissue near the ventricular surface, had also taken up fat in one or two large droplets. Even after 33 weeks, the atheromatous plaque had not spread beyond the pressure-bearing aortic face of the leaflet, leaving the redundant portion unencumbered.

The aortic pressure was 90/80 and 85/65 in the two rabbits. The thicknesses of the valve leaflets in these two rabbits were 0.1 mm and 0.12 mm. A study of silicone rubber casts of the closed aortic valves indicated that the pressure gradient was sustained by only part of the leaflet (Fig. 2.11); the rest of the leaflet (its redundant area) was under no pressure gradient. Because the redundant part of one leaflet came in contact with similar parts of the other two leaflets (Fig. 2.12a), diastolic pressure existed on both sides of this part, thereby eliminating the diastolic pressure gradient across it. Casts of the valves of the rabbits

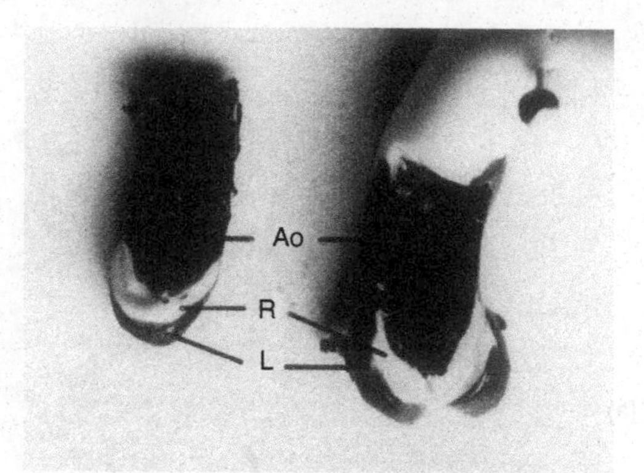

FIGURE 2.11. Photograph of a silicone rubber cast of the aortic valve from a rabbit. In order to visualize the leaflets clearly, a leaflet, sinus, and part of the ascending aorta (Ao) were cut away from the cast and displayed on the left. Also, parts of the casts were painted dark to clarify details of the valve geometry. One leaflet in its entirety and halves of the other two leaflets can be seen. L and R indicate load-bearing and redundant portions of a leaflet, respectively.

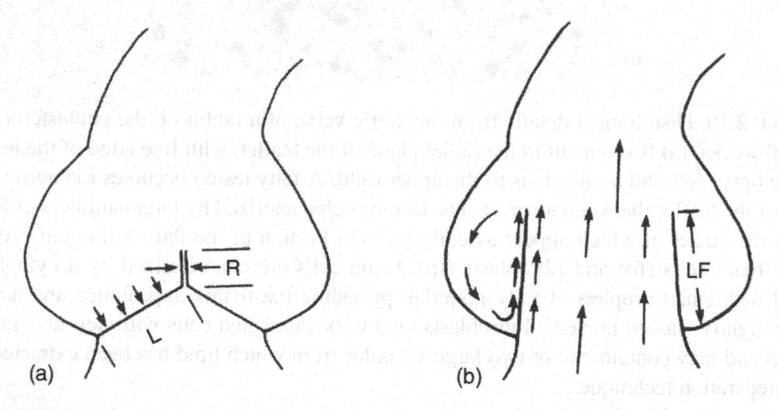

FIGURE 2.12. Schematic drawings of the rabbit aortic valve in the closed (a) and open (b) positions. (a) In the closed valve, only the load-bearing part (L) of the leaflet sustains a pressure gradient across it. The redundant part (R) of the leaflet has pressure on both sides of it and therefore sustains no pressure gradient. L and R in this figure also correspond with L and R of Figure 2.11. (b) Due to forward flow of the blood (central arrows), shear stress occurs on the entire ventricular surface of the leaflet (LF). Similarly, due to vortex formation in the sinus behind the leaflet, shear stress occurs on the entire aortic surface of the leaflet. Only two leaflets are shown for clarity although the valve has three leaflets.

on the cholesterol diet demonstrated that the area occupied by the lesion was the same as the load-bearing area of the leaflet. The silicone rubber casts of the closed aortic valves also showed that the leaflets in the load-bearing part are cylindrical. The radii of the cylindrical leaflet in the two rabbits were 2.3 mm and 2.5 mm.

3.3. *Stress Determination*

In previous studies using dogs, Thubrikar et al. have demonstrated that the orifice of the aortic valve during systole is circular (29, 30). If we assume that aortic valves in dogs and rabbits open in the same way, then the orifice of the opened valve in rabbits should also be circular. Hence, in rabbits, the leaflets of the valve are cylindrical in systole, as their load-bearing part was in diastole. Furthermore, because these leaflets are quite thin, they are likely to behave as a membrane. Therefore, intramural stress (wall stress) in the leaflet may be determined using the equation for stress in a cylindrical membrane (31).

$$\text{Hoop stress (circumferential)} = \frac{PR}{t} \qquad (1)$$

where P is the pressure gradient across the membrane, R is the radius of a cylinder, and t is the thickness of the membrane. The stress in the leaflet is 23 g/mm^2 during diastole (using $P = 70$ mmHg, $R = 2.4$ mm, and $t = 0.1$ mm) and 1.6 g/mm^2 during systole (using $P = 5$ mmHg, $R = 2.4$ mm, and $t = 0.1$ mm). This approach to stress determination has been used previously for dog valves (32), and the values obtained here compare favorably with those for the dog. In fact, it is of interest to note that in spite of the difference in size between valves of the rabbit and the dog, the stress in valve leaflets is similar and is highest in the load-bearing part of the leaflet. A significant amount of stress also occurs along the area of attachment of the leaflet, primarily due to flexion of the leaflets as the valve opens and closes. Although this stress has not been determined in the current study, it has been determined previously for dog valves where it was shown that compressive stresses occur along this area (33).

3.4. *Correlation with Stress*

Concerning the localization of atherosclerotic lesions, it has been thought for some time (34, 35) that the occurrence of these lesions at arterial branch points can be attributed to a disturbance in the flow of blood at the branches. Such an explanation applied to aortic valve leaflets does not account for the observations that fatty lesions occur only in load-bearing areas of the leaflets. For example, from Figure 2.12b it can be seen that in systole, shear stress occurs on the entire ventricular surface of the leaflets as the blood flows out of the ventricle. Similarly, shear stress occurs on the entire aortic surface of the leaflet due to vortex formation in the aortic sinuses (36). In diastole, eddy formation from backflow of blood can produce shear stress on the entire surface of the leaflet, and turbulent flow in the ventricle as it fills can produce shear stress on the ventricular surface of the load-bearing part of the leaflet. Thus, at no time in the cardiac cycle is there shear stress present only on the load-bearing part of the aortic surface of the leaflet where fatty lesions preferentially developed.

Occurrence of the lesions can, however, be explained by the intramural stress that occurs due to the pressure gradient across the valve leaflets. The intramural stress in the load-bearing area in diastole is approximately 15 times greater than the stress in the entire leaflet during systole. Furthermore, occurrence of early stages of the lesions in the area of leaflet attachment can be correlated with the large degree of flexion there. Even after a long period of high-cholesterol diet (33 weeks), the atherosclerotic lesions of the valve were confined to the load-bearing area of the leaflet, reinforcing the implication that wall stress due to pressure is important in this process.

4. How Stress and Stretch Might Influence Atherosclerosis

Endothelial injury, lipid accumulation, and smooth muscle cell proliferation are essential steps in atherosclerosis. There are several observations that link arterial wall stress and stretch to these processes. For example, endothelial cell morphology was observed by us (37) and by others (38, 39) to be different in the branch regions compared with that in the nonbranch regions (see Chapter 8). Cell culture studies on endothelial cells have established that cells orient in response to cyclic stretch so that their long axis is perpendicular to the direction of stretch (40, 41). The cells also proliferate at a higher rate (42) and show development of stress fibers (41) in response to stretch.

Low-density lipoprotein uptake in the artery also has been studied by us (43) and by others (44). We observed that low-density lipoprotein accumulation in the artery is greater in the branch region than in the nonbranch region (see Chapter 8).

The effect of cyclic stretch on smooth muscle cell proliferation also has been studied. We have observed smooth muscle cell proliferation of severalfold when the artery was de-endothelialized and stretched by a balloon compared with when the artery was only de-endothelialized but not stretched (45) (see Chapter 9). Similar observations have been reported by others (46). Cell culture studies on smooth muscle cells also have established that the cell orientation (47) and other cell functions are influenced by the cyclic stretch (48).

These observations suggest a strong relationship between wall stress, wall stretch, cyclic stress, cyclic stretch, and many cellular processes that are an essential part of atherosclerosis.

5. Existing Concepts of Pathogenesis of Atherosclerosis

The development of a coherent theory of origin and progression of atherosclerosis is compromised by lack of agreement about what constitutes the early stage and to some extent by a lack of consensus on the true nature of the lesions (49). Some see lipid in the lesions as the key to the initiation and progression of the disease, whereas others are more impressed by the proliferative aspects, especially the proliferation of SMCs, and yet others see the inflammatory nature of lesions as a central process. Still others attempt to combine all of these aspects, maintaining that the diversity of the disease results from its multifactorial nature.

5.1. Response-to-Injury Hypothesis

Currently, the most widely accepted theory explaining the initiation and development of the disease is the so-called modified response-to-injury hypothesis (49). Table 2.1 shows both, the response-to-injury and modified response-to-injury hypotheses. In the original hypothesis, endothelial injury or dysfunction is the primary event, which is followed by the migration and proliferation of medial SMCs, which form an intimal plaque, which subsequently accumulates lipid. In the modified hypothesis (lipid hypothesis), lipoprotein entry into the intima is the initial event. The monocytes are then attracted to the areas of lipid deposition. The uptake of lipoprotein by monocytes then results in macrophage-rich fatty streak lesion.

The problem with this explanation is twofold: first, not all fatty streaks become lesions; second, many patients with significant atherosclerosis do not have elevated blood lipid levels. Furthermore, early human lesions and fatty streaks contain an overwhelming amount of SMCs. The theory seems more applicable to experimental atherosclerosis created by feeding a high-cholesterol diet to rabbits, for example (49).

TABLE 2.1. Comparison of response-to-injury hypothesis and modified response-to-injury hypothesis.

Response to injury	Modified response to injury
Endothelial injury or stimulus causes release of PDGF or EDGF	Monocytes adhere to endothelium and enter the intima
↓	↓
Smooth muscle cells migrate from media and proliferate in intima	Lipid accumulates in macrophages in the intima
↓	↓
Lipoproteins form complexes with structural proteins, e.g., proteoglycans	Fatty streak disrupts endothelium
↓	↓
Lipoprotein-proteoglycan complexes taken up by smooth muscle cells and macrophages	Platelets adhere and release PDGF
↓	↓
Plaque composed mainly of lipid-laden smooth muscle cells	Smooth muscle cells migrate from media and proliferate in intima
	↓
	Plaque composed mainly of macrophage foam cells with smooth muscle cells at base

EDGF, endothelial cell-derived growth factor; PDGF, platelet-derived growth factor.

6. A New Hypothesis for Pathogenesis of Atherosclerosis: Smooth Muscle Cell Injury Hypothesis

A new hypothesis is proposed by Dr. Thubrikar (the author) in this section. This hypothesis is being proposed for the first time. It is a relatively straightforward hypothesis for a "complex" disease. The disease of atherosclerosis is complex in biological terms; however, from the point of view of vascular mechanics, the disease is really "simple." In mechanics, the failure of a pressure vessel occurs from fatigue in the region of stress-concentration. Similarly, we may expect that the failure of the artery (the pressure vessel) also occurs from fatigue in the region of stress-concentration. Furthermore, in case of nonbiological material, the failure presents itself as a rupture of the pressure vessel; however, in the case of the artery, we may expect failure to present itself as a pathological process. Thus, the hypothesis proposed is, *atherosclerosis is a result of failure of the artery from fatigue in the region of stress-concentration*. Now, we will expand on this concept so that atherosclerosis can be understood in terms of biology of the artery. For this, we need to recall the essential features of the artery and of atherosclerosis.

6.1. Essential Features

1. The artery has cyclic pressure, which produces cyclic stress in the artery wall.
2. The artery has branch ostia and bifurcation regions where stress-concentration occurs.
3. Cyclic load and region of stress-concentration are the two main requirements for fatigue failure.
4. Atherosclerotic plaques occur at branches and bifurcation and that, as proposed, represents fatigue failure of the artery (pressure vessel) there.
5. Atherosclerosis increases with mean pressure and pulse pressure just as fatigue failure increases (occurs sooner) with mean pressure and pulse pressure.
6. Atherosclerosis is inhibited by decreased transmural pressure, for example in the externally supported artery, which has less transmural pressure (less wall stress), just as fatigue failure is prevented by substantially reduced wall stress.
7. Atherosclerosis is reduced by reduced heart rate (as with β-blocker treatment) just as fatigue failure is reduced (delayed) with a reduced number of cycles imposed.
8. Atherosclerosis does not occur in the pulmonary artery and in the vein (unless substantially higher pressure is imposed) just as fatigue failure does not occur unless the stress is above a certain threshold value.
9. Atherosclerosis occurs at a pressure much lower than the bursting pressure of the artery, just as fatigue failure occurs at a stress much lower than the breaking stress of the material.
10. Atherosclerosis represents a process that produces a plaque, which grows with time just as fatigue represents a process where small damages accumulate over time to produce a failure.

Now, let us consider the cellular mechanism of atherosclerosis (fatigue failure).

6.2. Cellular Mechanism of Atherosclerosis

1. Stress in the wall of the artery, induced by pressure, is sustained primarily by the media (i.e., by smooth muscle cells and connective tissue).

2. Due to stress-concentration at the branches and bifurcations, there exist both higher pulsatile stress and higher pulsatile strain (stretch) in the media at these locations. Thus, SMCs are under higher cyclic tension and cyclic strain at these locations.
3. The wall stress is generally highest on the inner surface of the artery and decreases through the thickness toward the outside. Therefore, SMCs of the inner media are under the highest cyclic tension and cyclic strain.
4. Because of higher cyclic tension and cyclic strain, the SMCs of the inner media at the branches and bifurcations are easily "injured"; their cell-to-cell junction can be easily damaged and their cell-to-cell contact can be broken.
5. Injury to SMCs at these particular locations resulting from fatigue failure of the artery is the key to arterial disease including atherosclerosis.
6. SMCs are stimulated to proliferate both by cyclic strain and by removal of cell-to-cell contact inhibition.
7. The same forces produce changes in the endothelial cell morphology and in the connective tissue at these locations.
8. The same forces, through their influence on both the cells and the stretching of the tissue, promote increased penetration of LDL particles in this region.
9. LDL particles then serve as a promoting factor for SMC proliferation there.
10. Other factors that promote SMC proliferation at that location are hypertension, diabetes, cigarette smoking, and so forth.
11. Hence, SMCs located at the inner media, at the branches and bifurcations, become injured due to fatigue damage from pulsatile pressure, and this is the initial step in the disease of the artery.
12. When they proliferate due to injury stimulus, then atherosclerosis begins. This initial step does not require but can be aided by the presence of other agents like LDL.
13. The presence of a higher level of LDL, hypertension, diabetes, smoking, and so forth, is not required for the initiation or growth of atherosclerotic lesion; however, these factors can accelerate the disease process through enhancement of SMC proliferation.

7. Pathogenesis of Atherosclerosis and Aneurysm

Now, let us consider both atherosclerosis and aneurysm, the most common diseases of the artery. The basic premise used for the diseases of the artery is that all pressure vessels fail from fatigue in the region of stress-concentration. This premise was developed further to the point that fatigue failure presents itself as an injury to SMCs at the inner media in the branch and bifurcation regions. Obviously then, the injury to SMC could also result in cell death (apoptosis), which is the initial step in the development of aneurysms. Thus, we may express the unified mechanism of the arterial diseases as follows:

7.1. Proposed Pathogenesis of Atherosclerosis and Aneurysm

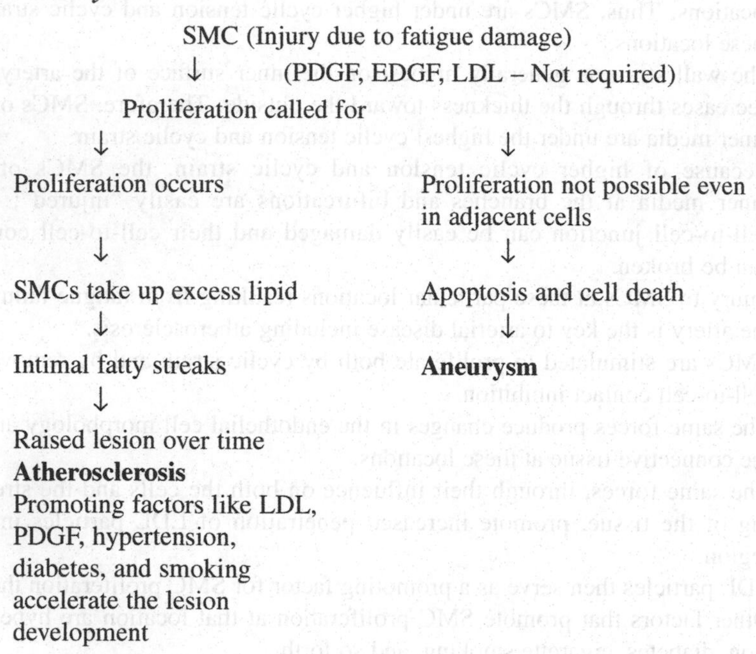

The main distinguishing feature of the proposed mechanism for atherosclerosis, from the previous mechanisms, is that the proposed mechanism does not require EDGF, PDGF, or LDL either for the initiation or for the growth of atherosclerotic lesion. It does not depend on the endothelial injury/dysfunction or lipid infiltration. The primary factor for both initiation and growth of the lesion is the same, and that is pulsatile pressure producing fatigue damage in the region of stress-concentration. It may be recalled that, in the artery, with each arriving pulse of pressure, the branch region stretches double the amount of the straight region (6% vs. 2.8%) (see Chapter 6). This differential stretching of SMC is the key event. It is important to remember also the arrangement of SMCs and connective tissue at the branch. In particular, not all of the SMCs are oriented in the same direction, and this imposes an additional challenge to the tissue when it has to stretch and even a bigger challenge when it has to stretch more. Through the thickness of the wall, when different layers of SMCs have different cell orientations, then the stretching of the wall imposes completely different forces on the cells, the forces that can only be more damaging to the cells (see Chapter 3). Furthermore, we have also noted that SMCs show a higher rate of cell replication at the branch in the normal artery, which would be expected in light of the challenges described above (see Chapter 9). According to the proposed mechanism, atherosclerotic lesions will begin and grow in all cases but perhaps at a slow rate. The presence of risk factors such as high level of cholesterol, high blood pressure, diabetes, smoking, and so forth, increases the rate of SMC proliferation and thereby accelerates the disease process.

7.2. *Examples*

The above mechanism explains many observations described in this book and is illustrated by the following examples:

1. Why atherosclerotic plaques are most often eccentric? Because atherosclerosis begins on the side of the artery near the ostium or bifurcation.
2. How does one explain atherosclerosis in the straight segment of the artery, though an infrequent occurrence? Because there can be regions of small defects or weaknesses in the straight segment, and these regions then act like regions of high local stress and strain.
3. Why there is a close relationship between the two diseases (atherosclerosis and aneurysm), such as both occurring together in the abdominal aorta (see Chapter 15)? Because they share a similar mechanism for their genesis.
4. Why do intracranial aneurysms occur at the branch sites while extracranial arteries develop atherosclerosis at the branch sites?
5. Why in the intracranial arteries do aneurysms and atherosclerosis occur in close proximity to each other (see Chapter 14)?
6. Why do anastomotic aneurysms and anastomotic intimal hyperplasia occur at the same location (see Chapter 13)?
7. Why are many of the risk factors the same for both diseases?
8. Why do many of the drugs that protect against one disease also protect against the other?
9. This is the known mechanism of failure for nonbiological material.

We may also want to examine the stretch aspect of the coronary arteries in light of the proposed hypothesis. We know that coronary arteries are subjected to more cyclic stretch due to cardiac contraction than all other arteries. During the growth period, up to 20 years of age, atherosclerosis is not noted in any of the arteries except in coronaries. It could be speculated that the growth processes are able to mitigate the processes of atherosclerosis in all of the arteries except in coronaries, because the coronaries are subjected to more stretch than all others. The same reasoning suggests that atherosclerosis would occur more in coronaries than in other arteries.

References

1. DeBakey ME, Lawrie GM, Glaeser DH: Patterns of atherosclerosis and their surgical significance. Ann Surg 1985;201:115-131.
2. Cornhill JF, Herderick EE, Stary HC: Topography of human aortic sudanophilic lesions. In: Liepsch DW (ed). Blood Flow in Large Arteries: Applications to Atherogenesis and Clinical Medicine. Basel, Karger, 1990;15:13-19.
3. Fry DL: Acute vascular endothelial changes associated with increased blood velocity gradients. Circ Res 1968;22:165.

4. Texon M: The hemodynamic basis of atherosclerosis. Further observations: The ostial lesion. Bull N Y Acad Med 1972;48:733-740.
5. Caro CG, Fitz-Gerald JM, Schroter JM: Arterial wall shear and distribution of early atheroma in man. Nature 1969;223:1159.
6. Ku KN, Giddens DP, Zarins CK, Glagov S: Pulsatile flow and atherosclerosis in the human carotid bifurcation. Arteriosclerosis 1985;5:293-302.
7. LoGerfo FG, Nowak MD, Quist WC, Crawshaw HM, Bharadvaj BK: Flow studies in a model carotid bifurcation. Arteriosclerosis 1981;1:235-241.
8. Rodkiewicz CM: Localization of early atherosclerotic lesions in the aortic arch in the light of fluid flow. J Biomech 1975;8:149-156.
9. Asakura T, Karino T: Flow patterns and spatial distribution of atherosclerotic lesions in human coronary arteries. Circ Res 1990;66:1045-1066.
10. Harvey JF: Theory and design of modern pressure vessels. Van Nostrand Reinhold Company, New York, 1974:251, 316, 338.
11. Thubrikar MJ, Roskelly SK, Eppink RT: Study of stress concentration in the walls of the bovine coronary arterial branch. J Biomech 1990;23:15-26.
12. Thubrikar MJ, Manuel L, Eppink RT: Intramural stress at arterial bifurcation *in vivo* [Abstract]. Proceedings of the 40th Annual Conference on Engineering in Medicine and Biology, Niagara Falls, 1987;29:208.
13. Chuong CI, Fung YC: Three dimensional stress distribution in arteries. J Biomech Eng 1983;105:268-274.
14. Velican C, Velican D: Natural History of Coronary Atherosclerosis. CRC Press, Boca Raton, 1989:191.
15. Willis GC: Localizing factors in atherosclerosis. Can Med Assoc J 1954;70:1-9.
16. Hemodynamic contribution to atherosclerosis. In: Wolf S, Werthessen NT (eds). Dynamics of Arterial Flow. Advances in Experimental Medicine and Biology, Vol. 115. Plenum Press, New York, 1976:357.
17. Burton AC: Physical principles of circulatory phenomena: The physical equilibria of heart and blood vessels. In: Handbook of Physiology, Vol. 1, Circulation. 1961: 85-106.
18. Harvey JF: Theory and Design of Modern Pressure Vessels. Van Nostrand Reinhold Company, New York, 1974:39.
19. Thubrikar M, Salzar R, Eppink R, Nolan S: Pressure induced mechanical stress in carotid artery atherosclerosis [Abstract]. 9th International Symposium on Atherosclerosis, Rosemont, IL, 1991:73.
20. Zarins CK, Glagov S, Giddens DP: What do we find in human atherosclerosis that provides insight into the hemodynamic factors in atherogenesis? In: Glagov S, Newman WP, Schaffer SA (eds). Pathobiology of the Human Atherosclerotic Plaque. Springer Verlag, New York, 1990:317-332.
21. Robicsek F: Atherosclerotic occlusive disease of the innominate and subclavian arteries. In: Robicsek F (ed). Extracranial Cerebrovascular Disease. Macmillan Publishing Company, New York, 1986:360.
22. Born GVB: Determinants of mechanical properties of atherosclerotic arteries [Abstract]. Ann Biomed Eng Abstr Suppl 1993;21:29.
23. Burton AC: Physiology and Biophysics of the Circulation. Year Book Medical Publishers, Chicago 1972:99.
24. Velican C, Velican D: Natural history of coronary atherosclerosis. CRC Press, Boca Raton, pg. 185-186.

25. Lyon RT, Runyon-Hass A, Davis HR, Glagov S, Zarins CK: Protection from atherosclerotic lesion formation by reduction of artery wall motion. J Vasc Surg 1987; 5:59-67.
26. Bomberger RA, Zarins CK, Glagov S: Subcritical arterial stenosis enhances distal atherosclerosis. J Surg Res 1981;30:205-212.
27. Mohler ER III: Are atherosclerotic processes involved in aortic-valve calcification? Commentary. Lancet 2000;356:524-525.
28. Thubrikar MJ, Deck JD, Aouad J, Chen JM: Intramural stress as a causative factor in atherosclerotic lesions of the aortic valve. Atherosclerosis 1985;55:299-311.
29. Thubrikar MJ, Harry RR, Nolan SP: Normal aortic valve function in dogs. Am J Cardiol 1977;40:563-568.
30. Thubrikar MJ, Bosher LP, Nolan SP: The mechanism of opening of the aortic valve. J Thorac Cardiovasc Surg 1979;77:863.
31. Harvey JF: Theory and Design of Modern Pressure Vessels. Van Nostrand Reinhold, New York, 1974:32.
32. Thubrikar MJ, Piepgrass WC, Deck JD, Nolan SP: Stresses of natural vs prosthetic aortic valve leaflets in vivo. Ann Thorac Surg 1980;30:230.
33. Deck JD, Thubrikar MJ, Schneider PJ, Nolan SP: Structure, stress, and tissue repair in aortic valve leaflets, Proc Annu Conf Eng Med Biol 1979;21:169.
34. Nerem RM: Hemodynamic contribution to atherosclerosis. In: Wolf S, Werthessen NT (eds). Dynamics of Arterial Flow. Plenum Press, New York, 1979:384.
35. Mohnssen HM, Kratzer M, Baldauf W: Microthrombus formation in models of coronary arteries caused by stagnation point flow arising at the predilection sites of atherosclerosis and thrombosis. In: Nerem RM, Cornhill JF (eds). The Role of Fluid Mechanics in Atherogenesis. Ohio State Univ. Exp. Sta., Columbus, 1978:12.
36. Bellhouse BJ: The fluid mechanics of the aortic valve. In: Ionescu MI, Ross DN, Wooler GH (eds). Biological Tissue in Heart Valve Replacement. Butterworths, London, 1972:23.
37. Baker JW, Thubrikar JM, Parekh JS, Forbes MS, Nolan SP: Change in endothelial cell morphology at arterial branch sites caused by a reduction of intramural stress. Atherosclerosis 1991;89:209-221.
38. Reidy MA, Langille BL: The effect of local blood flow patterns on endothelial cell morphology. Exp Mol Pathol 1980;32:276.
39. Zarins CK, Taylor KE, Bomberger RA, Glagov S: Endothelial integrity at aortic ostial flow dividers. Scanning Electron Microsc 1980;3:249.
40. Ives CL, Eskin SG, McIntire LV: Mechanical effects on endothelial cell morphology: In vitro assessment. In Vitro Cell Develop Biol 1986;22:500-507.
41. Sumpio BE, Banes AJ, Buckley M, Johnson G: Alteration in aortic endothelial cell morphology and cytoskeletal protein synthesis during cyclic tensional deformation. J Vasc Surg 1988;7:130-138.
42. Sumpio BE, Banes AJ, Levin LG, Johnson G Jr: Mechanical stress stimulates aortic endothelial cells to proliferate. J Vasc Surg 1987;6:252-256.
43. Thubrikar MJ, Keller AC, Holloway PW, Nolan SP: Distribution of low density lipoprotein in the branch and non-branch regions of the aorta. Atherosclerosis 1992;97:1-9.
44. Schwenke DC, Carew TE: Initiation of atherosclerotic lesions in cholesterol-fed rabbits. II. Selective retention of LDL vs. selective increases in LDL permeability in susceptible sites of arteries. Arteriosclerosis 1989;9:908.

45. Thubrikar MJ, Moorthy RR, Deck JD, Nolan SP: Smooth muscle cell proliferation: Is it due to endothelial injury or aortic stretch? [Abstract]. 10th International Symposium on Atherosclerosis, Montreal, 1994;100, 109.
46. Clowes AW, Clowes MM, Reidy MA: Role of acute distension in the induction of smooth muscle proliferation after endothelial denudation [Abstract]. Fed Proc 1987;46:720.
47. Sumpio BE, Banes AJ: Response of porcine aortic smooth muscle cells to cyclic tensional deformation in culture. J Surg Res 1988;44:696-701.
48. Sottiurai VS, Kollros P, Glagov S, Zarins CK, Mathews MB: Morphologic alteration of cultured arterial smooth muscle cells by cyclic stretch. J Surg Res 1983;35:490-497.
49. Damjanov I, Linder J (eds). Anderson's Pathology, Tenth Ed. Mosby, Chicago, 1996:1406.

3 Structure and Mechanics of the Artery

Michel R. Labrosse

1. Introduction

The circulatory system is made up of the heart (a central pump) and a vast array of tubes (arteries) that carry the blood away from the heart to the periphery and tubes (veins) that carry the blood from the periphery back to the heart. The cardiovascular system provides a continuous circulation of the blood in a system that is virtually closed. The aorta is a single systemic artery emerging from the heart. It gives origin, by successive branching, to hundreds of arteries of progressively smaller caliber. In an adult man, the aorta is about 30 mm in the outer diameter. Along the arterial tree, the arteries successively decrease in size, increase in number, undergo structural changes, and finish in arterioles that are as little as 10 μm in diameter. Roughly three-quarters of the blood volume resides in veins at low transmural pressure and one-quarter resides in arteries at high transmural pressure. Veins are therefore considered as a volume reservoir whereas arteries are a pressure reservoir.

The arteries have a dual function: they are conduits for blood flow and reservoirs for blood pressure. Also, due to the pumping action of the heart, both blood pressure and blood flow are pulsatile. In the cross section, the artery has a circular profile and generally a uniform thickness (Fig. 3.1). The structure of the artery is quite complex, as seen in Figure 3.1. The main components of the vessel wall are endothelium, smooth muscle cells, elastic tissue, collagen, and connective tissue. Arteries are prime targets for diseases such as *atherosclerosis* or *aneurysms* that each year claim the lives of scores of people worldwide. For these reasons, blood vessels have been the focus of active research for decades. Arteries present fascinating challenges related to their complex structure and behavior and the variability inherent to living tissues. Many studies (1) point out that cardiovascular disease may be triggered or aggravated by mechanical stimuli, such as wall stress or stretch resulting from the blood pressure, or shear stress resulting from the blood flow.

It is important to know how the artery is affected by the blood pressure or the blood flow. The heart offers at least two typical examples of such evolutions: the left ventricle is thicker than normal in patients with high blood pressure, while it is larger than normal in patients with a leaking aortic valve, to ensure that the net

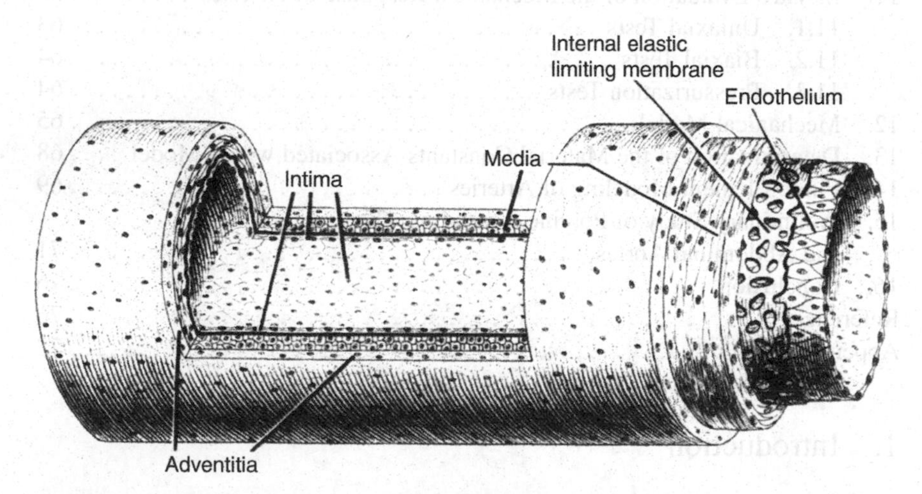

FIGURE 3.1. Drawing of a medium-sized (muscular) artery showing its layers. (Reproduced from Junqueira LC, Carneiro J: Basic Histology, 1980, pg 241, LANGE Medical Publications, Los Altos, Ca, 94022, with permission)

blood flow supplied by the heart is near normal. Similarly, it is known that the artery wall becomes thicker in patients with high blood pressure and the artery diameter becomes larger when it has to carry more blood volume. Along these lines we may ask, could the mechanical stimuli that induce adaptation response also cause damage to the artery? For instance, the artery is a pressure vessel subjected to cyclic loading; therefore, could there be a parallel between the mechanical failure of nonbiological pressure vessels and the occurrence of arterial diseases? In other words, if arteries were not afflicted by atherosclerosis or aneurysms, how would problems typical of pressure vessels be expected to manifest themselves in living arteries? Inquiries of this kind make the mechanics of arteries a very relevant area of investigation.

This chapter presents an overview of the most significant aspects of arterial mechanics. Central to the arteries' mechanical properties and response to mechanical stimuli are the composition and structure of the arterial wall (Fig. 3.1). Therefore, histological details will be presented. The changes observed between straight segments and arterial branches or bifurcations will be illustrated because these aspects have received little attention, although they may be extremely important with respect to mechanically induced pathogenesis. The baroreceptors, effect of posture, tethering force, residual stress, and smooth muscle cell activity will be discussed as key factors to keep in mind when studying arterial mechanics. The changes along the arterial tree and the influence of aging on the arteries' mechanical behavior will be described. The in vivo and in vitro evaluation of the mechanical response of arteries will be addressed, leading to the establishment of an anisotropic hyperelastic model. Details on the identification of the material constants and numerical implementation of the model will be given, along with recommendations for finite element modeling using commercial software. Finally, many of the topics discussed will be illustrated by a preliminary in vitro study we carried out on porcine aortas.

Unless noted otherwise, the main sources of information in this chapter are the excellent works by Humphrey (2), Fung (3), Dobrin (4), Silver (5), and McDonald (6). For more material on arterial histology and mechanics, additional references are provided.

2. Composition of Arteries

The principal organized constituents of the arterial wall are the smooth muscle cells and the extracellular matrix, composed of elastin, collagen, and ground substance (Fig. 3.1). These components are present in various quantities, orientations, and interconnections along the vascular tree. Because the mechanical properties of a material depend on its composition, structure, and microstructure, the arterial wall exhibits a wide range of behaviors depending on the location in the body.

Smooth muscle cells are spindle-shaped, typically 100 μm long and about 5 μm in diameter (Fig. 3.1). Each cell's nucleus reflects this long, narrow shape, which is an indicator of the principal direction of the individual muscle cell. As involuntary muscle, the cells can undergo partial contraction for long periods of time. In vivo, smooth muscle cells in most vessels tend to be oriented helically, almost circumferentially, but there are local variations, as will be seen. Such arrangements allow smooth muscle cell activity to alter the distensibility of large arteries and to control the luminal diameter of medium and small arteries. Another essential feature of arterial smooth muscle cells is that they synthesize elastin and collagen, thereby being primarily responsible for the maintenance or remodeling of the arterial extracellular matrix. For remodeling, they can also proliferate and change the wall thickness.

The extracellular matrix gives the arterial tissue its strength and basic shape as elastin and collagen are arranged into reinforcement structures. Elastic fibers consist of two components: 1) an amorphous central core of elastin surrounded by 2) a sheath of fibrillin microfibrils. The principal component (up to 90%) of mature elastic fibers is an amorphous material elastin and the fibers can be from 0.2 to 5.0 μm in diameter. These fibers are arranged in networks or sheets and can be stretched by more than 150% without breaking. Elastic fibers can remain stable for years in the body, unless prematurely broken down by increased proteolytic activity as observed in aortic aneurysms (2). Collagen fibers are made of collagen microfibrils and fibrils, ranging from 0.5 to 20 μm in diameter. The collagen fibers are usually undulated at physiologic loads but can only stretch by about 10% without breaking when straight and are much stiffer than elastic fibers. The half-life of collagen in the cardiovascular system varies between 0.5 and 3 months as a result of a balance mechanism between synthesis and degradation. This regulatory system may be shifted toward collagen synthesis in response to injury or disease (1).

3. Different Types of Arteries

Although the individual diameter of the arteries decreases as they travel from the heart toward the periphery, their combined diameter increases because the number of branches becomes very large. This leads to a progressive fall in average blood pressure and flow velocity. Parallel to these conditions, the arterial thickness and microstructure vary along the vascular tree. Although these variations are gradual, and some arteries exhibit mixed features, one distinguishes three types of arteries by properties and sizes (7):

i) *Elastic arteries* are larger vessels located close to the heart. They include the aorta, the branches originating from the aortic arch, and the pulmonary artery. The passive elasticity of these arteries allows them to expand when the heart ejects blood under high pressure and to retract when the heart relaxes. This phenomenon is crucial in augmenting the forward flow of blood and maintaining the flow throughout the cardiac cycle. It also plays a major role in the coronary circulation of the heart muscle itself.

ii) *Muscular arteries* (or conducting arteries) are essentially medium-diameter vessels located closer to the tissues they supply. They include the carotid, brachial, and iliac arteries, as well as the coronary arteries. Muscular arteries can regulate the local blood flow by vasoconstriction or vasodilatation, and this ability comes largely from the smooth muscle cells, which are the dominant elements in their walls.

iii) *Arterioles* are the smallest type of arteries; they feed the capillary beds of organs and tissues. Arterioles have a relatively small lumen and a thick muscular wall that makes them the major site of control of the peripheral resistance to blood flow, again by vasoconstriction or vasodilatation. In particular, an elevated tone in the smooth muscle cells of the arteriolar wall directly raises arterial pressure in the vascular system while relaxation of the tone lowers the pressure.

4. Structure of Arteries

All arteries are made of three concentric layers, or tunics: the innermost tunica intima, the middle tunica media, and the outermost tunica adventitia (Fig. 3.1).

The *intima* consists of one layer of 0.2- to 0.5-μm-thick endothelial cells held on a 1-μm-thick basal lamina mostly made of collagen. The intima is essentially similar in all arteries. However, in the aorta and larger muscular arteries including the coronaries, it may also include a subendothelial layer of connective tissue and some axially oriented smooth muscle cells. The normal intima has a negligible structural role because it is very thin. Yet, intimal hyperplasia or other conditions can increase its mechanical contribution. In any case, the intimal layer has a tremendous importance as a nonclotting interface with blood and a gateway for molecule transport to and from the bloodstream. A variably thick internal elastic lamina usually serves as a prominent boundary between the intima and the media.

The *media* begins at the internal elastic lamina that lines the intima and extends to a much less prominent external elastic lamina next to the adventitia. Both these laminae are perforated sheets of melded elastic fibers that permit the transmural transfer of water, nutrients, and electrolytes as well as cellular communication between the adjacent arterial tunics. The external elastic lamina is less marked in muscular arteries and does not exist in cerebral arteries. The media consists of concentric patterns of elastic fibers and smooth muscle cells. In an elastic artery, the pattern consists of elastic lamellar units or musculo-elastic fascicles whose number and thickness vary along the vascular tree (up to 60, about 15 μm thick in the thoracic aorta; up to 30, about 20 μm thick in the abdominal aorta). Each unit can be seen as layers of smooth muscle cells separated by 3-μm-thick fenestrated sheets of elastic fibers similar to the internal and external elastic laminae. Woven in between the elastin sheets are bundles of tiny collagen fibrils. As one proceeds from elastic to muscular arteries, the concentric layers become predominantly muscular, including up to three dozen layers of smooth muscle cells, although elastic layers also tend to be present in arteries of the arms and legs. Like elastic layers, collagenous fibers are reduced in favor of muscle.

Lastly, the *adventitia* makes up about 10% of the wall thickness in an elastic artery and considerably more in a muscular artery. The adventitial layer is essentially a dense network of collagen fibers interspersed with fibroblasts, elastic fibers, nerves, and vasa vasorum, tiny vessels providing the vessel's own blood supply. The collagen fibers in the adventitia are longitudinally oriented; along with those in the media, they provide the arterial wall with enough strength to prevent overdilatation under physiologic loads. Interestingly, the adventitia is almost absent in cerebral arteries. In arterioles, the vessels are reduced to the lining endothelium, a prominent internal elastic lamina and several layers of smooth muscle.

5. Macro- and Microstructural Changes Associated with Arterial Branches and Bifurcations

The general description given thus far dealt with the histology of straight cylindrical arterial segments. However, significant changes in the macro- and microstructure of the arterial wall can be observed where arteries branch off. Although branches have been the focal point of numerous studies on the blood flow and its potential role in pathogenesis, the detailed geometry and histology at the branches have received little attention. Figure 3.2 shows a longitudinal section at the branch region where a canine iliac artery splits into deep femoral and superficial femoral arteries (fixed in formaldehyde). In the figure, the main vessel gives rise to a side branch at about a 55-degree angle. While the arterial wall thickness is uniform away from the branch, there is a noticeable increase in wall thickness at the branch region. The wall is thicker at the proximal region of the branch ostium. In the distal region of the branch ostium, which is also the flow divider, the wall thickness increases significantly to a point of producing almost a sharp wedge. Such a sharp increase in the wall thickness at the main artery-branch intersection is a reminder of the reinforcements used at pressure vessel intersections in order to reduce stress-concentration (see Chapter 5). Such patterns of nonuniform thickness distribution at branches are not isolated, rather they are present throughout the body.

Parallel to the wall thickness changes observed at branches and bifurcations, there are significant variations in the arterial wall microstructure. One sees this in longitudinal sections of the aortic bifurcation of rabbits. (These observations and

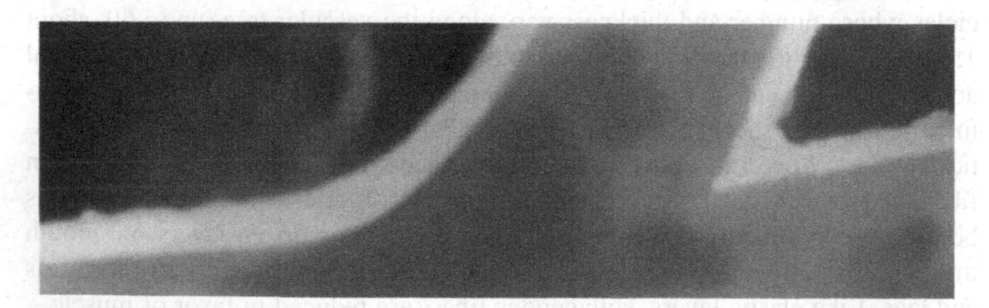

FIGURE 3.2. Longitudinal section of a canine iliac artery, branching into deep and superficial femoral arteries, showing wall thickening at the flow divider and across from it.

the material on which they are based are part of an on-going research project by Dr. J.D. Deck, Department of Cell Biology, University of Virginia, Charlottesville, VA, USA.) In the straight segment of the artery, smooth muscle cells and other elements of the arterial tunica media are wrapped circumferentially around the vessel in a tight helix (Fig. 3.3). The cells therefore can be seen to be cut across their long axis, in longitudinal histological sections, and they appear as small triangle-to-circular reddish structures (Fig. 3.3). Many of the smooth muscle cells are cut

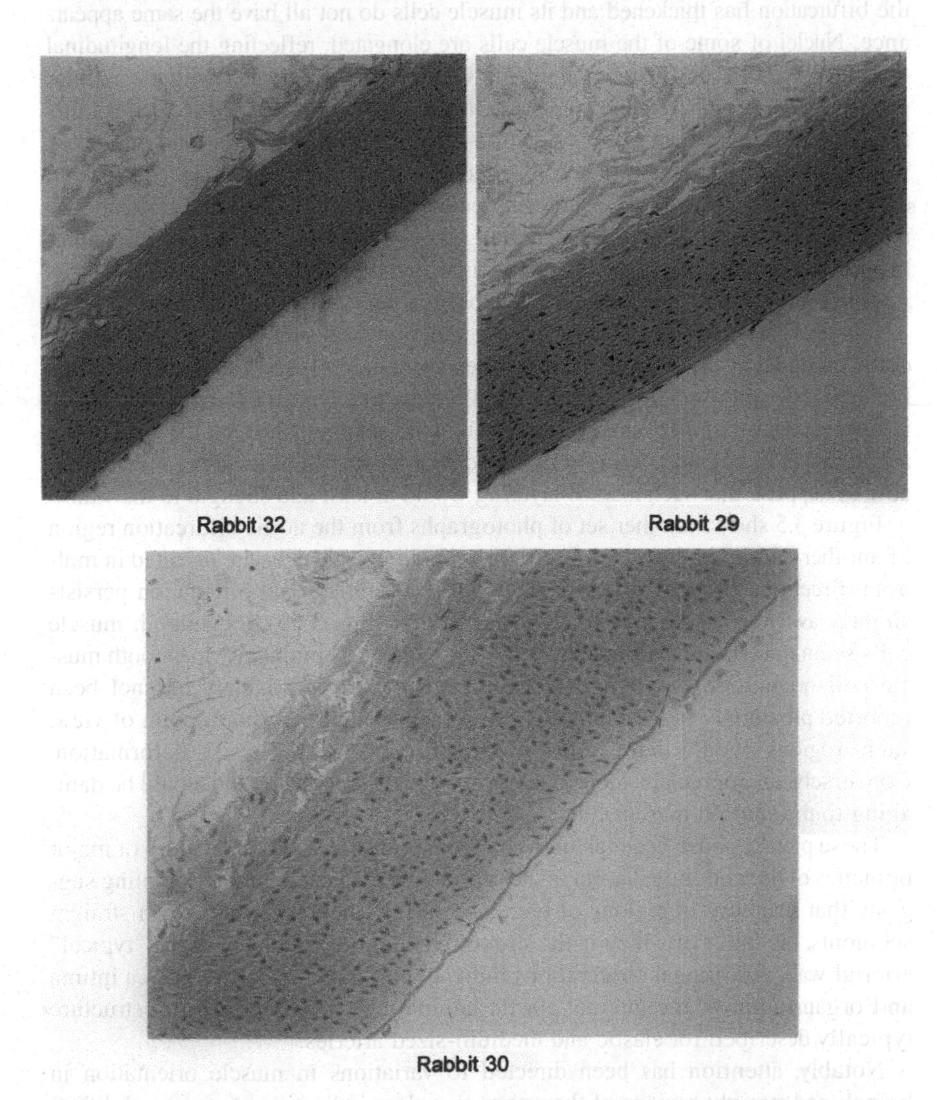

FIGURE 3.3. Longitudinal hematoxylin and eosin–stained sections of the rabbit abdominal aorta slightly away from the aortic bifurcation. The smooth muscle cells have been cut along their long axes as they are wound around the vessel. (*Please see color version on CD-ROM.*)

through the cell center and display the slightly smaller round, purplish nucleus. The significant fact to note is that almost all the smooth muscle cells have the same orientation (i.e., they have been cut across their long axis uniformly as they are wound around the vessel).

If the histology of these straight arterial segments is now compared with the structure of the same arteries where they have bifurcated to enter the legs (aortic bifurcation; Fig. 3.4), a different picture emerges: the muscular wall within the bifurcation has thickened and its muscle cells do not all have the same appearance. Nuclei of some of the muscle cells are elongated, reflecting the longitudinal orientation of the cells that contain them. Other nuclei are smaller with the circular profile typical of transected nuclei and cells. In other words, the muscle cells of this region run in more than one direction and may even lie in more than one layer. Figure 3.4 shows smooth muscle cells oriented at the branch (crotch) in at least three different directions: one layer contains cells oriented along the left arm of "V," another layer contains cells oriented along the right arm of "V," while yet another layer contains cells that appear perpendicular to the plane of the photographs. This interweaving of cells at the crotch of the bifurcation is seen in all the rabbits.

It is also apparent that the stratification and mixed orientation of smooth muscle cells, observed at the bifurcation, continues for some distance into the walls of the branches before the solely circumferential orientation, typical of straight segments, is once again restored. Whatever else might be accomplished by the interlacing arrangement of smooth muscle cells at and near an arterial bifurcation, it is reasonable to suppose that such a cross-layering would at least add strength to the wall.

Figure 3.5 shows another set of photographs from the aortic bifurcation region of another rabbit. At the crotch itself, the smooth muscle cells are oriented in multiple directions (Figs. 3.5e and 3.5e'). This multidirectional orientation persists slightly away from the crotch. In some regions (Fig. 3.5b), the smooth muscle cells seem to be braided or interwoven. This degree of complexity in smooth muscle cell orientation is not only very impressive but surprisingly has not been reported previously in the literature. From the vascular mechanics point of view, such regions would have greater strength and would resist deformation. Conversely, an appreciable deformation, particularly in this region, could be damaging to the smooth muscle cells.

These pictures have been taken from an ongoing study of the structure of major branches of the abdominal aorta in the rabbit, but even such limited sampling suggests that structure in regions of branching differs notably from that in straight segments, the latter of which is the structure commonly depicted as the "typical" arterial wall. Additional observations indicate that thickness of the tunica intima and organization of the internal elastic lamina may also deviate from structure typically described for elastic and medium-sized arteries.

Notably, attention has been directed to variations in muscle orientation in branch and straight regions of the artery as a clear indication of structural differences in these regions. One might then ponder what less readily visible distinctions also characterize sites of arterial branching. Therefore, thorough histological analyses are required to determine if and how smooth muscle cell

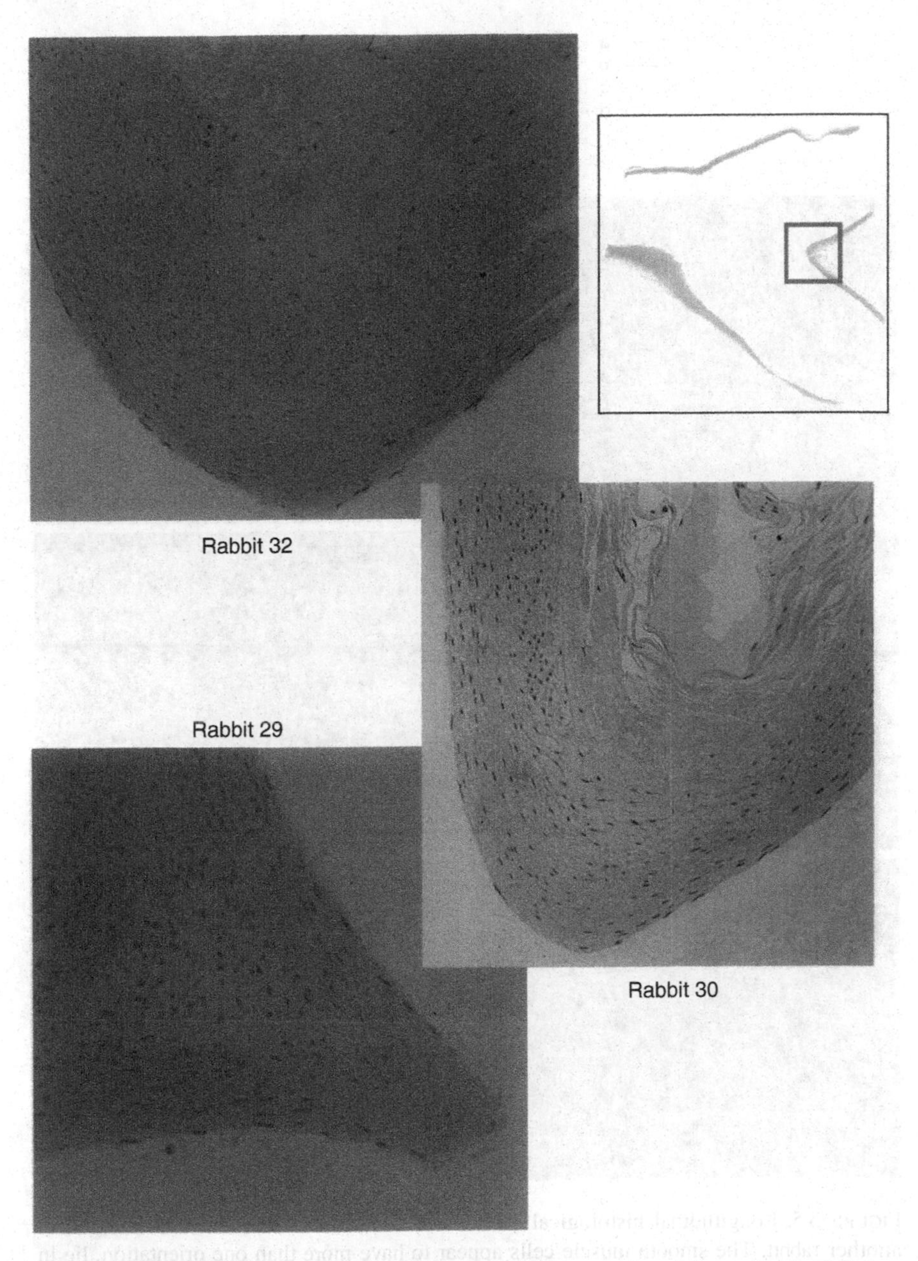

FIGURE 3.4. Longitudinal histological sections of the same aortas (shown in Fig. 3.3) but at the aortic bifurcation. The smooth muscle cells at the flow divider run in more than one direction and may even lie in more than one layer. (*Please see color version on CD-ROM.*)

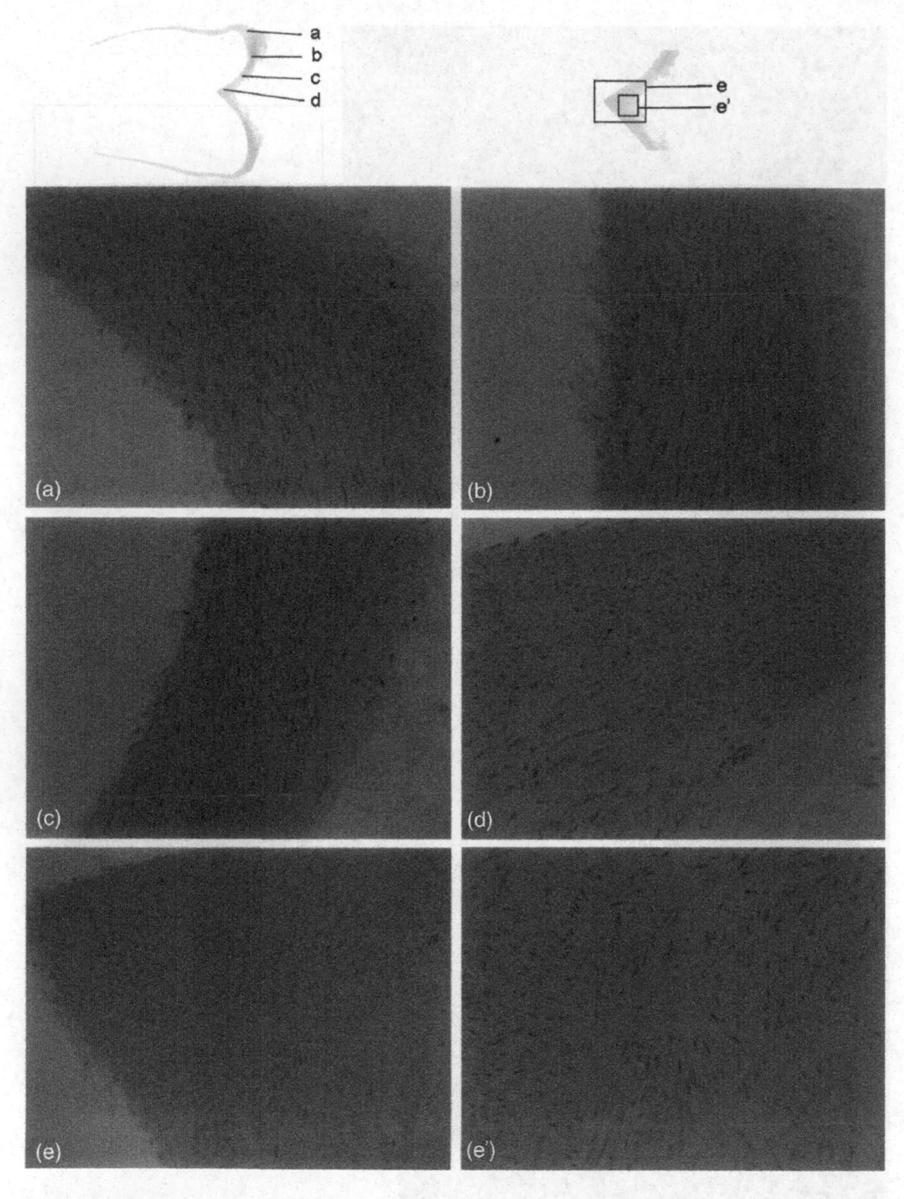

FIGURE 3.5. Longitudinal histological sections at and close to the aortic bifurcation in another rabbit. The smooth muscle cells appear to have more than one orientation, lie in multiple layers, and appear interwoven or braided. (*Please see color version on CD-ROM.*)

size, membrane composition, and junctions are different at bifurcations from what is known in straight regions. The detailed histology combined with the detailed mechanical studies of the branch region appears critical to the understanding of the development of vascular pathology at that location.

6. Function of Arteries

The primary function of the arterial system is to carry blood from the heart to the organs and tissues of the body. The high-pressure (120/80 mmHg) arterial system carries oxygen-rich blood from the left ventricle of the heart to the capillary beds present throughout the body (8). The capillary beds form extremely vast networks of microscopic tubules connecting the arterial side to the venous side of the circulation. This is where the exchange of metabolites (oxygen, nutrients, carbon dioxide, and other waste products) takes place between the blood and the tissue. Oxygen-depleted blood then flows through the venous system back to the right atrium of the heart. The shorter, low-pressure (25/8 mmHg) pulmonary arterial system carries oxygen-depleted blood from the right ventricle of the heart to the capillaries of the lungs where the blood is re-oxygenated and carried by the pulmonary veins to the left atrium of the heart (8).

Two basic principles governing the function of arteries can be delineated as follows. First, the blood needs to circulate efficiently throughout the body and therefore requires hollow tubes or pipes. Second, to deliver blood flow continuously, the arteries need to have a pressure-storage mechanism (elastic recoil, or *windkessel* effect—German for "elastic reservoir"). In other words, the heart ejects blood for one-third of the cardiac cycle, and the arteries supply blood for the remaining two-thirds of the cycle by using the energy stored in their stretched walls (6). Hence, the arteries perform a dual function: 1) as a conduit for the blood flow (pipes) and 2) as a storage of blood pressure (elastic pressure reservoir). Obviously, such a mechanism relies on the compliance or elasticity of arteries. Furthermore, the flexibility necessary in the body also requires that the arteries be elastic.

6.1. Baroreceptors

Special segments of arteries, primarily the carotid sinuses near the bifurcation of the common carotid arteries in the neck, and the aortic arch, have an additional function directly related to their elasticity. These locations harbor baroreceptors, which are important for keeping the mean arterial pressure almost constant. Baroreceptors are nerve endings in the wall of the arteries, and they provide reflex impulses proportionate to the vessel wall stretch (see Chapter 2). In turn, these impulses are factored in to control the heart rate, modify the force of atrial contraction, and change smooth muscle contractility in small vessels of the limbs and viscera (9).

6.2. Effect of Posture

Figure 3.6 shows the estimated pressure differences between the supine and erect positions in a 6-foot-tall man, for both the arterial and venous systems. With reference to the level of the heart, the weight of the blood column adds 88 mmHg to the arterial pressure in the feet of the standing man and subtracts 44 mmHg in his

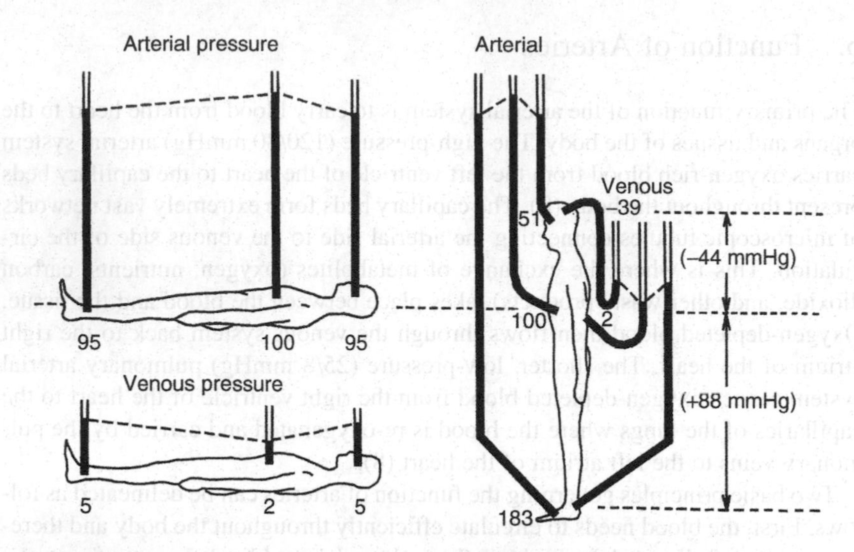

FIGURE 3.6. Effects of posture on levels of arterial and venous pressures (9).

head (10). The effect of posture on the blood pressure in the lower limbs is particularly important in the following two examples from the point of view of vascular mechanics. The *veins of the lower leg* sustain considerable hydrostatic pressure and consequently have thicker walls. This recognition has led to their most frequent use as arterial graft of choice in the coronary artery bypass surgery. The other example is that the *arteries of the lower extremities* have a greater propensity to develop atherosclerosis. Once again, these arteries have a much higher blood pressure due to the hydrostatic pressure head in the upright position. This exposure to a high pressure could be related to their enhanced tendency to develop atherosclerosis (see Chapter 2).

6.3. Adaptation to Pressure and Flow

Regardless of the advantages that elasticity confers to blood vessels under pressure, too much elasticity can incur exaggerated deformation and stress, resulting in elevated risks of failure. It is important to note that stretch and stress affect the whole vessel wall. Blood pressure is not the only potential source of injury to the vessel. Blood flow can also lead to erosion phenomena, which, although limited to the innermost layer of vessels, can elicit a powerful defense mechanism. As a result, blood vessels constantly adapt and optimize their thickness and radius with respect to the amount of pressure-related stress and stretch and the level of flow-related wall shear stress. For example, in case of hypertension, elevated wall stretch can trigger pressure adaptation whereby the wall thickness may increase so that the normal stress level returns. In case of flow alterations in an artery, the changes in wall shear stress can trigger flow adaptation causing the artery to change its radius to restore normal values of shear stress.

7. Anchoring of Arteries, Residual Stress, and Smooth Muscle Activity

Anchoring of arteries: For most vessels, perivascular tissue is contiguous with the adventitial layer and could provide some support. However, different conditions exist in some locations. For example, the vertebral arteries are partially encased in the spine, while the cerebral arteries are surrounded by cerebrospinal fluid. In the ascending aorta, the aortic arch and the supra-aortic vessels, the contraction of the heart during each cardiac cycle imposes significant downward displacement of the aortic root base. The displacement was found to range between 0 and 14 mm in patients and to considerably enhance the longitudinal stress in the ascending aorta (11). Similarly, knee and head motions can also impose large displacements and stresses on arteries in that region.

Alongside of these special conditions, the mechanical behavior of all arteries in vivo is strongly influenced by the fact that they are under a significant longitudinal force, known as *tethering force* (see Chapter 4). The wall volume of an arterial segment being approximately constant, the longitudinal stretch due to tethering causes the wall thickness to decrease, which in turn influences the circumferential distensibility of the artery. When the tethering force is released by excision, elastic and muscular arteries retract in young people by up to 30% and 50%, respectively (12). Such retraction can decrease by more than half or even disappear in older individuals or patients with atherosclerosis. Some studies report tortuosity or kinks in arteries, suggesting the presence of compressive forces. This clearly creates totally different load conditions for the arterial wall.

Residual stress: Residual stress is defined as the stress existing in the absence of externally applied loads. A significant amount of residual stress exists in the circumferential direction of some arteries. It is possible that the differential growth of the inner wall with respect to the outer wall creates residual stress that is responsible for distributed tension or compression across the wall thickness. This phenomenon is believed to homogenize the stress and/or strain within the arterial wall in vivo (see Chapter 4). If an artery is sliced circumferentially into small rings, these rings can spring open or contract upon radial excision, while the faces of the cut also exhibit shear deformation across the thickness (13). Such changes in geometry occur in various degrees according to the location along the arterial tree. The residual stress in the circumferential direction is primarily attributed to the response of elastin, although it is also influenced by the level of activity of the smooth muscle cells.

Experimentally, an excised artery can be cut circumferentially into 2- to 3-mm-wide rings for measurement of residual stress and no-load dimensions. These rings are immersed in saline and photographed (Fig. 3.7; see Fig. 3.16). The pictures are fed into image analysis software where the inner and outer circumference of the artery are traced and measured. Thus, one obtains the inner and outer radii A and B, respectively, of the unpressurized artery. The rings are then cut radially and left for 20 min before being photographed again. Using image

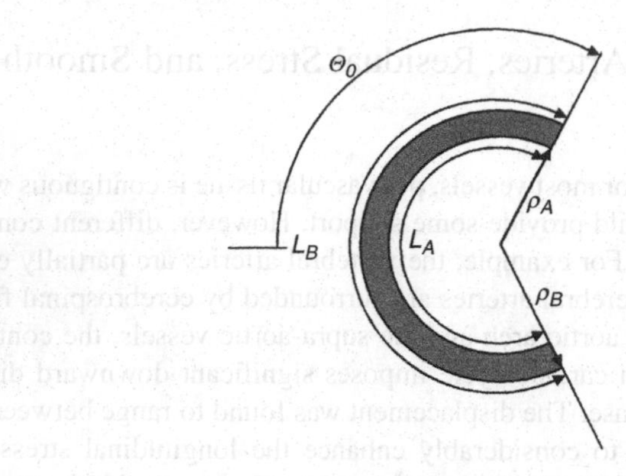

FIGURE 3.7. Schematic presentation of an arterial ring that springs open when it is cut. The opening angle Θ_0 and the inner and outer radii ρ_A and ρ_B of the cut artery can be determined from the lengths L_A and L_B and from the inner and outer radii of the uncut artery.

analysis software, one can measure the inner and outer perimeters L_A and L_B, respectively, of the cut artery. Then, the opening angle shown in Figure 3.7 can be determined as

$$\Theta_0 = \frac{1}{4\pi} \frac{L_B^2 - L_A^2}{B^2 - A^2}, \tag{1}$$

and the inner and outer radii of the radially cut artery can be calculated respectively as

$$\rho_A = 2\pi L_A \frac{B^2 - A^2}{L_B^2 - L_A^2} \text{ and } \rho_B = 2\pi L_B \frac{B^2 - A^2}{L_B^2 - L_A^2}. \tag{2}$$

Smooth muscle activity: As noted previously, the contractile state of an artery governs its luminal area and distensibility, and the changes can be significant depending on the location in the arterial tree. Therefore, when measuring the properties of arteries, including the state of residual stress, it is important to report the chemical environment of the vessel, which may affect smooth muscle cell activity (4). Potassium chloride or norepinephrine can be used to evoke contraction of the smooth muscle cells. In contrast, ethylene glycol tetra-acetic acid (EGTA) or potassium cyanide may suppress smooth muscle cell activity. Used in vitro, saline is expected to leave the smooth muscle cells in the contractile state that they were in upon excision. Aside from these considerations, it is naturally best to test the tissue as fresh as possible to minimize possible alterations of the wall components (see Chapter 4).

8. Changes Along the Arterial Tree

There are two important changes, among others, that occur along the length of the arterial tree. One is related to *pressure wave reflection* and the other to *arterial elasticity*. We mentioned earlier that, as one goes away from the heart to the periphery, the average blood pressure drops due to a small but finite energy loss. Interestingly, however, the pulse pressure increases in the thoracic and abdominal aorta (Fig. 3.8), due to reflected pressure waves arising from the peripheral terminations. We may note from the figure that the pulse pressure can be significantly greater in the abdominal aorta and femoral artery compared with that in the ascending aorta, as a result of the superposition of forward and reflected pressure waves. This is an important observation from the vascular mechanics point of view, because mechanical damage to the artery is more likely to occur where the peak pressure is higher, and/or the pulse pressure (cyclic load) is higher. In this regard, we may recall that abdominal aneurysms are more frequent compared with thoracic or ascending aortic aneurysms, and this may be in part due to the contribution of pressure wave reflection from the aortic bifurcation. Along the arterial tree, the blood flow decreases due to branching as one travels away from the heart (Fig. 3.8).

The elasticity of the artery in vivo can be evaluated as a change in the diameter of the vessel with each pressure pulse. Table 3.1 shows the radial dilation for a change of pressure of 1 mmHg. The dilation decreases from the ascending aorta to the iliac artery, indicating that the elasticity decreases gradually as one goes away from the heart. The same data can be converted to represent the *elastic modulus* in the circumferential direction (Table 3.2). The modulus increases (elasticity decreases) as one goes from the thoracic aorta to the femoral artery.

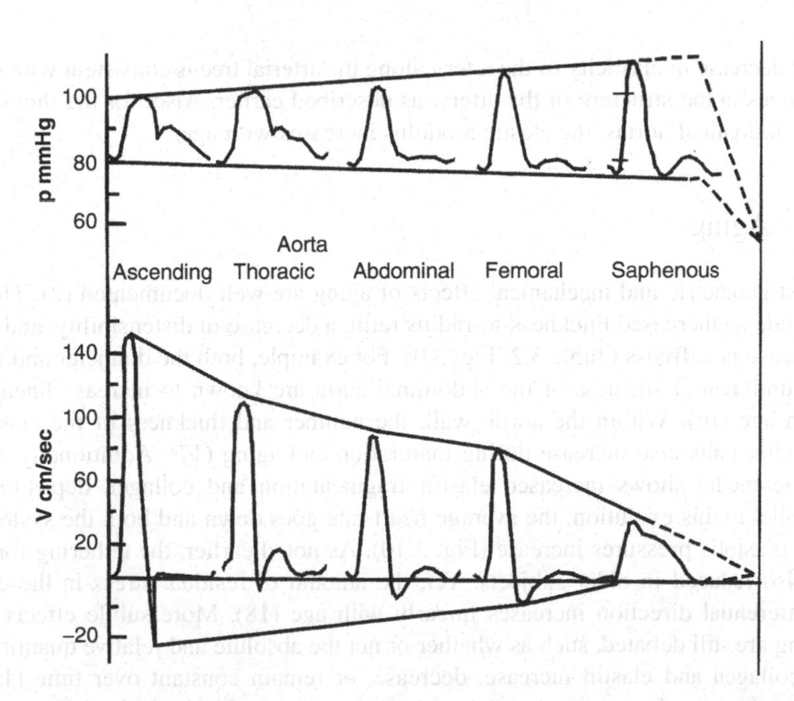

FIGURE 3.8. Schematic presentation of changes in pressure and flow contours as one moves away from the heart (6).

TABLE 3.1. Radial dilatation determined as $\frac{\Delta R}{R}$ in typical arteries, where R is the vessel radius and ΔR is the variation in radius under pressure increment ΔP (here, $\Delta P = 1$ mmHg). Adapted from Ref. 14.

Artery (dog)	Radial dilatation (/mmHg)
Ascending aorta	19.89×10^{-4}
Thoracic aorta (high)	13.36×10^{-4}
Thoracic aorta (middle)	4.58×10^{-4}
Thoracic aorta (low)	4.07×10^{-4}
Abdominal aorta (high)	2.85×10^{-4}
Abdominal aorta (low)	1.70×10^{-4}
Iliac artery	0.36×10^{-4}
Brachiocephalic trunk	2.82×10^{-4}

TABLE 3.2. Elastic modulus (MPa) determined as $E = \frac{\Delta P \cdot R^2}{\Delta R \cdot t}$, where R is the vessel radius, t the wall thickness, and ΔR is the variation in radius under pressure increment ΔP. Adapted from Ref. 15.

Artery	Young (11–20 years)	Old (36–52 years)
Thoracic aorta	0.6	2.0
Abdominal aorta	1.0	1.2
Iliac artery	2.9	0.7
Femoral artery	3.1	1.4
Carotid artery	0.8	1.1

The decrease in elasticity of the artery along the arterial tree is consistent with the changes in the structure of the artery, as described earlier. Also, for the thoracic and abdominal aortas, the elastic modulus increases with age.

9. Aging

Most geometric and mechanical effects of aging are well documented (2). They include an increased thickness-to-radius ratio, a decrease in distensibility, and an increase in stiffness (Table 3.2; Fig. 3.9). For example, both the diameter and the circumferential stiffness of the abdominal aorta are known to increase linearly with age (16). Within the aortic wall, the number and thickness of the elastic lamellar units also increase during maturation and aging (17). Additionally, the aortic media shows increased elastin fragmentation and collagen deposition. Parallel to this evolution, the average heart rate goes down and both the systolic and diastolic pressures increase (Fig. 3.10). As noted earlier, the tethering force is also reduced in older subjects. Yet, the amount of residual stress in the circumferential direction increases linearly with age (18). More subtle effects of aging are still debated, such as whether or not the absolute and relative quantities of collagen and elastin increase, decrease, or remain constant over time (12). From the vascular mechanics point of view, we note that both the increase in

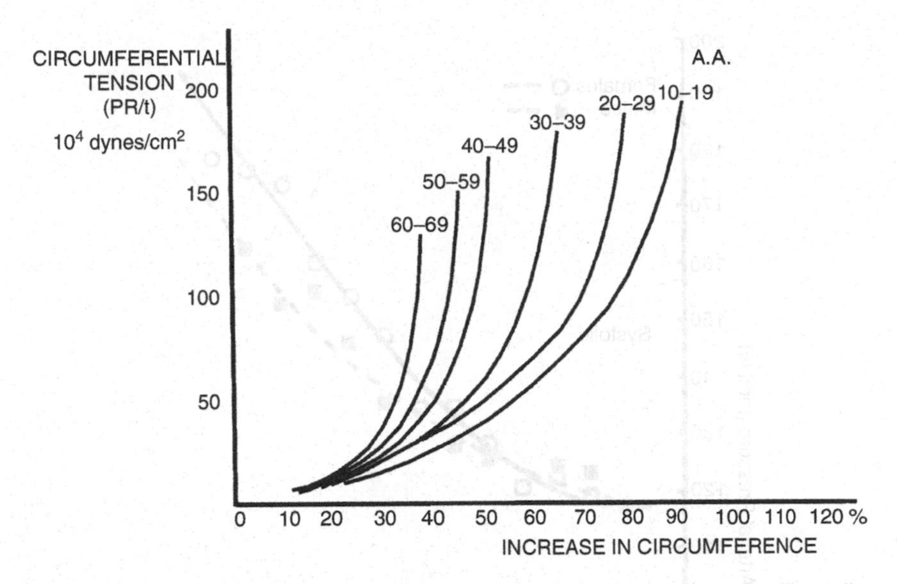

FIGURE 3.9. Aortic distensibility markedly decreases with age. P, pressure; R, radius; t, wall thickness. (Reproduced from Nakashima T and Tanikava J. Angiology 1971;22: 477-490, with permission of Westminster Publications)

pressure load and the decrease in elasticity with aging must make the artery more vulnerable because it is subjected to ever increasing mechanical stress and decreasing stress relief by deformation. Therefore, it may not be a coincidence that the occurrence of the two most prominent diseases of the artery—atherosclerosis and aneurysm—increases with age.

10. In Vivo Evaluation of the Mechanical Response of Arteries

There is currently no way to determine the state of strain or stress of an artery in vivo with respect to its undeformed or unstressed configuration. This is unfortunate because such information would be very valuable for establishing possible correlations with pathogenesis. It is however possible to make some measurements in vivo. For example, measuring the relative increase in the arterial diameter between diastole and systole allows one to evaluate the circumferential compliance (or stiffness, its inverse) of the vessel for comparison between healthy and diseased vessels (Table 3.1). Early studies in animals used invasive techniques, with calipers sutured to the surface of the arterial wall and fitted with strain gauges to measure relative circumferential and longitudinal strains. Significantly less invasive techniques have since been proposed, using optical sensors, cineangiography, intravascular ultrasound, echography, and magnetic resonance imaging

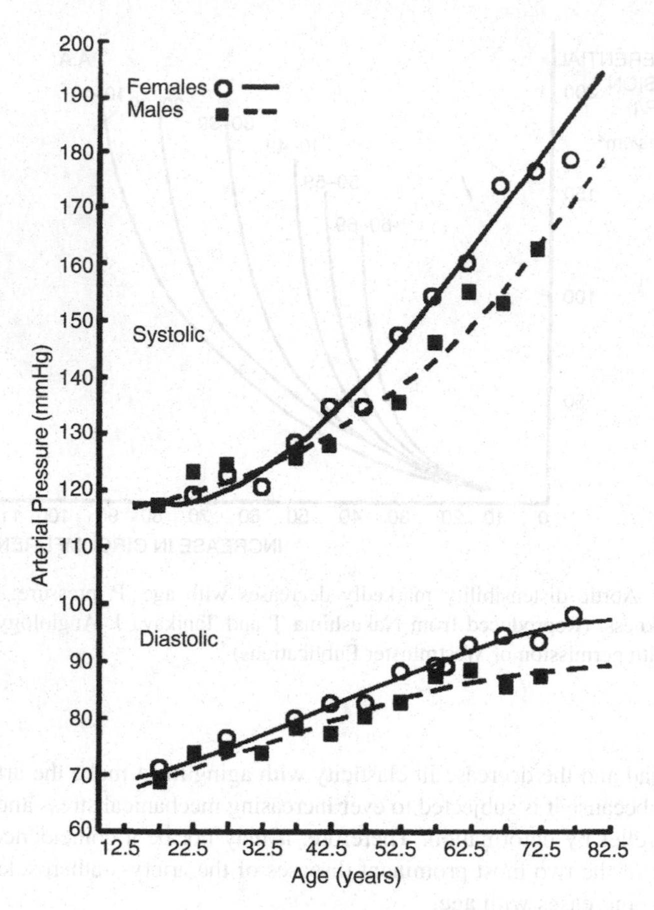

FIGURE 3.10. Arterial pressure as a function of age in the general population. (Reproduced from Hamilton et al. The aetiology of essential hypertension. I. The arterial pressure in the population. Clin Sci 1954;13:11, with permission of the Biochemical Society)

alone or combined with tagging or phase contrast. It was consistently found that the in vivo length of arteries changes very little, except due to heart or limb motion. Extensions induced by blood pressure are typically of the order 1–2%, except in the pulmonary artery where they average 8%. Changes in diameter during the cardiac cycle usually reach 10–15% in the pulmonary artery, 6–10% in the carotids, and 2–5% in the aorta (2). These are average values. Indeed, using an elaborate treatment of data obtained from phase-contrast MRI in a normal subject, Draney et al. (19) reported nonuniform deformations and circumferential variations in the cyclic strain of the thoracic aorta, with a peak average strain of 8 ± 11%. Overall, in vivo measurements are useful for comparative studies of the artery. They do not establish the comprehensive behavior of the material of the arterial wall. This can only be attempted in vitro.

11. In Vitro Evaluation of the Mechanical Response of Arteries

11.1. Uniaxial Tests

Uniaxial tensile tests on arterial strips provide an easy way to study the mechanical response of excised tissues and obtain stress-strain curves. This approach has been followed widely in the literature for comparative studies of different soft tissues (12). Such tissues typically exhibit a nonlinear response (Fig. 3.11), with first a low-stress, high-strain regime (also called low-modulus region) and then a high-stress, low-strain regime (high-modulus region). To approximate the nonlinear stress-strain curve using the local slopes of the curve at certain points, some investigators have carried out incremental measurements. They have described the material response by different values of elastic moduli pertaining to different strain levels resulting from stepwise loading and unloading (4). In addition to providing basic information on the mechanical response of the tissue, uniaxial tests yield interesting data about what level of mechanical stimulus is required for permanent damage to occur. Values of yield strain and yield stress are usually reported in that regard, along with the ultimate strain and stress describing tissue failure (20). Nevertheless, whether incremental or not, uniaxial tests only probe one direction at a time. Therefore, the cross-talk between the different directions

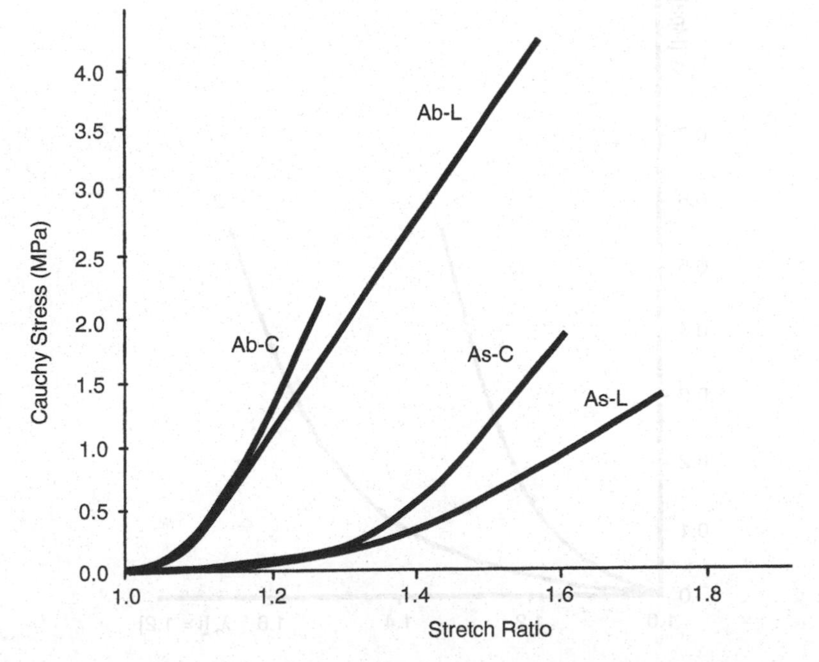

FIGURE 3.11. Average uniaxial stress-strain curves for the ascending aorta (As) and the abdominal aorta (Ab), from pig, in the circumferential (C) and longitudinal (L) directions.

(e.g., circumferential, longitudinal, and radial) of the tissue is not revealed, although potentially very significant. This is a shortcoming in regard to establishing constitutive equations usable in other situations, for example in three-dimensional finite element analysis.

11.2. Biaxial Tests

Biaxial testing machines allow one to stretch a small piece of arterial tissue in two perpendicular directions at the same time, possibly with different values of stretch on each side (Fig. 3.12). This represents a great improvement over uniaxial tests. Yet, because the sample is stretched in a plane, the natural curvature present in the arterial wall is not accounted for (18).

11.3. Pressurization Tests

In view of the limitations of the uni- and biaxial testing procedures, investigators have turned to sophisticated experimental setups that allow the accurate and independent control of dilation, extension, and twist of the whole artery (2). Let us recall at this point that the objectives of mechanical characterization are usually threefold: 1) to determine the best way to conduct experiments that yield useful

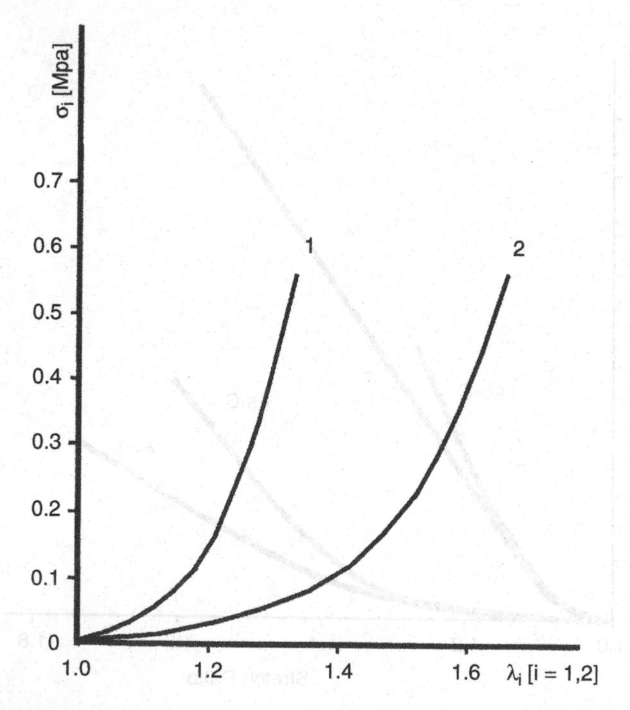

FIGURE 3.12. Biaxial stress-strain relationship for the human abdominal aorta: 1 and 2 denote the longitudinal and circumferential directions, respectively.

observations; 2) to provide a mathematical description of the material properties; 3) to predict the response of the material in situations where the parameters of interest are in the same range as during the experiment. For a model to achieve predictive quality obviously, it requires more than a mere curve fit of the experimental results. There must be a theoretical framework underlying the description of the material properties, and within which experiments are designed to produce specific information. Constitutive equations are typically not designed to represent the mechanical interaction between the components of the arterial wall at the cellular level. This is why phenomenological models have been favored, although recent studies have followed a more structural approach, wherein the media and adventitia are treated as thick-walled circular composite tubes reinforced by two families of collagen fibers (21). Still, the characterization of these fibers is essentially phenomenological.

For an illustrative purpose, let us consider a simpler experimental setup that does not involve twist. In this arrangement, the artery is cannulated at both ends and placed horizontally in a saline bath (Fig. 3.13). The distal cannula is blocked with a rubber stopper and let free to expand, while the proximal end is fixed and connected to a pressure head of saline. An ultrasound catheter is run through the proximal cannula to image the mid cross-section of the artery, allowing the measurement of the inner diameter as the luminal pressure is increased. Pressure measurements are made using a pressure transducer. A video camera positioned directly above the artery records the displacement of potassium permanganate markers placed on the surface of the artery (Fig. 3.13). Using image analysis software, one can measure the relative distances between the markers to determine the longitudinal stretch of the pressurized artery. Preconditioning typically includes five cycles of loading and unloading over the pressure range of interest.

12. Mechanical Model

The model presented below describes the arterial wall as an axisymmetric, thick-walled cylinder with homogeneous hyperelastic anisotropic and incompressible material properties. These assumptions can be justified as follows:

1. *Axisymmetry:* In general, an artery is approximately a straight circular cylinder subjected to deformations and loads (internal pressure and axial extension) that are symmetric.

2. *Thick-walled cylinder:* The thickness-to-radius ratio of arteries under physiological pressure usually varies between 1/8 and 1/10 (6), which would warrant the use of the thin-walled cylinder theory. However, doing so makes it impossible to determine the intramural distribution of stress and strain, while the possible variation in the circumferential, longitudinal, and radial stresses across the aortic wall is believed to be extremely relevant to pathogenesis. Because the unpressurized vessel is not a thin-walled cylinder by any means, the thick-walled theory is used as the

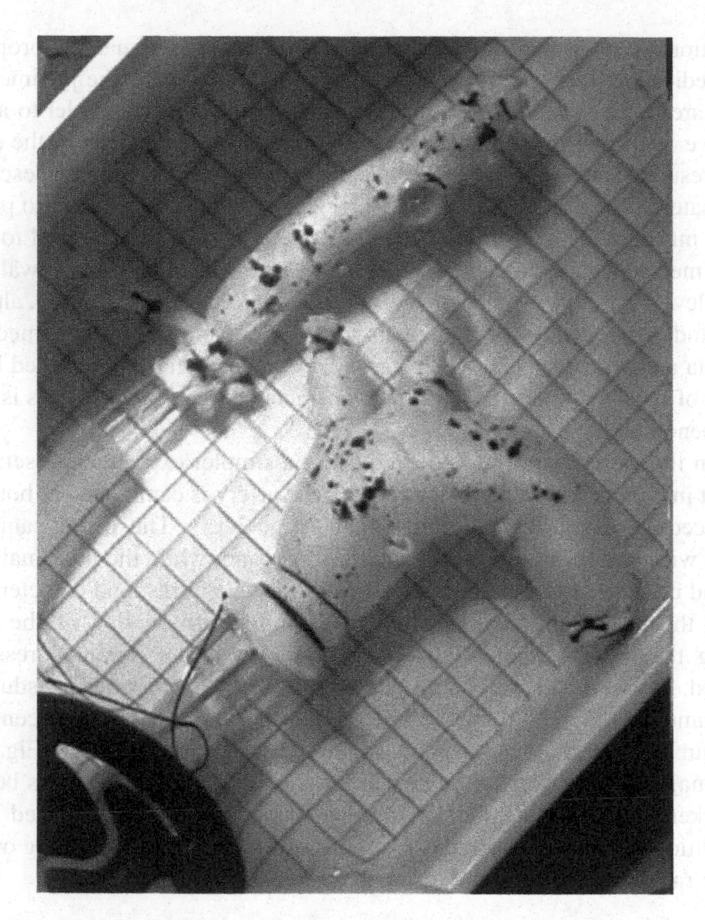

FIGURE 3.13. Photograph taken during a pressurization experiment on thoracic (bottom) and abdominal (top) segments of porcine aorta at 120 mmHg pressure. The potassium permanganate markers on the surface of the vessels were used to measure the longitudinal stretch, while ultrasound catheters inside the aortas were used to monitor the changes in the diameter.

most refined approach short of considering the actual heterogeneity of the wall and layers of different material properties (22).

3. *Homogeneity:* As detailed earlier, the arterial wall has three concentric layers: intima, media, and adventitia. The intima is extremely thin and can be ignored in a structural model. Two-layer models have been proposed by Von Mahltzan et al. (23) and more recently by Holzapfel et al. (21). For simplicity here, the arterial wall is assumed to have homogenized properties representing both the media and adventitia.

4. *Hyperelasticity:* A material is elastic when its state of stress only depends on the state of strain at a given time. Therefore, the stress-strain curve of an

elastic material is exactly the same during loading and unloading (recoverable deformation). Among elastic materials, hyperelastic materials are characterized by the strain energy function W describing the mechanical energy stored in the material under elastic deformation. It has been recognized (3) that, as long as the loading does not go excessively above physiological conditions, the passive response of arterial tissue exhibits repeatable hyperelastic behavior after preconditioning. Many mathematical expressions of the strain energy function have been proposed (2), but let us keep the definition of W general for now.

5. *Anisotropy:* The arterial wall has anisotropic material properties because its distensibility is different depending on the direction considered. Yet, experiments by Patel and Fry (24) showed that simultaneous axial extension and pressurization of arteries results in negligible shearing strains indicating that arterial tissue has elastic symmetry with respect to cylindrical coordinates. Studies by Humphrey et al. (25), Deng et al. (26), and Schulze-Bauer et al. (27) confirmed these findings. Under such loading, the general anisotropic response becomes orthotropic (three orthogonal preferred directions: radial, circumferential, and longitudinal).

6. *Incompressibility:* Barring exudation of internal fluid under extreme loads, studies based on different techniques indicate that the volume of the arterial wall remains constant during deformation.

It is assumed here for simplicity that 1) no residual stress is present when the artery is unloaded in vitro; 2) only inflation and extension are considered. Because no twist is introduced, one can postulate that the radial, circumferential, and longitudinal directions are principal and that the material response is orthotropic. Under such conditions, the principal directions of stress and strain do coincide, and significant simplifications ensue in the mathematical derivation. Due to large deformations undergone by the arterial wall, the model is built in the context of finite deformations.

In cylindrical coordinates, let a material particle located at (r, θ, z) in the deformed vessel be mapped to (R, Θ, z) in the undeformed body such that

$$r = r(R), \theta = \Theta, \text{and } z = z(Z),$$

in the radial, circumferential, and longitudinal directions, respectively. In these notations, $A \leq R \leq B$, where A and B are respectively the inner and outer radii in the unloaded geometry. Similarly, $a \leq r \leq b$, where a and b are respectively the inner and outer radii when the vessel is under internal pressure p and applied axial force F_z. The external pressure is assumed to be zero. One can demonstrate (see Appendix of this chapter) that the constitutive equations relating the material response to the global loads p and F_z are

$$p = \int_a^b \frac{1}{r} \lambda_{\theta\theta}^2 \frac{\partial W}{\partial E_{\theta\theta}} dr \text{ and } F_z = \pi \int_a^b \left(2\lambda_{zz}^2 \frac{\partial W}{\partial E_{zz}} - \lambda_{\theta\theta}^2 \frac{\partial W}{\partial E_{\theta\theta}} \right) rdr + \pi a^2 p (1 - \gamma), \quad (3)$$

with the stretch ratios and Green-Lagrange strains defined as

$$\lambda_{\theta\theta}=\frac{r}{R},\ \lambda_{zz}=\frac{\partial z}{\partial Z},\ \text{and}\ E_{ii}=\frac{1}{2}\left(\lambda_{ii}^{2}-1\right),\ i=\theta, z,\ \text{respectively},$$

and $\gamma = 1$ or 0 if the artery is treated as a closed or open cylinder, respectively.

If residual stress is introduced in the model through the opening angle Θ_0 determined in Eq. (1), the global constitutive equations (3) of the artery are unchanged, but the circumferential stretch ratio becomes

$$\lambda_{\theta\theta}=\frac{\pi r}{\Theta_0 \rho},\ \text{where}\ \rho=\sqrt{\frac{\pi\lambda_{zz}}{\Theta_0}(r^2-a^2)+\rho_A^2}, \tag{4}$$

and ρ_A is the inner radius of the unpressurized, radially cut artery determined in Eq. (2). The longitudinal stretch ratio is unchanged if we assume that the opening of the artery due to the radial cut occurs without measurable change in width of the ring sample. This was verified experimentally on multiple occasions. The reader is referred to (2) for complete detail on the equations for a cylindrical artery under inflation, extension, and twist, along with the restrictions of application relative to the model presented herein.

13. Determination of the Material Constants Associated with a Model

A suitable mathematical form of the strain energy function must be chosen to represent the mechanical response of the artery as determined by the experimental data. Following Takamizawa and Hayashi (28), let us assume a logarithmic form such that

$$W=-C_1\ln\left(1-\frac{1}{2}C_2E_{\theta\theta}^2+\frac{1}{2}C_3E_{zz}^2+C_4E_{\theta\theta}E_{zz}\right),$$

where C_1,..., C_4 are the material constants of the model, $E_{\theta\theta}$ and E_{zz} being the Green-Lagrange strains defined previously. If constants C_1,..., C_4 are known as well as the geometry of the unpressurized artery, then the mechanical response of the vessel can be completely determined under any pressure and/or longitudinal load, using Eqs. (A.1 to A.14, Appendix). Conversely, to identify the material constants, one uses a mathematical optimization procedure such that the predicted response of the artery closely matches the experimental data. Toward this goal, one may choose to minimize the objective function

$$\chi^2=\sum_{i=1}^{m}\left\{\left(e_p\right)_i^2+\left(e_f\right)_i^2\right\} \tag{5}$$

where m is the number of experimental points. On the one hand, $e_p = (p_{pred} - p_{\exp})/p_{\exp}$ is the error between the predicted pressure computed using Eq. (3) and the experimentally measured pressure. On the other hand, $e_f = (F_{z\,pred} - F_{z\,\exp})/F_{z\,\exp}$ is the error between the predicted longitudinal force computed from Eq. (3) and the experimentally applied force. Note that in the experimental setup described earlier, the expansion of the artery is free and therefore $F_{z\,exp}$ is zero. In this case, the definition of e_f can be changed to $e_f = F_{z\,pred}$. The nonlinear least-squares problem in Eq. (5) can be solved using numerical subroutines based on the Levenberg-Marquardt method for multivariate nonlinear regression (2).

Many different mathematical forms have been proposed for the strain energy function W, depending on the arterial tissue considered. However, they do not all perform equally well and none has been recognized as universal (27). Strain energy functions inspired from models for rubber and involving the strain tensor invariants I_1, I_2, I_3 such as

$$W = C_1(I_1 - 3) + \frac{C_2}{2}(I_1 - 3)^2 + \frac{C_3}{3}(I_1 - 3)^3 + f(I_2, I_3)$$

(29) are not capable of describing anisotropic materials. Indeed, due to the use of the strain tensor invariants, such functions are insensitive to the existence of preferred directions in the tissue and are inherently isotropic. Alternatively, strain energy functions have been proposed as functions of the principal stretch ratios or the components of Green-Lagrange strain tensor. Vaishnav et al. (30) defined a polynomial form

$$W = C_1 E_{\theta\theta}^2 + C_2 E_{\theta\theta} E_{zz} + C_3 E_{zz}^2 + C_4 E_{\theta\theta}^2 + C_5 E_{\theta\theta}^2 E_{zz} + C_6 E_{\theta\theta} E_{zz}^2 + C_7 E_{zz}^3,$$

where $C_1, \ldots, C_7$ are the material constants. Fung et al. (31) proposed an exponential model that many investigators have adopted under various forms,

$$W = \frac{C_1}{2}\left[\exp\left(C_2 E_{\theta\theta}^2 + C_3 E_{zz}^2 + 2C_4 E_{\theta\theta} E_{zz}\right) - 1\right]$$

where $C_1, \ldots, C_4$ are the material constants. More recently, Guccione et al. (32) introduced a modification of this model, a transversely isotropic (one preferred direction) version of which has been implemented in finite element code LS-Dyna (LSTC, Livermore, CA, USA) under the form

$$W = \frac{C_1}{2}\left[\exp\left(C_2 E_{\theta\theta}^2 + C_3\left(E_{zz}^2 + E_{rr}^2 + E_{rz}^2 + E_{zr}^2\right) + C_4\left(E_{\theta z}^2 + E_{\theta z}^2 + E_{\theta r}^2 + E_{r\theta}^2\right)\right) - 1\right] + \frac{1}{2}P(J - 1) \qquad (6)$$

wherein P is a Lagrange multiplier in charge of numerically enforcing material near-incompressibility ($J \cong 1$), and the Green-Lagrange strain components E_{ij} are modified to only include the effects of volumetric work. Although this model was created originally for heart tissue, it also has potential for applications with arterial tissue as will be shown below.

14. Finite Element Modeling of Arteries

The underlying principle in the finite element analysis for mechanical stress and strain is to divide a structure into a number of simple pieces (e.g., brick or shell elements) for which analytical solutions exist. The equilibrium equations for each element are then assembled into a set of equations that describe the whole structure. Although dynamic models are available, stress analyses of arteries are usually carried out with pressure implemented as a static load. For example, by using the systolic value of blood pressure, one can determine the highest stresses present in the artery under physiologic conditions. Because the equations are solved by computer, structures with complex shapes, boundary conditions, and loads can be considered and residual stress can be dealt with (33). Therefore, not only straight cylindrical arteries but also branches, aneurysms, and stenoses can potentially be addressed. However, the implementation of realistic material properties for arterial tissues is still under development. Thus, simplifications such as those listed below can be useful in carrying out first-order analyses. Examples of finite element studies using various simplified material models will be presented in Chapters 6 and 7.

Linear material models can be used to study an artery in vivo, because in the physiologic range, the mechanical response of the artery is described by the high modulus region of the stress-strain curve (from uniaxial, or better yet, biaxial tests). This means that starting with the geometry of the artery under 80 mmHg pressure, one can use a linear material model (Young's modulus and Poisson's ratio) to compute the stresses reached at 120 or 160 mmHg pressure with reasonable accuracy. Naturally, the strains are obtained with respect to the geometry at 80 mmHg pressure. Orthotropic linear models are refinements of isotropic linear models and theoretically need nine elastic constants to describe the arterial tissue instead of two.

Following the approach taken in incremental experiments, finite element studies can be designed that use incremental loads in combination with incremental linear material properties. However, such a technique is limited to cases where only small strains are reached, because it is implemented outside the mathematical framework of finite deformations.

Nonlinear material models for rubber or soft tissues available in finite element packages are based on the theory of hyperelasticity as shown in the previous section. When applying these models, it is important to use the unloaded geometry as a starting point because large deformations occur in the arterial wall before blood pressure reaches the physiological range of 120/80 mmHg. Otherwise, unrealistic strains and stresses are obtained. Another potential pitfall of such models is the incorrect determination of the material constants. As more and more finite element packages provide relatively easy ways to identify these constants, say for a Mooney-Rivlin material, great care must be taken to check that the behavior of the structure studied is correct. For example, if the constants are determined on the basis of uniaxial tests only, the three-dimensional response of the model will probably be wrong, because the necessary positive-definiteness of the stiffness matrix used in the computation is not guaranteed. The complete determination of the Mooney-Rivlin constants theoretically calls for uniaxial, biaxial, and shear tests.

Finally, because anisotropic hyperelastic material models have been available only recently, some investigators have gone around that difficulty by deliberately averaging the experimental results obtained in the circumferential and longitudinal directions to determine the "homogenized" material constants of an isotropic model (34).

The details of programming and implementing linear and nonlinear material models in finite element codes is beyond the scope of this chapter. The reader is referred to Bathe (35) and Bonet and Wood (36).

15. Preliminary Study of Porcine Ascending Thoracic and Abdominal Aortas[1]

The question at the center of the study presented below comes from the observation that in humans, the ascending aorta is more prone to dissections, while the abdominal aorta is more prone to aneurysms. Granted, the ascending aorta is subjected to cyclic pressurization and a cyclic longitudinal pull, while the abdominal aorta experiences cyclic pressurization of larger amplitude—due to pressure-wave reflection at the aortic bifurcation—and a static longitudinal pull. Still, regardless of these obviously dissimilar loading conditions, what differences in composition, structure, and mechanical properties could justify such different outcomes?

Histological differences have long been reported between the thoracic and abdominal regions of the aorta (17). Proportionately, collagen and elastin are the major components of the ascending thoracic aorta, whereas smooth muscle cells are the major components of the abdominal aorta. Moreover, the thoracic aorta contains more elastin and the abdominal aorta contains more collagen (17). Using X-ray diffraction techniques to image the orientation of collagen fibers in porcine aortic medial layers, Roveri et al. suggested that the extra amount of collagen going from the aortic arch downwards is deposited in a preferred circumferential orientation (37). As a preliminary study to explore how these variations in tissue composition and microstructure translate in terms of mechanical properties, we experimented on four fresh pig aortas and prepared sections for histological analyses. Figure 3.14 shows Verhoeff van Gieson's stained sections from the wall of the ascending and abdominal porcine aorta in transverse and longitudinal directions. Close to the lumen, one can distinguish the darker internal elastic lamina separating the intima from the media. The elastic fibers (or sheets) are stained in black and, as expected, their number is vastly greater in the ascending aorta than in the abdominal aorta. The media exhibits a regular arrangement of concentric lamellar units in both aortas. Overall, the ascending aorta has many more elastic lamellae and a much thicker media than the abdominal aorta. The adventitia shows a primary longitudinal orientation of elastin and collagen fibers in the abdominal aorta and occupies a much larger proportion of wall thickness than in the ascending aorta.

---

[1] The experiments on pig aortas were carried out by Joshua Newton at the Heineman Laboratory during his undergraduate research work. He was assisted by Brett Fowler. Dr. Geoffrey Gong prepared the microscopic pictures of the aortas. Their contributions are gratefully acknowledged.

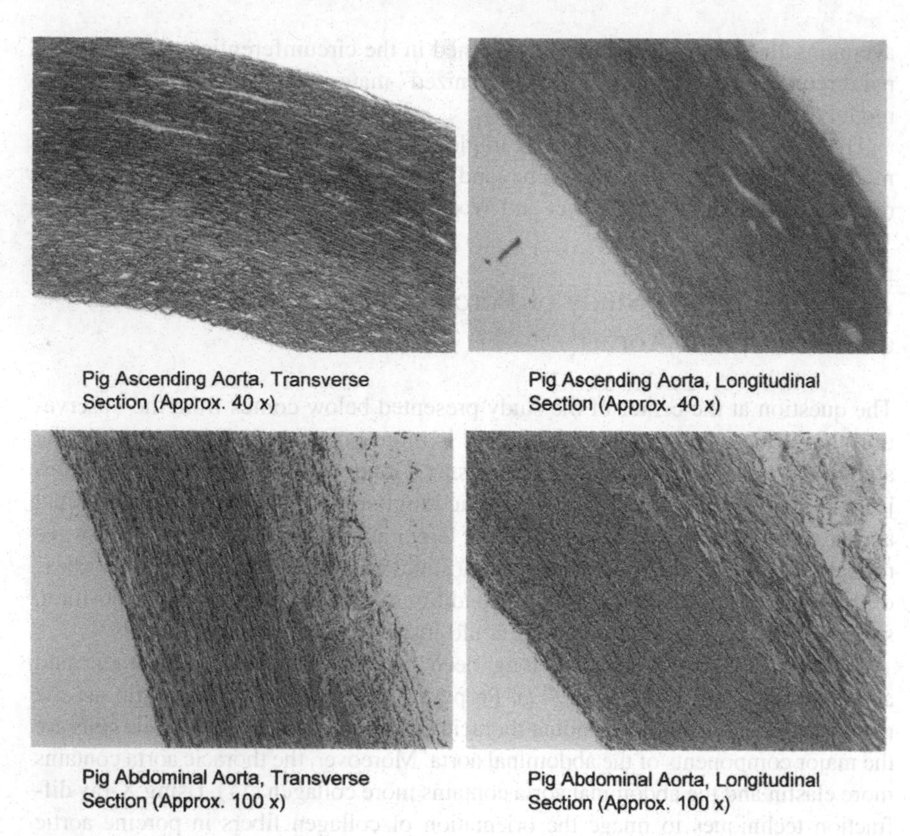

Pig Ascending Aorta, Transverse
Section (Approx. 40 x)
Pig Ascending Aorta, Longitudinal
Section (Approx. 40 x)
Pig Abdominal Aorta, Transverse
Section (Approx. 100 x)
Pig Abdominal Aorta, Longitudinal
Section (Approx. 100 x)

FIGURE 3.14. Verhoeff van Gieson–stained sections of porcine aortas. The elastin is stained black while smooth muscle cells are gray (usually yellow) and collagen fibers are red. (*Please see color version on CD-ROM.*)

Our in vitro measurements determined that the inner diameter and wall thickness were 17.0 ± 1.7 mm and 3.1 ± 0.43 mm, respectively, for the ascending aorta versus 8.7 ± 1.2 mm and 1.9 ± 0.43 mm, respectively, for the abdominal aorta. Pressure-diameter experiments were run on segments whose distal ends were tied and free to elongate, using the setup described previously (Fig. 3.13). In the circumferential direction, the dilation of the ascending aorta increased steadily up to 95% at 120 mmHg and rose to about 100% at 140 mmHg pressure; that of the abdominal aorta increased steadily to 50% at 60 mmHg, then to 60% at 140 mmHg pressure (Fig. 3.15). In the longitudinal direction, all the segments stretched quasi-linearly up to 80% at 140 mmHg pressure. Thus, the aorta was more distensible in the ascending region than in the abdominal region, which is consistent with the role of a compliant reservoir expected from the proximal aorta. Rings from the ascending aortas were found to spring wide open when cut (Fig. 3.16a), with $\Theta_0 = 57^\circ \pm 26^\circ$, while the abdominal aortas only showed little sign of opening or even pulled inward on themselves (Fig. 3.16b), with

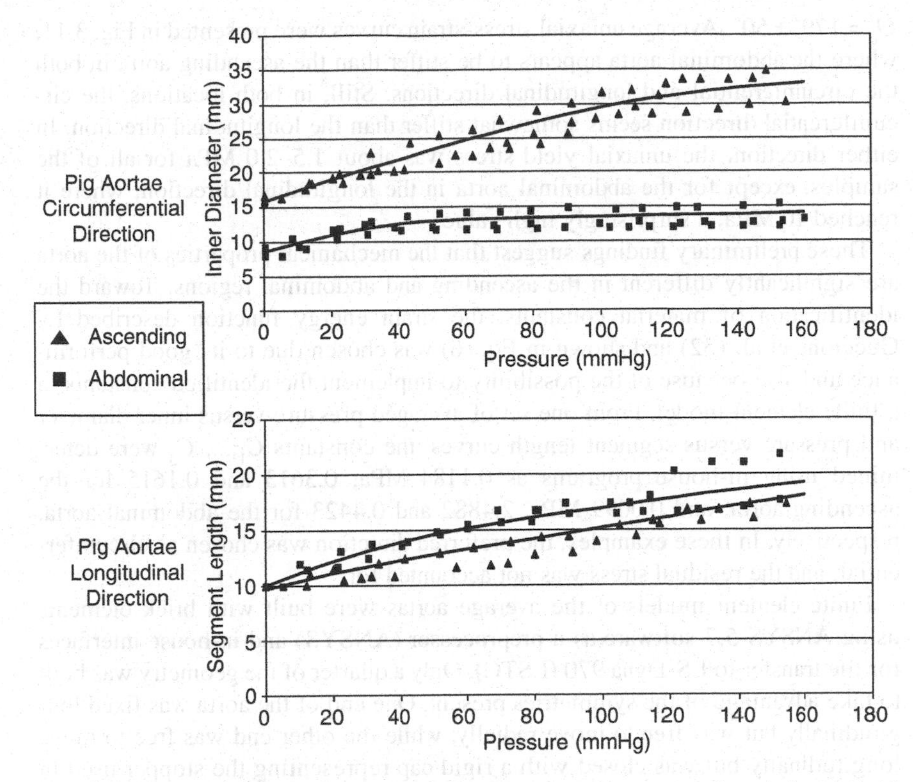

FIGURE 3.15. Measurements of the inner diameter (top) and a segment length (bottom) of porcine ascending and abdominal aortas during pressurization from 0 to 160 mmHg pressure. The solid lines represent manual curve fits used to identify the material constants. See text for details.

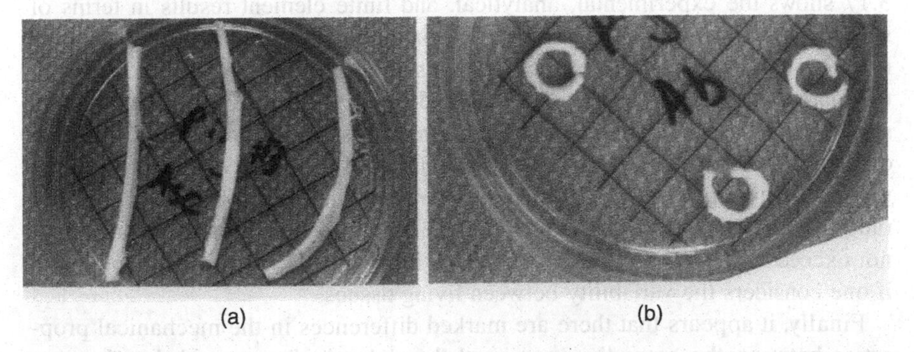

FIGURE 3.16. Photographs of three aortic rings from porcine ascending (a) and abdominal (b) aortas after they were cut radially and left for 20 min in saline. The ascending aorta rings opened up widely while the abdominal aorta rings opened very little or curled back on themselves.

$\Theta_0 = 179° \pm 50°$. Average uniaxial stress-strain curves were presented in Fig. 3.11, where the abdominal aorta appears to be stiffer than the ascending aorta in both the circumferential and longitudinal directions. Still, in both locations, the circumferential direction seems somewhat stiffer than the longitudinal direction. In either direction, the uniaxial yield stress was about 1.5–2.0 MPa for all of the samples, except for the abdominal aorta in the longitudinal direction, where it reached 10 MPa, a surprisingly high value.

These preliminary findings suggest that the mechanical properties of the aorta are significantly different in the ascending and abdominal regions. Toward the identification of material constants, the strain energy function described by Guccione et al. (32) and shown in Eq. (6) was chosen due to its good performance and also because of the possibility to implement the identified constants in a finite element model. From one set of averaged pressure versus inner diameter and pressure versus segment length curves, the constants C_1,..., C_3 were determined using in-house programs as 0.1184 MPa, 0.3613 and 0.1615 for the ascending aorta, and 0.0099 MPa, 2.4882 and 0.4423 for the abdominal aorta, respectively. In these examples, the preferred direction was chosen as circumferential, and the residual stress was not accounted for.

Finite element models of the average aortas were built with brick elements using ANSYS 5.7 software as a preprocessor (ANSYS) and in-house interfaces for file transfer to LS-Dyna 970 (LSTC). Only a quarter of the geometry was built to take advantage of the symmetries present. One end of the aorta was fixed longitudinally but was free to move radially, while the other end was free to move longitudinally but was closed with a rigid cap representing the stopper used in the experiment. For the analysis, the luminal pressure was ramped between 0 and 160 mmHg over a period of 0.16 s. Because the constant C_4 cannot be determined from a simple pressurization test where no twist is involved, it was given the same value as C_3 for the analysis. Its actual value does not influence the results however, except where shear occurs between the longitudinal and the circumferential or radial directions (i.e., at the interface between the cap and the aorta). Figure 3.17 shows the experimental, analytical, and finite element results in terms of stretch ratios in the circumferential and longitudinal directions. The analytical results were determined from the equations established in the Appendix, given the material constants and the unpressurized geometry of the average aortas. The stretch ratios in the circumferential direction were evaluated from the aortic inner diameter. As can be seen in Fig. 3.17, the analytical and finite element results are within a maximum of 5% from each other over the 0 to 160 mmHg pressure range. The relative error between the finite element and experimental results does not exceed 15% over the same range. This is acceptable for predictive purposes if one considers the variability between living tissues.

Finally, it appears that there are marked differences in the mechanical properties between the ascending aorta and the abdominal aorta and that they are deeply rooted in variations in the aortic wall microstructure and composition. Although such a result may seem obvious, this preliminary study provided some quantitative information and testing ground for experimental and numerical techniques.

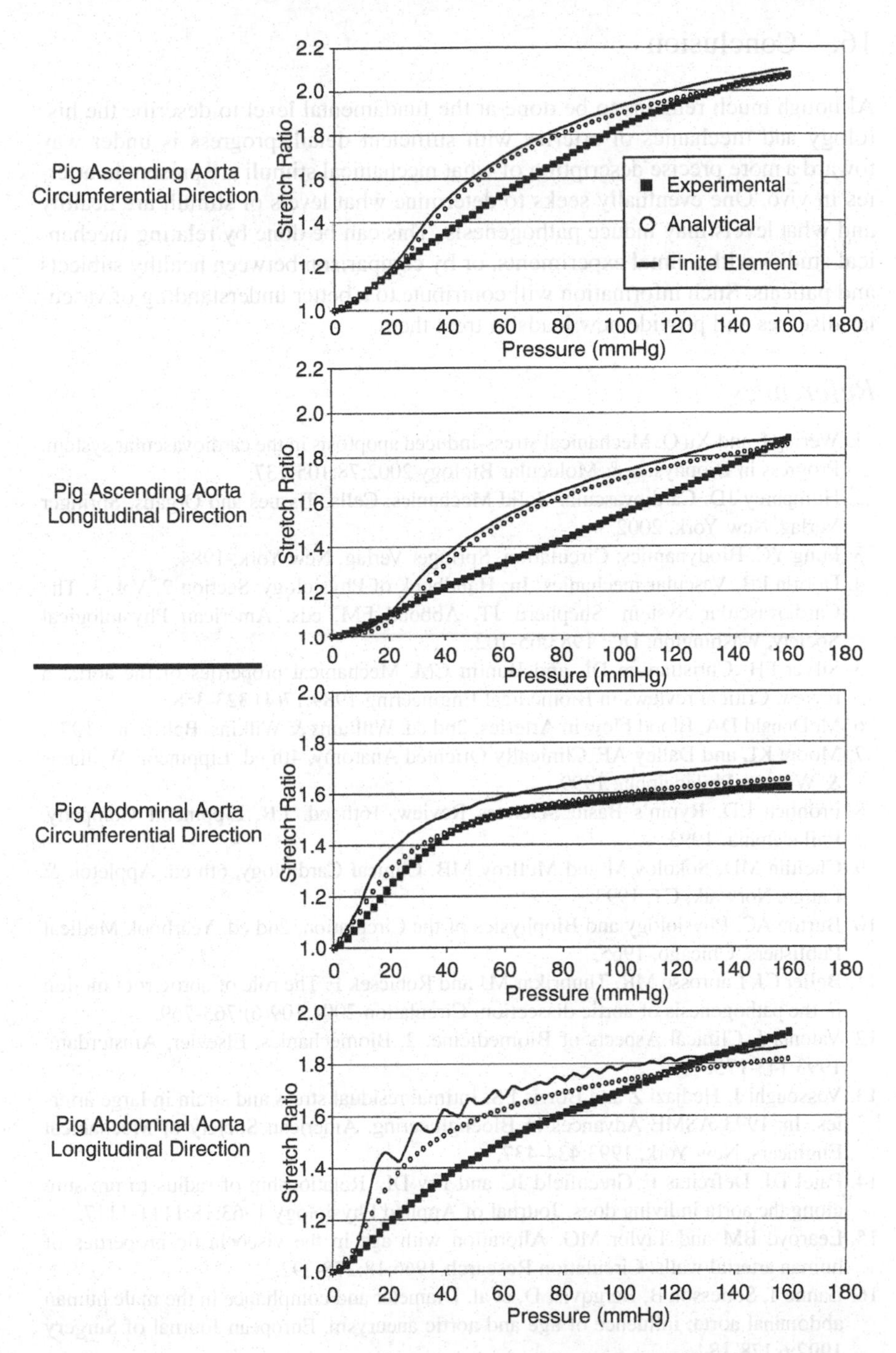

FIGURE 3.17. Comparison between experimental, analytical, and finite element results of pressurization of average porcine aortas in the ascending and abdominal regions. See text for details.

16. Conclusion

Although much remains to be done at the fundamental level to describe the histology and mechanics of arteries with sufficient detail, progress is under way toward a more precise description of what mechanical stimuli are present in arteries in vivo. One eventually seeks to determine what levels of stimuli are healthy and what levels may induce pathogenesis. This can be done by relating mechanical studies with animal experiments, or by comparison between healthy subjects and patients. Such information will contribute to a better understanding of vascular diseases and provide new leads to treat them.

References

1. Wernig F and Xu Q. Mechanical stress-induced apoptosis in the cardiovascular system. Progress in Biophysics & Molecular Biology 2002;78:105-137.
2. Humphrey JD. Cardiovascular Solid Mechanics: Cells, Tissues and Organs. Springer Verlag, New York, 2002.
3. Fung YC. Biodynamics: Circulation. Springer Verlag, New York, 1984.
4. Dobrin PB. Vascular mechanics. In: Handbook of Physiology. Section 2, Vol. 3: The Cardiovascular System. Shepherd JT, Abboud FM, eds. American Physiological Society, Washington, DC: 1983:65-102.
5. Silver FH, Christiansen DL and Buntin CM. Mechanical properties of the aorta: a review. Critical reviews in Biomedical Engineering 1989;17(4):323-358.
6. McDonald DA. Blood Flow in Arteries, 2nd ed. Williams & Wilkins, Baltimore, 1974.
7. Moore KL and Dalley AF. Clinically Oriented Anatomy, 4th ed. Lippincott Williams & Wilkins, Philadelphia, 1999.
8. Frohlich ED. Rypin's Basic Sciences Review, 16th ed. J.B. Lippincott Company, Philadelphia, 1993.
9. Cheitlin MD, Sokolov M and McIlroy MB. Clinical Cardiology, 6th ed. Appleton & Lange, Norwalk, CT, 1993.
10. Burton AC. Physiology and Biophysics of the Circulation, 2nd ed. Yearbook Medical Publishers, Chicago, 1965.
11. Beller CJ, Labrosse MR, Thubrikar MJ and Robicsek F. The role of aortic root motion in the pathogenesis of aortic dissection. Circulation 2004;109(6):763-769.
12. Valenta J. Clinical Aspects of Biomedicine, 2: Biomechanics. Elsevier, Amsterdam, 1993:143-175.
13. Vossoughi J, Hedjazi Z and Borris FS. Intimal residual stress and strain in large arteries. In: 1993 ASME Advances in Bioengineering. American Society of Mechanical Engineers, New York, 1993:434-437.
14. Patel DJ, Defreitas F, Greenfield JC and Fry DL. Relationship of radius to pressure along the aorta in living dogs. Journal of Applied Physiology 1963;18:1111-1117.
15. Learoyd BM and Taylor MG. Alteration with age in the viscoelastic properties of human arterial walls. Circulation Research 1966;18:278-292.
16. Lanne T, Sonesson B, Bergqvist D, et al. Diameter and compliance in the male human abdominal aorta: influence of age and aortic aneurysm. European Journal of Surgery 1992;6:178-184.
17. Schlatmann TJM and Becker AE. Histologic changes in the normal aging aorta: implications for dissecting aortic aneurysm, The American Journal of Cardiology 1977; 39(1):13-20.

18. Okamoto RJ, Wagenseil JE, DeLong WR, et al. Mechanical properties of dilated human ascending aorta. Annals of Biomedical Engineering 2002;30:624-635.
19. Draney MT, Herfkens RJ, Hughes TJR, et al. Quantification of vessel wall cyclic strain using cine phase contrast magnetic resonance imaging. Annals of Biomedical Engineering 2002;30:1033-1045.
20. Raghavan ML, Webster MW and Vorp DA. Ex vivo biomechanical behavior of abdominal aortic aneurysm: assessment using a new mathematical model. Annals of Biomedical Engineering 1996;24:573-582.
21. Holzapfel GA, Gasser TC and Ogden RW. A new constitutive framework for arterial wall mechanics and a comparative study of material models. Journal of Elasticity 2000;61:1-48.
22. Matsumoto T and Sato M. Analysis of stress and strain distribution in the artery wall consisted of layers with different elastic modulus and opening angle. JSME International Journal, Series C 2002;45(4):906-912.
23. Von Maltzahn WW, Warriyar RG and Keitzer WF. Experimental measurements of elastic properties of media and adventitia of bovine carotid arteries. Journal of Biomechanics 1984;17:839-847.
24. Patel DJ and Fry DL. The elastic symmetry of arterial segments in dogs. Circulation Research 1969;24:1-8.
25. Humphrey JD, Kang T, Sakarda P and Anjanappa M. Computer-aided vascular experimentation: a new electromechanical test system. Annals of Biomedical Engineering 1993;21:33-43.
26. Deng SX, Tomioka J, Debes JC and Fung YC. New experiments on shear modulus of elasticity of arteries. American Journal of Physiology 1994;266:H1-H10.
27. Schulze-Bauer CAJ, Moerth C and Holzapfel GA. Passive biaxial mechanical response of aged human iliac arteries. Journal of Biomechanical Engineering 2003; 125:395-406.
28. Takamizawa K and Hayashi K. Strain energy density function and uniform strain hypothesis for arterial mechanics. Journal of Biomechanics. 1987;20(1):7-17.
29. Tickner EG and Sacks AH. A theory for the static elastic behavior of blood vessels. Biorheology 1967;4:151-168.
30. Vaishnav RN, Young JT, Janicki JS and Patel DJ. Nonlinear anisotropic elastic properties of the canine aorta. Biophysics Journal 1972;12:1008-1027.
31. Fung YC, Fronek K and Patitucci P. Pseudoelasticity of arteries and the choice of its mathematical expression. American Journal of Physiology 1979;237:H620-H631.
32. Guccione JM, McCulloch AD and Waldman LK. Passive material properties of intact ventricular myocardium determined form a cylindrical model. Journal of Biomechanical Engineering 1991;113:42-55.
33. Delfino A, Stergiopulos N, Moore JE and Meister J-J. Residual strain effects on the stress field in a thick wall finite element model of the human carotid bifurcation. Journal of Biomechanics 1997;30(8):777-786.
34. Prendergast PJ, Lally C, Daly S, et al. Analysis of prolapse in cardiovascular stents: a constitutive equation for vascular tissue and finite element modeling. Journal of Biomechanical Engineering 2003;125:692-699.
35. K.J. Bathe. Finite Element Procedures. Prentice Hall, Englewood Cliffs, NJ, 1996.
36. Bonet J and Wood RD. Nonlinear Continuum Mechanics for Finite Element Analysis. Cambridge University Press, New York, 1997.
37. Roveri N, Ripamonti A, Pulga C, et al. Mechanical behavior of aortic tissue as a function of collagen orientation. Makromol Chem 1980;181:1999-2007.

Appendix

In cylindrical coordinates, let a material particle located at (r, θ, z) in the deformed vessel be mapped to (R, Θ, Z) in the undeformed body such that

$$r = r(R), \theta = \Theta, \text{ and } z = z(Z),$$

in the radial, circumferential, and longitudinal directions, respectively. In these notations, $A \le R \le B$, where A and B are respectively the inner and outer radii in the unloaded geometry. Similarly, $a \le r \le b$, where a and b are respectively the inner and outer radii when the vessel is under internal pressure p and applied axial force F_z. The external pressure is assumed to be zero.

A fundamental measure of deformation is provided by the deformation gradient tensor $\boldsymbol{F}$ whose components in cylindrical coordinates, under the circumstances, reduce to

$$[\boldsymbol{F}] = \begin{bmatrix} \partial r/\partial R & 0 & 0 \\ 0 & r/R & 0 \\ 0 & 0 & \partial z/\partial Z \end{bmatrix}.$$

In turn, the components of Cauchy-Green strain tensor defined as $\boldsymbol{C} = \boldsymbol{F}^T\boldsymbol{F}$ are

$$[\boldsymbol{C}] = \begin{bmatrix} \lambda_{rr}^2 & 0 & 0 \\ 0 & \lambda_{\theta\theta}^2 & 0 \\ 0 & 0 & \lambda_{zz}^2 \end{bmatrix}.$$

Here, $\boldsymbol{C}$ is populated by the squares of the stretch ratios in the principal directions, namely

$$\lambda_{rr} = \frac{\partial r}{\partial R}, \lambda_{\theta\theta} = \frac{r}{R}, \text{ and } \lambda_{zz} = \frac{\partial z}{\partial Z}. \tag{A.1}$$

Another useful description of strain is given by the Green-Lagrange strain tensor defined as $\boldsymbol{E} = \frac{1}{2}(\boldsymbol{C} - \boldsymbol{I})$. Here,

$$[\boldsymbol{E}] = \begin{bmatrix} E_{rr} = \frac{1}{2}(\lambda_{rr}^2 - 1) & 0 & 0 \\ 0 & E_{\theta\theta} = \frac{1}{2}(\lambda_{\theta\theta}^2 - 1) & 0 \\ 0 & 0 & E_{zz} = \frac{1}{2}(\lambda_{zz}^2 - 1) \end{bmatrix},$$

where obviously the components $E_{r\theta}$, $E_{\theta z}$, and E_{rz} are zero because no shear deformation is present.

In the most familiar description of stress, Cauchy stress characterizes the current force per unit deformed area. More specifically, Cauchy stress tensor $\boldsymbol{\sigma}$ defines the traction vector $\boldsymbol{t}$ acting on the *current* material unit area da of normal vector $\boldsymbol{n}$ such that $\boldsymbol{t} = \boldsymbol{\sigma n}$. Because it is often necessary to refer to the undeformed geometry, the second Piola-Kirchoff stress tensor is also used, although it has no physical meaning. The second Piola-Kirchoff stress tensor $\boldsymbol{S}$ defines the traction vector $\boldsymbol{T}$ acting on the *undeformed* material unit area dA of normal vector $\boldsymbol{N}$ such that $\boldsymbol{T} = \boldsymbol{SN}$. Both the Cauchy and the second Piola-Kirchoff stress tensors are connected to each other by

$$\boldsymbol{\sigma} = J^{-1}\boldsymbol{FSF}^T, \text{ where } J \equiv \det \boldsymbol{F}.$$

Because no moment per unit volume is present to load the vessel, both tensors are symmetric.

In cylindrical coordinates, the static equilibrium of a material volume element is given in general by

$$\begin{cases} \partial\sigma_{rr}/\partial r + \partial\sigma_{r\theta}/r\partial\theta + \partial\sigma_{zr}/\partial z + (\sigma_{rr} - \sigma_{\theta\theta})/r = 0 \\ \partial\sigma_{r\theta}/\partial r + \partial\sigma_{\theta\theta}/r\partial\theta + \partial\sigma_{z\theta}/\partial z + 2\sigma_{r\theta}/r = 0, \\ \partial\sigma_{rz}/\partial r + \partial\sigma_{\theta z}/r\partial\theta + \partial\sigma_{zz}/\partial z + \sigma_{rz}/r = 0 \end{cases}$$

where σ_{ij} are the components of the Cauchy stress tensor. Then, due to the assumption of structural and loading axisymmetry, the only nontrivial equation is

$$\frac{\partial\sigma_{rr}}{\partial r} + \frac{\sigma_{rr} - \sigma_{\theta\theta}}{r} = 0. \tag{A.2}$$

Now that the strains and stresses have been defined, let us introduce the material constitutive equation that connects them. By definition of the strain energy function W for an hyperelastic material, $dW = \boldsymbol{S}d\boldsymbol{E}$ and specifically here,

$$dW = S_{rr}\, dE_{rr} + S_{\theta\theta}\, dE_{\theta\theta} + S_{zz}\, dE_{zz}, \tag{A.3}$$

where S_{ij} are the components of the second Piola-Kirchoff stress tensor.

From the assumption of material incompressibility, the ratio of deformed to undeformed volume is one. In other words,

$$dv/dV \equiv J \equiv \det \boldsymbol{F} = 1,$$

and as a consequence,

$$\det \boldsymbol{C} = 1, \text{ and } \lambda_{rr}\lambda_{\theta\theta}\lambda_{zz} = 1. \tag{A.4}$$

Using this relationship, one stretch ratio can be expressed in terms of the other two. Similarly, if we initially assumed the strain energy function to be a function of E_{rr}, $E_{\theta\theta}$, and E_{zz}, it can now be written as a function of $E_{\theta\theta}$ and E_{zz} only. Then, by differentiation,

$$dW = \frac{\partial W}{\partial E_{\theta\theta}}\, dE_{\theta\theta} + \frac{\partial W}{\partial E_{zz}}\, dE_{zz}, \tag{A.5}$$

and the terms in Eqs. (A.3) and (A.5) can be identified. Accordingly,

$$\begin{cases} S_{\theta\theta} - \dfrac{\lambda_{rr}^2}{\lambda_{\theta\theta}^2} S_{rr} = \dfrac{\partial W}{\partial E_{\theta\theta}} \\ S_{zz} - \dfrac{\lambda_{rr}^2}{\lambda_{zz}^2} S_{rr} = \dfrac{\partial W}{\partial E_{zz}} \end{cases},$$

and using the equation $\boldsymbol{\sigma} = J^{-1}\boldsymbol{FSF}^T$ with $J = 1$, one gets

$$\begin{cases} \sigma_{\theta\theta} - \sigma_{rr} = \lambda_{\theta\theta}^2 \dfrac{\partial W}{\partial E_{\theta\theta}} \\ \sigma_{zz} - \sigma_{rr} = \lambda_{zz}^2 \dfrac{\partial W}{\partial E_{zz}} \end{cases}. \tag{A.6}$$

Substituting $\sigma_{\theta\theta} - \sigma_{rr}$ from Eq. (A.6$_1$) into the equilibrium equation (A.2) and integrating with boundary conditions $\sigma_{rr}(a) = -p$ and $\sigma_{rr}(b) = 0$, one obtains the intramural radial stress as

$$\sigma_{rr}(r) = \int_a^r \frac{1}{r} \lambda_{\theta\theta}^2 \frac{\partial W}{\partial E_{\theta\theta}} dr - p, \tag{A.7}$$

where the internal pressure p satisfies

$$p = \int_a^b \frac{1}{r} \lambda_{\theta\theta}^2 \frac{\partial W}{\partial E_{\theta\theta}} dr. \tag{A.8}$$

On the other hand, the resultant of the longitudinal stresses in the wall must balance the external axial force applied to the artery and the action of pressure on the closed end, assuming that the artery is closed. Namely,

$$2\pi \int_a^b \sigma_{zz} r dr = F_z + \gamma \pi a^2 p, \tag{A.9}$$

where $\gamma = 1$ if the artery is treated as a closed cylinder ($\gamma = 0$ for an open cylinder, respectively).

Noting that $\sigma_{zz} \equiv (\sigma_{zz} - \sigma_{rr}) + \sigma_{rr}$, Eq. (A.9) can be rewritten with $\sigma_{zz} - \sigma_{rr}$ from Eq. (A.6$_2$) as

$$2\pi \int_a^b \lambda_{zz}^2 \frac{\partial W}{\partial E_{zz}} r dr + 2\pi \int_a^b \sigma_{rr} r dr = F_z + \gamma \pi a^2 p. \tag{A.10}$$

Finally, by integrating by parts $2\pi \int_a^b \sigma_{rr} r dr$ in Eq. (A.10) using Eq. (A.7), one obtains

$$F_z = \pi \int_a^b \left(2\lambda_{zz}^2 \frac{\partial W}{\partial E_{zz}} - \lambda_{\theta\theta}^2 \frac{\partial W}{\partial E_{\theta\theta}} \right) r dr + \pi a^2 p (1 - \gamma). \tag{A.11}$$

Equations (A.6) and (A.7) can be seen as the local constitutive equations of the artery, while Eqs. (A.8) and (A.11) are its global constitutive equations, relating the material response to the global loads p and F_z.

Before these equations can be implemented, other useful relationships need to be established. First, from Eqs. (A.1$_1$), (A.1$_2$), and (A.4), one can derive $r dr = \dfrac{1}{\lambda_{zz}} R dR$ where λ_{zz} is assumed to be independent of r, which is true at least near the connections to the cannulae. Therefore, $\lambda_{zz} r^2 + k = R^2$ where k is an arbitrary constant. This constant can be determined by realizing that under zero pressure, $r = a$ when $R = A$, wherein A is the (known) inner radius of the unpressurized artery. Similarly, under zero pressure, $r = b$ when $R = B$, wherein B is the (known) outer radius of the unpressurized artery. Thus, $k = A^2 - \lambda_{zz} a^2 = B^2 - \lambda_{zz} b^2$, which finally yields the relationship

$$\lambda_{zz}(b^2 - a^2) = B^2 - A^2 \tag{A.12}$$

relating the longitudinal stretch ratio and current outer and inner radii to the unpressurized radii. In the experimental setup described earlier, the inner radius a can be measured directly for each desired pressure. On the other hand, the longitudinal stretch ratio can be calculated using the distance between two markers in the longitudinal direction. Let d_0 denote the relative distance between two markers along the unpressurized artery and d denote the current relative distance under a given pressure. Then the longitudinal stretch ratio is $\lambda_{zz} = \frac{d}{d_0}$, and from Eq. (A.12), the outer radius is determined as

$$b = \sqrt{a^2 + \frac{B^2 - A^2}{\lambda_{zz}}}. \tag{A.13}$$

Finally, the circumferential stretch ratio across the arterial wall can practically be determined as

$$\lambda_{\theta\theta} \equiv \frac{r}{R} = \sqrt{\frac{r^2}{A^2 + \lambda_{zz}(r^2 - a^2)}}. \tag{A.14}$$

Altogether, given the geometry of the unpressurized artery (radii A and B), the vessel geometry and deformation under any pressure and/or longitudinal load can be described completely by just experimentally determining the current inner radius a and the current longitudinal stretch ratio λ_{zz}.

4
Pressure Vessel Principles

1. Equilibrium in the Artery Under Internal Pressure

1.1. Determination of Wall Stress in the Artery: The Basic Approach

The stress in the wall of the artery can be determined by first considering the wall to be very thin compared with the radius (Fig. 4.1). At the equilibrium, the artery has a radius r and thickness t at the internal pressure P. The free body diagram suggests the following for the equilibrium of forces in the circumferential direction.

The distending force due to pressure

$$= P \cdot area$$

$$= P \cdot 2r \cdot l$$

where l is the length under consideration. The internal force produced in the wall of the artery in response to the distending force

$$= \sigma_c \cdot t \cdot l \cdot 2$$

where σ_C is the stress in the circumferential direction. Because these two forces balance each other, we have

$$P \cdot 2r \cdot l = \sigma_c \cdot t \cdot l \cdot 2$$

or

$$\sigma_C = \frac{Pr}{t}. \tag{1}$$

This expression is similar to another expression given by the "law of Laplace," which describes wall tension for a very thin wall geometry such as soap bubbles.

The wall tension T_C in this case is given by:

$$T_C = Pr.$$

The wall tension T_C represents the circumferential tension per unit length in a very thin wall geometry. For a finite wall thickness, then

$$\sigma_C = \frac{T_C}{t}.$$

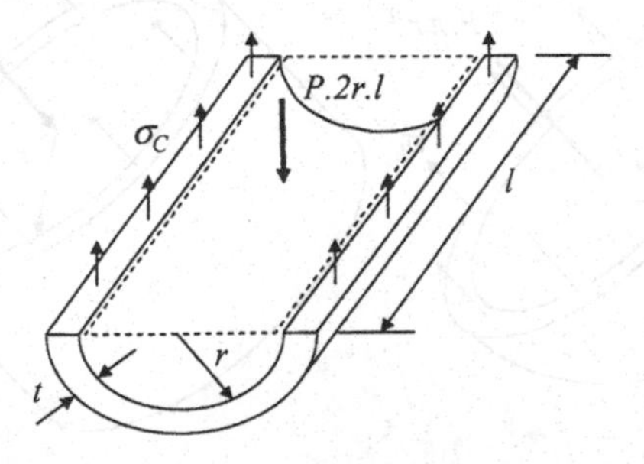

FIGURE 4.1. Schematic presentation of a section of the artery of length (ℓ), thickness (t), and radius (r) showing force due to internal pressure (p) and that due to circumferential wall stress σ_c.

In a clinical setting both parameters, the wall tension and the wall stress, continue to be used to describe the state of the artery because the internal pressure and the diameter are more readily available than the wall thickness.

Using a similar approach, we can determine the stress in the wall of the artery in the longitudinal direction (Fig. 4.2). From the free body diagram, we note that the distending force in the longitudinal direction due to internal pressure is

$$= P \cdot \pi r^2.$$

The internal force in the wall of the artery in the longitudinal direction produced in response to the above distending force

$$= \sigma_L \cdot 2\pi \left(r + \frac{t}{2} \right) \cdot t$$

where σ_L is the stress in the longitudinal direction.
At equilibrium, these two forces are balanced.

$$P \cdot \pi r^2 = \sigma_L \cdot 2\pi \left(r + \frac{t}{2} \right) t.$$

Because r is large compared with t, often the above expression is simplified as

$$P \cdot \pi r^2 = \sigma_L \cdot 2\pi r \cdot t$$

then

$$\sigma_L = \frac{Pr}{2t}. \tag{2}$$

Comparison of Eqs. (1) and (2) indicates that the longitudinal stress is only half the circumferential stress in the artery. This is important information in the context of strength of the artery in these two directions and for the development of pathology.

From the law of Laplace, as described earlier, the wall tension in the longitudinal direction is

$$T_L = \frac{Pr}{2}.$$

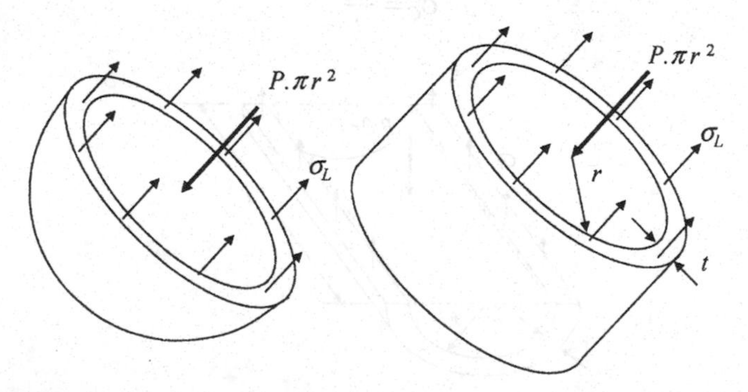

FIGURE 4.2. Schematic presentation of equilibrium of forces in the longitudinal direction of the artery (right). P, pressure; r, internal radius; t, wall thickness; and σ_L, longitudinal wall stress. Similar equilibrium of forces illustrated in the spherical geometry (left).

In case of a *spherical geometry* of the radius r and thickness t, the equilibrium between the distending force from internal pressure P and the wall stress developed as a restoring force can be written as follows (Fig. 4.2):

$$P \cdot \pi r^2 = \sigma \cdot 2\pi r \cdot t$$

$$\sigma = \frac{Pr}{2t} \tag{3}$$

that is, the wall stress in a spherical geometry is the same as the longitudinal stress in the cylindrical geometry and it is half the circumferential stress in the cylindrical geometry. Thus, when a cylindrical artery develops an aneurysm, the spherical shape of the aneurysm serves to reduce the circumferential stress. This aspect will be examined again in Chapter 15.

1.2. Wall Stress in Different-Size Arteries

The magnitude of stress in the artery is dependent on the internal pressure and on the ratio r/t [Eqs. (1), (2)]. It would be expected in the natural development that the artery would maintain a constant stress as it changes *size* from the largest at the heart (aorta) to the smallest at the periphery. To do so it would have to maintain the ratio r/t to be constant or nearly constant. Table 4.1 shows the measured values for the large-size (about 25 mm in diameter) and medium-size (about 6 mm in diameter) arteries and it is noted that the arteries maintain a constant ratio r/t of about 8–10 at normal body pressure (1).

This principle of constant stress is one of the most important governing principles in the development and adaptation of the artery. For instance, if the blood pressure increases, as in conditions of chronic *hypertension*, then the artery would adapt either by decreasing the radius or by increasing the wall thickness to restore the normal value of stress. In hypertension, it is known that the artery wall becomes thicker, once again to maintain a constant level of stress.

Similarly, if the artery curves for instance as in case of the *aortic arch*, then it is expected that the inner curvature of the arch will have a different thickness than the outer curvature because the stresses are different in these two locations. This will be described in detail a little later.

TABLE 4.1. R/t for the arteries.

	Bergel (1960)	Gow & Taylor (1968)
Thoracic aorta	9.52	7.14
Abdominal aorta	9.52	8.33
Femoral artery	8.93	7.69
Carotid artery	7.58	—
	(at 100 mmHg pressure)	(mean pressure 105–124 mmHg)

1.3. *Magnitude of Wall Stress*

For the artery the circumferential stress

$$\sigma_C = \frac{Pr}{t}$$

and the longitudinal stress

$$\sigma_L = \frac{Pr}{2t}.$$

The ratio $\frac{r}{t}$ is about 9 from Table 4.1.

Then for a pressure of 105 mmHg (0.0139 N/mm²),

$$\sigma_C = 0.126 \text{ N/mm}^2 = 12.6 \text{ N/cm}^2$$

and

$$\sigma_L = 0.063 \text{ N/mm}^2 = 6.3 \text{ N/cm}^2.$$

Thus, for most of the arteries in the body, the wall stress is about 13 N/cm² circumferentially and 6 N/cm² longitudinally for a mean arterial pressure of 105 mmHg.

1.4. *Wall Stress in the Aortic Arch and Tortuous Arteries*

The aortic arch has a special geometric feature that is important to consider for development of stress in the wall. This geometric feature is also present in small-size arteries in numerous places in the body. Therefore, the analysis of the wall stress in this curved geometry is applicable also to other curved or tortuous arteries.

Figure 4.3 shows a typical curved geometry of an aortic arch (for example). The upper surface is called a *synclastic surface* where the two centers of curvature lie on the same side (lower) of the surface. The lower surface is called an *anticlastic surface* where the two centers of curvature lie on opposite sides of the surface (one above and one below). A horse saddle is an example of an anticlastic surface.

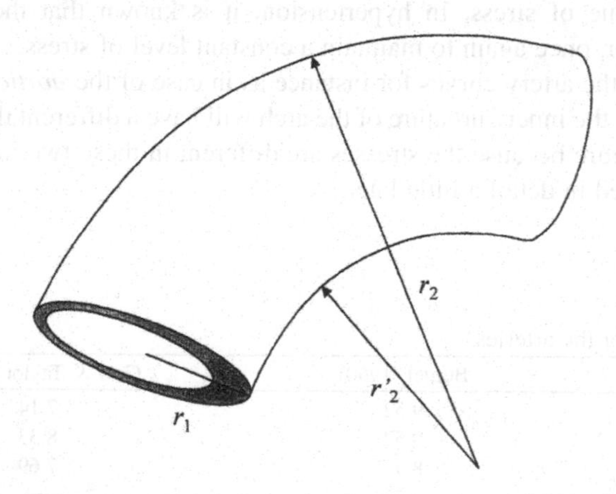

FIGURE 4.3. Schematic presentation of the curved geometry of an aortic arch. r_1, radius of the aorta; r_2, radius of the outer curvature of the arch; and r_2', radius of the inner curvature of the arch.

In classical textbooks on pressure vessels, the stress at any given point on a curved surface can be described in general terms where the curved surface is described to have two principal radii of curvature at that point. In this case, the equilibrium equation yields:

$$\frac{\sigma_1}{r_1} + \frac{\sigma_2}{r_2} = \frac{P}{t} \tag{4}$$

where σ_1 and σ_2 are first and second principal stresses and r_1 and r_2 are first and second principal radii of curvature at a given point. σ_1 is along the direction of r_1 and σ_2 is along r_2. For example, in a cylinder σ_1 will be circumferential stress if the radius of a cylinder is r_1.

The principal radii are defined as follows. At any point on a given surface (say surface of the heart), we may draw the normal to a tangent plane. Then from two points somewhere on this normal we may describe two arcs, which touch the given surface in two different planes, at right angles to each other. A unique pair of radii exist at this point, where one radius is minimum and the other is maximum and these two radii are called the principal radii of curvature.

Using Eq. (4) for the upper surface (outer curve) of the aortic arch, we have

$$\frac{\sigma_1}{r_1} + \frac{\sigma_2}{r_2} = \frac{P}{t}$$

or

$$\frac{\sigma_1}{r_1} = \frac{P}{t} - \frac{\sigma_2}{r_2}. \tag{5}$$

Using the same equation for the lower surface (inner curve), we have

$$\frac{\sigma_1}{r_1} + \frac{\sigma_2}{(-r_2')} = \frac{P}{t}$$

or

$$\frac{\sigma_1}{r_1} = \frac{P}{t} + \frac{\sigma_2}{r_2'}. \tag{6}$$

Comparison of Eqs. (5) and (6) indicates that the hoop stress σ_1 is much larger on the lower curvature (inner curvature) of the arch than on the upper curvature (outer curvature). Accordingly, to maintain the constant state of stress, the artery would be expected to have a larger wall thickness on the inner curvature than on the outer curvature. In case of the aortic arch, it has been observed that the wall is much thicker at the bottom of the arch than at the top (2). As mentioned earlier, this principle of different stresses or different wall thicknesses to maintain constant wall stress must hold true also in curved arteries.

1.5. Stress Distribution from Inside to Outside of the Artery Wall

In pressure vessels—and the artery is a pressure vessel—the stress distribution in the wall is different depending upon whether the wall is thin or thick in proportion to the radius. In a thin-walled pressure vessel, the ratio r/t is 10 or more, and the stress in the wall is relatively constant through the thickness. The stress determinations done so far for the artery have been for this condition. However, when the cylinder is thick (i.e., the wall of the artery is thick in proportion to the radius), then the variation in the stress from the inner wall to the outer wall becomes appreciable, and the average stress equations are not satisfactory. A gun barrel is an example of a thick-wall cylinder. In general, the ratio r/t of less than 10 can be considered to represent a thick-wall cylinder. Because the artery has a ratio of 8 to 10, it can fall under the category of either thick or thin cylinder. Also, when the artery is under zero pressure, the ratio r/t can be as small as 2–4, which then makes the artery a thick-wall cylinder. Therefore, it is important to understand how the stress may vary for a thick-wall cylinder (i.e., for an artery when the r/t is in the range of a thick cylinder).

The equilibrium of forces for a thick cylinder under internal pressure have been described in several standard texts on pressure vessels (3). The circumferential (hoop) stress and the radial stress (in the direction of the radius) are given as follows:

$$\sigma_t = \frac{r_i^2 \cdot P}{r_o^2 - r_i^2}\left(1 + \frac{r_o^2}{r^2}\right) \tag{7}$$

and

$$\sigma_r = \frac{r_i^2 \cdot P}{r_o^2 - r_i^2}\left(1 - \frac{r_o^2}{r^2}\right) \tag{8}$$

where σ_t is circumferential stress, σ_r is radial stress, r_i is inner radius, r_o is outer radius, P is internal pressure (external pressure is zero), and r is the radius where the stresses σ_t and σ_r exist. These equations are also called *Lamé equations* or *thick-cylinder formulas*.

The above equations show that both the stresses are maximum at the inner surface where r is minimum and that σ_r is always a compressive stress and smaller than σ_t . σ_t is a tensile stress that is maximum at the inner surface. At the inner surface ($r = r_i$),

$$\sigma_{t_{inner}} = P\frac{\left(r_o^2 + r_i^2\right)}{\left(r_o^2 - r_i^2\right)}. \tag{9}$$

The minimum σ_t occurs at the outer surface ($r = r_o$), and it is given by

$$\sigma_{t_{outer}} = \frac{2r_i^2 \cdot P}{r_o^2 - r_i^2}. \tag{10}$$

Comparison of Eqs. (9) and (10) shows that the circumferential stress on the outer surface is less than that on the inner surface by the value of internal pressure P, that is,

$$\sigma_{t_{inner}} = \sigma_{t_{outer}} + P. \tag{11}$$

Figure 4.4 shows the variation of σ_t through the wall for the ratio

$$K = \frac{r_o}{r_i} = 2.$$

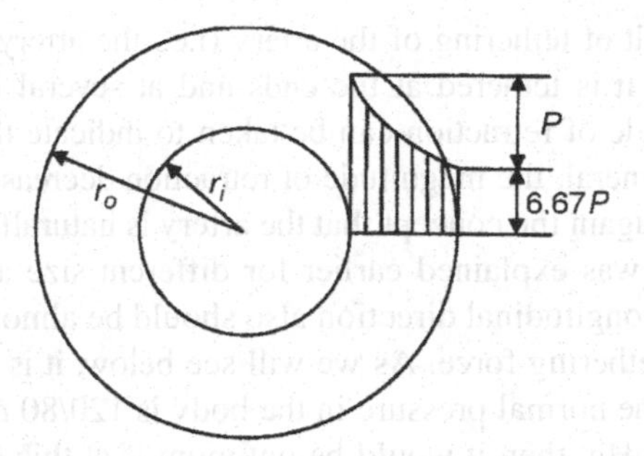

FIGURE 4.4. Schematic presentation of the distribution of wall stress in a thick cylinder. P, inside pressure; r_o, outer radius; and r_i, inner radius. The circumferential stress is highest on the inner surface of the wall and decays through the wall thickness to the outer surface.

TABLE 4.2. Ratio of maximum stress to average stress ($\sigma_{max}/\sigma_{ave}$) for a cylinder for various values of (r_o/r_i).

r_o/r_i	1.1	1.14	1.2	1.6	2.0
$\sigma_{max}/\sigma_{ave}$	1.05	1.07	1.1	1.37	1.67

It may be noted that the inner surface has considerably higher stress than the average stress, and this is an important consideration in the design of a pressure vessel and in the design of the artery. It is of interest to compare the maximum stress value obtained by the thick-cylinder formula [Eq. (9)] and that obtained by the thin-cylinder formula [Eq. (1)] for various values of K (ratio r_o/r_i) shown in Table 4.2 (3). For small wall thickness ($K = 1.1$), there is only a little difference. For a wall thickness of 10–14% of the radius (i.e., for the artery), the maximum stress is only 5–7% higher than the average stress. For a wall thickness equal to the inside radius ($K = 2$), the maximum stress is 67% higher than the average stress.

The radial stress σ_r on the outer surface is zero, as seen from Eq. (8), and on the inner surface it is $-P$ or equal to pressure but compressive. The average radial stress through the thickness is $P/2$. These equations can be applied to the arteries when the r/t ratio is in the range 2–8 particularly as the artery is pressurized beginning at a zero pressure.

2. Other Factors Affecting Wall Stress in the Artery

2.1. *Tethering*

In the body, it is known that if the artery is cut in the circumferential direction, then the cut ends immediately retract and the artery shortens in the longitudinal direction. Thus, the artery wall has a tensile force present in the longitudinal direction even when the internal pressure is zero. This force is called *tethering force* because it is present as a result of tethering of the artery (i.e., the artery exists in its original location because it is tethered at the ends and at several locations along its length). The magnitude of retraction can be taken to indicate the amount of tethering force and, in general, the magnitude of retraction decreases with age.

Considering once again the concept that the artery is naturally designed to have a constant stress, as was explained earlier for different-size arteries, we expect that the stress in the longitudinal direction also should be almost constant even in the presence of the tethering force. As we will see below, it is so.

Considering that the normal pressure in the body is 120/80 mmHg, or an average of about 105 mmHg, then it would be optimum if at this pressure the artery were at a length equal to that which it has in the body without any imposition of tethering force. The longitudinal force present in this condition would then be the constant force the artery would have even if the internal pressure was to change. For instance, if the pressure becomes lower than 105 mmHg, then the longitudinal force is reduced and the artery would tend to shorten lengthwise. In this condition, the tensile tethering force of appropriate magnitude can be applied to restore the artery length. If the pressure becomes higher than 105 mmHg, then the longitudinal force will be larger and the artery would be longer. In this condition, the compressive tethering force of appropriate magnitude can be applied to restore the artery length. In other words, the artery length is maintained constant with a constant total force in the longitudinal direction where the total force is made up of that due to internal pressure and that due to tethering.

The interaction between the longitudinal stress due to pressure (σ_P) and that due to tethering (σ_T) is illustrated in Figure 4.5. The net longitudinal stress (σ_N) is given by

$$\sigma_N = \sigma_P + \sigma_T.$$

Let us consider a carotid artery, held at its in situ length to represent a tethered artery in vivo under internal pressure (4). All stresses are plotted as a function of circumferential strain where the unstressed circumference was defined as that observed in the unstretched, unpressurized vessel. At zero pressure, as the tethering force stretches the artery to in vivo length, the circumferential strain becomes negative and the tethering force is maximum (data points on the far left). When the internal pressure is increased in steps, the longitudinal stress due to pressure increases, the circumference of the artery increases, and the tethering force decreases. The net longitudinal stress remains nearly constant until high pressures and large circumferential strains (greater than 25%) are encountered. At normal pressures, circumferential strains can be 40–50% and in such cases, the force due to pressure is a dominating factor for determination of the wall stress compared with the force due to tethering.

2.2. *Active versus Passive Conditions of the Artery*

The stress in the wall exists to balance the force of internal pressure plus other forces such as those arising from tethering. In nonbiological materials, this stress

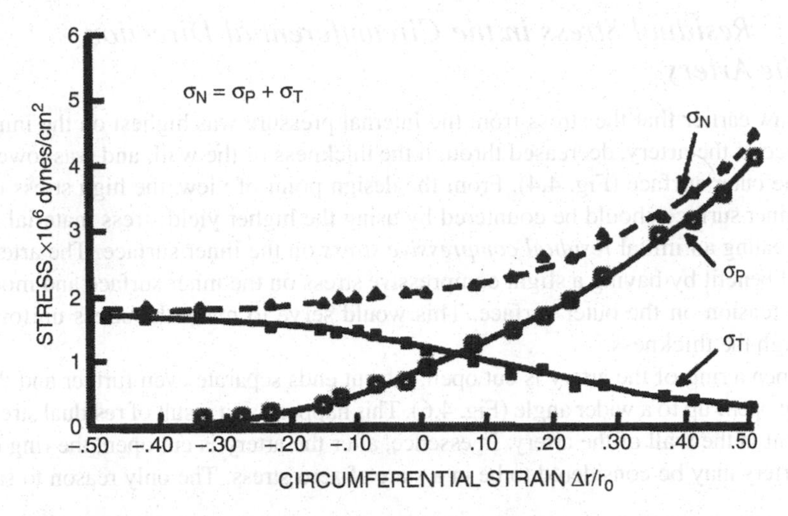

FIGURE 4.5. Net longitudinal stress (σ_N) as a function of circumferential strain in a dog carotid artery. Zero strain indicates vessel circumference of relaxed, excised, totally unloaded vessel. Net longitudinal stress is a sum of the stress due to tethering (σ_T) and that due to pressure (σ_P). Interaction between these components results in almost constant net longitudinal stress up to large circumferential strains and high pressures. (From Dobrin PB, Hand book of Physiology, pg. 82, Lippincott Williams & Wilkins, with permission.)

will exist purely by "passive" mechanisms, the example being a steel pipe under pressure or a rubber band under stretch. In the artery, however, there are both the passive and the active components. The passive components would be concerned with the stretching of all of the cellular and noncellular components of the artery wall. Because the artery wall consists of significant amounts of contractile cells, the contractile state of the cells will generate some force (active component), which will participate in bearing a portion of the wall stress. Thus, the total stress in the wall can be written as:

$$\text{Total stress}\,(\sigma_T) = \text{active stress}\,(\sigma_A) + \text{passive stress}\,(\sigma_P)$$

or

$$\sigma_T = \sigma_A + \sigma_P.$$

Although the total stress can be determined, its partition into active and passive components is not easily known. The passive stress would be that present due to elasticity of the wall in a "dead" artery. The active component is due to contraction of the living smooth muscle cells in the wall. In general, passive stress is a function of the degree of stretch in the wall, while the active stress is independent of stretch (by definition) and depends on physiological activity of the tissue, such as vasomotor activity. The active tone of the artery is greatly influenced by vasoactive drugs.

2.3. *Residual Stress in the Circumferential Direction of the Artery*

We saw earlier that the stress from the internal pressure was highest on the inner surface of the artery, decreased through the thickness of the wall, and was lowest on the outer surface (Fig. 4.4). From the design point of view, the high stress on the inner surface should be countered by using the higher yield stress material or by creating an initial *residual compressive stress* on the inner surface. The artery could benefit by having a slight compressive stress on the inner surface and moderate tension on the outer surface. This would serve to make the stress uniform through the thickness.

When a ring of the artery is cut open, the cut ends separate even further and the artery opens up to a wider angle (Fig. 4.6). This happens as a result of residual stress present in the wall of the artery. In essence, after the artery is cut open, the ring of the artery may be considered to be in a state of zero stress. The only reason to say

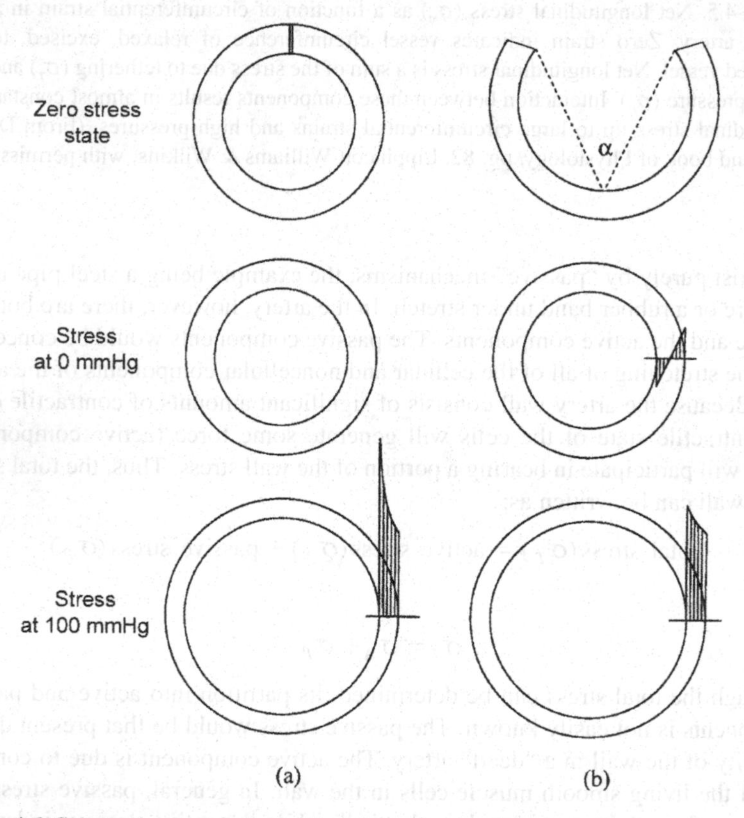

FIGURE 4.6. Schematic presentation of the effect of residual stress on the stress distribution in the artery under pressure. (a) In the absence of residual stress, the stress in the artery under pressure varies significantly through the wall thickness. (b) In the presence of residual stress, the stress in the artery under pressure is more uniform through the wall thickness.

"may be" in this case is because the artery is a complex material and therefore one must consider the possibility that if the ring is cut on the opposite side, the opening angle may be different, thereby implying that the residual stress may be different. In general, it is a good approximation to assume that one cut will produce a zero stress state in the artery wall. Fung's experiments support this hypothesis (5).

Starting with a zero stress state, then for a given opening angle one can determine the stresses that would be induced when the two ends are joined together to form a ring. As shown in Figure 4.6, we will have compressive stress on the inner surface and tensile stress on the outer surface for an opening angle α but only zero stress if the opening angle is zero. When the artery is under pressure, its wall becomes thin as its diameter increases. The distribution of stress in the wall is quite nonuniform and skewed when there is no residual stress. However, the stress distribution becomes more uniform when the residual stresses are combined with the stresses that occur with pressurization. The artery may desire more uniform distribution of stress through its wall, and this will be achieved through the presence of residual stress. Residual stress, in general, does not change the average stress in the wall or the stress in the middle of the wall thickness (midplane stresses) but rather changes the stress distribution through the wall thickness. It may be important to note other implications of residual stress. For instance, residual stress, overall, decreases with age, whereas the arterial diseases increase with age. The residual stress may change with conditions such as acute or chronic hypertension and with the presence of certain drugs in the system, which may affect the state of the cells in the arterial wall.

Some experimental data and the effect of residual stress on the stress in the artery in vivo have been discussed in detail by Fung (5).

3. Circular Hole in a Plate Under Tension

Here we begin to analyze the geometry that is "a hole in a plate" and develop this to a geometry that is "a hole in a cylinder," which would be developed even further to a geometry describing "an artery with a branch." Although many of the theories for stress determination have been developed for metals, these theories can be applied to biological tissues, such as arteries, particularly to gain qualitative information.

The stress distribution in the vicinity of a small circular hole of radius a in a plate stretched elastically by a uniform tensile stress σ can be derived as follows: For the element at a distance r (Fig. 4.7), the stress in the radial direction (σ_r) and that in the tangential (circumferential) direction (σ_t) are given by following equations:

$$\sigma_r = \frac{\sigma}{2}\left(1 - \frac{a^2}{r^2}\right) + \frac{\sigma}{2}\left(1 + 3\frac{a^4}{r^4} - 4\frac{a^2}{r^2}\right)\cos 2\theta \tag{12}$$

$$\sigma_t = \frac{\sigma}{2}\left(1 + \frac{a^2}{r^2}\right) - \frac{\sigma}{2}\left(1 + 3\frac{a^4}{r^4}\right)\cos 2\theta. \tag{13}$$

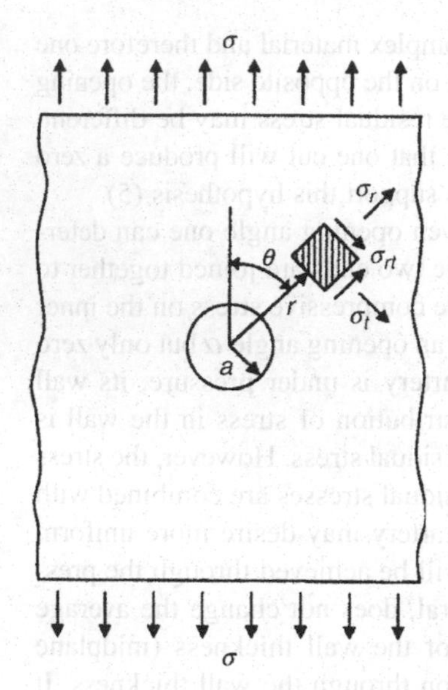

FIGURE 4.7. Hole in a plate subjected to tension.

At the circumference of the hole, $r = a$, and $\sigma_r = 0$.

$$\sigma_t = \sigma(1 - 2\cos 2\theta). \tag{14}$$

The tangential stress σ_t is maximum at points where $\theta = 90$ degrees and -90 degrees located on the circumference of the hole and on the axis perpendicular to the direction of the applied tension. At these points, the stress

$$\sigma_t = 3\sigma. \tag{15}$$

The tangential stress is minimum at points where $\theta = 0$ degrees and 180 degrees, and at these points

$$\sigma_t = -\sigma. \tag{16}$$

Thus, it may be noted that a small hole in a plate subjected to tension causes an increase in the stress at the hole to a maximum value of 3 times that in the normal undisturbed region of the plate.

If the plate is stretched further so as to cause yielding, the first regions to yield are two small regions, at the edge of the hole and on the axis perpendicular to the direction of the load. This yielding is caused by the high (3σ) tangential stress and extends only to a short distance from the edge of the hole.

3.1. Stress-Concentration Around a Circular Hole in a Plate Under Tension

The tangential stress σ_t is of importance around the hole as was seen above. The distribution of this stress on the axis perpendicular to the direction of the load at $\theta = 90$ degrees is given by the following equation:

$$\sigma_t = \frac{\sigma}{2}\left(2 + \frac{a^2}{r^2} + 3\frac{a^4}{r^4}\right). \tag{17}$$

As can be seen in Figure 4.8, the stress is maximum at the edge of the hole where $r = a$ and $\sigma_t = 3\sigma$, and it decays quite rapidly. At $r = 2a$,

$$\sigma_t = 1.22\sigma. \tag{18}$$

Figure 4.8 shows that the effect of a small hole is very limited and damps out rapidly. For practical purposes, Eq. (17) can be used for plates of a width more than 5 times the hole diameter. For plates of a smaller width than this, or for proportionally larger holes, experimental results may be more important than theoretical calculations, particularly in the case of arteries.

3.2. *Lüder Lines*

In metals, particularly when the tensile stress reaches the value of yield point, the material undergoes plastic flow, which is the sliding of a portion of the material along the planes where shearing resistance has been overcome. Such sliding produces slip bands over the surface of the material, and these bands can be visible as dark bands on a shiny surface and/or the bands produce a wavy surface that can be felt with the fingertips. These lines (bands) are known as Lüder lines (Fig. 4.9).

Lüder lines generally begin at points of stress-concentration, as the yield point stresses are first reached there and their propagation direction is influenced by the direction and intensity of these localized high stresses. Lüder lines therefore can be taken as indicators of the highest stresses present in the region and of the direction in which these stresses were acting.

Figure 4.9 shows an example of when a plate with a circular hole is subjected to tension in the vertical direction, the Lüder lines appear at the edge of the hole in two locations and propagate in the direction consistent with the stress

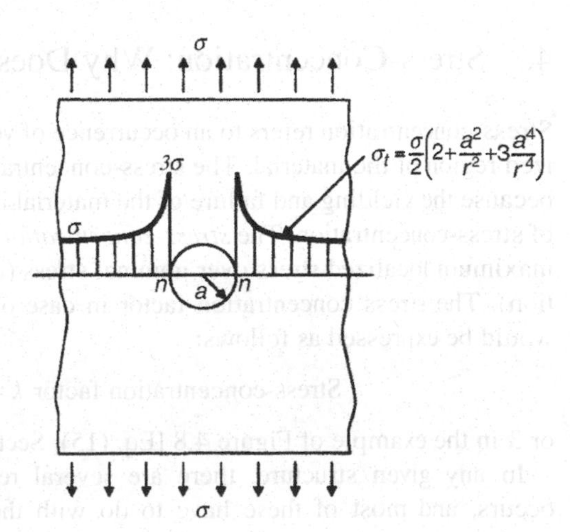

FIGURE 4.8. Variation in stress in a plate containing a circular hole and subjected to uniform tension.

FIGURE 4.9. Lüder lines at the edge of a hole in a plate subjected to tension in the vertical direction.

concentration described in the previous section. The highest tensile stresses acting in the vertical direction reach the yield point at two locations on the horizontal axis at $\theta = 0$ degrees angle and cause yielding of the material there producing Lüder lines. The lines accurately show the points of maximum stress concentration. Conversely, Lüder lines have been used to determine stress-concentration factors.

4. Stress-Concentration: Why Does It Occur?

Stress-concentration refers to an occurrence of very high stresses in a small localized region of the material. The stress-concentration is very important in practice because the yielding and failure of the material is expected to start at the location of stress-concentration. The *stress-concentration factor* (k) is defined as a ratio of maximum localized stress over nominal stress (i.e., a stress away from this location). The stress concentration factor in case of a circular hole in a plate then would be expressed as follows:

$$\text{Stress-concentration factor } k = \frac{\sigma \max}{\sigma} = \frac{\sigma_t}{\sigma}$$

or 3 in the example of Figure 4.8 [Eq. (15), Section 3].

In any given structure, there are several reasons why stress-concentration occurs, and most of these have to do with the practical nature in which the material is used. Why the stress-concentration occurs may be best illustrated by the following examples.

4.1. Hole in a Plate

In the case of a hole in a plate, the primary reason for the occurrence of stress-concentration is the "absence of the material, i.e., the presence of the hole." As long as this fundamental geometry exists, the stress-concentration in unavoidable, and furthermore, the occurrence of stress-concentration does not depend upon whether the material is a biological tissue, such as the artery, or steel. The degree of stress-concentration may vary with the material properties, but stress-concentration will always be present in this geometry.

In Figure 4.10, one can see that the tensile stress acting in the vertical direction in regions away from the hole has the material of the plate available to sustain it. For instance, let us consider the condition along the axis *nn*. The stress acting along C_1D_1 is sustained by the material along *CD*, the stress along B_1C_1 is sustained by the material along *BC*, but the stress acting along A_1B_1 has no material along axis *nn* to sustain it. This entire stress can be considered to be concentrated at point *B*, in which case the stress at point *B* will be infinite. Or, one can think of an "influence length," and that is the stress along A_1B_1 together with that along B_1C_1 is sustained by the material along *BC*. Once again, a very high stress will be present in region *BC*. In fact, the stress in the *BC* region will be increased by a factor A_1C_1/BC or the ratio of the area over which the applied force is acting to the area of the material to which the force is transferred.

In essence, there will always be stress-concentration at the hole in a plate as a result of the "absence of the material or presence of a hole."

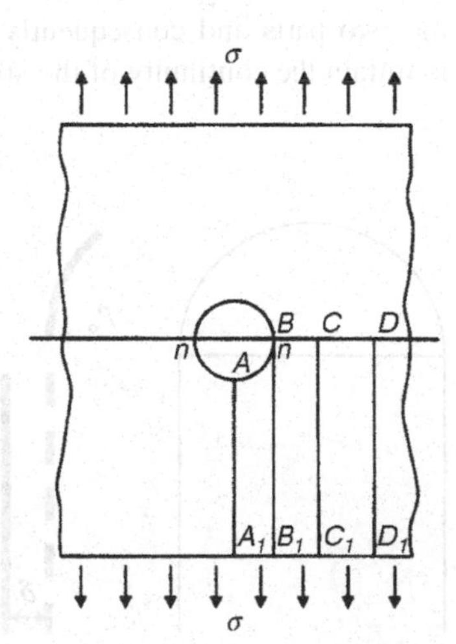

FIGURE 4.10. Schematic drawing to illustrate the absence of material (presence of a hole) as a cause of stress increase at the hole when the plate is in tension.

4.2. *Geometric Discontinuity Stresses*

Primary stresses are those that occur in response to the load and are present over the entire material where the loads are acting. These stresses are quite important. There are also secondary stresses that occur as a result of various considerations such as the presence of a hole in a plate or changes in the geometry of the structure. Such stresses are localized in nature but they are still very important because they can reach a high value and can cause yielding or rupture to begin there.

Let us consider the following example for changes in the geometry. Let us consider a cylindrical pressure vessel with a spherical head covering one end of a cylinder (Fig. 4.11). We can see how this change in the geometry can induce discontinuity stresses. The internal pressure will produce larger circumferential stress in a cylinder, thereby causing larger expansion at the top of a cylinder, whereas the same pressure will produce smaller circumferential stress in the sphere, thereby causing smaller expansion at the bottom of the hemisphere. At the juncture of these two parts, the continuity of the vessel still has to be preserved, and this will be achieved by local bending. This localized bending will set up additional stresses called *discontinuity stresses*. To preserve continuity, an additional force say P_o will act in the opposite directions on the two parts at the juncture as shown in Figure 4.11.

Usually, the discontinuity stresses are the combination of the membrane stresses and the bending stresses. At the junction, the membrane stresses also change because there is a change in the radius of the cylinder. To determine the total stress at the junction, it is necessary to find 1) the primary membrane stress, 2) the change in the membrane stress, and 3) the bending stress. The *total stress* can then be obtained by *superposition* of these three stresses. Further elaboration of this principle can be found in Ref. 6.

Similar discontinuity stresses are set up also in the case of a cylinder closed with a flat plate.

Overall then, geometric discontinuity results in a differential deformation in the two parts and consequently causes localized stress to increase, in order to maintain the continuity of the structure.

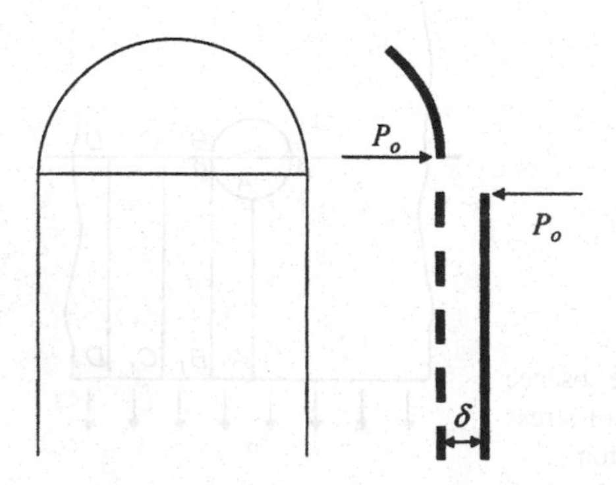

FIGURE 4.11. Discontinuity at the hemispherical head and cylindrical shell juncture. Under pressure, the difference in expansion between the two geometries results in discontinuity stresses at the juncture.

In keeping with the principle, a spherical geometry of an aneurysm in the artery will set up discontinuity stresses at the neck of the aneurysm, where the spherical geometry meets with the cylindrical geometry, resulting in stress concentration at the neck.

4.3. Material Discontinuity Stresses or Stresses in a Bimetallic Joint: Stresses When the Material Changes While the Geometry Remains the Same

Once again, as shown in Figure 4.12, when a cylinder made of one material for one part of the length and of another material for the other part is subjected to internal pressure, the discontinuity stresses will be produced. Under pressure, one part will expand more than the other due to the differences in their material properties. To maintain the continuity in the structure, local bending stresses will occur, which will increase stresses in the region of the joint (i.e., there will be stress-concentration present at the joint).

This phenomenon is quite important in the artery because changes in the material properties in the vasculature occur for several reasons. For instance, end-to-end arterial anastomosis causes localized stresses at the anastomosis because the suture material creates a band of increased stiffness. Artery-to-synthetic graft anastomosis and artery-to-vein anastomosis are other examples that will also result in stress-concentration at the region of anastomosis.

The relevance of this will be discussed again in Chapters 12 and 13 where we consider anastomotic aneurysms and anastomotic intimal hyperplasia.

5. Hole in a Cylinder Under Internal Pressure

The stress-concentration around a circular hole in a cylinder or sphere, with forces applied by internal or external pressure, can be obtained from the cases of simple tension or compression by using the method of superposition. In the case

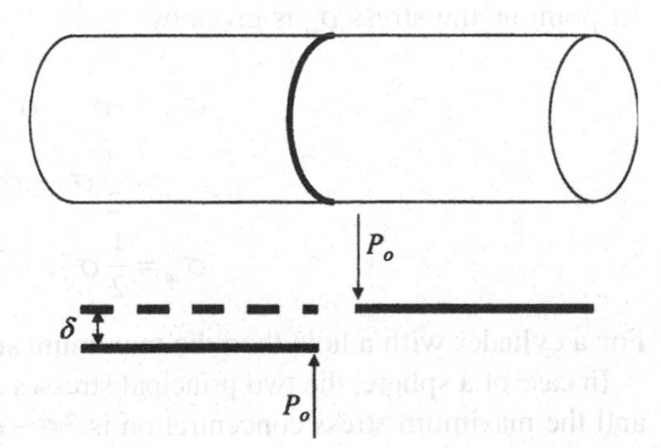

FIGURE 4.12. Discontinuity stresses at the juncture of two different materials. Under pressure, two different materials expand differently producing discontinuity stresses.

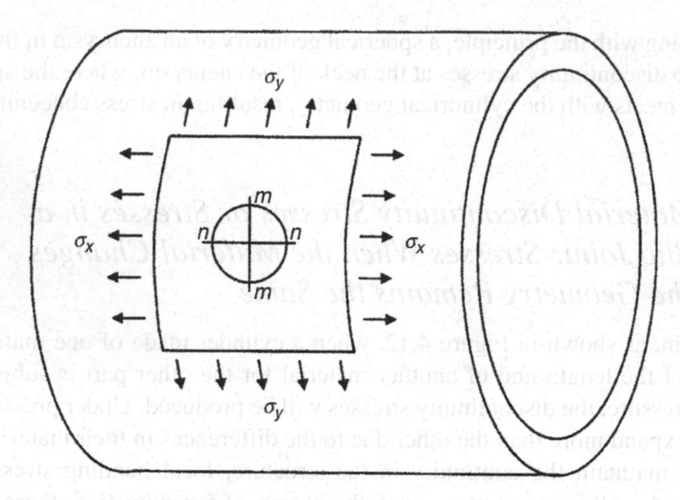

FIGURE 4.13. Hole in a cylinder under internal pressure.

of a cylinder stressed by internal pressure, as in the case of the artery, the longitudinal stress is half of the hoop stress (Fig. 4.13). Therefore, maximum stress will occur at points *n,n* on the longitudinal axis and will be given by (σ_n) as follows [refer to Eqs. (15) and (16), Section 3]:

$$\sigma_n = 3\sigma_y - \sigma_x. \tag{19}$$

We arrive at this equation using the principle of superposition. For instance, the stress σ_y alone produces a stress of $3\sigma_y$ at point *n*, and the stress σ_x alone produces a stress of $-\sigma_x$ at point *n*. By superposition then, under the combined action of σ_y and σ_x, the stress at point *n* is given by

$$\sigma_n = 3\sigma_y - \sigma_x$$

or

$$\sigma_n = 3\sigma_y - \frac{1}{2}\sigma_y$$

$$\sigma_n = 2.5\sigma_y. \tag{20}$$

At point m, the stress σ_m is given by

$$\begin{aligned}\sigma_m &= 3\sigma_x - \sigma_y\\ &= \frac{3}{2}\sigma_y - \sigma_y\\ \sigma_m &= \frac{1}{2}\sigma_y.\end{aligned} \tag{21}$$

For a cylinder with a hole then the maximum stress concentration is $2.5\sigma_y$.

In case of a sphere, the two principal stresses are equal and we have $\sigma_x = \sigma_y = \sigma$ and the maximum stress concentration is $3\sigma - \sigma = 2\sigma$.

For several arteries with branches, we will explore the stress concentration in Chapters 6 and 7, and we may expect that the stress-concentration may be close to value $2.5\sigma_y$ because this phenomenon is primarily dependent on the geometry and only secondarily dependent on the material properties.

5.1. Elliptical Opening in a Plate

There are several occasions when instead of a circular opening an elliptical opening is desired. Particularly in a cylindrical structure, such as an artery, as we will see below, an elliptical opening may be more desirable than a circular opening. In this section, we will analyze stress-concentration around an elliptical opening.

As in the case of a circular hole, the problem of an elliptical hole in a plate (Fig. 4.14) subjected to uniform tensile stress applied at the ends of a plate has been solved mathematically (7). When the major axis of an ellipse is perpendicular to the direction of tension (Fig. 4.14a), the maximum stress occurs at the ends of the major axis (e.g. at points *n,n*). This stress is given by the following equation:

$$\sigma_n=\sigma\left(1+2\frac{a}{b}\right) \tag{22}$$

where *a* is the major axis and *b* is the minor axis of an ellipse. The stress at points *m,m* is

$$\sigma_m=-\sigma. \tag{23}$$

As can be noted, the stress increases with the ratio *a/b* (i.e., with the degree of ellipticity).

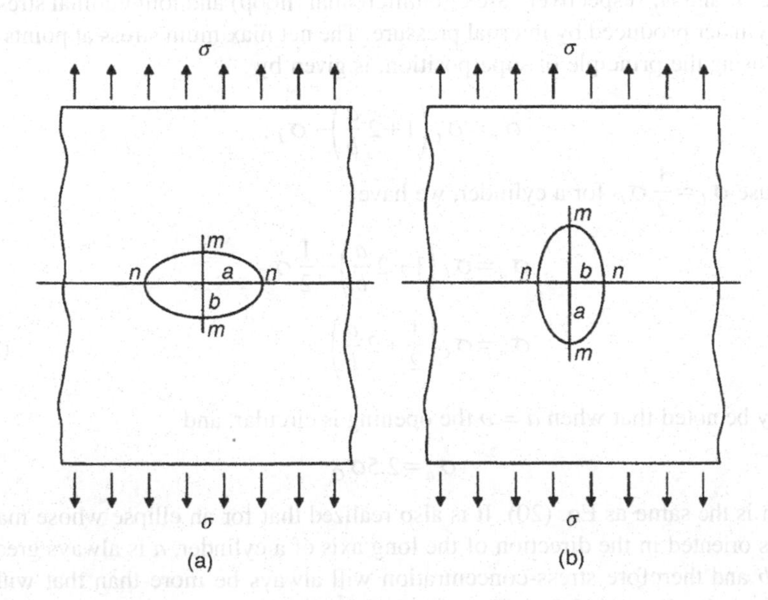

FIGURE 4.14. Elliptical hole in a plate in tension.

When the major axis is parallel to the direction of applied force (Fig. 4.14b), the stress at the ends of the minor axis (i.e., at point *n,n*) is given by:

$$\sigma_n = \sigma\left(1 + 2\frac{b}{a}\right) \tag{24}$$

and that at the ends of the major axis (i.e., at points *m,m*) is given by

$$\sigma_m = -\sigma. \tag{25}$$

The maximum stress occurs at points *n,n* and the stress value approaches σ when the b/a ratio is small (i.e., the ellipse is very slender).

5.2. Elliptical Opening in a Cylinder

Once again, stress concentration around an elliptical hole in a cylinder under internal pressure can be found by the principle of superposition. Let us first consider a case where the major axis of the ellipse is in the longitudinal direction (Fig. 4.15a). The maximum stress at points *n,n* arising as a result of only the hoop stress acting in a cylinder, is given by [refer to Eq. (15), Section 3]

$$\sigma_{n_1} = \sigma_C\left(1 + 2\frac{a}{b}\right). \tag{26}$$

The stress at points *n,n* arising as a result of only longitudinal stress acting in a cylinder is given by [refer to Eq. (16), Section 3]

$$\sigma_{n_2} = -\sigma_L \tag{27}$$

where σ_C and σ_L respectively are circumferential (hoop) and longitudinal stresses in a cylinder produced by internal pressure. The net maximum stress at points *n,n* then, using the principle of superposition, is given by

$$\sigma_n = \sigma_C\left(1 + 2\frac{a}{b}\right) - \sigma_L.$$

Because $\sigma_L = \frac{1}{2}\sigma_C$ for a cylinder, we have

$$\sigma_n = \sigma_C\left(1 + 2\frac{a}{b}\right) - \frac{1}{2}\sigma_C$$

$$\sigma_n = \sigma_C\left(\frac{1}{2} + 2\frac{a}{b}\right). \tag{28}$$

It may be noted that when $a = b$ the opening is circular, and

$$\sigma_n = 2.5\sigma_C$$

which is the same as Eq. (20). It is also realized that for an ellipse whose major axis is oriented in the direction of the long axis of a cylinder, a is always greater than b and therefore stress-concentration will always be more than that with a circular hole, that is, $>2.5\sigma_C$ [Eq. (28)].

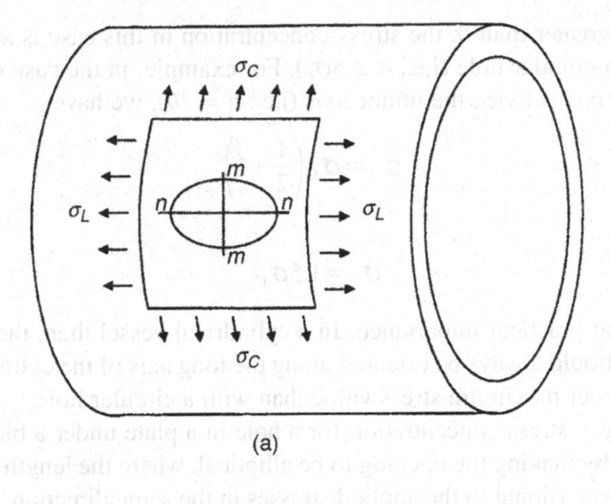

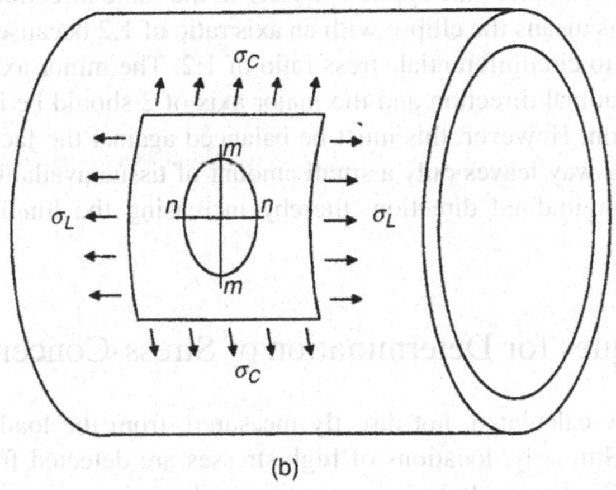

FIGURE 4.15. Elliptical opening in a cylindrical pressure vessel.

Let us now consider a case where the major axis of an ellipse is oriented in the hoop direction (Fig. 15b). Once again using the principle of superposition, we have the following.

Stress at points n,n [refer to Eqs. (24) and (25)]:

$$\sigma_n = \sigma_C\left(1 + 2\frac{b}{a}\right) - \sigma_L$$

$$= \sigma_C\left(1 + 2\frac{b}{a}\right) - \frac{1}{2}\sigma_C$$

$$= \sigma_C\left(\frac{1}{2} + 2\frac{b}{a}\right). \tag{29}$$

Because a is greater than b, the stress-concentration in this case is always lower than that for a circular hole (i.e., $< 2.5\sigma_C$). For example, in the case of an ellipse whose major axis is twice the minor axis (i.e., $a = 2b$), we have

$$\sigma_n = \sigma_C\left(\frac{1}{2}+1\right)$$

or

$$\sigma_n = 1.5\sigma_C. \tag{30}$$

This is of great practical importance. In a cylindrical vessel then, the minor axis of an ellipse should always be oriented along the long axis of the cylinder, thereby obtaining a lower maximum stress value than with a circular hole.

The minimum stress concentration for a hole in a plate under a biaxial load is thus obtained by making the opening to be elliptical, where the lengths of the axes are directly proportional to the applied stresses in the same direction. For a cylindrical vessel, this means the ellipse with an axis ratio of 1:2 because we have longitudinal stress to circumferential stress ratio of 1:2. The minor axis of 1 should be in the longitudinal direction and the major axis of 2 should be in the circumferential direction. However, this must be balanced against the fact that a larger hole created this way leaves only a small amount of tissue available to carry the stress in the longitudinal direction, thereby increasing the longitudinal stress significantly.

6. Techniques for Determination of Stress-Concentration

Usually stress is calculated, not directly measured, from the load and a cross-sectional area. Similarly, locations of high stresses are detected from strains or deformations. To detect stress-concentration and to determine stress values, advantage is taken of the fact that strain is proportional to stress and thus by measuring strain, stress can be known.

There are several methods available to determine stress, and we will deal here with some of them.

6.1. Strain Measurement

This method consists of measuring surface strains on an actual vessel or structure with a device such as electrical resistance strain gauge.

6.2. Photoelastic Method

This method consists of optically measuring the principal stress differences in isotropic transparent material models, which become doubly refractive when polarized light is passed through the model (8). Normally, these models are relatively small and inexpensive. An extension of this method has also been made to permit analyses of the actual structure. It consists of bonding thin sheets of photoelastic plastic to the part of a structure to be analyzed. The surface strains of the part under consideration are transferred to the plastic coating and measurements made by reflected light (8).

6.3. *Moiré Method*

The moiré fringe technique is one of the most adaptable experimental stress analysis methods for evaluation of thermal, as well as pressure, stresses (9). [*Moiré* is the French word for "watered"]. It can be applied directly to metals in models or actual structures. When grids with periodic rulings are made to overlap, interference patterns called moiré fringes are produced. To use this method for stress analysis, fine regularly spaced lines are placed on the undeformed test specimen. The same lines are also placed on a transparent screen. The specimen is then deformed. Moiré fringes are formed when the transparent master grid is superpositioned upon the deformed grid. From the fringes produced, the displacements and strains can be determined.

Figure 4.16 shows the moiré fringe pattern in the region of a hole in a plate subjected to the tensile load in the vertical axis. The regular spacing of vertical fringes at relatively short distances away from the hole indicates that the effect of the hole is local. The closeness of these fringes adjacent to the hole edge at its horizontal axis indicates high strains in this area. Thus, moiré fringe pattern can be used very effectively to visually demonstrate the stress-concentration at the hole in a plate.

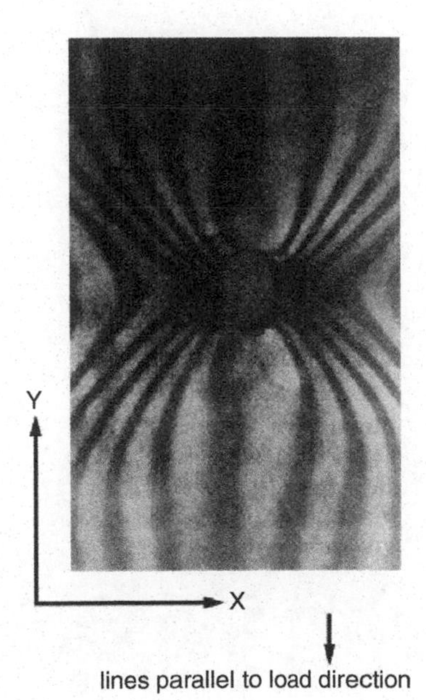

FIGURE 4.16. Moiré fringe pattern in the region of a hole in a plate subjected to tension.

Some of these methods have been used for determination of stress-concentration in bones. However, for arteries almost none of these methods have been applied. The major difficulty is a large deformation that the artery goes through under pressure. Most of the methods are applicable in a small strain range (perhaps <4%).

References

1. McDoald DA: Blood Flow in Arteries. Williams & Wilkins, Baltimore, 1974:258-259.
2. Burton AC: Physical principles of circulatory phenomena; In: The physical equilibria of heart and blood vessels, in Handbook of physiology, Vol. 1, Circulation 1961, pgs. 88-89.
3. Harvey JF: Theory and Design of Modern Pressure Vessels, Second ed. Van Nostrand Reinhold, New York, 1974:56-57.
4. Dobrin PB: Vascular mechanics. In: Handbook of Physiology – The Cardiovascular System III. pgs. 65-102.
5. Fung YC: Biomechanics – Motion, Flow, Stress and Growth. Springer-Verlag, New York, 1990:384-390.
6. Harvey JF: Theory and Design of Modern Pressure Vessels, Second ed. Van Nostrand Reinhold, New York, 1974:151-156.
7. Harvey JF: Theory and Design of Modern Pressure Vessels, Second ed. Van Nostrand Reinhold, New York, 1974:319-322.
8. Harvey JF: Theory and Design of Modern Pressure Vessels, Second ed. Van Nostrand Reinhold, New York, 1974:14-15.
9. Harvey JF: Theory and Design of Modern Pressure Vessels, Second ed. Van Nostrand Reinhold, New York, 1974:18-21.

5
Pressure Vessel Intersections

1. Stress-Concentration at "T"-Branch Pipe Connection: The Area Method

The arteries have branches that come off at various angles and some of them do come off at 90 degrees making it similar to a "T" connection. There are different degrees of stress-concentration for different branch angles. We will therefore consider two cases, one where the branch angle is 90 degrees and the other where it is different from 90 degrees.

For metallic pipes, theories have been worked out that allow us to determine the stress-concentration at a "T"-branch pipe connection reasonably accurately. We will not consider these theories in detail here because they may not be applicable to the blood vessels, where the elastic properties are quite different from those of the metals and where the deformations are very large in comparison with those in metals. However, it is of great benefit to consider some of the approaches used in the analysis of "T"-connections in metals, because these approaches suggest that *the phenomenon of stress-concentration must occur in the arteries also*, although quantitative determination must be done using the actual data obtained from the arteries. It is with this point of view that we will consider the determination of stress-concentration in a "T"-branch connection.

The so-called area method allows the determination of stress-concentration in an approximate sense. For metals, the results of the area method have been compared with those of the other more mathematical approach, and they have been found to be satisfactory, considering particularly the simplicity of the area method. We will consider this approach below.

Let us consider the "T"-branch pipe connection (Fig. 5.1) defined by the diameter ratio for the branch to the main vessel of d/D and the thickness ratio of t/T, made of linear elastic material and subjected to the internal pressure P. There are two regions of high stress in this configuration: high bending stress at point A, where the hoop stress is curtailed and equilibrium in the direction normal to the vessel must be maintained by bending action alone; and point B, where large hoop stresses occur as a result of the removed material. At A, bending stresses are localized over a fraction of thickness of the pipe, and for the deformable material, the capacity for stress relief by deformation is very high. Consequently, the stress at point A is not as high as that at point B. Furthermore, at point B, the highly stressed zone is extensive, penetrating through the entire thickness of the cylinder, and

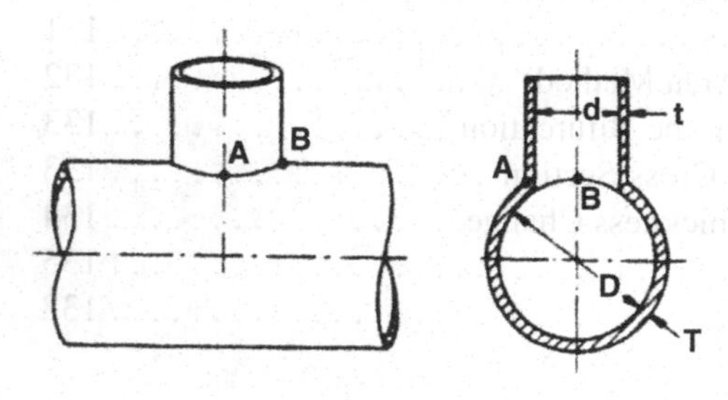

FIGURE 5.1. A straight branch pipe connection, or "tee" or "T."

subjected to only limited stress relief under local deformation. Thus, stress at point B is of most significance for design and/or rupture considerations (1).

Figure 5.2 shows the origin of the high local stresses at point B. The hoop stress in the pipe is obtained as follows (see Chapter 4): $\sigma_c = \frac{PR}{T}$ for the main pipe; or $\sigma_c = \frac{P \cdot R \cdot \ell}{T \cdot \ell}$. The term $R \cdot \ell$ equals area C and $T \cdot \ell$ equals area C'. Thus $\sigma_c = P\frac{C}{C'}$. In other words, the hoop stress is given by the force of the pressure acting on lumen area C, divided by the corresponding wall area C'. Thus, for most of the length of the main pipe and also for most of the length of the branch pipe away from the hole, the hoop stress is given by the pressure times the ratio of areas such as C divided by areas such as C'. In fact, the pressure may be considered as transferred radially to the vessel wall, and therefore, wherever the vessel wall is present, the stress from the pressure is the same.

If we continue to work this process toward the junction for both, the main vessel and the branch vessel, we see that the pressure force over an area such as F is not accounted for because there is no corresponding wall of the pipe present. This force is called the *discontinuity hoop force* and it gives rise to the *discontinuity hoop stress* field (1).

Experimental tests have shown that the discontinuity hoop stress is highest at point B and it decays rapidly in the direction of the main vessel as well as the branch vessel. Because the total hoop stress field (normal hoop stress plus discontinuity hoop stress) is continuous, the stress gradient is high toward the branch at the junction as the main vessel hoop stress is higher than the branch vessel hoop stress, because the branch is small.

From the viewpoint of the shell theory, it may be said that additional high bending stresses are superimposed on the membrane stresses at the juncture. For the first approximation, however, these bending stresses may be neglected because they do not make a net contribution to the discontinuity hoop stresses. Doing so can yield a value called maximum average hoop stress at the juncture.

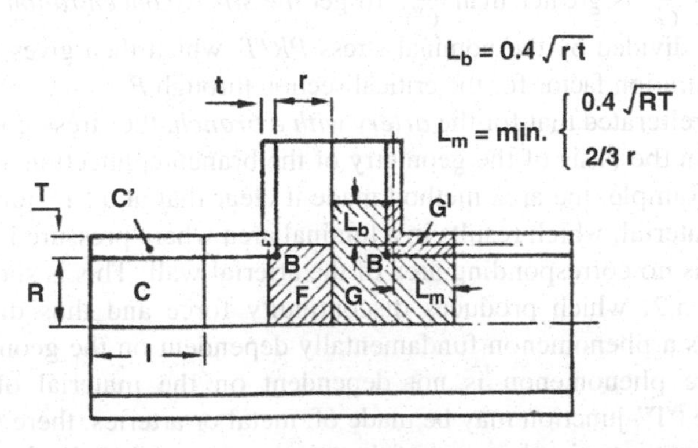

FIGURE 5.2. Section through the plane of symmetry, showing the main points, lengths, and areas of the area method for a "T"-branch connection.

This stress is approximately equal to the value of the hoop stress at the midsurface at the junction. For more accuracy, one can consider stress variation, due to bending, through the thickness of the vessel in a separate analysis and then add this stress to the total maximum average stress so that the maximum stress at point B can be known.

Let us consider the case of a small thin nozzle (Fig. 5.2). In the nozzle, the rate of decay of the localized hoop stress with distance from the juncture would be expressed by the attenuation length (influence length), which is some multiple of $\sqrt{rt}$ (1). It has been determined by detailed analysis that the maximum discontinuity hoop stress is the same as that obtained when the resultant discontinuity force was distributed uniformly over a nozzle length equal to $0.4\sqrt{rt}$, that is, $L_b = 0.4\sqrt{rt}$. To put it simply, the influence length (L_b) (the length affected by the discontinuity stresses) in the nozzle equals $0.4\sqrt{rt}$. Using the same approach, we can say that the influence length (L_m) in the main vessel equals $0.4\sqrt{RT}$. For the main vessel, another consideration is necessary, particularly if the branch vessel is small. The main vessel can be treated as though it is behaving similar to a cylinder with a hole (see Chapter 4). From these considerations, it was observed that the influence length in the main vessel equals $\frac{2}{3}r$. Thus, for the "T"-branch the influence length for the main vessel is taken as the minimum of $0.4\sqrt{RT}$ or $\frac{2}{3}r$. In other words, the influence length is dependent on the size of the branch. Once again, the influence length in the artery is likely to be different from that in metallic pipes.

To obtain average discontinuity hoop stress, the force over area F must be distributed in part over the branch pipe length L_b and over the main pipe length L_m. The total stress at point B is obtained by adding the above discontinuity stress to the nominal hoop stress. Thus, the total stress at the juncture is obtained by using area G for the pressure and area G' for the vessels (Fig. 5.2). Hence, the hoop stress at the junction is $\sigma_c = P\frac{G}{G'}$. If we recall that the nominal hoop stress is $\sigma_c = P\frac{C}{C'}$, then it is obvious that the stress at the junction is considerably higher because the area ratio $\frac{G}{G'}$ is greater than $\frac{C}{C'}$. To get *the stress-concentration factor*, this stress can be divided by the nominal stress PR/T, which then gives the average stress-concentration factor for the critical section through B.

It may be reiterated that for the *artery with a branch*, the stress-concentration must occur on the basis of the geometry of the branch connection as described above. For example, the area method made it clear that at a "T"-junction there is missing material, which results in a luminal area where pressure is acting and where there is no corresponding area of the arterial wall. This is shown as area F in Figure 5.2, which produces discontinuity force and thus discontinuity stress. This is a phenomenon fundamentally dependent on the geometry at the junction. The phenomenon is not dependent on the material of the pipe. Whatever the "T"-junction may be made of, metal or arteries, there will always exist stress-concentration because of the missing material in the form of a hole or the ostium.

In the artery, force F may be distributed over a larger influence length, both in the main vessel and in the branch, because the artery is easily deformable compared with metal. This will lower the stress-concentration factor, but it will not eliminate stress-concentration.

The bending stresses in the region of the juncture are determined by using a complex mathematical procedure, however, they can be accounted for in the determination of stress-concentration by a simplified procedure, that is, by using the so-called bending correction factor.

The bending correction factor K_b is given as follows (1):

$$K_b = 1 + \frac{T}{D} \Big/ \sqrt{\frac{r}{t} \Big/ \frac{R}{T}} \tag{1}$$

For the vast majority of practical branch pipe connections, this factor lies between 1 and 1.1.

The accurate stress-concentration factor (K) then is $K_a \cdot K_b$, where K_a is determined on the basis of the area method described above.

Thus, $K = K_a \cdot K_b$.

Finally, it may be said that the area method is based on generalization and simplification of the complex mechanical behavior of a juncture, while it correctly reflects the most important aspect of overall equilibrium. The area method may be employed for the artery, particularly if the experimental results are available for comparison.

1.1. Sample Calculation Using Area Method: An Example

Let us consider an example shown in Figure 5.3 (1).

$$\frac{d}{D} = \frac{r}{R} = 0.5;$$

$$\frac{T}{D} = \frac{1}{13.4}; \text{ and } \frac{r/t}{R/T} = 0.99.$$

Corner and fillet radii are given by

$\frac{\rho_c}{T} = 0.578$ and $\frac{\rho_f}{T} = 0.461$, where ρ_c and ρ_f represent the corner radius and fillet radius, respectively.

$$L_b = 0.4\sqrt{rt} = 0.4\sqrt{1.69 \times 0.28} = 0.275 \text{ in.}$$

$$L_m = \min \begin{bmatrix} 0.4\sqrt{RT} = 0.4\sqrt{3.38 \times 0.54} = 0.54 \\ 0.667r = 1.13 \end{bmatrix} = 0.54 \text{ in.}$$

The measured areas then are

$G = 8.94$ in^2, and $G' = 0.377$ in.^{2}.

The average stress concentration factor K_a is,

$$K_a = P\frac{G}{G'} \Big/ P\frac{R}{T} = \frac{23.7}{6.7} = 3.54.$$

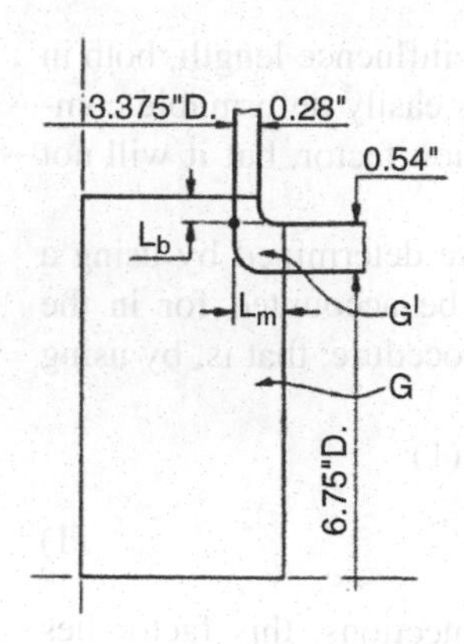

FIGURE 5.3. Example of "T"-branch connection. (Reproduced from Lind NC, ASME paper 67-WA/PVP-7, Winter Annual Meeting and Energy Systems Exposition, Nov. 1967, American Society of Mechanical Engineers, with permission from ASME International.)

From Eq. (1), the bending correction factor K_b is

$$K_b = 1 + \frac{0.746}{0.98} = 1.076.$$

The overall stress concentration factor K is

$$K = K_a \cdot K_b = 3.54 \times 1.076 = 3.81.$$

Hence, the stress at the junction is 3.81 times greater than the nominal stress in the larger pipe. The actual value of the stress concentration factor determined for this model, using stress photoelasticity technique, is 3.50. The difference between the results of the two methods is 9%.

2. Stress-Concentration of a Nozzle in a Spherical Vessel

2.1. *Area Method*

The problem of determining stress-concentration produced by an outlet placed in the wall of a pressure vessel has long been the subject of considerable engineering efforts (2). Theoretical methods have been applied in the solution of such problems, notably the shell theory. The shell theory usually concerns itself with the mid-surface. In many cases, the stresses in a certain region of a complex geometry can be indeterminate by the shell theory, and this is besides the difficulties in the practical solution of shell problems. Also, the results can be less accurate than the expectations in engineering.

It is therefore necessary to consider other approaches. One approach can be to consider simpler concepts, such as static equilibrium, continuity of stresses, and so forth. One such simpler concept is the area method, which provides results that may be semiempirical but suitable for practical design. It may be noted that quite often, because of the nature of the complex problem, experimental solutions are sought such as those obtained by photoelastic methods.

The area method considered below usually yields results equal to, or superior in accuracy to, the shell theory. Also, the method is applicable to reinforced configurations that are difficult to solve by any other method. The results of the area method have been compared with those from photoelastic tests.

The first step in the area method is to locate the mid-surfaces of the nozzle and the main vessel (Fig. 5.4) (2). The next step is to locate point B by passing a line perpendicular to the nozzle's mid-surface. The average hoop stress across the section at B depends on the *attenuation (influence) length* for the nozzle and that for the sphere. In the area method, the discontinuity hoop stress at the juncture is computed by finding the discontinuity hoop force and then by allowing this force to be uniformly distributed over the attenuation lengths in the nozzle and the sphere. The attenuation length is shown below to be $0.4\sqrt{rt}$ *or* $0.4\sqrt{RT}$. For small openings, this approximation is invalid and the stresses are instead distributed over the attenuation length in the main vessel equal to the inner radius of the nozzle. According to the area method then, the maximum average hoop stress at the juncture is $P\,\frac{G}{F}$, where P is the internal pressure, and G and F are the areas shown in Figure 5.4.

2.2. *Example*

Consider the geometry in Figure 5.5 (2). We first locate point P on the mid-surface of the nozzle by drawing a horizontal line from the root of the fillet. Then points A and B are located as follows:

$$PA = 0.4\sqrt{rt} = 0.4\sqrt{1.74 \times 0.6} = 0.41 \text{ in.}$$

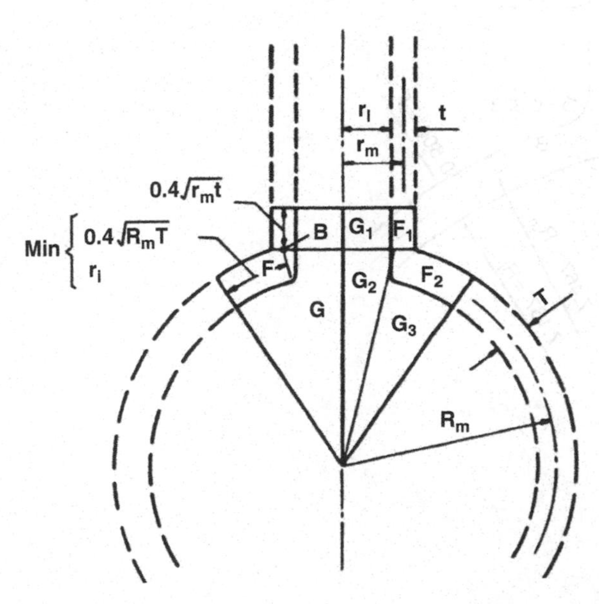

FIGURE 5.4. Areas used in the analysis of a nozzle in a spherical pressure vessel.

$$PB = \min \begin{bmatrix} 0.4\sqrt{RT} = 0.4\sqrt{7.47 \times 0.6} = 0.85 \\ r_i = 1.43 \end{bmatrix}$$

$$\text{So, } PB = 0.85 \text{ in.}$$

Next, lines AN and BO are drawn. Then areas F and G are determined. $F = 0.85$ in.^{2} and $G = 9.85$ in.^{2}.

The stress concentration factor K_s for the average hoop stress equals:

$$K_s = \frac{\text{maximum stress}}{\text{nominal stress}} = \frac{PG/F}{PR_i/2T}$$

or
$$K_s = \frac{2GT}{R_i F} = \frac{2 \times 9.85 \times 0.6}{7.17 \times 0.85} = 1.94.$$

The stress raising effect of the fillet (bending correction) is $K_b = 1.02$.

Hence, the maximum stress-concentration factor is $K = 1.94 \times 1.02 = 1.98$.

In this model, K was determined photoelastically to be 1.93. These two values differ by only 3%.

2.3. *Influence Length*

We saw earlier that the discontinuity force is concentrated at the vessel intersection. It is important to know how far the effect of such a localized force extends into the vessel. The length of the vessel influenced by this force is called the *influence*

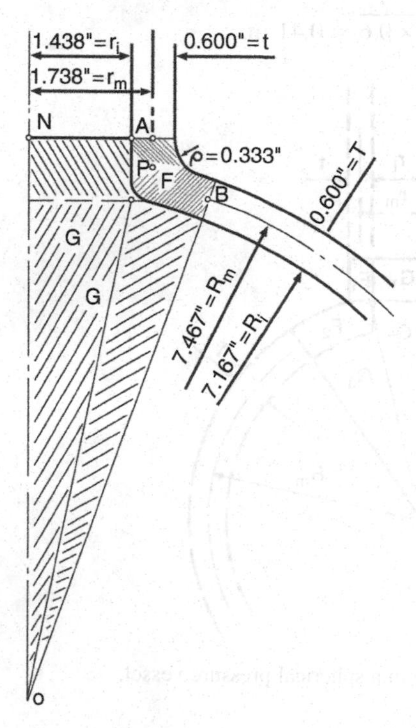

FIGURE 5.5. Example of analysis of a nozzle in a spherical pressure vessel.

length. It would be desirable to use the reinforcement of the vessel intersection only up to this length. In the area method, the discontinuity force was considered to be uniformly distributed over the influence lengths in the nozzle and the sphere. One simple approach used to derive this length is to consider the case of a cantilever beam loaded at a free end with the same discontinuity force (3). Let L represent the distance on the cantilever beam, farthest from the point of application of the load (discontinuity force) at which a significant effect is registered. Then from the equations for the end deflection and the slopes of a beam, it was found that $L = \frac{1}{\beta}$, where β is *the attenuation factor* described earlier (3),

or
$$\beta = \frac{\left[3(1-\nu^2)\right]^{1/4}}{\sqrt{rt}};$$

or
$$\beta = \frac{1.285}{\sqrt{rt}} \text{ for } \nu = 0.3 \text{ for steel.}$$

In the case of vessel intersections, the load is considered applied not at the free end of the beam but in the middle of the beam, which is held at both ends. Consequently, influence length L is distributed on both sides of the load, either side being $L/2$. Hence, influence length on any one side is given by $L/2$.

Then, $\frac{L}{2} = \frac{1}{2\beta} = \frac{\sqrt{rt}}{2 \times 1.285} = 0.4\sqrt{rt}$. Thus, the influence length is $0.4\sqrt{rt}$ on the nozzle side and $0.4\sqrt{RT}$ on the sphere side, as in the example above.

2.4. *Effect of Bending (Stress-Concentration Correction Factor K_b)*

Stress raising factor (K_b) from bending is a function of radius of fillet (ρ) at the junction (4). The value of K_b varies with ρ/T; it decreases as ρ/T increases, and it is usually in the range of 1.3 to 1.0. It is understood that the problem of stress raising effect from bending is oversimplified here. However, the goal is to show that the stress raising effect of the bending is small and has little effect on the total stress-concentration factor. In other words, as a first approximation, the area method is adequate to establish the order of stress-concentration.

In fact, it may be stated, in the words of Lind (1), that "a complicated mathematical analysis tends to suggest to the unwary that the procedure is a highly exact method of stress analysis, and is thus undesirable."

2.5. *Comparison with Shell Theory*

Waters (5) has described an analysis by shell theory that introduces only a small discontinuity at the nozzle-sphere juncture. It is of interest to compare Waters' results with those of the area method for very thin shells. Waters' results are based on Geckeler's second approximation (6). It is valid for very small central angles

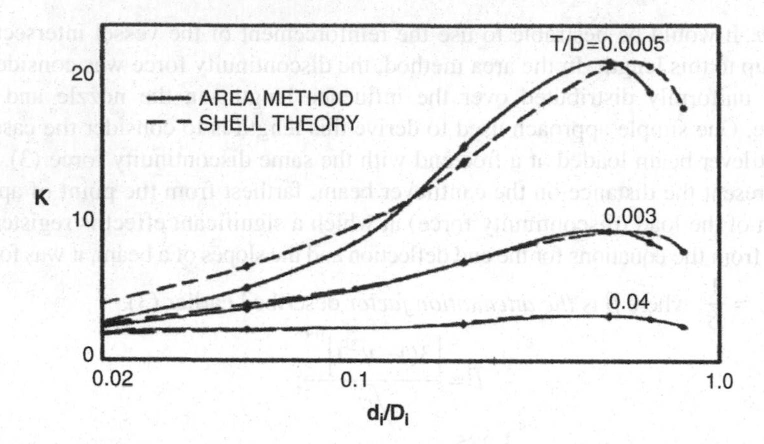

FIGURE 5.6. Stress concentration factor (K) versus ratio of nozzle diameter (d_i) to spherical vessel diameter (D_i). Comparison with shell theory for unreinforced nozzle.

of the opening. Figure 5.6 shows the comparison for an equal nominal stress level in the nozzle and the sphere. The agreement is very good.

3. An Elastic Shell Analysis of Stress-Concentration of a Pressurized "T"-Branch Pipe Connection

In brief, we will consider one of the approaches used by Lind (7) in the determination of stress-concentration at a "T"-branch pipe connection. A simplified version of the analysis will be considered here, because it was found that a number of special approximations could be made without impairing the practical accuracy of the method. These approximations also serve to bring out the principal features of the approach. Let us refer back to Figure 5.2.

Overall equilibrium between the discontinuity force and the discontinuity stresses in the branch region can be expressed as

$$\frac{a_1 h_1}{2\, b_1}\cdot \sigma_{c_1} + \frac{a_2 h_2}{2\, b_2}\cdot \sigma_{c_2} = P\left(a_1 - \frac{h_1}{2}\right)\left(a_2 - \frac{h_2}{2}\right) \tag{2}$$

where

$$b = \left[3\left(1-\nu^2\right)\frac{a^2}{h^2}\right]^{1/4}$$

and a = mid-surface radius, h = wall thickness, σ_c = hoop stress, P = internal pressure, ν = Poisson's ratio, and subscripts 1 and 2 denote main vessel and branch vessel, respectively. Note slightly different symbols from Figure 5.2 for radius and wall thickness (a and h instead of R and T).

Considering that the hoop strain and the total hoop stress are continuous at the juncture, we have

$$\text{Total hoop stress} = \sigma_{c_1} + P\frac{a_1}{h_1} = \sigma_{c_2} + P\frac{a_2}{h_2}. \tag{3}$$

From Eqs. (2) and (3) we obtain the solution for the total hoop stress at B as follows (7):

$$\sigma_{c_1} + P\frac{a_1}{h_1} = P\frac{2a_1a_2 + \left(a_1^2/b_1\right) + \left(a_2^2/b_2\right)}{\left(\dfrac{a_1h_1}{b_1}\right) + \left(\dfrac{a_2h_2}{b_2}\right)} \tag{4}$$

where certain terms in the order of h/a have been neglected in comparison to unity.

Equation (4) can be divided by the nominal stress in the main vessel $\left(P\dfrac{a_1}{h_1}\right)$ to obtain *stress-concentration factor* K as follows:

$$K = \frac{2\cdot a_2\cdot h_1 + \left(a_1h_1/b_1\right) + \left(a_2^2/b_2\right)\Big/\left(a_1/h_1\right)}{\left(\dfrac{a_1h_1}{b_1}\right) + \left(\dfrac{a_2h_2}{b_2}\right)}$$

K determined as above has been compared with the experimental values of K for a variety of geometries, and the agreement between them has been found to be good as seen from Table 5.1 (7).

TABLE 5.1. Stress-concentration factor K for "T"-branch connection for various models: comparison between theory and experiment.

Diameter ratio $\frac{d_2}{d_1}$	Stress ratio d_2t_1/d_1t_2	Main-vessel thickness ratio d_1/t_1	K (theory)	K (exp.)
0.50	0.98	13.4	3.74	3.50
0.50	0.99	13.2	3.73	3.50
0.50	0.98	13.3	3.73	3.74
0.80	1.00	13.0	4.32	4.10
1.00	1.00	13.0	4.40	3.90
0.28	1.23	26.9	3.54	4.00
0.63	0.92	19.0	4.52	5.00
1.00	1.00	19.0	5.06	5.43
0.34	1.02	98.0	6.53	5.70
0.29	0.56	13.1	2.66	3.32
0.29	0.59	12.9	2.67	2.90
0.29	0.58	13.0	2.67	3.28
0.29	0.56	13.3	2.67	3.36
0.57	0.41	13.1	2.56	2.70
0.30	0.48	26.9	3.24	3.00
0.50	0.512	30.0	9.02	8.33

1, main vessel; 2, branch.

3.1. *Structural Analysis of "T"-Branch Shell Intersection*

Prince (8) performed a finite element analysis of "T"-branch pipe connections considering the geometry to be made of thin elastic shells. The results are illustrated in the case of the following example (Fig. 5.7). The geometry was divided into triangular shell elements. The stress at the junction is shown in Figure 5.8. Once again, very high circumferential stress on the outer surface of the main vessel is seen. The circumferential stress at the intersection is almost 6.5 times that in the main body of the cylinder. Axial stress is also increased at this location but by a small amount. The agreement between the experimental values obtained from the strain gauge method and analytical values is very good.

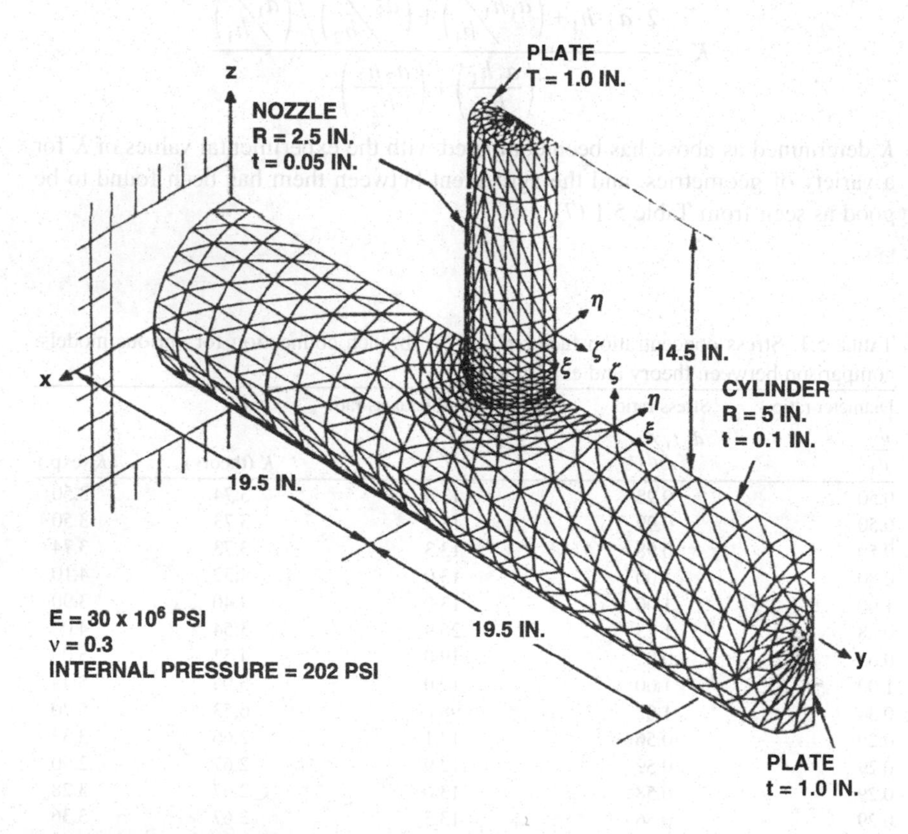

FIGURE 5.7. Nozzle to cylinder intersection. (Reproduced from Prince N et al., Proceedings of 1st International Congress on Pressure Vessel Technology, American Society of Mechanical Engineers, 1969:245-254, with permission from ASME International.)

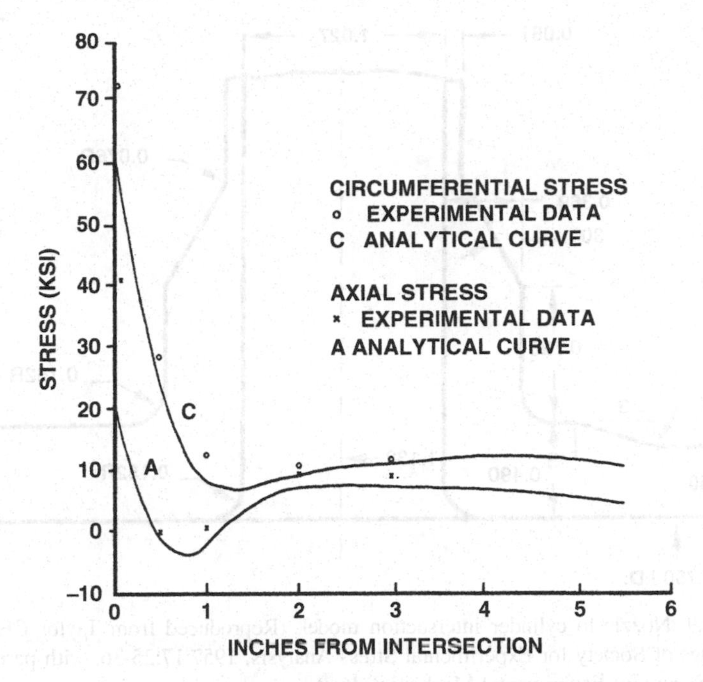

FIGURE 5.8. Stresses at the outer surface of the nozzle to cylinder intersection for the internal pressure of 202 PSI. Comparison of analytical solution with the experimental data obtained with strain gauge. (Reproduced from Prince N et al., Proceedings of 1st International Congress on Pressure Vessel Technology, American Society of Mechanical Engineers, 1969:245-254, with permission from ASME International.)

4. Photoelastic Study of Stresses Around Reinforced Outlets in Pressure Vessels

Because the stresses at the branch regions are quite high, these regions are often *reinforced* by adding extra material. Determination of stresses at a complex geometry becomes more challenging by analytical methods. Therefore, analytical results must be confirmed experimentally. *The photoelastic method* is used quite often to experimentally determine the stresses at the branch. We will consider this method for determination of stresses at a "T"-branch connection, which has been reinforced.

4.1. Photoelastic Method

Let us consider a "T"-branch under internal pressure shown in Figure 5.9, where we want to determine the stresses (9). The "frozen stress" method of three-dimensional photoelasticity will be described here. The details of the method can be found in several places (9, 10), and we will consider only a brief description here. First, a model

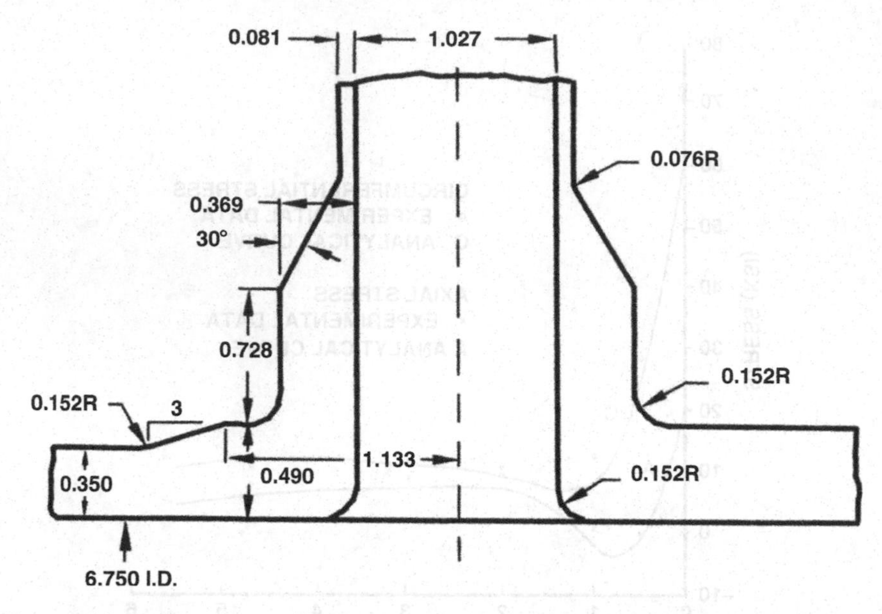

FIGURE 5.9. Nozzle to cylinder intersection model. (Reproduced from Taylor CE et al., Proceedings of Society for Experimental Stress Analysis, 1957:17:25-36, with permission from the Society of Experimental Mechanics, Inc.)

of the "T" is made from an epoxy resin. Normally this can be a scaled-down version of the actual geometry to be analyzed. Usually, the model is made from component parts and then cemented together to form a complete pressure vessel. The cemented joints and connections supplying the internal pressure are located remotely from the area to be analyzed. The best procedure consists of a cycle in which the material is slowly heated to 150°C, then loaded by internal pressure and maintained for 30 min to reach equilibrium. Then the model is cooled under pressure load at a rate of 2°C per hour to room temperature. This way, the photoelastic patterns are "frozen" into the model. The model is subsequently sliced with care so as not to heat it. When polarized light is passed through a slice, fringe patterns such as those shown in Figure 5.10 can be seen (9).

Several points need consideration in the method.

a) The three-dimensional photoelastic procedures yield highly reproducible results.
b) Overall errors in the method are about 5%.
c) The similarity between photoelastic models and metal prototypes is very good. Even though the unit strain in the model is greater than the unit strain in the metal prototype, the stress-strain relationship for the epoxy resin is linear in the range of stress used. Because the large strains do not appreciably change the geometry of the model, these strains do not introduce significant errors.
d) Poisson's ratio may also be different for the model compared with that for the metal (steel). However, this also does not affect the maximum principal stresses significantly.

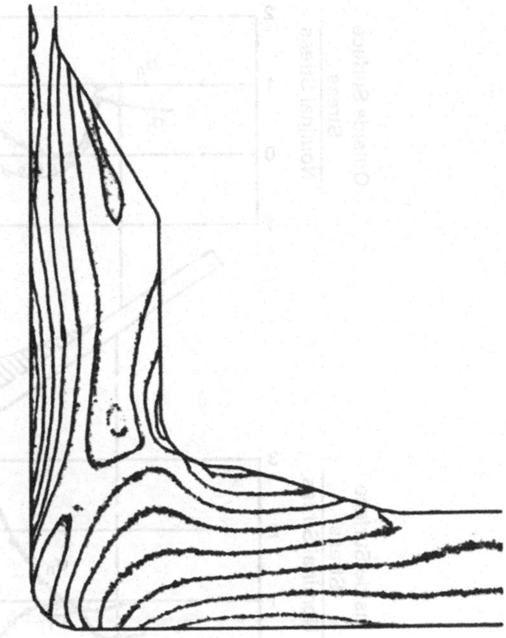

FIGURE 5.10. Fringe photograph of a slice taken from the model in Figure 5.9. (Reproduced from Taylor CE et al., Proceedings of Society for Experimental Stress Analysis 1957;17: 25-36, with permission from the Society of Experimental Mechanics, Inc.)

e) The values for the stresses are obtained in the range where the material behaves elastically with a linear stress-strain relationship. Therefore, the stress-concentration factor is regarded as *elastic stress-concentration factor*.

Table 5.2 shows the parameters of the model and also stress values in the model (10). Figure 5.11 shows the distribution of stresses (circumferential and tangential) on the inner and the outer surfaces of the pressure vessel (10). All of the stress

TABLE 5.2. Dimensions and photoelastic stress results of the model.

Model	Longitudinal plane	Transverse plane
d/D	0.418	0.418
h/D	0.087	0.087
t/d	0.087	0.087
t/h	0.418	0.418
Total reinforcement	55%	85%
Stresses on inside surface		
σ *n*.max/σ nom	2.84	—
σ *t*.max/σ nom	0.90	—
Stresses on outside surface		
σ *n*.max/σ nom	1.23	1.65
σ *t*.max/σ nom	0.78	2.10

σ_n, circumferential stress; σ_t, tangential (longitudinal) stress; σ_{nom}, nominal stress; *D*, inside diameter of a main vessel; *d*, inside diameter of a nozzle; *h*, wall thickness of a main vessel; *t*, wall thickness of a nozzle.

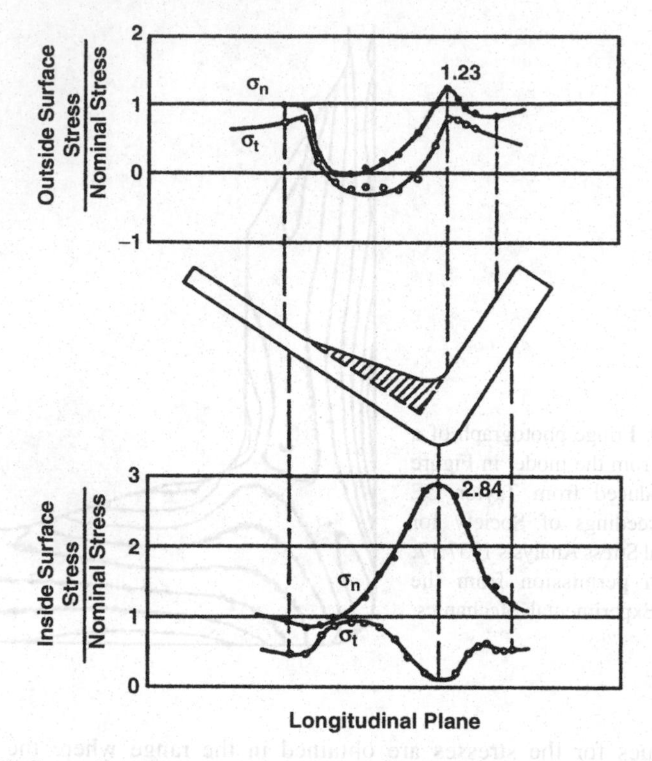

FIGURE 5.11. Stress distribution at the nozzle-cylinder intersection when the intersection is reinforced. Stress values are obtained using photoelasticity method.

values are obtained from the photoelastic patterns. As can be seen, the stresses on the inner surface at the branch region are very high (stress concentration factor of 2.84) even after the reinforcement.

5. Nonradial Branch (Branch at an Angle)

We saw in Table 5.1 what the stress-concentration factor is for a "T"-branch connection for various models. In the body, however, several branches of the artery come off at an angle different from 90 degrees to the main vessel. Thus, it is necessary to determine what the stress-concentration will be in such a geometry.

To start with, we notice that a nonradial branch makes a hole in the main vessel, which is not circular but rather elliptical. Just as an elliptical hole in a plate produces a higher stress-concentration than does a circular hole (see Chapter 4), a nonradial branch also has a higher stress-concentration than its radial counterpart. Analytical and experimental investigations have shown that maximum stresses occur in the vessel on the major axis of the elliptical opening close to the branch (11,12). Furthermore, the stresses are greater for branches of increased nonradiality. In cylindrical vessels, the stress-concentration of the nonradial branch can be approximately determined by relating it to that for the same but radial branch by the following equations:

1. For the nonradial branch coming off in a circumferential plane as shown in Figure 5.12, we have (13),

$$K_{nr} = K_r\left(1 + 2\sin^2\phi\right).$$

where K_{nr} = nonradial nozzle stress concentration factor, K_r = radial nozzle stress concentration factor, and ϕ = angle between the axes of the (assumed) radial branch and the nonradial branch.

2. For the nonradial branch coming off in a longitudinal plane as shown in Figure 5.13,

$$K_{nr} = K_r\left[1 + \left(\tan\phi\right)^{4/3}\right]$$

The stress-concentration factor goes up sharply with the angle of nonradiality. In these cases, smooth transition in the geometry is even more important than that with the radial branch. This is especially applicable to the acute internal lip and external crotch (Fig. 5.13), where it has been found that the maximum stress occurs and the fatigue failure originates. As we have seen in Chapter 2, the arterial disease is also more common at these sites.

Although thus far we focused on the maximum stress in the geometry, let us now consider another region, which also has high stress even though it may not be the highest stress. That is region A, which we may call the proximal lip of the

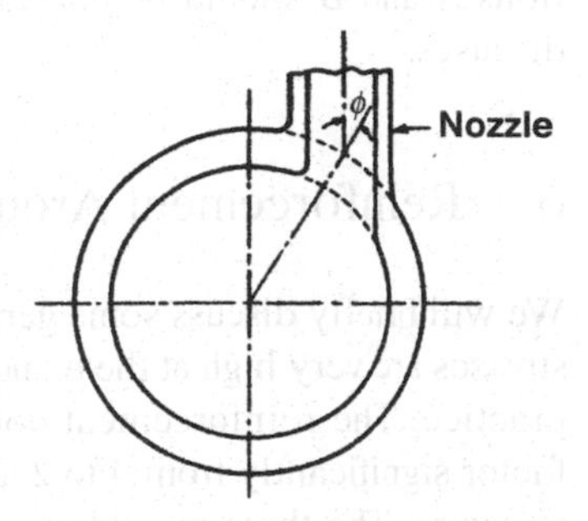

FIGURE 5.12. Nonradial nozzle. Hillside nozzle in a sphere or cylinder.

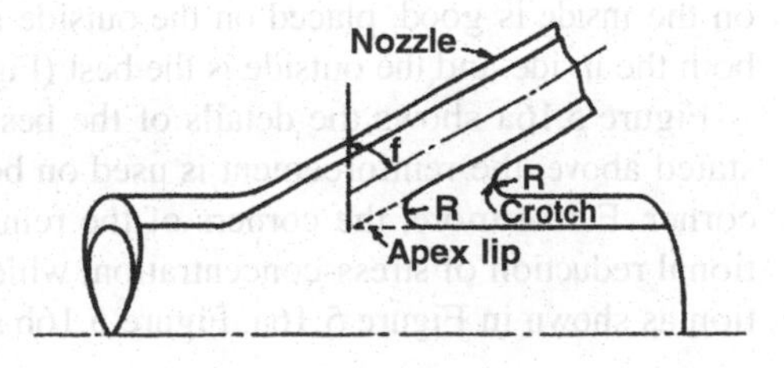

FIGURE 5.13. Nonradial nozzle. Lateral nozzle in a cylinder.

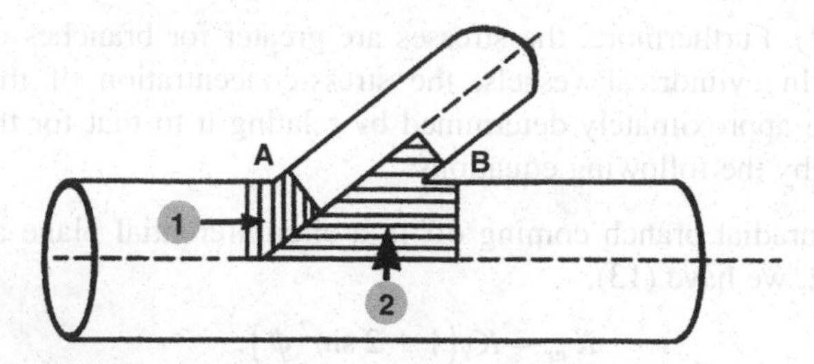

FIGURE 5.14. Schematic presentation of the area method for determination of stress-concentration at the arterial branch. Pressure load proportional to areas 1 and 2 acts on the tissue in regions A and B, respectively.

branch, and we will call region *B* the distal lip of the branch (Fig. 5.14). If we use the principles of the area method described earlier (see Section 1), we can see that the pressure areas unaccounted for are area 1 and area 2. Because area 2 is larger, it will produce larger stress in region *B*. Area 1 will increase the stress in region *A*. The influence length for these two areas will have to be determined experimentally in case of the arteries. However, it is clear that these two pressure areas will cause stress increase at locations *A* and *B*. Even though the stress at *A* is lower than that at *B*, it could be important because the arterial disease may occur from the mechanisms that require synergy between different types of forces. For instance, the arterial disease could be influenced by the forces of blood flow and blood pressure and these two forces could have different interactions at locations *A* and *B*. Therefore, the stress-concentration at both the locations *A* and *B* should be considered important for the development of vascular diseases.

6. Reinforcement Around the Branch

We will briefly discuss some general principles of reinforcement (14). Because the stresses are very high at the branch areas, reinforcement in these areas is a common practice. The reinforcement done properly can reduce the stress-concentration factor significantly from 4 to 2 and thus have a profound influence on the life of a structure. The three possible ways of placing the reinforcement at the branch are illustrated in Figure 5.15. Among them it is noted that the reinforcement placed on the inside is good, placed on the outside is even better, and that distributed on both the inside and the outside is the best (Fig. 5.15).

Figure 5.16a shows the details of the best way the reinforcement is used. As stated above, the reinforcement is used on both the inside corner and the outside corner. Furthermore, the corners of the reinforcement are rounded off for additional reduction of stress-concentration, which then results in the final configuration as shown in Figure 5.16a. Figure 5.16b shows the drawing of the artery with

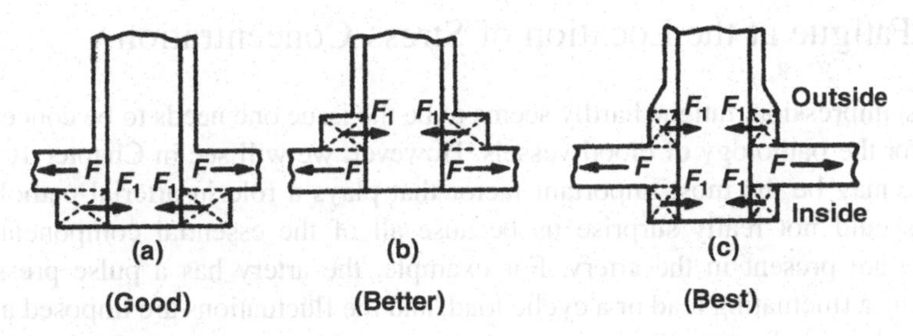

FIGURE 5.15. Diagrammatic location of nozzle opening reinforcement: (a) unbalanced inside, (b) unbalanced outside, (c) balanced.

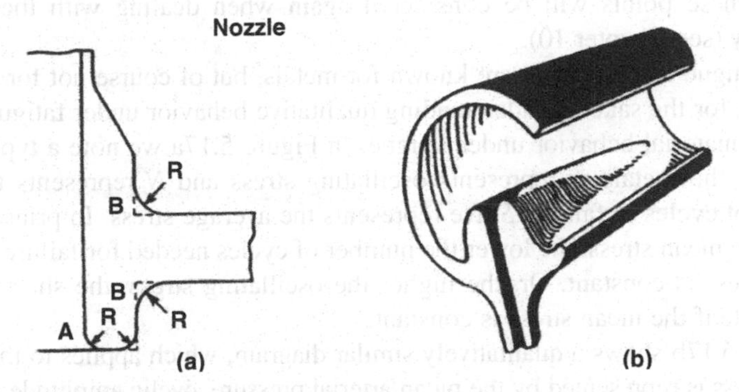

FIGURE 5.16. (a) Details of nozzle reinforcement to reduce stress-concentration. R is radius at sharp corners. (b) Drawing illustrating an arterial cushion where a side branch leaves its parent artery. Similarity between A and B should be noted.

the originating side branch. The similarity between the Figures 5.16a and 5.16b is inescapable, and it indicates that the artery branch follows the principles of reinforcement required on the basis of stress-concentration.

The dominant principle, present in the artery-branch area, has been totally overlooked in the medical literature. Such a protrusive branch geometry was speculated to be good for the blood flow so that the branch may capture the blood cells as it avoids the cell free zone near the wall of the main vessel. In contrast, the prominent protrusive feature seen at the origin of the branch is more in line with the reinforcement needed to reduce stress-concentration due to internal pressure load than with its function to capture blood cells. Once again, this special branch geometry emphasizes the important role of stress-concentration in the vasculature. It may be recalled that the stresses at the branch locations are local in nature, nonetheless they are important in determining the fatigue life of the structure.

7. Fatigue at the Location of Stress-Concentration

At first impression, fatigue hardly seems to be the issue one needs to be concerned with for the pathology of blood vessels. However, we will see in Chapter 10 that fatigue may be the most important factor that plays a role in arterial pathology. This should not really surprise us because all of the essential components of fatigue are present in the artery. For example, the artery has a pulse pressure, meaning a fluctuating load or a cyclic load, and the fluctuations are imposed at the rate the heart is beating (i.e., so many beats or cycles per minute). Of course, the pulse pressure as well as the heart rate fluctuates during the course of one day.

The most important point here is that in a pressure vessel virtually all failures occur as a result of fatigue—fatigue in the area of stress-concentration. Furthermore, eliminating or minimizing fatigue prolongs the life of a pressure vessel. These points will be considered again when dealing with the arterial pathology (see Chapter 10).

The fatigue life diagrams are known for metals, but of course not for arteries. However, for the sake of understanding qualitative behavior under fatigue, let us consider material behavior under fatigue. In Figure 5.17a we note a typical *S-N* curve for the metals. *S* represents oscillating stress and *N* represents the total number of cycles to failure. *S*-ave represents the average stress. In principle, the higher the mean stress, the lower the number of cycles needed for failure if oscillating stress is constant. Or, the higher the oscillating stress, the shorter is the fatigue life if the mean stress is constant.

Figure 5.17b shows a qualitatively similar diagram, which applies to the artery. Mean stress is represented by the mean arterial pressure, cyclic amplitude of stress is represented by the pulse pressure, and the life of the artery is represented by say the heart rate times the time, that is, the total number of cycles to *failure or injury*. The last parameter (artery life) can be looked at as that represented by the degree of arterial disease and/or by the extent of cellular injury. In other words, the artery may not fail like metal does but may become diseased, which is similar to failure.

Once again, the artery is expected to become diseased sooner if the mean blood pressure is higher or if the pulse pressure is higher. These parameters will be considered again in Chapter 10 on arterial pathology. In summary, the pressure vessel is most likely to fail from fatigue at a location of stress-concentration. Figure 10.12 in Chapter 10 shows a failure of one such pressure vessel from fatigue at a location of stress-concentration.

8. Effect of Size of the Artery and Size of the Branch on Stress-Concentration

In the vascular system, the arteries occur in various sizes, and therefore it is important to understand how the size of the artery may influence the development of the disease. Similarly, branches of different sizes may originate from the artery of a given size. This will be similar to having holes (ostia) of different sizes in the

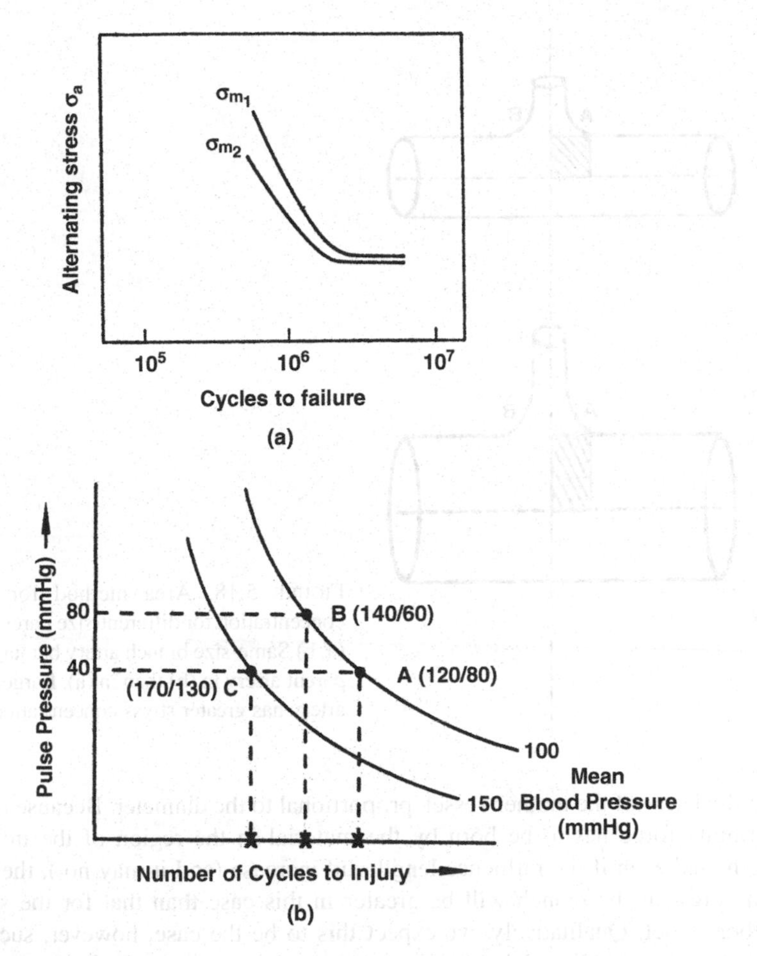

FIGURE 5.17. Schematic representation of fatigue data. (a) Average stress $\sigma m_2 > \sigma m_1$. (b) A similar representation of fatigue data for the artery.

artery, and it is important to know how that may influence the development of vascular disease.

8.1. *The Effect of Size of the Artery*

Figure 5.18 shows two examples where the same-size branch originates from a mother artery of two different sizes. First, on the basis of the area method described earlier, we note that the area unaccounted for, upon which the pressure force is acting, is larger in the case where the mother artery is larger. This larger area represents a larger discontinuity force and thus produces larger discontinuity stress at points *A* and *B*. This will be the case even if there is an increase in the

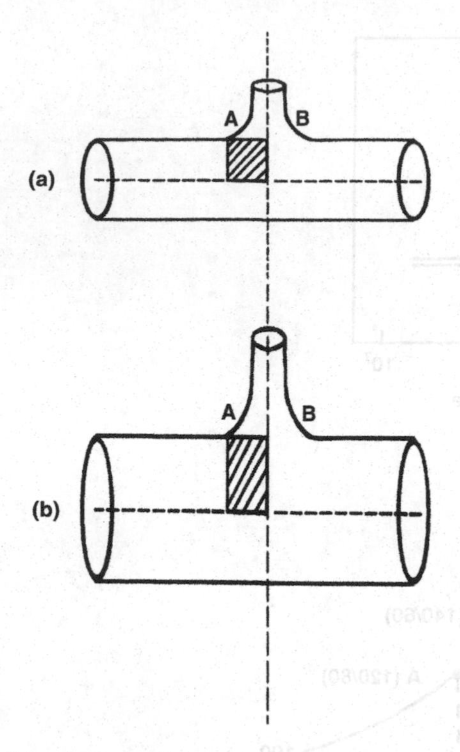

FIGURE 5.18. Area method for stress-concentration for different-size parent artery. (a, b) Same-size branch artery but larger size parent artery in (b) than in (a). Larger parent artery has greater stress concentration.

wall thickness of the mother vessel, proportional to the diameter. Because the discontinuity force has to be born by the material in the region of the influence length, and even if the influence length did increase (and it may not), the maximum stress at the branch will be greater in this case than that for the smaller mother vessel. Qualitatively, we expect this to be the case, however, such data remains to be established for the artery.

8.2. The Effect of Size of the Branch (Hole)

Figure 5.19 shows two examples where the smaller or the larger size branch originates from the same mother vessel. Once again, on the basis of the area method we note that a larger branch will have a larger discontinuity force and therefore a greater discontinuity stress.

Thus, if the high stress-concentration contributes to vascular disease, then we may expect that the disease will be worse in the larger vessels and at the larger branch sites.

8.3. Analytical Approach

To understand how the size of the main artery or the size of the branch affects the stress-concentration in the branch region, let us now consider the analytical

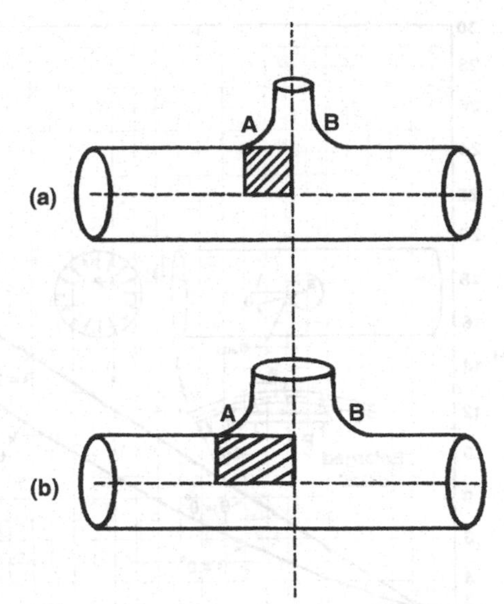

FIGURE 5.19. Area method for stress concentration. Branch artery is smaller in (a) than in (b). Larger branch has greater stress-concentration.

approach, even though we have explored the area method above. For the "T"-branch geometry, the stress-concentration factor (K_t) has been worked out for a cylindrical shell having a circular hole under internal pressure. Figure 5.20 shows one such chart for different-size holes (or branches) (15). For pressure loading, the analysis assumes that the force representing the total pressure, corresponding with the area of the hole, is carried as a perpendicular shear force distributed around the edge of the hole. Thus, we first compute the force acting in the area of the hole, which is equal to pressure times the area of the hole. This force is assumed to act around the edge of the hole as a perpendicular shear force as shown schematically in the chart (Fig. 5.20).

In Chapter 4, we saw that for a circular hole in a cylinder under internal pressure, the stress concentration factor is 2.5. This stress concentration occurred under the influence of only circumferential stress and longitudinal stress without any attachment of a structure (like a nozzle) to the hole. When there is another structure attached to the hole, additional discontinuity stresses are produced as described earlier. Therefore, in the chart (Fig. 5.20) we note that the stress-concentration factor begins at a value of 2.5 and increases with the attachment of a structure (nozzle).

Because the major stress rise occurs in the main vessel, sometimes the problem can be analyzed by ignoring the attached structure (nozzle) and considering the shear stress acting only at the hole in the main vessel. This is the approach presented in the chart. The results are presented as a function of dimensionless parameter β, where

$$\beta = \frac{\sqrt[4]{3\left(1 - \nu^2\right)}}{2}\left(\frac{a}{\sqrt{Rh}}\right)$$

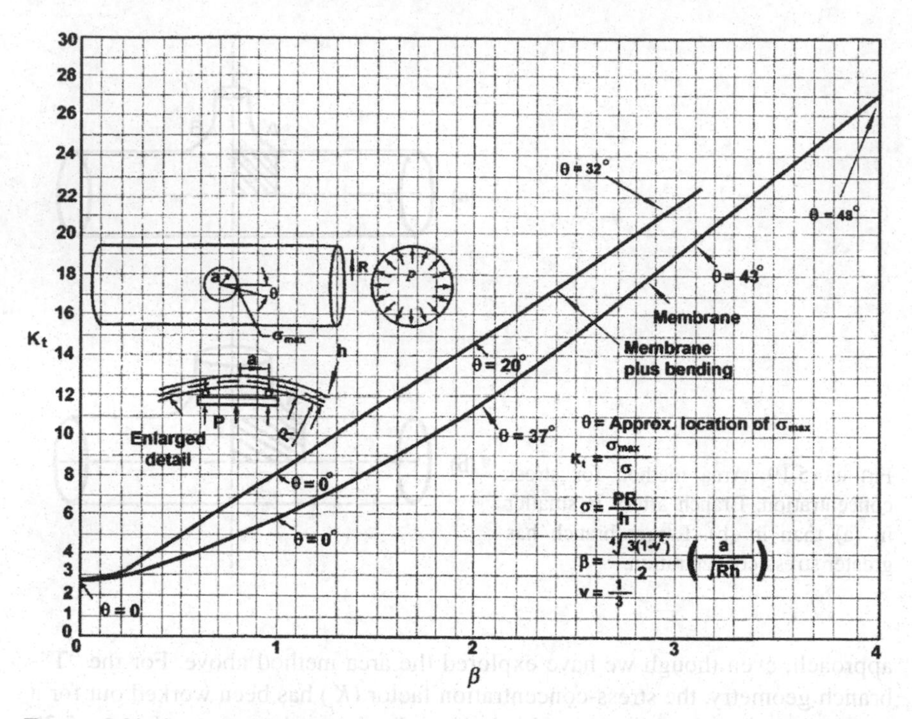

FIGURE 5.20. Stress-concentration factors for a circular hole in a cylindrical shell with internal pressure. (Reproduced from Pilkey WD: Peterson's Stress Concentration Factors. John Wiley & Sons, New York, 1997:260, with permission.)

where R is the mean radius of the cylinder, h is its thickness, a is the radius of a hole, and ν is Poisson's ratio. The analysis is valid for small h/R and small a/R (see Ref. 15 for details). For steel, one may use $\nu = 1/3$. Then

$$\beta = 0.639\left(\frac{a}{\sqrt{Rh}}\right) \tag{5}$$

or

$$\beta = 0.639\,\frac{\left(a/R\right)}{\sqrt{h/R}}.$$

From Figure 5.20, one can see that the stress concentration factor (K_t) increases as β increases. From the above Eq. (5) we note that β increases with a (i.e., as the size of the hole increases). Thus, the stress concentration factor increases with the size of the hole (or branch).

Now, let us examine the maximum stress at the hole.

$$K_t = \frac{\sigma_{\max}}{\sigma_{\text{nominal}}}.$$

Also, from the chart we may approximately anticipate the following relationship:

$$K_t = m \times \beta + c$$

where m is the slope and c is the intercept.
Then

$$m\beta + c = \frac{\sigma_{\text{max}}}{\sigma_{\text{nom}}}$$

$$\begin{aligned}\sigma_{\text{max}} &= \sigma_{\text{nom}} \times m\beta + \sigma_{\text{nom}} \times c \\ &= \sigma_{\text{nom}} \times m \times 0.639 \frac{a}{\sqrt{Rh}} + \sigma_{\text{nom}} \times c \\ &= \frac{PR}{h} \cdot \frac{a}{\sqrt{R}\sqrt{h}} \times 0.639m + \sigma_{\text{nom}} \times c.\end{aligned}$$

Because

$$\sigma_{\text{nom}} = \frac{PR}{h} \text{ (see Chapter 4)}$$

Then,

$$\sigma_{\text{max}} = \frac{P}{h} \cdot \sqrt{R} \cdot \frac{a}{\sqrt{h}} \cdot 0.639\, m + \frac{PR}{h} \cdot c. \tag{6}$$

So, as can be noted from the above Eq. (6), the maximum stress at the hole increases as the radius (R) of the main vessel increases.

It may be pointed out that Eq. (5) suggests that K_t may decrease as the vessel radius increases because β decreases. However, the decrease in K_t is due to an increase in the nominal stress (σ_{nom}). Actually, the maximum stress (σ_{max}) increases with the radius (R) of the vessel.

In summary, both, the larger size of the hole (branch) and the larger size of the main vessel increases the maximum stress at the branch. We can expect this to be the case even for the artery, although most of the analysis has been carried out for metals.

There is one more important aspect we need to consider regarding the *location of maximum stress* at the branch. Figure 5.20 shows that the location of maximum stress changes around the branch ($\theta = 0$, 37, 43 degrees, etc.) as the size of the hole (β) changes. The location moves away from the original site ($\theta = 0$ degrees) as the size of the hole (branch) increases. The reason being, with a larger hole the stress in the longitudinal direction increases because there is less material available to sustain the stress. This aspect of the size of the hole affecting the location of the maximum stress was addressed in Chapter 4. This information is very important when we consider the location of the origin of vascular disease around the branch ostium. For example, intimal hyperplasia at the end-to-side anastomosis may be maximum at the heel (see Chapter 13) or at the lateral wall depending upon the size of the branch in comparison with the size of the main vessel. The larger the branch ostium, the more the involvement of the lateral wall.

9. Arterial Bifurcation

In the body, there are several places where arteries bifurcate, such as the aortic bifurcation into iliac arteries or carotid artery bifurcation into internal carotid and external carotid arteries. At the bifurcation, we have two special locations, which

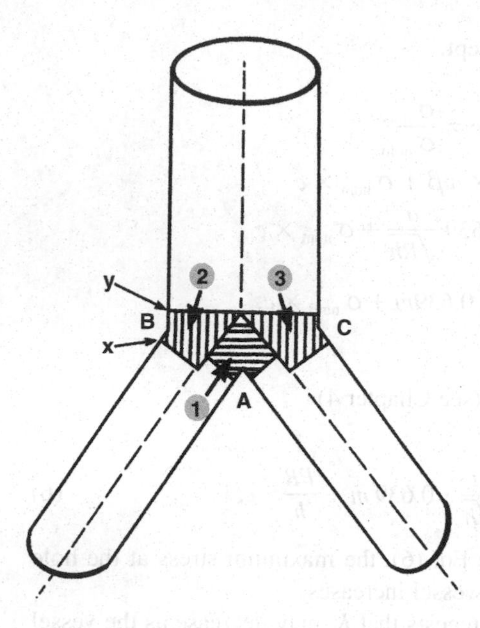

FIGURE 5.21. Area method for stress-concentration at the artery bifurcation. Stress-concentration exists at the crotch (A) and at the hip (B and C). XY defines the region influenced at location B. Pressure load proportional to areas 1, 2, and 3 acts on the arterial tissue at locations A, B, and C, respectively.

are the crotch (point A in Fig. 5.21) and the hip (points B and C in Fig. 5.21) where the stress increase can be significant. Also at these locations, vascular pathologies are frequent (see Chapter 2). Therefore, we will consider how the stress may be distributed at the bifurcation and the parameters that may influence it. The bifurcation considered here as an example is not a specific one and therefore general principles described are applicable to aortic, carotid, or any other arterial bifurcation in the body. There are at least four parameters in the bifurcation geometry that have an influence on the stress at the bifurcation; they are described below.

9.1. Discontinuity Stresses (Area Method)

First and foremost, the bifurcation represents discontinuity in the geometry and thus discontinuity stresses are produced that concentrate in the local region of the bifurcation. Once again, using the area method we can establish the regions of stress-concentration. As depicted in Figure 5.21, at the crotch (point A) one needs to determine the influence length. Then the force equivalent to the pressure acting on area 1 can be taken to act on the material of the length given by the influence length. As explained earlier, the ratio of the area acted upon by the pressure to the area of the material (influence length × wall thickness) gives an average stress in the crotch region. Because this ratio is obviously greater than the similar ratio in the straight segment of the artery, the crotch region has much larger stress than does the straight region.

The hip region (points *B* and *C*) also has larger stress than the straight region of the vessel on the basis of the area method. For example, the force acting in this region is equivalent to the pressure acting on area 2. The material sustaining this force is that whose length is given by the influence length (*xy* in Fig. 5.21). Once again, considering the ratio of the area acted upon by the pressure to the area of the material, we note that this ratio is greater than the similar ratio in the straight segment. Therefore, the stress at the "hip" of the bifurcation is also larger than that in the straight segment.

9.2. Elliptical Cross Section at the Bifurcation

Figure 5.22 shows that the cross section of the artery in region *A* is basically circular and that at the bifurcation (region *B*) is almost elliptical. Furthermore, the minor axis of the ellipse is the same as the diameter of the main artery slightly away from the bifurcation, whereas the major axis of the ellipse is considerably larger than the diameter. The major axis of the ellipse is located at the "hip" of the bifurcation.

Also, as explained in the figure, the circumferential stress at the hip is proportional to the major axis of the ellipse. The circumferential stress, slightly away from the bifurcation is proportional to the diameter of the main artery. Because the major axis is larger than the diameter, the stress at the hip is greater than that in the straight segment of the main vessel.

9.3. Bending in the Elliptical Cross Section

The artery is quite elastic and deforms relatively easily under internal pressure. Therefore, any elliptical cross section of the artery will have a tendency to deform

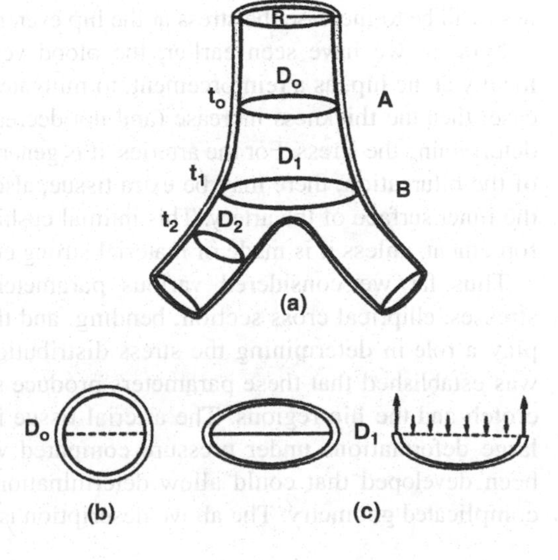

FIGURE 5.22. (a) Change in diameter and wall thickness near the bifurcation. $t_o > t_1 > t_2$, whereas $D_1 > D_o > D_2$. (b) Cross section at level A. (c) Cross section at level B. Wall tension is higher at the hip of the bifurcation because the major axis of the ellipse D_1 is greater than the diameter D_o. Both the smaller wall thickness and larger diameter increase wall stress at the hip compared with that in the main artery.

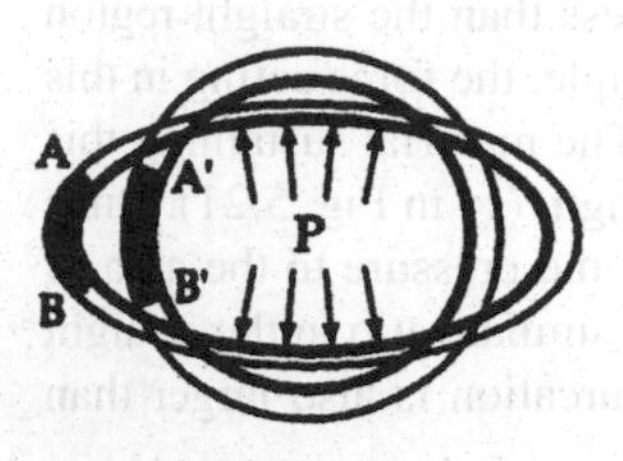

FIGURE 5.23. A pressure increase tends to change the cross section from elliptical to circular thereby causing larger bending stresses at the hip of the bifurcation.

to become circular. The example in Figure 5.23 shows how an elliptical cross section may deform to become circular. In the hip region, the bending deformation will tend to flatten the curvature in this region. Thus, the bending will produce tensile stress on the inner surface of the artery and compressive stress on the outer surface. In principle then, the bending deformation will increase the stress on the inner surface in the hip region.

9.4. Stress Increase Due to Thickness Change

It is known that the thickness of the artery is proportional (approximately) to its diameter (see Chapter 4). As can be seen from Figure 5.22, the main vessel has a thickness t_o, which is significantly larger than the thickness of the branch t_2. Quite possibly, the ratio t_o/t_2 could equal the ratio D_o/D_2. Now, because the artery wall is continuous from the main vessel to the branch, it is likely that the wall thickness changes gradually from t_o in the main vessel to t_1 in the hip region to t_2 in the branch. If this is the case, then the thickness t_1 in the hip is smaller than the thickness t_o in the main vessel. This thickness reduction alone will increase the circumferential stress at the hip. Also, as explained in the previous section, the diameter at the hip is larger. The combined effect then of a larger diameter and a smaller thickness will be to increase the stress at the hip even more.

Now, as we have seen earlier, the blood vessel may increase its thickness locally at the hip, as a reinforcement, to mitigate the stress increase. If this is the case, then the thickness increase (and not decrease as above) will play a role in determining the stress. For the arteries, it is generally known that in the hip region of the bifurcation, there may be extra tissue, also know as "intimal cushion," on the inner surface of the artery. This intimal cushion may not serve as a true reinforcement, unless it is made of material strong enough to bear the stress.

Thus far we considered various parameters, for example, discontinuity stresses, elliptical cross section, bending, and thickness variation, all of which play a role in determining the stress distribution in the arterial bifurcation. It was established that these parameters produce stress-concentration in both the crotch and the hip regions. The arterial tissue is complex and undergoes very large deformations under pressure compared with metals. Theories have not been developed that could allow determination of stress in the artery with a complicated geometry. The above description is given to establish qualitatively that the stress-concentration must occur at the bifurcation. The quantitative determination of stress-concentration must be carried out for the particular arterial bifurcation of interest. Such determination must include not only the modeling and analytical approach but also experimental data to validate the results. In Chapters 6 and 7, we will describe the actual determination of stresses performed for the artery branches and bifurcation.

10. Curved Arteries

We saw in Chapter 4 that in the case of the aortic arch, the stress in the aortic wall at the inner curvature of the arch is greater than that on the outer curvature. The same principle applies to the curved arteries. In the vascular system, there are numerous places where the arteries are curved and sometimes even tortuous, exhibiting an acute curvature. It is of interest to know what the stress distribution is like in the artery especially when it exhibits an acute curvature.

In this section, we will consider the stress distribution in curved arteries in detail. For the geometry, which is doughnut-shaped or torus or part thereof, the equilibrium of forces can be obtained as shown in Figure 5.24. Consider the

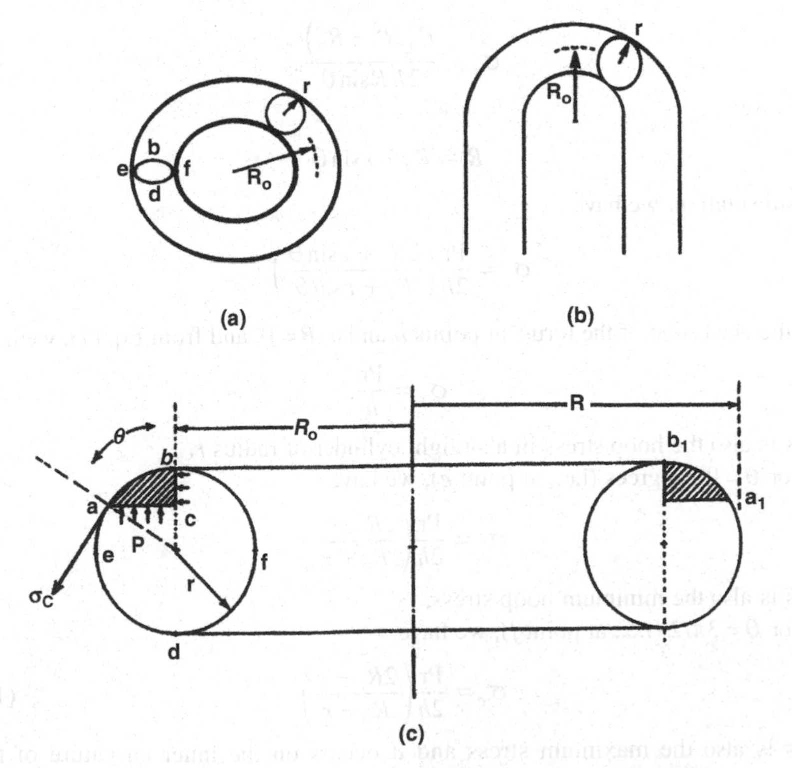

FIGURE 5.24. (a) A torus. (b) A bend. (c) The equilibrium of forces around the bend.

segment aba_1b_1, all around where the forces need to be balanced. The stress in the circumferential direction is σ_c at point *a* all around. There is another value of circumferential stress at point *b* all around. Now the force due to pressure can be resolved to be acting on the two surfaces represented by line *ac* all around and line *bc* all around. In other words, here we are concerned with the volume of area *abc* as it rotates all around and the equilibrium of forces related to this volume. R_o is the radius of the centerline of the torus, *r* is the radius of the pipe, *h* is the wall thickness, and *P* is the internal pressure.

Vertical force of pressure on the annular plane

$$ac = P \times \left(\pi R^2 - \pi R_o^2\right)$$
$$= P\pi\left(R^2 - R_o^2\right).$$

This force has to be balanced by the vertical component of stress σ_c because there is no vertical stress component at point *b*.

The balancing force due to σ_c then is

$$= 2\pi R \times h \times \sigma_c \times \sin\theta$$

or

$$\sigma_c \sin\theta\, h 2\pi R = P\pi\left(R^2 - R_o^2\right)$$

$$\sigma_c = \frac{P\left(R^2 - R_o^2\right)}{2hR\sin\theta}.$$

Because

$$R = R_o + r\sin\theta$$

by substitution, we have

$$\sigma_c = \frac{\Pr}{2h}\left(\frac{2R_o + r\sin\theta}{R_o + r\sin\theta}\right). \tag{7}$$

On the centerline of the torus, at points *b* and *d*, $\theta = 0$, and from Eq. (7), we have

$$\sigma_c = \frac{\Pr}{h}. \tag{8}$$

This is also the hoop stress in a straight cylinder of radius *r*.

For $\theta = 90$ degrees (i.e., at point *e*), we have

$$\sigma_c = \frac{\Pr}{2h}\left(\frac{2R_o + r}{R_o + r}\right). \tag{9}$$

This is also the minimum hoop stress.

For $\theta = 3\pi/2$ (i.e., at point *f*), we have

$$\sigma_c = \frac{\Pr}{2h}\left(\frac{2R_o - r}{R_o - r}\right). \tag{10}$$

This is also the maximum stress and it occurs on the inner curvature of the bend. The stress distribution around the circumference is given by Eq. (7).

Maximum stress occurs on the inner curvature (at point f), minimum stress occurs on the outer curvature (at point e), and regular hoop stress (straight cylinder) occurs at points b and d. The stress increase is strongly dependent on R_o, meaning the more acute the bend (smaller the R_o), the more is the stress increase at the inner curvature and the stress decrease at the outer curvature. Figure 5.25 shows the variation in the maximum and minimum stress as a function of R_o/r. For the R_o/r of 3 or 2, we can note that the stress on the inner curve can be almost double that on the outer curve. In arteries then, particularly for smaller ratios of R_o/r, that is, for more acute bends, we may find that the wall at the inner curvature is thicker and the wall at the outer curvature is thinner than that at the lateral regions. This will be the compensatory mechanism to create uniform stress distribution around the circumference of the artery. As mentioned in Chapter 4, for the aortic arch such a difference in the wall thickness was found to be the case.

This difference in the wall stress could be very significant for the vein graft used in the coronary bypass procedure. Because a longer than needed segment is usually used in patients, the vein graft left inside the chest will generally have a bend in it. Around the bend, the vein wall may become considerably thicker on the inner curve as the vein adapts to the wall stress. This could contribute to the graft stenosis (see Chapter 11). Thus, one should avoid an acute bend in the vein graft as much as possible.

The longitudinal stress σ_ℓ at any cross section such as b, e, d, f in torus can be seen to be similar to that in a straight cylinder.

Thus, $\sigma_\ell = \dfrac{\text{Pr}}{2h}$, and it is the same at all locations in torus.

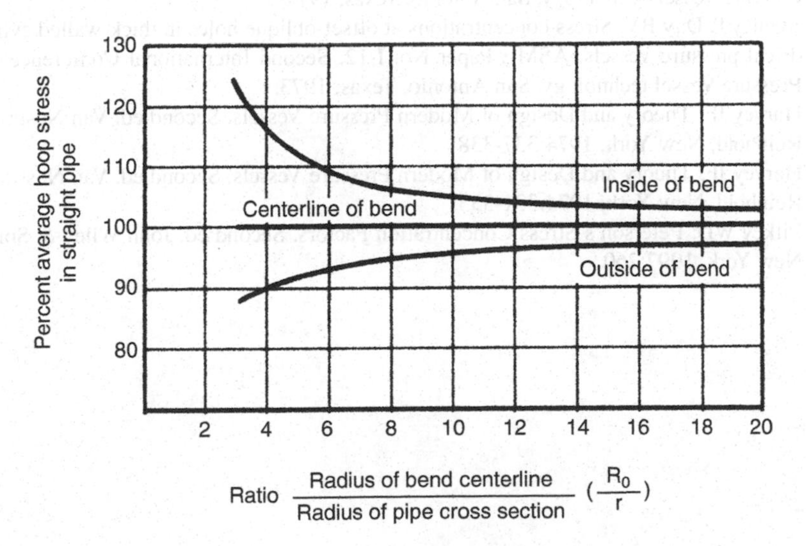

FIGURE 5.25. Variation in hoop stress as a function of radius of the bend.

References

1. Lind NC: Approximate stress-concentration analysis for pressurized branch pipe connections. ASME Paper 67-WA/PVP-7, 1967:951-958.
2. Lind NC: A rapid method to estimate the elastic stress concentration of a nozzle in a spherical pressure vessel. In: Nuclear Structural Engineering 2. North-Holland, Amsterdam, 1965:159-168.
3. Harvey JF: Theory and Design of Modern Pressure Vessels, Second ed. Van Nostrand Reinhold, New York, 1974:146-150.
4. Lind NC, Hradek R, Cook RD: Influence of fillet radii on stresses near outlets in pressure vessels. Univ. of Illinois TAM Report No. 167, 1961.
5. Waters EO: Stresses near cylindrical openings in a spherical vessel. Welding Research Council Bulletin No. 96, 1964.
6. Timoshenko S: Theory of Plates and Shells, First ed. McGraw-Hill, New York, 1940.
7. Lind NC: An elastic-shell analysis of the stress concentration of a pressurized "T" branch-pipe connection. Proceedings of the 1st International Congress on Pressure Vessel Technology (Delft 1969), American Society of Mechanical Engineers, 1969;269-276.
8. Prince N, Rashid YR: Structural analysis of shell intersections. Proceedings of the 1st International Congress on Pressure Vessel Technology (Delft 1969), American Society of Mechanical Engineers, 1969;245-254.
9. Taylor CE, Schweiker JW: A three-dimensional photoelastic investigation of the stresses near a reinforced opening in a reactor pressure vessel. Proceedings of the Society for Experimental Stress Analysis 1957;17(1):25-36.
10. Taylor CE, Lind NC, Schweiker JW: A three-dimensional photoelastic study of stresses around reinforced outlets in pressure vessels. Welding Research Council (Eng. Foundation, NY) Bulletin No. 51, 1959;26-41.
11. Yu JCM, Shaw WA: Stress distribution of a cylindrical shell non-radially penetrated into a spherical pressure vessel. ASME Paper No. I-9, Second International Conference on Pressure Vessel Technology, San Antonio, Texas, 1973.
12. Stanley P, Day BV: Stress concentrations at offset-oblique holes in thick-walled cylindrical pressure vessels. ASME Paper No. I-12, Second International Conference on Pressure Vessel technology, San Antonio, Texas, 1973.
13. Harvey JF: Theory and Design of Modern Pressure Vessels, Second ed. Van Nostrand Reinhold, New York, 1974:337-338.
14. Harvey JF: Theory and Design of Modern Pressure Vessels, Second ed. Van Nostrand Reinhold, New York, 1974:331-333.
15. Pilkey WD: Peterson's Stress Concentration Factors. Second ed. John Wiley & Sons, New York, 1997:260.

6 Stress-Concentration in the Artery I

1. Introduction

Determination of stress in the artery in vivo will continue to be a challenge in the field of biomechanics for years to come. This is so for many reasons: 1) The artery is a complex and inhomogeneous biological material, and it changes with age. The *connective tissue* responsible for bearing the major portion of the load is distributed differently in different layers of the artery wall. How much of the load is shared by the *contractile components* of the artery is not known. How much of the load is born by the *cellular components* is also not known, just to name a few parameters. 2) The forces acting on the artery in the body are complex. For example, the amount of tethering force in the artery, the presence of contractile force, the presence of residual stress, and the effect of the external organs in contact with the artery are some of the parameters whose contributions are not precisely known.

3) The artery undergoes very large deformations, a large increase in both diameter and length, and it has nonlinear stress-strain properties. All these conditions together make it extraordinarily difficult to develop theoretical approaches to this subject.

For these reasons, data on the stress distribution in the artery has been sparsely available. In this chapter, we will consider some of the observations made from the viewpoint of the artery as a pressure vessel. In other words, some of the principles of the pressure vessels, such as the occurrence of stress-concentration at the branch, described in Chapters 4 and 5, will be considered here for a limited number of cases. It will become obvious that much work needs to be done in this area. Although the studies in this field have been limited, they establish, as a matter of principle, that the results are qualitatively applicable to many other arteries in the vascular system.

2. Stress-Concentration in the Bovine Coronary Arterial Branch

2.1. *The Experiments*

The data on the stresses and strains in the bovine coronary artery branch were obtained by Thubrikar et al. (1). Both the analytical and experimental approaches were used in gathering the data. In the following pages, we will describe their methods and results.

The wall stress was determined in the bovine circumflex coronary artery and its first major branch in vitro. To determine the stress, a method of finite element analysis was used. In this method, it is necessary to know 1) the geometry of the arterial branch, 2) the material properties of the artery, 3) the loads on the artery (i.e., the internal pressure), and 4) the boundary conditions. The analytical results were compared with the experimentally measured strains in the branch region.

From cow hearts, the circumflex coronary artery was dissected, beginning at its junction with the left anterior descending coronary artery to well beyond the first major branch. The distal ends of the main vessel and branch were tied and the proximal end of the main vessel was cannulated. The arteries were cleaned and treated with chemicals to ensure that only the passive properties would be present. The arteries were placed in 205 mmol solution of EGTA (ethylene glycol tetraacetic acid) and then in an EGTA solution containing 0.01 mmol norepinephrine. In the first step, free calcium is combined, and in the second step, any calcium released upon cell contraction due to norepinephrine is combined with EGTA.

For measurements, markers were placed on the artery as follows. The artery was gently pressurized to assume its natural shape, and a fine powder of potassium permanganate was sprinkled on it. The powder reacted with the tissue, forming a permanent discoloration, which resulted in randomly scattered marks of about 0.1 mm diameter (see Chapter 7, Fig. 7.3). The artery was held fixed at the cannulated end, while the remainder of the artery was free to expand (Fig. 6.1).

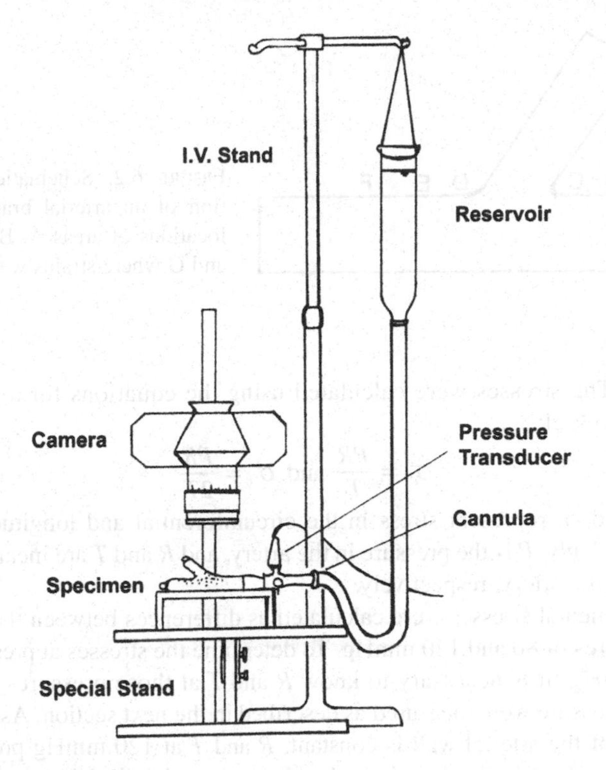

FIGURE 6.1. Experimental setup.

The cannula was connected to a saline reservoir, which could be raised or lowered to produce the desired pressure. A camera was positioned over the artery and focused on the marks. The artery was preconditioned by pressurizing it from 0 to 160 mmHg five times. First, the artery was subjected to 60 mmHg static pressure and photographed. Then the pressure was increased in steps of 20 mmHg until a pressure of 160 mmHg was reached and photographed at each step.

The measurements were made under magnification in seven areas shown in Figure 6.2. The distances between two marks oriented circumferentially and two marks oriented longitudinally were measured at each pressure. The distances were plotted against pressure for the entire pressure range, and a least square fit line was drawn. The distances at 80 and 120 mmHg pressures were read from this line. The incremental strain (ε) was given by:

$$\varepsilon = \frac{L_{120} - L_{80}}{L_{80}}$$

where L = the distance between the marks at 80 mmHg or at 120 mmHg.

Coefficients of elasticity: Measurements on the main artery were used to calculate the moduli of elasticity and Poisson's ratios in the circumferential and longitudinal

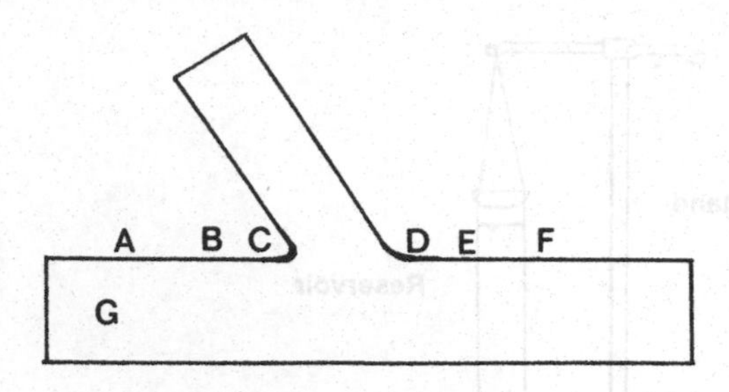

FIGURE 6.2. Schematic representation of an arterial branch showing locations of areas A, B, C, D, E, F, and G where strains were measured.

directions. The stresses were calculated using the equations for a thin-walled cylindrical vessel:

$$\sigma_c = \frac{PR}{T} \text{ and } \sigma_\ell = \frac{PR}{2T}$$

where σ_c and σ_ℓ represent stress in the circumferential and longitudinal directions, respectively, P is the pressure in the artery, and R and T are mean radius and thickness of the artery, respectively.

The incremental stresses were calculated as differences between the stress values at pressures of 80 and 120 mmHg. To determine the stresses at pressures of 80 and 120 mmHg, it is necessary to know R and T at those pressures. R and T at 80 mmHg pressure were measured as described in the next section. Assuming that the volume of the arterial wall is constant, R and T at 120 mmHg pressure were obtained by using measured strain in the circumferential direction to obtain R and measured strains in both circumferential and longitudinal directions to obtain T. The experimentally measured incremental strains were related to the incremental stresses by the familiar constitutive equations for orthotropic material:

$$\varepsilon_c = \frac{\sigma_c}{E_c} - \frac{\nu_\ell \sigma_\ell}{E_\ell} \tag{1}$$

and

$$\varepsilon_\ell = \frac{\sigma_\ell}{E_\ell} - \frac{\nu_c \sigma_c}{E_c} \tag{2}$$

where ε, E, and ν represent strain, modulus of elasticity, and Poisson's ratio, respectively; and c and ℓ represent circumferential and longitudinal directions.

For Poisson's ratio, Patel's (2) average value is about 0.3. Therefore,

$$\frac{\nu_c + \nu_\ell}{2} = 0.3 \tag{3}$$

was taken as a third equation. Finally, the expression

$$E_c \nu_\ell = E_\ell \nu_c \tag{4}$$

which holds true for orthotropic material was included. With these four equations, all four elastic constants (E_c, E_ℓ, ν_c, ν_ℓ) were calculated.

Geometry of the arterial branch: Geometric parameters necessary for the finite element model are shown in Figure 6.3. These include radii and thicknesses at various locations in the branch area. The numbers shown in the figure are representative values (in millimeters) for these dimensions for the coronary arteries studied. The arterial specimens were fixed in 4% buffered formaldehyde at 80 mmHg pressure. This configuration was defined as reference geometry. Cross-sectional cuts were made and the resulting rings photographed to measure the radii and the wall thicknesses of the main vessel and the branch. The specimens

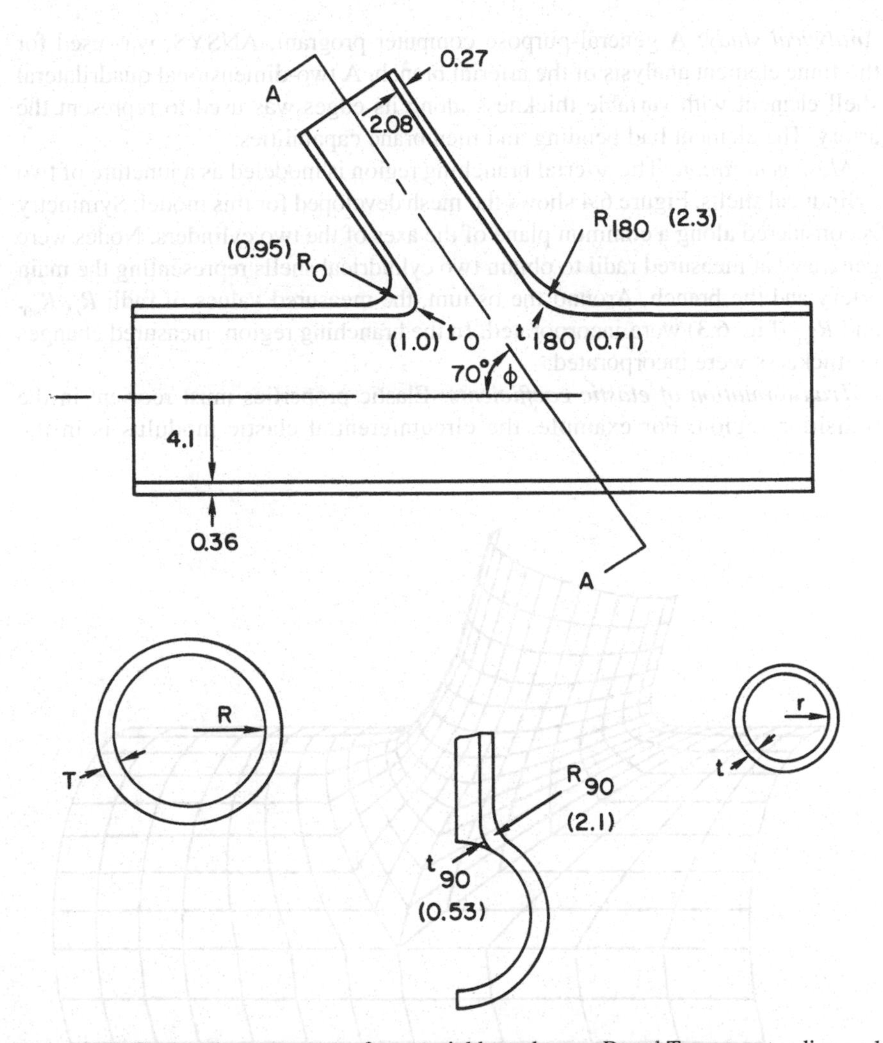

FIGURE 6.3. Geometric parameters of an arterial branch area. R and T represent radius and thickness of the main vessel; r and t represent radius and thickness of the branch. R_o, R_{90}, and R_{180} represent the radii of the transition region at, respectively, 0, 90, and 180 degrees around the ostium, and t_0, t_{90}, and t_{180} represent corresponding thicknesses of the arterial wall. ϕ is the branch angle. The numbers represent dimensions in millimeters for a representative coronary arterial branch.

were cut along the plane common to axes of both the arteries and the sectioned specimens photographed to measure the branch angle, increase in thickness at the branch, and the radii of curvatures in the transition region. The specimens were also cut along the axis of the branch corresponding with A-A in Figure 6.3 and turned on end so that the thickening of the wall in this plane could be measured.

2.2. *The Model*

Analytical study: A general-purpose computer program, ANSYS, was used for the finite element analysis of the arterial branch. A two-dimensional quadrilateral shell element with variable thickness along its edges was used to represent the artery. The element had bending and membrane capabilities.

Mesh generation: The arterial branching region is modeled as a juncture of two cylindrical shells. Figure 6.4 shows the mesh developed for this model. Symmetry is considered along a common plane of the axes of the two cylinders. Nodes were generated at measured radii to obtain two cylindrical shells representing the main artery and the branch. Around the ostium, the measured values of radii R_o, R_{90}, and R_{180} (Fig. 6.3) were incorporated. In the branching region, measured changes in thickness were incorporated.

Transformation of elastic coefficients: Elastic properties must reorient in the transition region. For example, the circumferential elastic modulus is in the

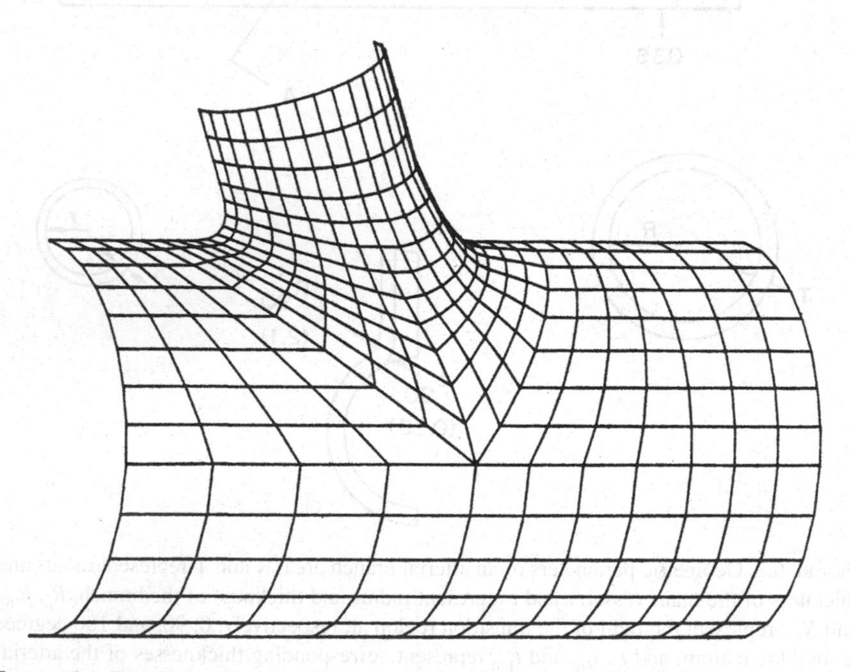

FIGURE 6.4. The finite element mesh representing the coronary circumflex artery and its first branch. For the artery, the direction of blood flow in vivo is from right to left.

circumferential direction of the main artery at the beginning of the transition region and in the circumferential direction of the branch artery at the end of the transition region. The reorientation of the elastic properties in the transition region was carried out (1).

Other conditions: One end of the main vessel in the finite element model was constrained. This corresponds with the cannulated end of the artery in the experiment. The model was loaded with a pressure of 40 mmHg (0.0053 M Pa), which corresponds with the increase in pressure from a diastolic of 80 mmHg to a systolic of 120 mmHg. The outer edges of the artery were loaded with nodal forces to reflect the longitudinal stress resulting from an internal pressure.

The following assumptions were made for the stress analysis of the arterial branch:

1) The artery is linearly elastic over the operating pressure range of 80 to 120 mmHg; thus, the incremental theory of elasticity is valid; 2) the artery is orthotropic; 3) stresses in the straight segment of the artery can be obtained by stress formulas for thin cylindrical shells; and 4) initial geometry of the artery was taken at 80 mmHg pressure.

The geometry of the coronary circumflex arterial branch, used in the analysis, was as follows (Fig. 6.3): branch angle $\phi = 70$ degrees, main vessel $R = 4.1$ mm and $T = 0.36$ mm, branch $r = 2.08$ mm and $t = 0.27$ mm. Around the ostium, radii were $R_o = 0.95$ mm, $R_{90} = 2.1$ mm, $R_{180} = 2.3$ mm, and thicknesses were $t_o = 1$ mm, $t_{90} = 0.53$ mm, and $t_{180} = 0.71$ mm. The coefficients of elasticity for the artery were determined to be $E_c = 1.8$ MPa, $E_\ell = 0.8$ MPa, $\nu_c = 0.42$, and $\nu_\ell = 0.19$.

2.3. The Observations

The results were obtained in the form of stress contour plots. Figures 6.5, 6.6, and 6.7 show maximum principal stress contours on the inner, middle, and outer surface of the artery, respectively. In the branch area, the stresses are not uniformly distributed, instead they are localized. This is indicated by the close spacings of the contour lines where the stress is increasing rapidly. On the inner and middle surfaces of the artery, the stress increases toward the ostium. The highest principal stress occurs on the plane of symmetry at the acute angle of the branch. The stress is greatest on the inner surface of the artery. The stress-concentration factor, K, is defined as the ratio of the highest stress in the branch region to the circumferential stress in the straight segment. Table 6.1 lists the highest principal stress and stress-concentration factor in the branch region. The stress increases by a factor of 3.4 to 2.8 at both the distal and the proximal regions of the ostium compared with that in the straight segment. The nominal circumferential stress in the straight segment is 0.06 MPa. It is also important to note that the stress gradients are steep on the inner surface of the artery. For example, the stress varies from 0.18 MPa to 0.06 MPa over a distance of less than 1 mm (Fig. 6.5). Hence, in the branch area the arteries have to sustain not only high stresses but also a high stress gradient. The stress concentration subsides over a distance of approximately 2 mm. Comparison of these stresses with those along the X-axis of the

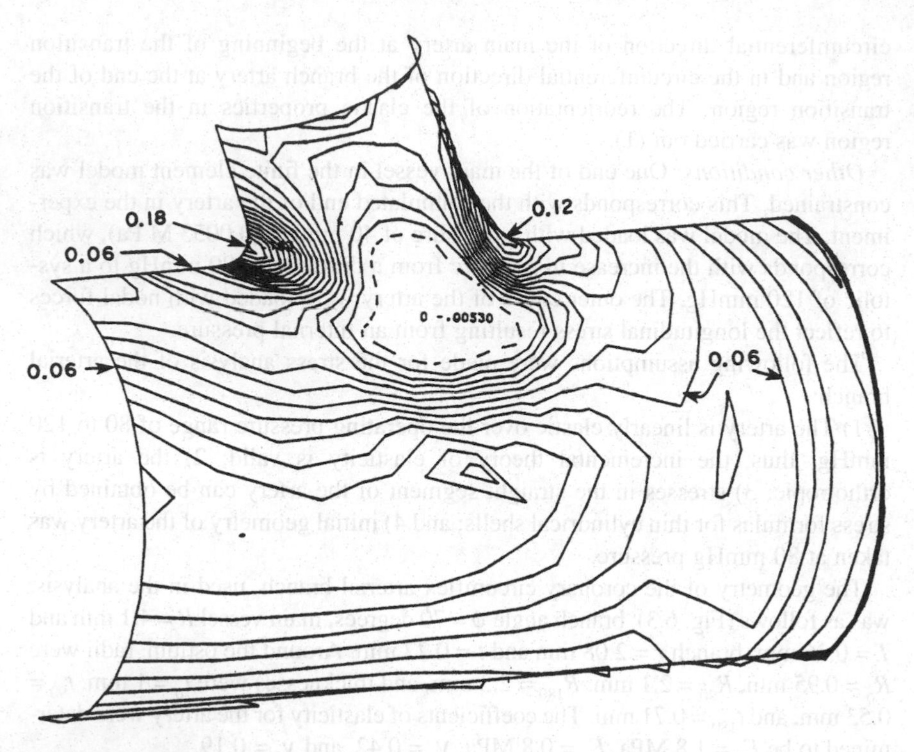

FIGURE 6.5. Maximum principal stress contours on the inner surface of the artery. These stresses are due to an incremental pressure loading of 40 mmHg. The stress increases from one contour to the next toward the ostium. The stress-concentration can be seen at both the proximal and distal regions of the ostium. The numbers represent stress values in MPa for various contours. Stress increment between two contours is 0.01 MPa.

elements (Fig. 6.4) reveals that the principal stresses are in nearly circumferential direction. The stresses are produced by membrane action as well as by a significant amount of bending.

It is of interest to compare analytical strains to the experimental data. Because the analytical strains correspond with the middle surface and the experimental strains correspond with the outer surface, their magnitudes may not be compared directly, but their trends can be compared. The circumferential strains along the plane of symmetry ($x = 0$ plane) are plotted as a function of their location and compared in Figure 6.8. Both the analytical strains and the experimental strains show the same trend. The strains increase toward the branch ostium and reach a maximum value of 5–7% at the branch compared with 2.5% away from the branch.

Parametric studies: The model was explored to determine the relative importance of some of the parameters. Effect of the parameters was studied in an isotropic model because it was observed that the results from an isotropic model were not significantly different than those from the orthotropic model.

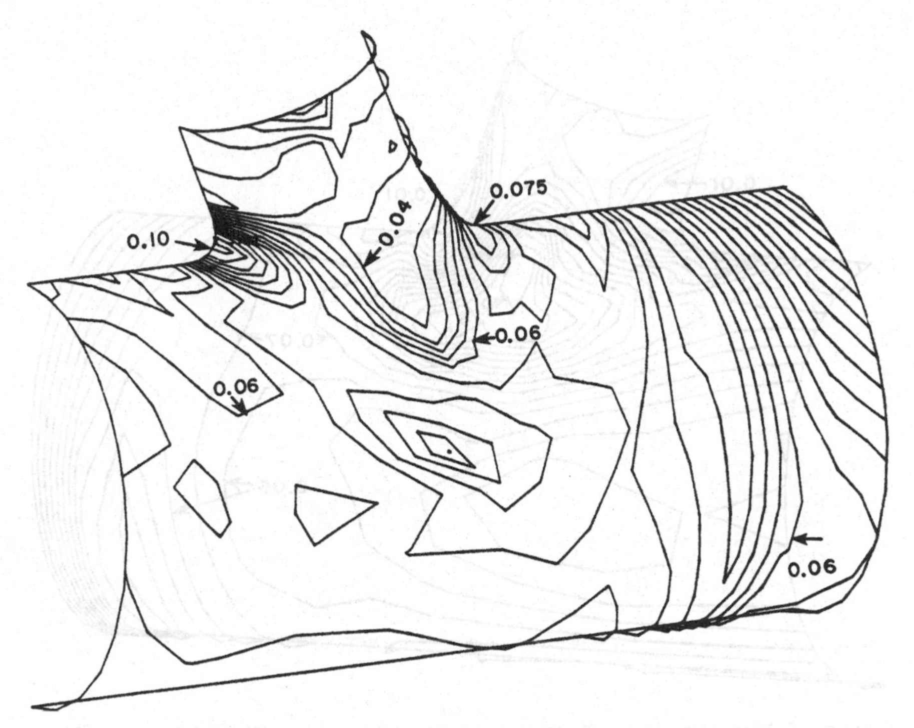

FIGURE 6.6. Maximum principal stress contours on the middle surface of the artery. The stress-concentration can be seen around the ostium. The numbers represent stress values in MPa for various contours. Stress increment between two contours is 0.005 MPa.

Isotropic model: The stress-concentration factor, K, at the inner surface was greater by only 6.7% for the orthotropic model than for the isotropic model (Table 6.2).

Relative thickening at the branch: The wall thickness away from the branch was left unchanged while that at the branch was made first larger and then smaller. An increase in the thickness ratio (t_o/t) by 11% decreased the stress concentration factor by 9.4% (Table 6.2).

Radius of curvature in the transition region (R_o)*:* This parameter had little effect on the stress-concentration. Apparently by making the transition smoother (increasing R_o), any effect to reduce the stress is more than offset by the increased hoop stress in the branch, due to the local increase in the radius. It is interesting to note that the decreased radius also decreased the stress, although only slightly.

These parametric studies clearly show that the stress increase in the branch area is primarily a result of the presence of the ostium (i.e., the unique geometry of the branching region). The stress increase is affected only secondarily by such factors as elastic properties, thickening at the branch site, curvature around the ostium, and length of the transition zone. Hence, the stress-concentration described in

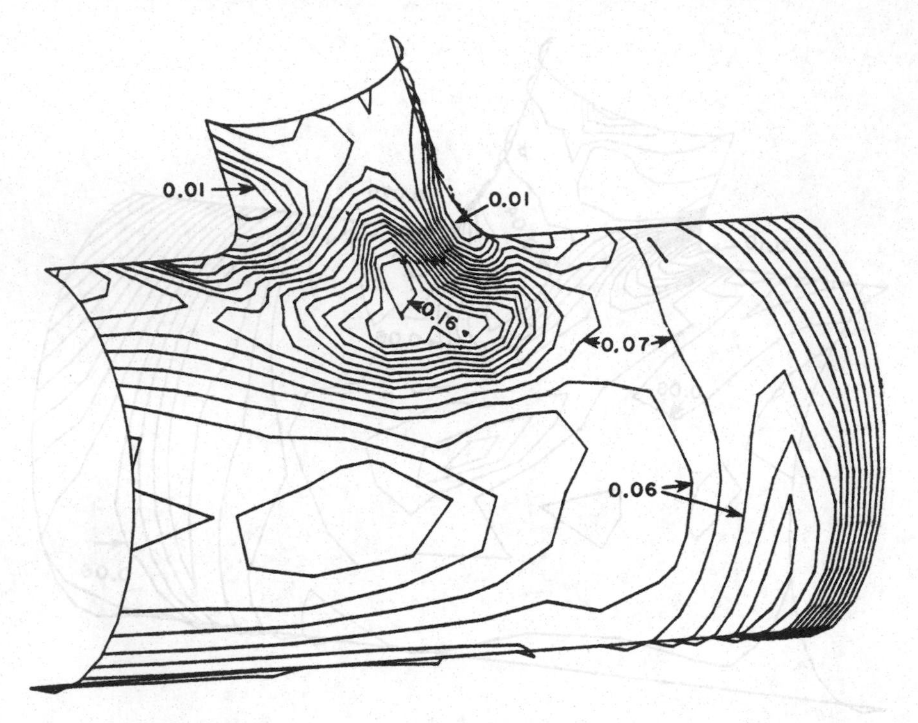

FIGURE 6.7. Maximum principal stress contours on the outer surface of the artery. Stress values are in MPa. Stress increment is 0.01 MPa.

TABLE 6.1. Intramural stress-concentration around the ostium.

	K	Distal max. stress (MPa)	K	Proximal max. stress (MPa)
Inner surface	3.38	0.205	2.8	0.169
Middle surface	1.7	0.103	1.5	0.09

Stress-concentration factor $K = \dfrac{\textit{maximum stress}}{\textit{circumferential stress } PR/T}$

Chapter 5 for pressure vessel intersections also occurs at the arterial branch region as expected.

3. Stress-Concentration in the Human Carotid Artery Bifurcation

The stress-concentration phenomenon has also been studied in the human carotid artery bifurcation. Here we describe the approach used by Salzar et al. (3). As one may expect, examination of several carotid bifurcations can reveal several different

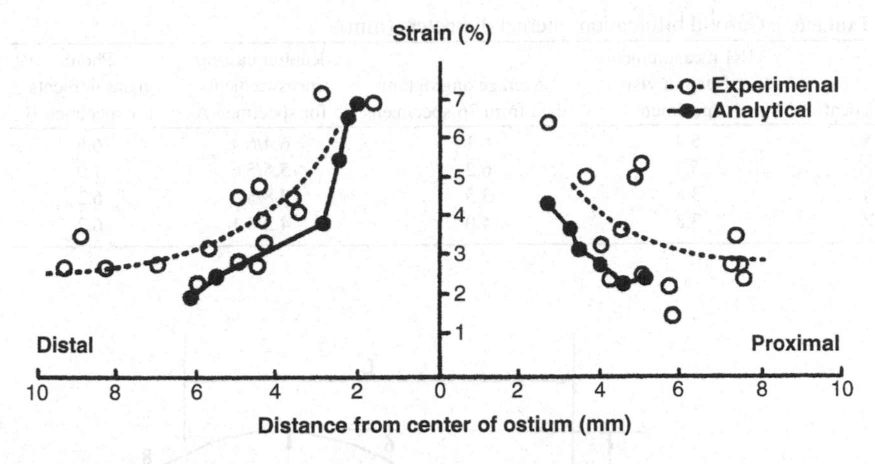

FIGURE 6.8. Comparison of experimental and analytical strains on the outer surface of the artery in the arterial branch area. Zero on the abscissa represents a point of intersection of the axis of the main artery with the axis of the branch (the branch radius is 2.08 mm). Strains correspond with a pressure increase from 80 mmHg to 120 mmHg.

TABLE 6.2. Effect of various parameters on stress-concentration factor K.

Parameter or model	Change in parameter (%)	Change in K from basic isotropic model (%)	
		Middle surface	Inner surface
Orthotropic model	—	+0.6	+6.7
t_o/t	+11.1	−9.7	−9.4
t_o/t	−11.1	+9.7	+5.7
R_o/r	+25	+1.3	+7.9
R_o/r	−12.5	−1.9	−3.8

geometries. Salzar found that among multiple geometries, two dominant patterns could be identified (3).

3.1. The Geometry

The internal diameters of the six in vitro specimens obtained from cadavers, the 76 angiogram specimens, and the MRI images are summarized and shown to display both the limited physiological variation and the consistent agreement between the measurement techniques (Table 6.3; Fig. 6.9). Generally, the common (proximal) carotid has the largest internal diameter (average 6.4 mm) followed by the diameter at the sinus bulb (average 6.0 mm). The internal carotid artery diameter is average 4.9 mm, and the external carotid artery diameter is average 4.5 mm. The internal diameters for the two bifurcations modeled (A and B) are also shown in the table.

TABLE 6.3. Carotid bifurcation internal diameters (mm).

Location	MRI measurements of one in vivo specimen	Average angiogram data from 76 specimens	Rubber casting measurements for specimen A	Photo measurements for specimen B
V	6.4	6.1	6.4/6.4	6.6
X	5.5	6.2	5.5/5.6	7.0
Y	3.8	3.5	5.8/5.4	6.2
Z	3.8	4.0	4.2/4.1	6.2

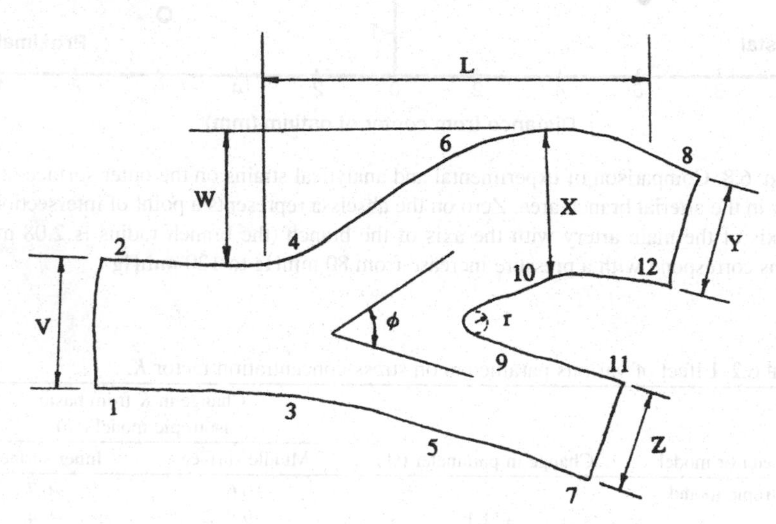

FIGURE 6.9. Typical carotid artery showing the locations of recorded internal diameters corresponding with Table 6.3, wall thickness locations for the six in vitro specimens corresponding with Table 6.4, and the overall geometric dimensions of the six in vitro specimens corresponding with Table 6.5.

From measurements of the six in vitro specimens, common geometric patterns were observed, with the most important being the thinning of the artery wall along the sinus bulb (Table 6.4; Fig. 6.9). Generally, the common carotid (locations 1 and 2) displays a circular cross section and a fairly uniform wall thickness (0.74–0.99 mm). In the region of the sinus bulb (location 6), considerable wall thinning occurs (0.43–0.68 mm). Note that in Table 6.2, only specimens D and E have actual wall thickness data at the sinus bulb. Because of severe lesions affecting the intima in this area, the disease-free wall thicknesses cannot be precisely determined. In order to extrapolate this disease-free wall thickness into the sinus bulb, a linear interpolation of the surrounding data was made. Also in this region, the cross section of the vessel becomes ellipsoidal. The internal carotid eventually obtains a more uniform wall thickness (0.22–0.45 mm) and a circular cross section. The transition from the bifurcation to the external carotid (location 5) does not undergo localized wall thinning unlike the internal carotid. The angle (ϕ)

TABLE 6.4. Arterial wall thickness (mm) and bifurcation angle from six in vitro specimens.

Location	Specimen					
	A	B	C	D	E	F
1	0.82	0.74	0.99	0.80	—	0.93
2	0.87	0.74	0.99	0.74	—	0.86
3	1.09	0.62	1.05	0.93	0.49	0.86
4	0.44	0.49	0.99	0.62	—	—
5	0.87	0.49	—	0.99	0.49	0.49
6	—	—	—	0.43	0.68	—
7	0.44	0.49	0.74	0.25	0.25	0.25
8	0.22	0.37	0.49	0.49	0.43	0.49
9	0.54	0.62	0.43	0.49	0.49	0.49
10	0.44	0.62	0.49	0.49	—	—
11	0.33	—	0.37	0.25	0.49	0.25
12	0.49	0.37	0.49	0.49	—	0.37
ϕ	42°	44°	50°	27°	26°	35°

TABLE 6.5. Overall geometry of the six in vitro artery specimens (mm).

Sample	Bulb length (L)	Bulb width (W)	Radius (r)
A	18.0	5.0	0.5
B	21.0	7.2	0.7
C	19.1	8.4	0.9
D	23.0	6.0	0.3
E	25.0	2.5	0.6
F	19.0	3.5	0.5

between the internal and external carotid varies between 27 degrees and 50 degrees. The overall dimensions of the six carotid bifurcations are also tabulated (Table 6.5; Fig. 6.9). The elevation of the sinus bulb (dimension W) varied in each specimen, ranging from 2.5 to 8.4 mm. The bulb length (dimension L) ranged from 18.0 to 25.0 mm. The radius (r) at the crotch varied from 0.3 to 0.9 mm.

3.2. *The Analysis*

Two representative carotid artery bifurcations (chosen from six samples) were modeled for the stress analysis using the finite element method. Figure 6.10 shows their overall geometry. Assuming that the modelled geometry represents the geometry during diastole (at 80 mmHg pressure), the stress was determined for the incremental pressure load of 40 mmHg (0.0053 N/mm^2). This load equals the pulse pressure when the normal pressure is 120/80 mmHg. The results are obtained as incremental stresses. The nominal circumferential stress in the common carotid artery is given by:

$$\sigma_c = \frac{\Pr}{t} = 0.0053 \times 3.2/0.8\ N/mm^2;$$

$$= 0.02\ N/mm^2.$$

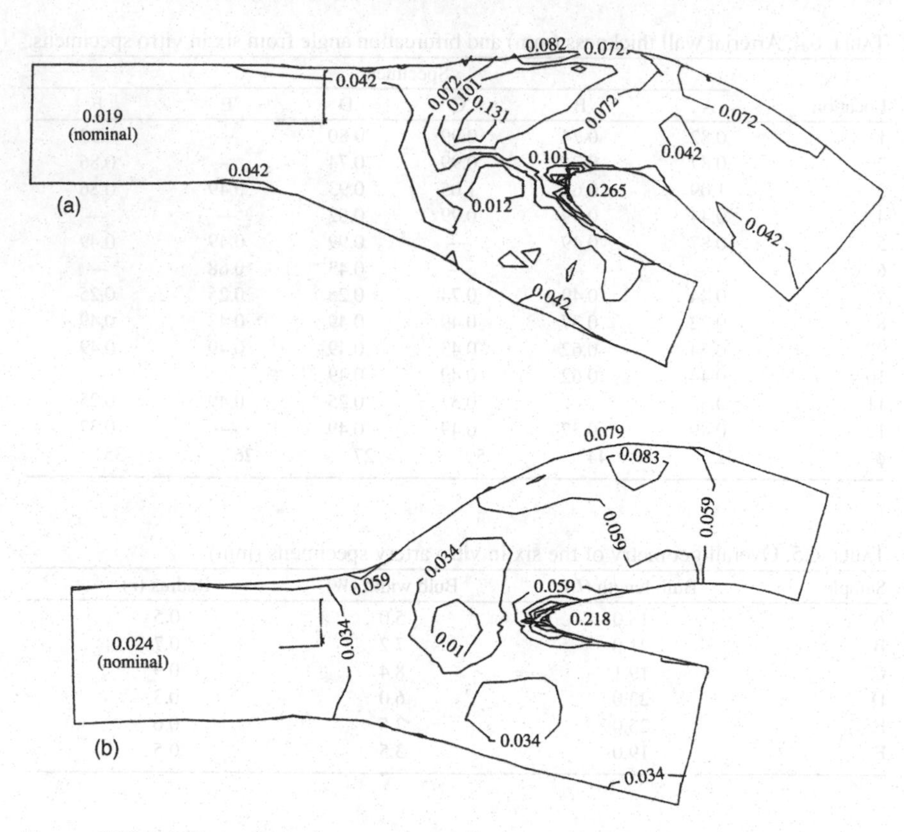

FIGURE 6.10. (a) First principal stress contours in the mid-wall for specimen (A). Note that stresses are highest and most concentrated at the point of bifurcation, while stresses at the sinus bulb are high, but more distributed. Stress increment between two contours is 0.0298 N/mm^2. (b) First principal stress contours in the mid-wall surface for specimen B. Similar to specimen A, elevated stresses are concentrated at the point of bifurcation and more distributed along the sinus bulb. Stress increment between two contours is 0.0245 N/mm^2.

In the carotid artery bifurcations, mid-wall stress-concentrations were found to occur at the point of bifurcation and along the sinus bulb. The resulting contour line plots (Figs. 6.10a and 6.10b) show the distribution of the mid-wall stresses. The stress contours at the point of bifurcation (crotch) are concentrated over a very small area of tissue and have stress-concentration factors of 9.2 (0.22 N/mm^2) to 14.2 (0.27 N/mm^2). The stress-concentration factor is the ratio of maximum stress to the nominal stress.

The stress contours along the sinus bulb are more distributed and occur over the larger area of wall thinning; with stress-concentration factors of 3.3 (0.08 N/mm^2) to 4.4 (0.82 N/mm^2). Also, the examination of the stress contours in the two models suggests that with the exception of the localized regions of stress-concentration, the bulk of the two arteries remains at the nominal stress.

At the inner wall, the finite element analysis showed the largest stress-concentration at the point of bifurcation and a more distributed stress in the sinus bulb (3). Though the stresses on the inner walls are higher than those along the mid-surface, due to the limitations of the thin shell finite element models used, these stress values should be considered important qualitatively. In general, the stresses on the inner wall surface will be higher than those on the mid-surface as seen previously in Chapter 4.

The variation in the mechanical properties ($E = 1.4$ to $6.2\ \mathrm{N/mm^2}$) and in the wall thickness did not significantly change the *locations* of the stress-concentrations, however, change in the wall thickness had a profound effect on the *magnitude* of the stress. By varying the wall thickness uniformly throughout the models, the stress magnitudes were altered. From Figure 6.11, it can be seen that the uniform reduction of wall thickness dramatically increased the wall stress-concentration, with a 50% decrease in wall thickness producing an approximate 200% increase in stress at both the crotch and the sinus bulb. Similarly, a uniform 50% increase in wall thickness leads to an approximate 40% decrease in stress in these areas. The pattern of stress distribution does not change with the wall thickness. In other words, even dramatic physiological variations in wall thickness will not alter the fact that localized areas of high mechanical stress will occur in approximately the same regions.

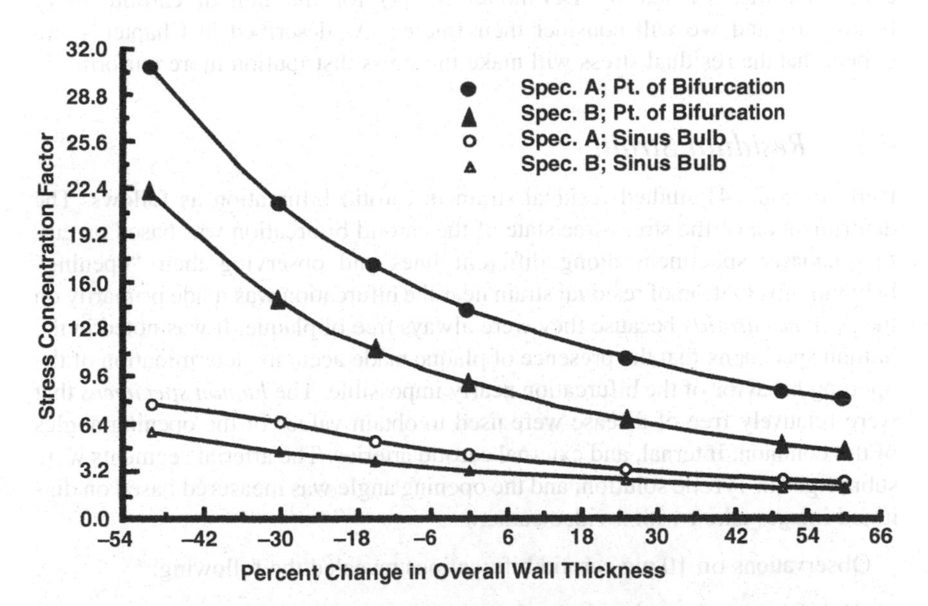

FIGURE 6.11. Stress-concentration factor at the mid-wall surface for point of bifurcation and for sinus bulb as a function of a uniform percent change in overall wall thickness from the in vitro measurements for specimens A and B. Increasing the wall thickness reduces the critical stresses but does not eliminate them.

In summary, in the human carotid artery bifurcation, the stress-concentration occurs at two locations: the crotch and the sinus bulb. Furthermore, at the crotch the stress-concentration occurs in a much smaller area of the arterial tissue, whereas at the sinus bulb it occurs in a much larger area of the tissue. For a pressure load of 40 mmHg, the stresses at the crotch, the sinus bulb, and the common carotid are 0.22–0.27, 0.08, and 0.02 N/mm^2, respectively.

If we compare the nominal stress in the carotid artery (0.02 N/mm^2) with that in the coronary artery described earlier (0.06 N/mm^2), we find that they are different primarily because the carotid artery is thicker than the coronary artery (0.8 mm vs. 0.36 mm). Also in Chapter 4, the nominal circumferential stress for the artery in general was reported to be 0.126 N/mm^2 for a pressure load of 105 mmHg. This is equal to the nominal stress of 0.05 N/mm^2 for the pressure load of 40 mmHg and is comparable to the value reported above for the carotid and coronary arteries, especially considering the different techniques used in thickness measurement.

4. Role of Residual Stress in Stress-Concentration at the Human Carotid Artery Bifurcation

As mentioned earlier, there are only a few examples where the stress-concentration in the artery has been studied. The role of residual stress in the stress-concentration was studied by Delfino et al. (4) for the human carotid artery bifurcation and we will consider their finding. As described in Chapter 4, we expect that the residual stress will make the stress distribution more uniform.

4.1. Residual Strain

Delfino et al. (4) studied residual strain in carotid bifurcation as follows. The determination of the stress-free state of the carotid bifurcation was based on cutting cadaver specimens along different lines and observing their "opening" behavior. Evaluation of residual strain near the bifurcation was made primarily on the *porcine carotids* because they were always free of plaque. It was noted in the human specimens that the presence of plaque made accurate determination of the opening behavior of the bifurcation nearly impossible. The *human specimens* that were relatively free of disease were used to obtain values of the opening angles of the common, internal, and external carotid arteries. The arterial segments were submerged in Tyrode solution, and the opening angle was measured based on digitized images taken with a video camera.

Observations on 10 pig carotid bifurcations revealed the following:

a) A first cut along the plane of symmetry separates the bifurcation into two essentially identical parts. Each branch of the two parts opens into a circular sector. It was noted that the edges of the common carotid and both branches remained in the same plane. The apex took on the configuration of a "bump" protruding toward the luminal side of the half-artery. The bifurcation angle at the apex did not change.

b) A second cut along the mid-line of the remaining common carotid artery separated the common carotid into two pieces and allowed the external and internal branches to be separated. The opening angles of the two branches increased slightly. The shape of the bump at the apex was not significantly affected. Additional cuts across the arteries did not change the geometry further.

From these observations, a representative model of the stress-free state of the carotid artery bifurcation was constructed.

4.2. *The Analytical Model*

Geometry of the carotid bifurcation: The construction of the geometry of the carotid bifurcation was based on the average angiographic measurements of Bharadvaj et al. (5). The diameter of the common carotid artery was 7.2 at 120 mmHg. The opening angle of the common carotid artery was 130 degrees, based on the measurement of eight human carotid arteries (standard deviation of 15 degrees). Equations described in Ref. 4 were used to find the radius and the opening angle of each branch. The values given in Figure 6.12 provide the radii and the opening angles of each branch. The radii and opening angles of the two branches are therefore calculated based on the value of the common carotid to

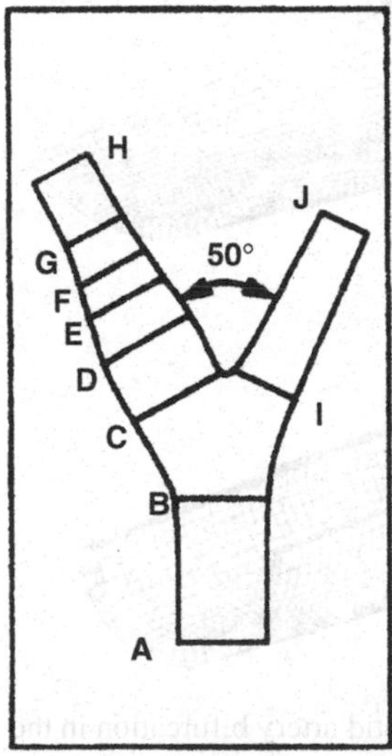

position	thickness [mm]	radius unloaded [mm]	radius stress-free [mm]	opening angle in degrees
A	0.9	3.1	4.46	130
B	0.6	3.1	4.42	129
C	0.6	3.3	4.62	132
D	0.55	3.3	4.61	132
E	0.52	3.15	4.46	130
F	0.49	2.7	4.03	124
G	0.43	2.2	3.54	115
H	0.4	2.2	3.54	115
I	0.6	1.83	3.23	108
J	0.4	1.83	3.19	107

FIGURE 6.12. Geometrical dimensions of the carotid artery bifurcation in free-strain state and in unloaded state corresponding with the positions shown in the schematic diagram at the left. (Reproduced from Delfino A et al., J Biomechanics 1997;30:777-786, with permission from Elsevier.)

satisfy the assumption that the edges stay in the same plane. The actual measured values of the opening angles for the internal and external carotid were 116 ± 20 degrees and 100 ± 18 degrees, respectively.

The residual strains were introduced by first deforming the bifurcation from the stress-free to the unloaded state (Fig. 6.13). The loading necessary to perform this deformation was then applied to the stress-free state and then the artery was pressurized to 120 mmHg. An axial stretch of 10% was applied at the ends of the external and internal carotid arteries, creating a constraint of motion in the local axial directions. The artery was free to move in the circumferential and radial directions. The axial stretch of 10% was based on average measurements made in situ, before extraction of the human carotids. To assess the effects of residual strain on the stress field a second finite element model was constructed assuming no residual strains. In this case the stress-free geometry is similar to the unloaded geometry (Fig. 6.13B). The same pressure and axial stretching were applied to this second model.

The wall material represented by the elements was assumed isotropic and homogenous. The following form of the strain energy density function W was applied:

$$W = \frac{a}{b}\left[\exp\left(\frac{b}{2}(I_1 - 3)\right) - 1\right].$$

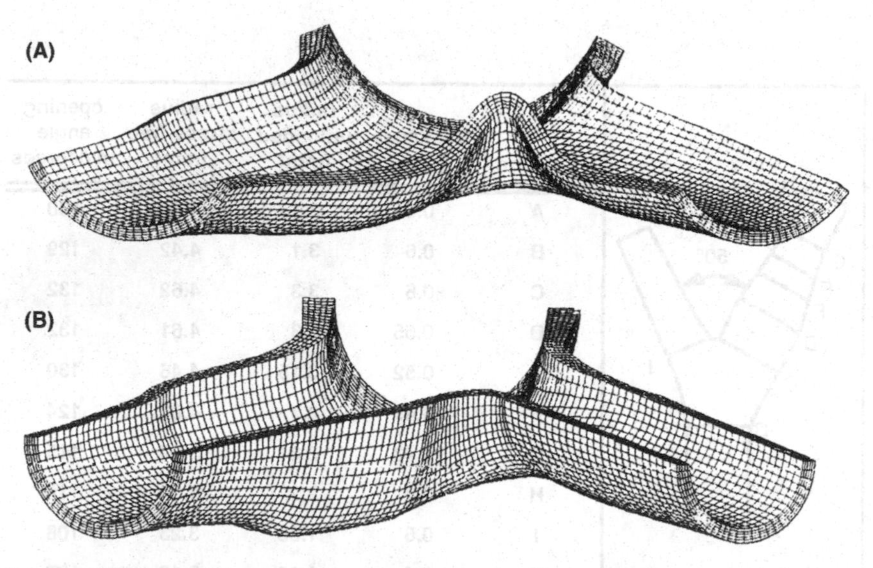

FIGURE 6.13. Illustration of the finite element model of the carotid artery bifurcation in the (A) stress-free and (B) unloaded states. The stress-free state was constructed from observations of the carotid artery segments shown in Figure 6.12. The unloaded state was based on published average angiographic measurements. For the model without residual strain, the unloaded configuration (B) was taken as the stress-free state. (Reproduced from Delfino A et al., J Biomechanics 1997;30:777-786, with permission from Elsevier.)

The parameters a and b are the constants representing the material properties. We assume that the material properties of each branch of the carotid artery bifurcation are the same.

Measurements of the pressure-radius relation for different axial stretching were performed on eight relatively disease-free human common carotid arteries obtained fresh from autopsy. The material constants a and b were calculated using a nonlinear least-squares algorithm to fit the average of the experimental measurements. The material constants found were $a = 44.2$ kPa and $b = 16.7$.

4.3. The Model Results

The results of the *model with residual strain* show that the maximum principal stresses are more uniformly distributed through the wall thickness at all locations, in comparison with the case where the residual strains were not included (Fig. 6.14). To characterize the stress distribution throughout the wall thickness, the stress uniformity factor (UF), defined as the ratio between the maximum principal stress at the inner wall and the maximum principal stress at the outer wall at a given anatomic location, was employed. In the common carotid artery, where the stress varied between 80 kPa at the inner wall and 72 kPa at the outer wall, the uniformity factor was *1.1 with the inclusion of residual strains* (Fig. 6.14A) and *3.0 without residual strains* (Fig. 6.14B). A plot of the variation in principal stress from the inner to the outer wall at several anatomic locations, with residual strain included, is shown in Figure 6.15A. The inner wall is at 0, the outer wall at 1, and the mid-surface at 0.5 on the X-axis. The figure shows that the location of the highest UF is at the lateral wall of the sinus bulb near the plane of symmetry. At this location, the UF is 4.1. Near the proximal part of the external carotid artery, the stress is higher at the outer wall than the inner wall, so the value of UF is 0.64. The value of UF at the apex is 1.16 with residual strain included and 8.8 in the model without residual strain (Fig. 6.15B). In the absence of residual strain (Fig. 6.15B), the variation of the principal stress throughout the wall shows that the location of highest UF is at the apex. At the other three locations (i.e., common, sinus, and external), the UF is 3.2, 2.5, and 2.7, respectively.

The inclusion of residual strains also affected the spatial distribution of inner wall maximum principal stresses throughout the model geometry. Figure 6.14A shows that with residual strains, the stress at the inner wall is very localized in the center of the apex and the principal stresses range from 132 to 221 kPa. For the model without residual strain, (Fig. 6.14B), the inner wall stresses are higher, ranging from 236 to 655 kPa, and distributed over a wider area of the apex. At the inner wall of the common carotid artery, the stress in the model with residual strain ranged from 65 to 85 kPa and the stress without residual strain was approximately 150 kPa. In the proximal portions of the internal and external carotid arteries, the stresses in both models are highest on the lateral walls where the local curvature is maximal. The inner wall stresses at these two locations were, respectively, 220 and 100 kPa with residual strain and 194 and 278 kPa without residual strain.

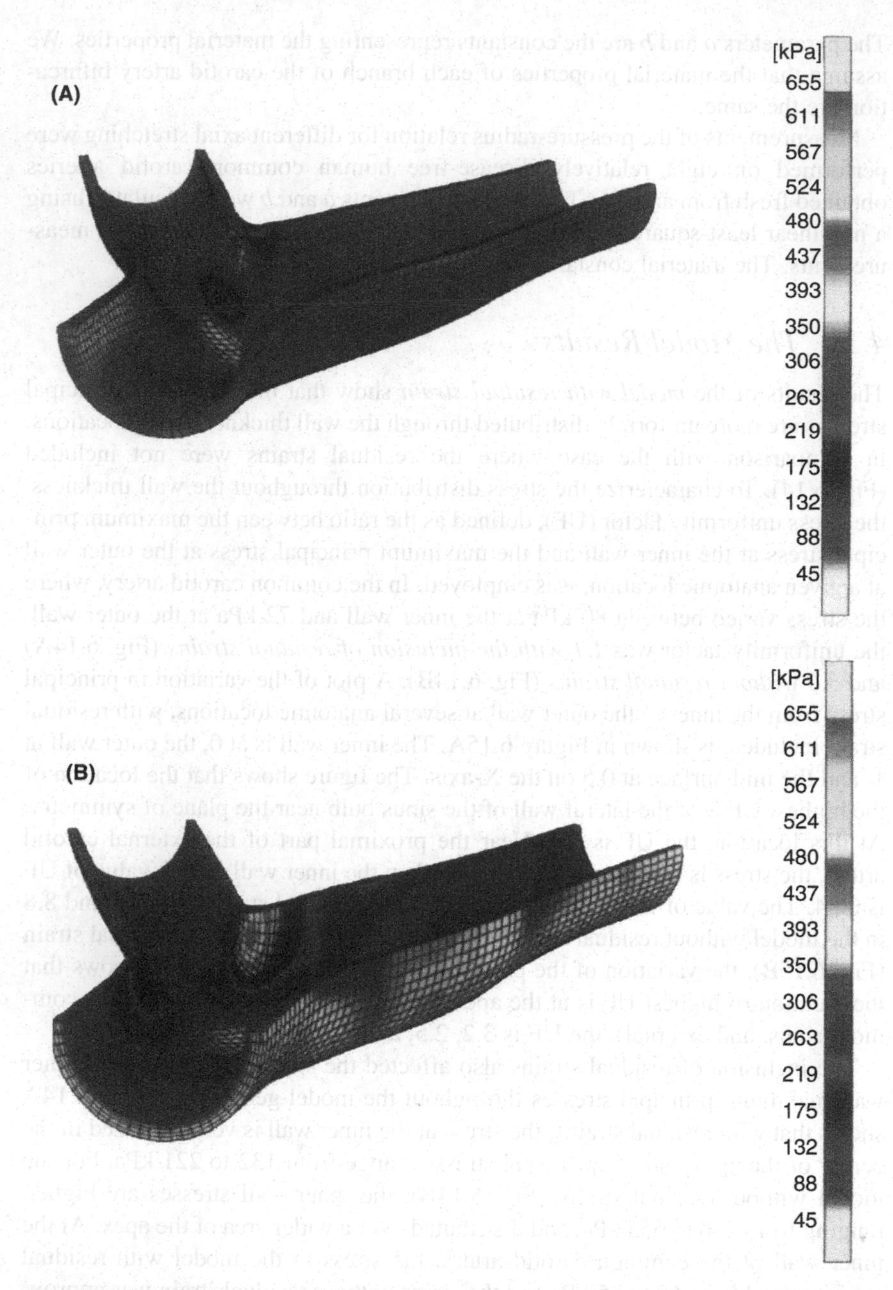

FIGURE 6.14. Maximum principal stress distribution at a pressure of 16 kPa (120 mmHg) with the residual strain (A) and without the residual strain (B) taken into account. The distribution of stress was made more uniform throughout the model geometry and through the wall thickness by the presence of residual strain (A). There is a concentration of stress at the apex, and the inner wall stresses are much higher without the residual strain (B). (Reproduced from Delfino A et al., J Biomechanics 1997;30:777-786, with permission from Elsevier.) (*Please see color version on CD-ROM.*)

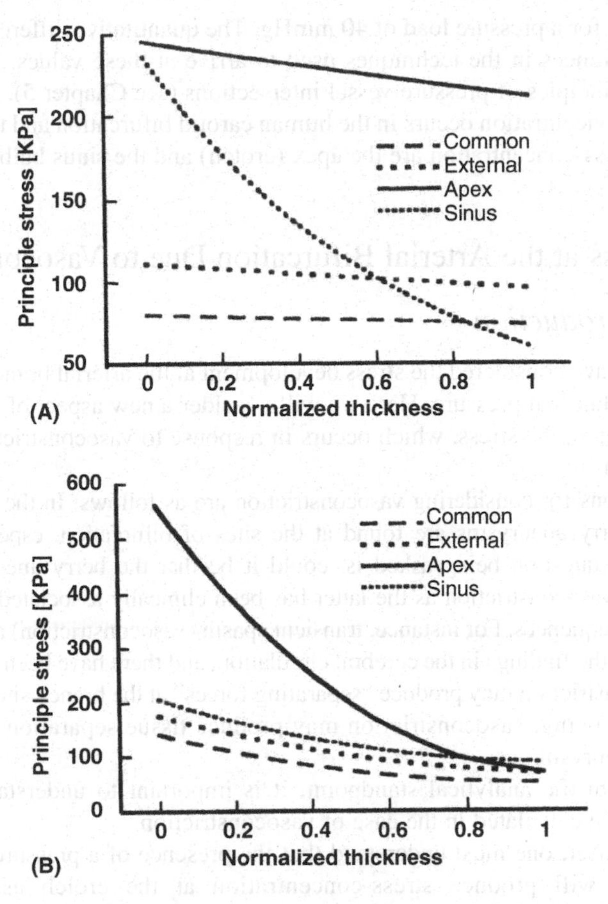

FIGURE 6.15. (A) Comparison of maximum principal stress at four locations in the carotid bifurcation, with residual strain taken into account. The highest stress was located at the apex, and the highest UF value was along the lateral wall of the carotid sinus. (B) Same as (A) but without residual strain. The highest UF value was at the apex. (Reproduced from Delfino A et al., J Biomechanics 1997;30:777-786, with permission from Elsevier.)

Overall, the inclusion of the residual strain in the model made the stress distribution more uniform throughout the carotid bifurcation except at the sinus bulb where it made the stress distribution less uniform. As for the stress-concentration, the highest stress occurred at the apex, followed by that at the sinus. Qualitatively, the results of the current study, which included the residual strain, and of the earlier study, which did not include the residual strain, are in agreement as far as the stress-concentration is concerned. The mid-wall stresses in the current study (including residual strain) at the apex, sinus, and common carotid are approximately 225, 120, and 75 kPa, respectively, for a pressure load of 120 mmHg. The mid-wall stresses in the earlier study (3) at the same locations were 220–270, 80, and 20 kPa, for a pressure load of 40 mmHg. The quantitative differences are due to the differences in the techniques used to arrive at these values. As expected from the principles of pressure vessel intersections (see Chapter 5), we note that the stress-concentration occurs in the human carotid bifurcation and the two locations of stress-concentration are the apex (crotch) and the sinus bulb.

5. Stress at the Arterial Bifurcation Due to Vasoconstriction

5.1. Introduction

So far, we have considered the stress development at the arterial branch due to the load of the luminal pressure. Here we will consider a new aspect of stress development, that is, the stress, which occurs in response to vasoconstriction (muscle contraction).

The reasons for considering vasoconstriction are as follows: In the cerebral circulation, berry aneurysms are found at the sites of bifurcation, especially at the crotch. The question being asked is, could it be that the berry aneurysms form because of vasoconstriction as the latter has been clinically associated with pathological consequences. For instance, transient spasm (vasoconstriction) and ischemia are some of the findings in the cerebral circulation, and there have been speculations that vasoconstriction may produce "separating forces" at the branch sites (6, 7). The implication being, vasoconstriction may produce tissue separation, which may produce aneurysm.

Also, from the analytical standpoint, it is important to understand how the stress may be calculated in the case of vasoconstriction.

At the outset, one must understand that the presence of a pressure load at the bifurcation will produce stress-concentration at the crotch as previously described. However, for the study of vasoconstriction, it is convenient to begin at the level of stress, which includes stress-concentration under pressure load, as a reference level. In other words, one may define the state of stress present under pressure, whatever the stress distribution may be, as the reference state.

Using the geometry under pressure as a reference geometry, one can determine the additional stress produced by vasoconstriction. Then, the total stress could be computed by adding the reference stress and the stress due to vasoconstriction using the principle of superposition. In biological problems of this kind, such segmental approaches for stress analysis are quite common, mainly because dealing with the entire problem in a single attempt is too complex.

5.2. Determination of Stress Due to Vasoconstriction

We will consider how the problem of stress analysis due to vasoconstriction is formulated and solved in the case of *canine renal artery bifurcation*. A general method of two-dimensional finite element stress analyses is used in which experimentally measured displacements, produced by vasoconstriction, are prescribed.

The problem was investigated by Fenton et al. (8). First, renal arteries with branches were removed from anesthetized dogs and placed in a bath of modified Krebs-Ringer solution at 37°C. Intraluminal pressure was maintained at 120 mmHg (16 kPa). The three ends of the arterial branch segment were tethered 1 cm from the branch point. The surfaces of the arteries were marked using graphite dust with a particle size of about 20 μm. A dissecting microscope (for magnification) was used to photograph the graphite markers at the branches. A typical experiment was to first photograph the graphite markers in a reference state, which in this case was with the internal pressure of 120 mmHg. Then, stimulate the smooth muscle cells by drugs such as dopamine hydrochloride or norepinephrine to cause vasoconstriction in the artery, and then photograph the markers again. The change in the marker position provides the displacements due to vasoconstriction. Another variation used in the experiment was to reduce the internal pressure from 120 to 50 mmHg (16 to 6.7 kPa), and then measure the displacements of the markers. Both the active vasoconstriction and the reduction of pressure produced displacements, which are qualitatively similar, that is, they both caused reduction in the arterial dimensions, and therefore, both conditions were analyzed.

Figure 6.16 shows the positions of graphite particles and the region of interest in the canine renal artery branch. In the example shown, the region of interest is defined by a quadrangle formed by four graphite particles 1-2-3-4. It is important to remember that localized effects in the branch are being studied and therefore the region of interest is very small. Each side of a quadrangle is only about *0.5 mm*. Displacements along the four boundaries of the quadrangle are presented by a linear interpolation between the measured displacements of their vertices. In

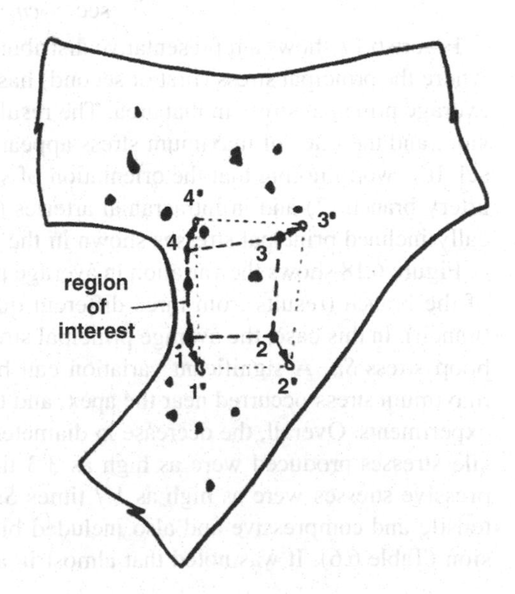

FIGURE 6.16. Tracing of a renal artery branch point, showing the graphite marks and a typical region of interest 1-2-3-4 on the adventitial surface. Characteristic displacements of the four markers of the quadrangle to 1′-2′-3′-4′ are shown; not to scale. (Reproduced from Fenton TR et al., J Biomechanics 1986;19:501-509, with permission from Elsevier.)

other words, displacements of points 1 and 2 to points 1′ and 2′ in the figure are used to describe the displacement of the entire boundary line 1–2 to a new boundary line 1′–2′. This is done for all four sides.

To begin with, a two-dimensional "plate" type finite element model is created, using the original positions of four graphite particles, which has the shape of a quadrangle. The quadrangle is then discretized into multiple triangles of constant strain. The material was considered linearly elastic and homogenous and a small strain analysis was used. The displacements of the four boundaries of the quadrangle are provided as input in the analysis and the stresses are determined. The results represent change in stress from the reference state, which was, in this case, the state under physiologic hydrostatic pressure of 120 mmHg. The reference stress Su was defined as $Su = \frac{PR}{t}$, where P = internal pressure, R = internal radius, and t = wall thickness. Su was determined for the three branches and then averaged. The analysis is a first-order approximation, and the authors estimate 5% error in the computational procedure and about 5% error between the small and large strain analysis. It is acknowledged that the assumption of isotropic and homogeneous material behavior is a simplification and that all derived stress applies to the passive components of the artery wall.

The results were based on the four renal artery branches in which both chemical vasoconstriction and pressure reductions were used. The average diameters of the parent and daughter vessels were 2.5 mm and 1.8 mm, respectively, and the mean wall thicknesses were 0.3 and 0.2 mm, respectively. The angle of the branch was 64 degrees on average. The value of the elastic modulus used was 1500 gmf/cm^2 and Poisson's ratio was 0.49. The elastic modulus of 1500 gms of force/cm^2 equals

$$1.5 \times 9.81 \text{ kg}. \frac{m}{\sec^2} \cdot \frac{1}{cm^2} = 14.7 N/cm^2.$$

Figure 6.17 shows a representative distribution of normalized principal stresses where the principal stress (first or second) has been divided by the corresponding average principal stress in that area. The results show the presence of biaxial tension, and the lines of maximum stress appear to follow the curvature of the vessel. It is worth noting that the orientation of smooth muscle cells in the coronary artery branch (7) and in intracranial arteries (9) appear to be similar to the helically inclined principal stresses shown in the figure.

Figure 6.18 shows the variation in average principal stress in the general region of the branch (results from three different quadrangles in a single renal arterial branch). In this case, the average principal stress has been divided by the average hoop stress Su. A significant variation can be noted in the branch region. The maximum stress occurred near the apex, and this is reported in Table 6.6 for four experiments. Overall, the decrease in diameter ranged from 0% to 26%. The tensile stresses produced were as high as 3.3 times the hoop stress Su. The compressive stresses were as high as 1.7 times Su. The resulting stresses were both tensile and compressive and also included biaxial tension and biaxial compression (Table 6.6). It was noted that almost in all cases, there was a change in the

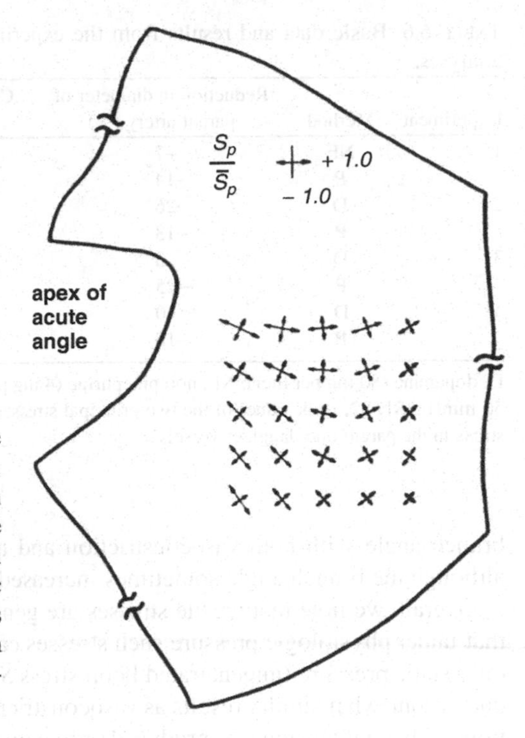

FIGURE 6.17. Principal stresses in one quadrangle in the apical area of case 1. Both principal stresses *Sp* have been divided by the corresponding average principal stress $\overline{Sp}$ in the area. Physical dimensions are not to scale. Elastic modulus = 1500 gmf cm^{-2}, Poisson's ratio = 0.49. (Reproduced from Fenton TR et al., J Biomechanics 1986;19:501-509, with permission from Elsevier.)

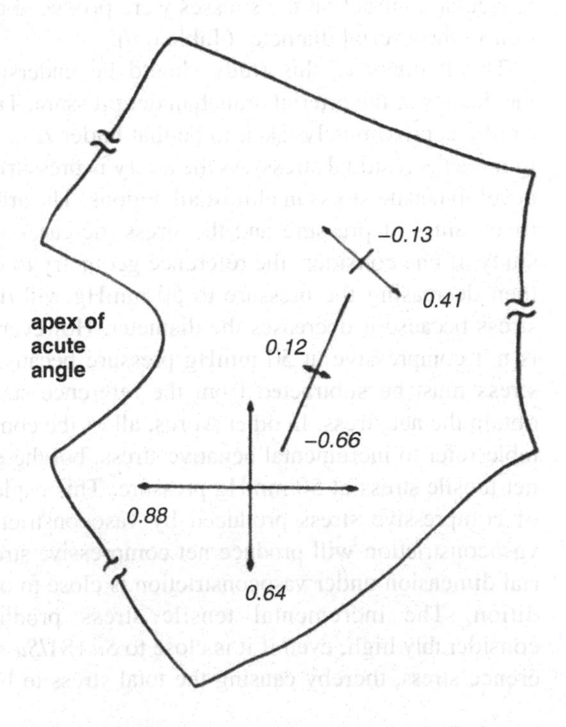

FIGURE 6.18. Variation of average principal stresses in three quadrangles in case 1. Stresses were divided by the average hoop stress in the three vessels of the branch. Physical dimensions are not to scale. Elastic modulus = 1500 gmf cm^{-2}, Poisson's ration = 0.49. (Reproduced from Fenton TR et al., J Biomechanics 1986; 19:501-509, with permission from Elsevier.)

TABLE 6.6. Basic data and results from the experimental procedures and finite element analyses.

Experiment	Method	Reduction in diameter of parent artery (%)	Change in branch angle (%)	S1/*Su*	S2/*Su*
1	NE	−7	−10	1.3	1.0
	P	−14	−14	−1.0	−1.3
2	D	−26	4	3.3	0.2
	P	−13	19	2.4	−1.7
3	D	0	0	−0.2	−0.5
	P	−25	−10	−0.6	−1.1
4	D	0	0	1.1	−1.7
	P	−15	−31	0.3	−0.9

D, dopamine (80 mg per liter); NE, norepinephrine (4 mg per liter); P, pressure reduction from 120 to 50 mmHg; S1, S2, peak values of the two principal stresses in any one quadrangle; *Su*, average hoop stress in the parent and daughter vessels.

branch angle with both vasoconstriction and reduction of pressure (Table 6.6), although the branch angle sometimes increased and sometimes decreased.

Overall, we note that tensile stresses are generated in a contracting artery, and that under physiologic pressure such stresses can be greater than those caused by only static pressure (unpenetrated hoop stress *Su*). The reduction in pressure produced somewhat similar results as vasoconstriction with the following two exceptions: i) biaxial tension was produced only with muscular contraction, and ii) with muscular contraction the stresses were produced even when there was no reduction in the arterial diameter (Table 6.6).

The findings of this study should be understood in the context of overall mechanics of the arterial branch under pressure. The zero stress state of the artery can be approximately taken to be that under zero internal pressure load, and this ignores the residual stress. As the artery is pressurized, it continues to expand and develop tensile stress in almost all regions. The artery has tensile stress at all positive values of pressure and the stress increases with pressure. As in the above study, if one considers the reference geometry to be that at 120 mmHg pressure, then decreasing the pressure to 50 mmHg will definitely produce compressive stress because it decreases the diameter. However, the actual stress in the artery is not compressive at 50 mmHg pressure because the incremental compressive stress must be subtracted from the reference stress at 120 mmHg pressure to obtain the net stress. In other words, all of the compressive stresses shown in the table refer to incremental negative stress, but the entire branch region is under a net tensile stress at 50 mmHg pressure. This explanation is true also in the case of compressive stress produced by vasoconstriction. The only situation when vasoconstriction will produce net compressive stress will be that when the arterial dimension under vasoconstriction is close to or less than that at no load condition. The incremental tensile stress produced by vasoconstriction is considerably high, even if it is close to *Su* (S1/*Su* ≡ 1), and it will add to the reference stress, thereby causing the total stress to be even higher. The study used the value of 14.7 N/cm^2 for the modulus of elasticity. This value is 5 to 10 times lower than that reported by us and others (1, 3). Because in the model the strain is used as an input to determine the stress, the use of a higher modulus will result in even higher values of stress.

The occurrence of berry aneurysms and other branch point pathology, particularly in cerebral arteries, was a stimulus to study the effect of vasoconstriction on the development of stress in that region. The finding that muscle contraction produces tensile stresses that can be additive to the stress-concentration due to pressure load is quite significant. The greater the total stress at the branch, the greater the likelihood of tissue damage. Therefore, contraction of the media in the artery is an important factor and should be considered when exploring vascular pathology at the branch sites.

References

1. Thubrikar MJ, Roskelley SK, Eppink RT: Study of stress concentration in the walls of the bovine coronary arterial branch. J Biomechanics 1990;23:15-26.
2. Patel DJ, Fry DL: The elastic symmetry of arterial segments in dogs. Circulation Res 1969;24:1-8.
3. Salzar RS, Thubrikar MJ, Eppink RT: Pressure-induced mechanical stress in the carotid artery bifurcation: A possible correlation to atherosclerosis. J Biomechanics 1995;28:1333-1340.
4. Delfino A, Stergiopulos N, Moore Jr. JE, Meister JJ: Residual strain effects on the stress field in a thick wall finite element model of the human carotid bifurcation. J Biomechanics 1997;30:777-786.
5. Bharadvaj BK, Mabon RF, Giddens DP: Steady flow in a model of the human carotid bifurcation Part I – flow visualization. J Biomechanics 1982;15:349-362.
6. Stehbens WE: Hemodynamics and the Blood Vessel Wall. Charles C. Thomas, Springfield, IL, 1981.
7. Boucek RJ, Takeshita R, Brady AH: Microanatomy and intramural forces within the coronary arteries. Anat Rec 1965;153:233-242.
8. Fenton TR, Gibson WG, Taylor JR: Stress analysis of vasoconstriction at arterial branch sites. J Biomechanics 1986;19:501-509.
9. duBoulay GH: The natural history of intracranial aneurysms. Am Heart J 1967;73: 723-729.

7 Stress-Concentration in the Artery II

1. Introduction

In the past chapter, we considered the stress-concentration at the coronary artery branch and at the carotid artery bifurcation. We also considered the stress-concentration, which may occur due to vasoconstriction in the branch region. In this chapter, we will consider the following topics: i) the stress-concentration at the cerebral artery bifurcation; ii) stress-concentration in vivo; and iii) stress-concentration where the absolute magnitude of the stress is determined. Understanding the stress-concentration at the cerebral artery bifurcation provides us with a common link between the aneurysms, which form more frequently in the intracranial arteries, and atherosclerosis, which occurs more frequently in the extracranial arteries. As hypothesized in Chapter 1, we can consider the stress-concentration as a common factor, which can lead to either pathology. Also, we would like to know whether the stress-concentration occurs in vivo. Finally, we must determine the absolute values of stress, beginning with the initial geometry of the artery at zero pressure and ending with a pressure load of 120 mmHg or greater.

2. Shape Change at the Apex of Human Cerebral Artery Bifurcation Under Pressure

As we continue to study the mechanics of the arterial branch, we keep in mind that different arterial branches may behave differently, depending upon their tissue structure, and accordingly the pathology they develop could be different. As we note in Chapter 14, the frequency of branch aneurysms is much greater in the cerebral arteries than in the arteries elsewhere in the body.

We have also observed that the study of the branch has been performed by different authors using different approaches. In this section, we will consider the measurements on the apex of the cerebral artery bifurcation at various internal pressures. In the bifurcation, the most interesting structure is the apex (or the crotch). There are only a few studies that have reported direct observations on the apex, and therefore, the study considered here is rare and important in providing an understanding of the development of the branch pathology.

2.1. *Study of the Apex of Cerebral Artery Bifurcation*

MacFarlane et al. (1) studied five cerebral arterial bifurcations obtained at autopsies from subjects ranging in age from 40 to 74 years. Two middle cerebral arteries, two anterior cerebral arteries, and one vertebral artery, all with branches, were studied (see Chapter 14, Figs. 14.3 and 14.4 for anatomical locations of these arteries). The study focuses on the examination of *apical curvature* in the two orthogonal planes as depicted in Figure 7.1. The artery bifurcations were cleared of adventitial connective tissue and perfused with 0.9% saline to remove blood. The bifurcations were cannulated, ligated, and mounted in a saline bath at 21°C. The branches were left free to lengthen. The daughter branches were aligned in a

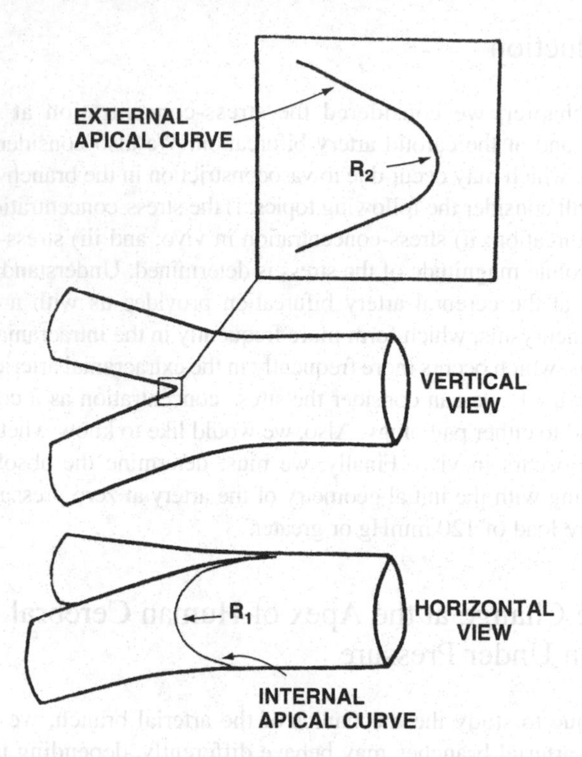

FIGURE 7.1. Schematic diagram of the apical region of a bifurcation showing the internal and external apical curves. The external apical curve is very small compared with the internal curve. (Reproduced from Macfarlane TWR et al., Stroke 1982;14:70-76, with permission from Lippincott, Williams & Wilkins.)

horizontal plane. A special Plexiglas light pipe was used to illuminate the apex of the bifurcation. The combination of high-intensity light and the semitransparent nature of the cerebral arteries made it possible to observe and photograph the apical curve by the camera aligned horizontally. Another camera photographed the external apical curve from the vertical view.

The bifurcations were pressurized in increments of 10 mmHg pressure and photographed. The photographs were magnified and digitized for the outline of the internal apical curve. From the configuration of the apical curve, the radius of curvature was calculated.

Figure 7.2 shows the contours of the apical curve at various luminal pressures. The contours clearly indicate that the shape of the apical curve changes with transmural pressure. Overall, the apical curve flattens in the central region and shoulder regions broaden with increasing pressure. The curve does not just become larger but it changes shape. Three of the five specimens showed that the internal apical curve flattened in the central region. The data in the central region of the curve was more accurate than that in the shoulder region because in the

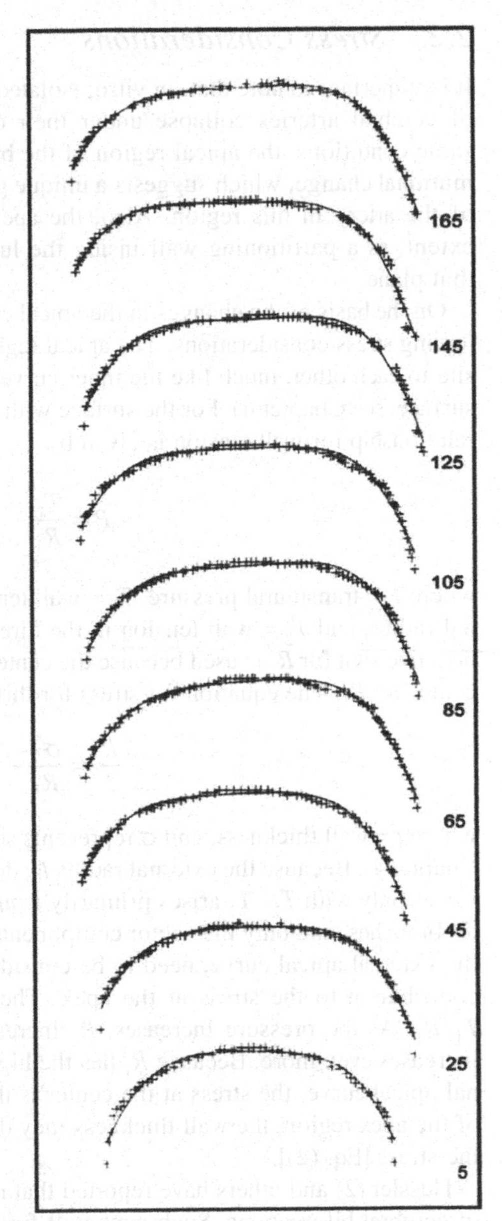

FIGURE 7.2. Computer-drawn serial array of apical curves indicating the digitized coordinates (crosses) and the coordinates generated by fourth-degree polynomial equations corresponding with each coordinate set (line). The specimen is the second bifurcation of the right middle cerebral artery from a 34-year-old male. The bifurcation was subjected in steps to transmural pressure ranging from 5 mmHg to 165 mmHg (numbers in the figure). (Reproduced from Macfarlane TWR et al., Stroke 1982;14:70-76, with permission from Lippincott, Williams & Wilkins.)

central region, the flow divider is sharply defined and away from the lateral walls of the bifurcation. In the shoulder region, the flow divider broadens and merges with the parent trunk and therefore it is less sharply defined.

In contrast with the internal apical curve, the external apical curve did not show a significant change in shape or size. The radius of the external curve was quite small compared with that of the internal curve.

2.2. *Stress Considerations*

It is important to note that, in vitro, isolated unpressurized cylindrical segments of cerebral arteries collapse under their own weight. In contrast, under the same conditions, the apical region of the bifurcation shows little or no conformational change, which suggests a unique geometry and structural organization of the artery in this region. Also, the apex may be expected to act, to some extent, as a partitioning wall inside the lumen, thereby imparting strength in that plane.

On the basis of the changes in the apical curvature, the authors propose the following stress considerations. The apical region has two curvatures that are opposite to each other, much like the inner curve region of the aortic arch (anticlastic surface, see Chapter 4). For the surface with two opposite curvatures, the Laplace relationship for wall tension is given by

$$P = \frac{T_1}{R_1} - \frac{T_2}{R_2} \tag{1}$$

where P = transmural pressure, T_1 = wall tension in the direction of R_1, the internal radius, and T_2 = wall tension in the direction of R_2, the external radius. The negative sign for R_2 is used because the center for R_2 is on the opposite side of the center for R_1. The equation for stress for this geometry is

$$\frac{P}{t} = \frac{\sigma_1}{R_1} - \frac{\sigma_2}{R_2} \tag{2}$$

where t = wall thickness, and σ represents stress in the respective directions (see Chapter 4). Because the external radius R_2 does not change much, the term T_2/R_2 varies only with $T_2 \cdot T_2$ arises primarily from longitudinal loading of the daughter branches, and only the vector components, acting tangentially at the center of the external apical curve, need to be considered. Thus, T_2/R_2 makes a minimal contribution to the stress at the apex. The major stress arises from the term T_1/R_1. As the pressure increases, R_1 increases and therefore the product PR_1 increases even more. Because R_1 has the highest value at the center of the internal apical curve, the stress at the center is the highest. Also with the expansion of the apex region, the wall thickness may decrease, which will further increase the stress [Eq. (2)].

Hassler (2) and others have reported that medial gaps are common at the apex of cerebral bifurcations. Such gaps will further decrease the wall thickness and increase the stress-concentration, thereby making it even more probable that the stress will reach the critical yield point at this location. Thus, it is not surprising that the aneurysms tend to arise from the center of the apical region where the stress is maximum, rather than from the edges (see Chapter 14). Campbell and Roach (3) have also observed larger fenestrations in the elastic membrane at the apex of the cerebral bifurcations. These fenestrations could have been produced by greater stress at that location.

3. Stress-Concentration at the Arterial Branch in Vivo

Determination of stress-concentration at the arterial branch in vivo is complicated. In vivo, the artery is under some degree of active tone (vasoconstriction) and therefore the dimensions it occupies are a combined result of i) passive expansion due to luminal pressure, ii) passive lengthening due to tethering force, and iii) active contraction caused by neural stimulation or by circulating chemical agents. The amount of tethering force and active contraction are not known. Any attempt to determine stress-concentration therefore involves some simplifications and assumptions.

Manuel and Thubrikar (4) carried out such a study. They determined the stress-concentration at the ilio-femoral branch in canines where the iliac artery bifurcates into deep femoral and superficial femoral arteries.

The approach used in the determination of stress-concentration involves multiple steps. First, the strains in the artery are measured experimentally in vivo over a given pressure range and used to determine the material properties. Then the experiments are repeated on the same artery in vitro to determine the tethering force, which is needed to determine the material properties mentioned above. Then the geometry of the arterial branch is determined to create a model for the finite element analysis. The results of the analysis are obtained in terms of stresses and strains at various key regions. The strains obtained from the analysis are compared with those measured experimentally to check the accuracy of the analysis. The next step is to explore the effect of those parameters, which are either not known or known less accurately, on the results by carrying out parametric studies.

3.1. *In Vivo Experiments*

Adult dogs weighing 18–20 kg were anesthetized and maintained on a respirator. The ilio-femoral arterial branch (not a true bifurcation) was surgically exposed to obtain the main vessel of about 6 cm length and the branch of about 2 cm length. The loose fatty tissue over the adventitia was removed. The artery was marked with finely ground potassium permanganate powder (Fig. 7.3). The markers were used for strain measurements. The experimental setup for recording the positions of the markers was as follows. A video camera with either a micro lens (giving magnification of 40×) or a macro lens (to capture the whole artery) was used for the recording. A micro lens was necessary to record marker positions in local regions.

To measure intra-arterial pressure, a 5F catheter, having only side holes, was connected to a pressure transducer, which was connected to a digital display unit to compute mean pressure. The arterial pressure was raised or lowered by intravenous infusion of angiotensin or nitroprusside. The marker positions were recorded at three locations proximal, at, and distal to the branch ostium.

First, at a site proximal to the branch, marker positions were recorded at a control systemic pressure. Then pressures were first lowered and then raised and the

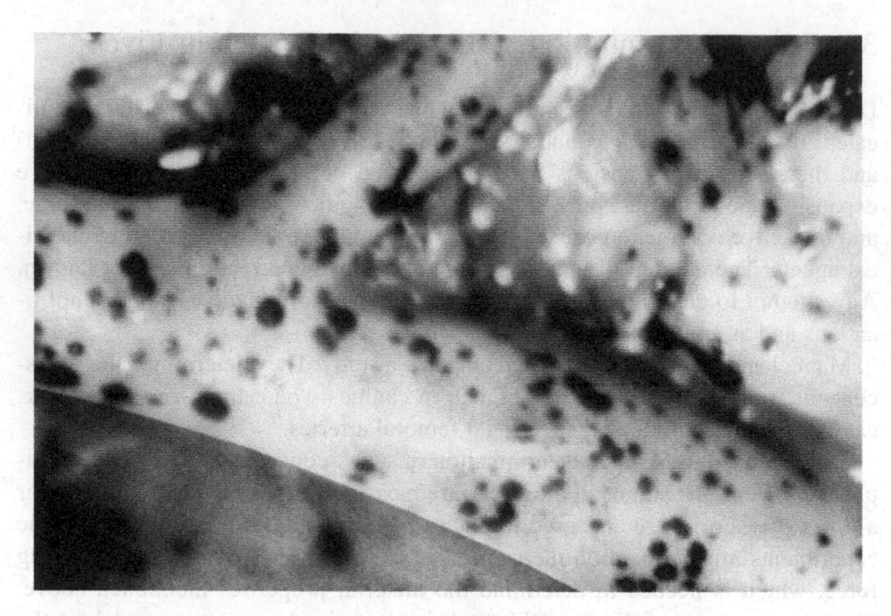

FIGURE 7.3. The marked ilio-femoral artery branch region in vivo (canine).

recording done close to 60, 80, 100, 120, and 140 mmHg mean pressure. The measurements at this site were used to determine the incremental moduli of elasticity and the Poisson's ratios in the longitudinal and circumferential directions. Similar measurements were repeated at the branch and distal to the branch. These measurements were used to determine circumferential and longitudinal strains for comparison with the analytical results.

For both the circumferential and longitudinal directions, a plot of pressure versus distance between markers is made (Fig. 7.4). Using the distance at a pressure of 80 mmHg as the initial length, the incremental strains for a pressure increase from 80 to 120 mmHg were obtained. It may be reiterated that the distances were related to the mean pressure because the change in the distance in response to pulsatile pressure was not measurable. The arterial branch was then removed for experiments in vitro.

3.2. *In Vitro Experiments*

The sole purpose of these experiments was to determine the longitudinal tethering force present in the artery in vivo. In the in vitro experiments, the arteries are not tethered. The experiments were performed using a procedure similar to that used in vivo. The iliac artery proximal to the bifurcation was cannulated and the other two branches were tied. The cannulated end was held fixed while the remainder of the artery was free to lengthen. The cannula was connected to a

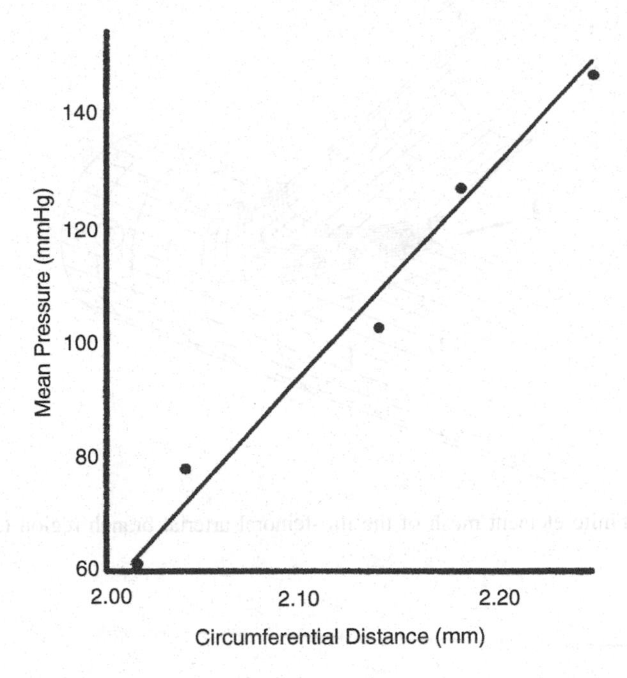

FIGURE 7.4. Mean pressure versus circumferential distance in the main artery.

saline reservoir, which could provide the desired pressure. Video recordings of the markers proximal to the branch region were made at pressures of 60, 80, 100, 120, and 140 mmHg. The same markers used in the in vivo experiments were used here. Two markers in the circumferential direction and two in the longitudinal direction were chosen at each of the three sites (Fig. 7.5). Distances between the markers were measured at each pressure. Strains were determined for an increase in pressure from 80 mmHg to 120 mmHg.

3.3. Geometry of the Branch

After the in vitro experiment, the artery was fixed with 4% formaldehyde under 80 mmHg pressure for 2 h. Cross-sectional cuts were made in the main vessel and two branches. The bifurcation was then split longitudinally along the common plane of axes of the main vessel and the branches. All of these sections were recorded with a video camera. Figure 6.3 of Chapter 6 shows various geometric measurements needed for the generation of a finite element model. The radii and thicknesses R and T for the main vessel and r and t for the branch were measured. Around the ostium, thickness and radii were measured at 0, 90, and 180 degrees from the common plane of the axis and denoted as t_0, t_{90}, and t_{180} and R_0, R_{90}, and R_{180} (Table 7.1).

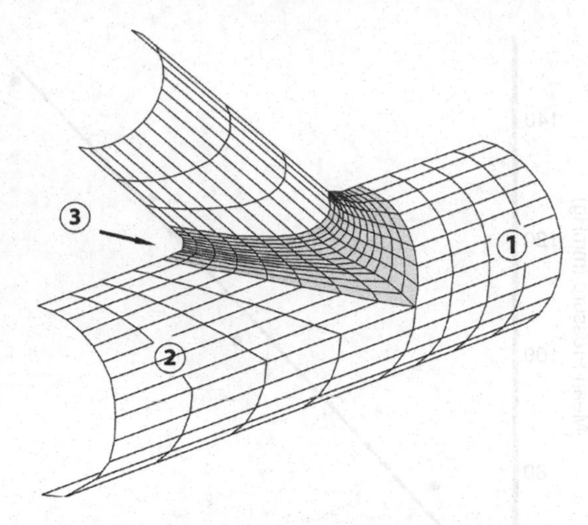

FIGURE 7.5. Finite element mesh of the ilio-femoral arterial branch region (3, transition region).

TABLE 7.1. Measured geometric parameters of ilio-femoral branch (mm).

Parameter	Artery 1	Artery 2	Artery 3
R	2.10	2.50	3.08
T	0.26	0.23	0.28
r	1.43	1.05	1.41
t	0.20	0.17	0.18
Φ (degrees)	39.75	50.50	61.00
R_o	0.20	0.51	0.45
R_{90}	2.68	2.30	3.11
R_{180}	3.80	4.01	3.89
t_o	0.34	0.46	0.68
t_{90}	0.25	0.17	0.70
t_{180}	0.20	0.14	0.53

3.4. *Theoretical Considerations*

A number of assumptions were made to determine stress distribution at the arterial branch in vivo. These are 1) stresses in the straight segment of the artery are obtained by formulas for a thin cylindrical shell; 2) artery wall is orthotropic; 3) artery is linearly elastic over the pressure range of 120/80 mmHg; and 4) initial geometry is taken at 80 mmHg pressure.

A) *Incremental stresses and strains:* Incremental strains were obtained by plotting distances between markers in the circumferential and longitudinal directions versus mean pressure (60–140 mmHg). A linear regression line was drawn and the distances at 80 and 120 mmHg pressures were obtained (Fig. 7.4). The incremental strains are given as:

$$\varepsilon_C = \frac{C_{120} - C_{80}}{C_{80}} \text{ and } \varepsilon_\ell = \frac{L_{120} - L_{80}}{L_{80}}$$

where ε_C and ε_ℓ are incremental strains in the circumferential and longitudinal directions, respectively. C and L are distances in the circumferential and longitudinal directions, respectively, at 80 or 120 mmHg pressure. These equations apply to both in vivo and in vitro experiments.

Incremental stresses due to an incremental pressure of 40 mmHg (pressure change from 80 to 120 mmHg) are given by the formulas for a thin-walled cylinder as follows:

$$\sigma_C = \frac{PR}{T} \text{ and } \sigma_\ell = \frac{PR}{2T} + F'$$

where σ_C and σ_ℓ are incremental stresses in the circumferential and longitudinal directions, respectively, P is the incremental pressure 40 mmHg, R is the mean radius of the main vessel, T is the thickness of the main vessel, and F' is the longitudinal tethering force per unit area of cross section of the main vessel. The radius R and thickness T were determined from the geometry at 80 mmHg pressure. The tethering force F', which occurs only in vivo, was determined from in vitro experiments as described later.

B) *Material properties:* For the artery, it was necessary to determine moduli of elasticity and Poisson's ratios in the circumferential and longitudinal directions. These four elastic constants were determined from the following equations for orthotropic material.

$$\varepsilon_c = \frac{\sigma_c}{E_c} - \nu_\ell \cdot \frac{\sigma_\ell}{E_\ell} \tag{3}$$

$$\varepsilon_\ell = \frac{\sigma_\ell}{E_\ell} - \nu_c \frac{\sigma_c}{E_c} \tag{4}$$

$$\varepsilon_c \cdot \nu_\ell = \varepsilon_\ell \cdot \nu_c \tag{5}$$

$$\text{and } \frac{\nu_c + \nu_\ell}{2} = 0.3 \tag{6}$$

where ε_c and ε_ℓ are incremental moduli of elasticity in the circumferential and longitudinal direction, respectively, and ν_c and ν_ℓ are corresponding Poisson's ratios. These equations are valid for both in vivo and in vitro experiments. Once the stresses and strains are known, these equations are used to calculate incremental elastic constants. A similar approach has been described in Chapter 6.

C) *Longitudinal tethering force:* To determine the tethering force present in vivo, a suitable experimental procedure and theoretical approach had to be developed. In the in vitro experiment, the artery was free to expand longitudinally, whereas in vivo the artery is tethered. The tethering force was determined as that force which would make the longitudinal strain in vitro the same as that in vivo. In the in vitro experiments, first all four elastic constants were determined from the measured strains and calculated stresses using Eqs. (3) to (6). Then using Eq. (4), the stress σ_ℓ was calculated by substituting the strain ε_ℓ measured in vivo. This longitudinal stress σ_ℓ has both the pressure component and the tethering force component, that is,

$$\sigma_\ell = \frac{PR}{2T} + F'.$$

From this, the tethering force F' was calculated.

Hence, for in vivo conditions, the stress, the strain, and the tethering force were known and all four elastic constants were determined by using Eqs. (3) to (6).

3.5. *The Finite Element Model*

A general-purpose computer program, ANSYS, was used for the finite element analysis of the arterial branch. A two-dimensional quadrilateral shell element with variable thickness along its edges was used to represent the artery.

A) *Mesh generation:* The Ilio-femoral arterial branch is modeled as two intersecting thin cylindrical shells. Figure 7.5 shows the finite element mesh for this model. A sharp intersection between two cylindrical shells was obtained first and then the nodes at the intersection were moved to new positions so that the radii of curvature could be drawn from the main vessel to the branch equal to the measured radii R_0, R_{90}, and R_{180}. To complete smoothing of the geometry around the ostium, the radius of curvature was varied linearly between these values. In the region of the branch, measured thicknesses were incorporated.

B) *Transition region:* At the ostium, the elastic properties in the circumferential direction of the main vessel must reorient themselves till they equal the elastic properties in the circumferential direction of the branch. The region over which this reorientation takes place is defined as a transition region (see Chapter 6). In all of the arteries studied, it was seen that the main artery began to thicken at a distance from the branch axis roughly equal to the radius of the main vessel. This thickening could be considered as reinforcement and the distance over which it occurs could be considered as an *influence length* described in Chapter 5. This distance was used to represent the boundary of the transition region (Fig. 7.5). The measured values of the thicknesses around the ostium t_0, t_{90}, and t_{180}, and a linear variation in the thickness were incorporated in the transition region. Also, the elastic properties were reoriented as described in Chapter 6.

C) *Loading and boundary conditions:* The model was loaded with an incremental pressure of 40 mmHg, which represents the difference between the systolic and diastolic pressure in vivo. All three ends of the arterial bifurcation were loaded with equivalent longitudinal nodal force to represent both the longitudinal stress, which results from an internal pressure, and the tethering force.

3.6. *Parametric Studies*

These studies were done to determine the relative importance of various parameters on the stress. The analysis was carried out for the following conditions: a) The artery was considered *isotropic* and its modulus of elasticity was taken as an average of ε_c and ε_ℓ; b) the *transition region* was made larger; c) at the junction, the *thickness* of the elements was increased by 100%; d) at the junction the *elastic modulus* was increased; e) at the junction the *radius of curvature* R_o was increased; and f) the *tethering force* was changed.

3.7. *Analytical Results*

3.7.1. Stress-Concentration

The geometric parameters of the three ilio-femoral branches studied are listed in Table 7.1. The elastic properties and tethering forces are listed in Table 7.2. The results are obtained in terms of iso-stress contours on the inner, middle, and outer surfaces of the artery wall. From these contours, the magnitude and location of the highest stress are known. The first principal stress is also the maximum stress at a given point. The stress-concentration factor, *K*, is defined as a ratio of the maximum stress to the nominal stress where the nominal stress is the circumferential stress (*PR*/*T*) in the main artery. Another stress quantity of interest is the "von Mises" stress. The "von Mises" stress is the equivalent stress and it is considered important because according to "von Mises" theory, yielding takes place when the equivalent stress equals the yield stress. It is given by the following expression:

$$\sigma_e = \frac{2}{\sqrt{2}}\left[(\sigma_1 - \sigma_2)^2 + (\sigma_2 - \sigma_3)^2 + (\sigma_3 - \sigma_1)^2\right]^{1/2}$$

where σ_e is the "von Mises" stress and σ_1, σ_2, σ_3 are the first, second, and third principal stresses, respectively.

Plots of the first principal stress on the inner, middle, and outer surfaces and "von Mises" stress on the inner surface of one branch are shown in Figures 7.6 and 7.7. In these plots, the stress increases from one contour to the next toward the ostium. The first principal stress on the inner surface of the artery is highly localized around the branch ostium (Figure 7.6). The stress is enhanced at both the distal and the proximal regions of the ostium but it is highest at the distal region. The stress-concentration reduces significantly over a small distance away

TABLE 7.2. Elastic properties and tethering force.

	Artery 1	Artery 2	Artery 3
ε_c (N/mm^2)	2.51	3.25	2.61
ε_L (N/mm^2)	1.58	2.46	2.10
ν_c	0.37	0.34	0.33
ν_ℓ	0.23	0.26	0.27
F (N/mm^2)	−0.059	−0.031	−0.132

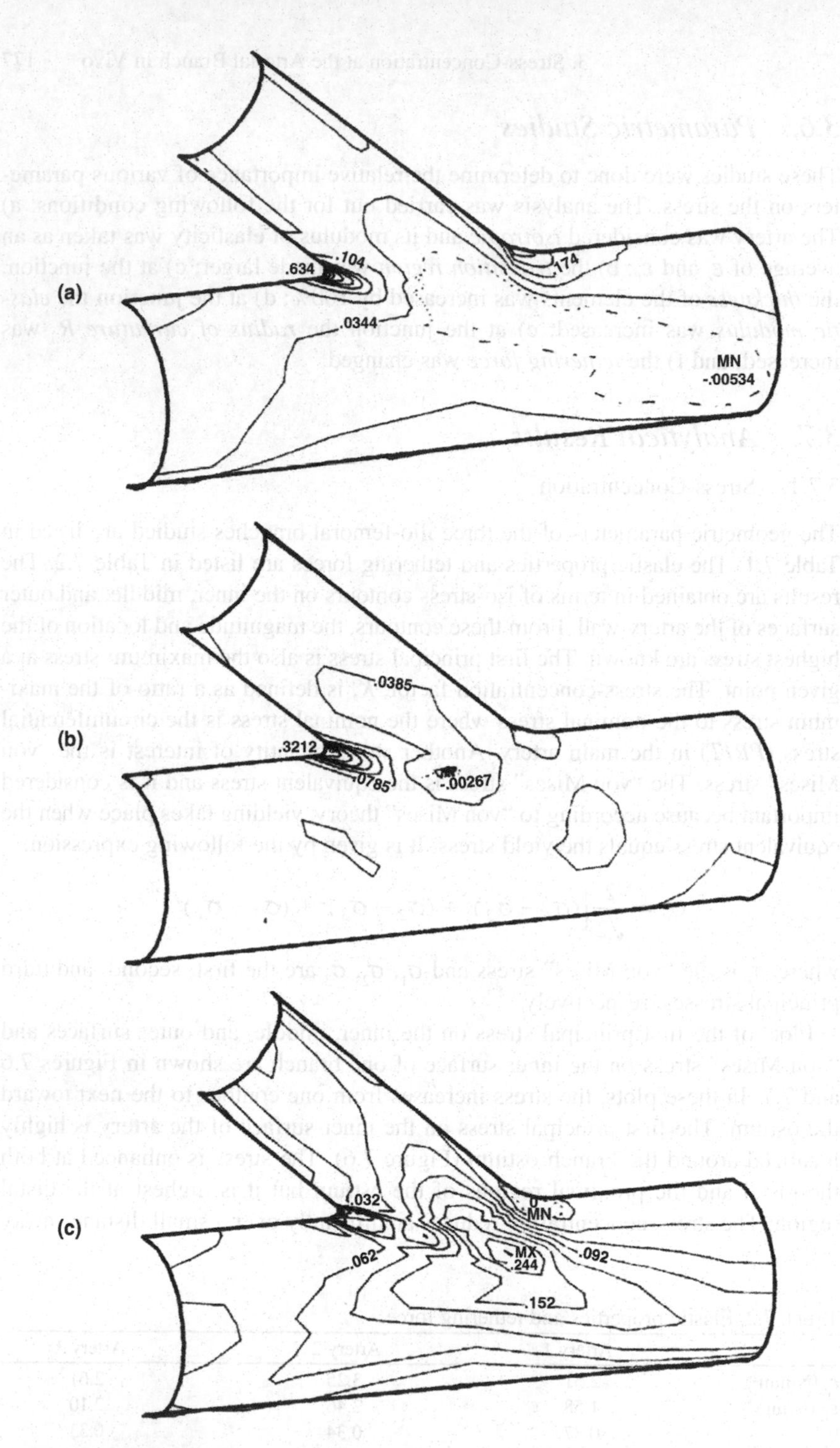

FIGURE 7.6. The first principal stress (N/mm²) on the inner (a), middle (b), and outer (c) surfaces of the ilio-femoral arterial branch. The stress-concentration is present at the distal lip and the proximal lip of the branch ostium on all three surfaces.

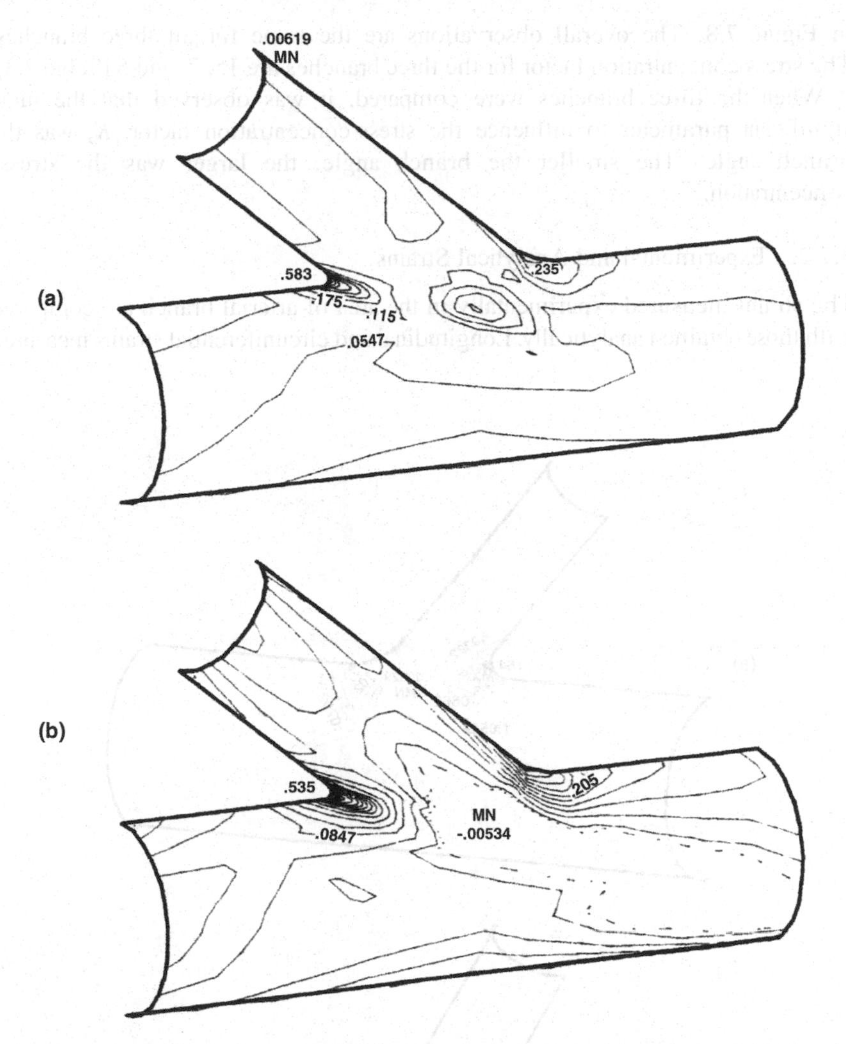

FIGURE 7.7. (a) Von Mises stress (N/mm^2) on the inner surface of the artery considering the orthotropic properties. (b) The first principal stress (N/mm^2) on the inner surface of the artery considering the isotropic properties.

from the branch point. When the element component stresses are studied, it is observed that the maximum stress is nearly circumferential. The stress gradient pattern (i.e., the manner in which the stress decays) has a pronounced V-shape around the crotch of the arterial branch. The first principal stress at the middle and outer surfaces of the artery (Figs. 7.6b and 7.6c) is also concentrated around the branch point.

Both the first principal stress (Fig. 7.6) and the "von Mises" stress (Fig. 7a) were largest at the distal branch point on the inner surface. The first principal stress on the middle surfaces of the other two artery branches are shown in Figure 7.8. The overall observations are the same for all three branches. The stress-concentration factor for the three branches are 15, 7, and 5 (Table 7.3).

When the three branches were compared, it was observed that the most significant parameter to influence the stress-concentration factor, K, was the branch angle. The smaller the branch angle, the larger was the stress-concentration.

3.7.2. Experimental and Analytical Strains

The strains measured experimentally in the area of arterial branch are compared with those obtained analytically. Longitudinal and circumferential strains measured

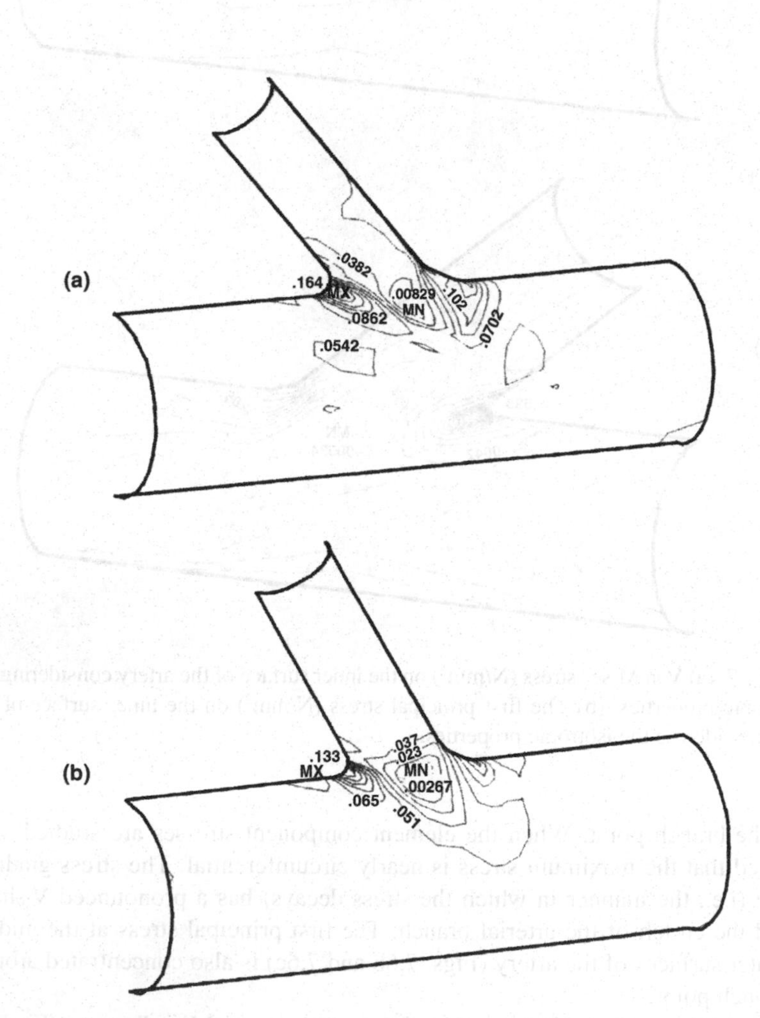

FIGURE 7.8. The first principal stress (N/mm²) on the middle surface of the other two (a, b) arteries.

TABLE 7.3. Stress-concentration factors (distal region).

	Artery 1		Artery 2		Artery 3	
PR/T (N/mm²)	0.043		0.058		0.060	
K_{inner}	14.54	Proximal 8.8	6.90	Proximal 4.26	4.79	Proximal 4.37
K_{mid}	7.34		2.83		2.22	
K_{outer}	5.60		4.26		2.57	

TABLE 7.4. Experimental and analytical strain comparison.

		Artery 1		Artery 2		Artery 3	
Site	Strain	Exp.	Anal.	Exp.	Anal.	Exp.	Anal.
1	ε_L	0.003	0.002	0.004	0.004	0	0.007
	ε_c	0.029	0.038	0.021	0.017	0.038	0.040
2	ε_L	0.007	0.009	0.011	0.014	0.004	0.0097
	ε_c	0.029	0.035	0.014	0.021	0.008	0.0372
3 Branch ostium	ε_L	0.039	0.041	0.024	0.018	0.041	0.051
	ε_c	0.040	0.081	0.028	0.054	0.051	0.082

at the three sites (Fig. 7.5) and the analytical strains at these sites are shown in Table 7.4. The analytical strains are at the middle surface of the artery whereas those measured experimentally are on the outer surface. Because the artery is thin walled, the strains at the middle surface do not differ much from those at the outer surface. In all of the experiments, the measured strains are much higher at the branch than at points away from the branch (Table 7.4). The analytical strains compare fairly well with the experimental strains except at the branch site. At this site, the nodes selected to compute analytical strains were not exactly in the circumferential or the longitudinal direction and therefore discrepancy is expected. The agreement between analytical and experiment strains at other sites is good, suggesting that the model is reasonable.

3.7.3. Parametric Studies

These studies reveal the relative influence of various parameters on the stress-concentration. The consideration of *isotropy* provided a stress distribution (Fig. 7.7b), which is very similar to that produced by the consideration of orthotropy (Fig. 7.6a) and it decreased K by only 15%. Extending the *transition zone* one row of elements on each side of the main vessel and the branch to produce a more gradual change in the geometry resulted in a 17% decrease in K. Doubling the *thickness* in the branch area resulted in 16% decrease in K. In an isotropic model, a 50% increase in the *modulus* produced a 7% increase in K. For the above conditions, the stress contour plots show that the nature of stress-concentration does not change much. The radius of curvature R_0 or the tethering force have an even smaller effect on K. Hence, the stress-concentration is a function primarily of the branch geometry, and it is affected only minimally by other parameters. Similar observations were made in Chapter 6.

3.8. Comments

The artery is a complex structure made of three layers: intima, media and adventitia. Each of these layers is composed of different tissue components and therefore has different mechanical properties. The orientation of connective tissue fibers and smooth muscle cells at the branch site is not well defined, which will govern the local material properties. Residual stresses for the canine ilio-femoral branch are not known. These and other such parameters make an accurate stress analysis forbiddingly complex. The work described above represents an initial step in the stress analysis of a complex system and provides results that are valid qualitatively and perhaps correct to an order of magnitude quantitatively. The initial geometry taken at 80 mmHg pressure represents an arbitrary state of reference stress, and the results give incremental stresses and strains. Parametric studies lend further credence to the results.

In conclusion, a stress-concentration as high as 2.2- to 7.3-fold exists on the mid-surface (Table 7.3) of the canine ilio-femoral arterial branch in vivo. The stress-concentration is primarily a function of the geometry of the branch and is minimally influenced by errors in the measurements of material properties or tethering force.

4. Determination of Absolute Values of Stress and Stress-Concentration at the Arterial Branch

4.1. Introduction

In Chapter 6 and previous sections, we have considered various approaches to the determination of stress-concentration at the arterial branch. Most of the approaches described have dealt with the incremental stress analysis where the starting geometry of the arterial branch is taken at 80 mmHg and the branch is loaded with an incremental pressure of 40 mmHg. These approaches have provided information on the change in the stresses between 80 and 120 mmHg pressure, the nature of stress distribution, the location of stress-concentration, and the stress-concentration factor. What they do not provide is the absolute values of the stress and the stress-concentration. These limitations have existed primarily because the artery has nonlinear orthotopic material properties and it undergoes large deformations. The analysis of the artery branch, which includes both nonlinearity and orthotropicity, still remains to be done. Because it is important to know the absolute values of stress and stress-concentration, Sharrets and Thubrikar (5) carried out an analysis that considers the artery to be nonlinear, isotropic, and undergoing large deformations. We will describe their approach because it is one of the most important steps in the stress analysis since it provides, although approximately, the absolute values of stress over a pressure range 0–140 mmHg. It also provides a good understanding of the arterial mechanics beginning with zero pressure load.

4.2. *Steps Used in the Analysis*

The multiple steps used in the analysis are as follows:

1. Experimentally determine the geometry and nonlinear orthotropic material properties over the entire pressure range from 0 to 140 mmHg.
2. Using nonlinear orthotropic material properties, determine *nonlinear isotropic* material properties over the entire pressure range.
3. Divide the problem into several small problems and use small pressure loads for each problem (e.g., 10 or 20 mmHg pressure load).
4. For each load step, determine the new geometry at the beginning of the load step. This means 10 geometries for 10 pressure load steps.
5. For each load step, determine the incremental modulus of elasticity thereby converting a nonlinear stress-strain curve into several small linear stress-strain segments. This provides a single value of elastic modulus for each load step.
6. Because the branch region has more stress and therefore a higher modulus than the straight region upon the application of load, determine the modulus for the branch region at each load step.
7. Allow the modulus to change gradually from that in the branch to that in the straight region.
8. Make sure that both the geometry and the material properties remain continuous from one model to the next, so that the results are cumulative.
9. Finally, compare the strains obtained from the analysis to those measured experimentally.

The ilio-femoral arterial branch from a canine is studied. The original geometry of the artery is taken at 0 mmHg pressure and used for the initial model for the finite element analysis. For each 10 mmHg pressure increment, using the experimental data, a new finite element model with new geometry is created. Each model is subjected to the stress analysis using an incremental pressure load. The resulting incremental stresses are then added to provide the absolute values of the stresses.

4.3. *Experimental Method*

The *ileo-femoral arterial branch* is isolated and removed from the canine. The branch is prepared for in vitro experiments using the marker technique described in the previous section. The marker positions in the circumferential and longitudinal directions are recorded at each step as the luminal pressure is increased from 0 to 140 mmHg in steps of 10 mmHg. The artery branch is then fixed at 0 mmHg pressure using 10% formalin. The branch is cut along various planes and photographed to obtain the entire geometry and the wall thickness (see Chapter 6, Fig. 6.3).

Using the markers in the circumferential direction, the circumferential strain is determined at various pressure steps from the following equation:

$$\epsilon_{cp} = \frac{D_{cp} - D_{co}}{D_{co}}$$

where D_{cp} is the distance at pressure P, and D_{co} is the distance at pressure 0 mmHg. Similarly, using the markers in the longitudinal direction, the longitudinal strain is determined as follows:

$$\in_{Lp} = \frac{D_{Lp} - D_{Lo}}{D_{Lo}}$$

where D_{Lp} and D_{Lo} represent the distance at pressure P and pressure 0, respectively. Figure 7.9 shows a typical plot of pressure versus strain for the circumferential and the longitudinal directions for the main artery.

The stress at each value of pressure is determined, assuming that the artery is a thin-walled shell structure and using the following equations:

$$\sigma_{cp} = \frac{PR}{T} \text{ and } \sigma_{Lp} = \frac{PR}{2T};$$

where σ_{cp} and σ_{Lp} represent the stress at pressure P in the circumferential and longitudinal direction, respectively; R is the radius, and T is the thickness of the main artery at pressure P. These stresses are calculated after the radius and thickness are determined at each value of pressure as follows.

Radius increases directly with circumferential strain, so

$$R_p = R_o\left(1 + \in_{cp}\right),$$

where R_p and R_o represent radius at pressure P and pressure zero, respectively.

The new thickness (T_p) at pressure P is calculated by *conservation* of volume as follows:

$$2\pi R_o T_o L_o = 2\pi R_p T_p L_p.$$

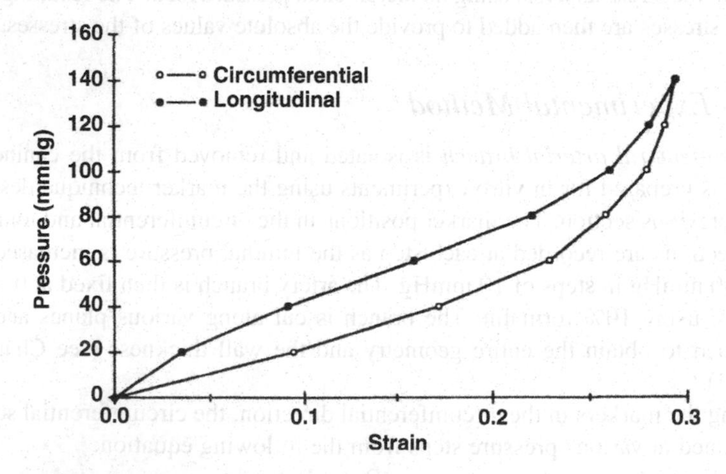

FIGURE 7.9. Experimental strains measured in one artery (ilio-femoral branch) as a function of pressure.

Noting that $L_p = L_o(1 + \in_{LP})$ and by substitution *of* R_p we get

$$2\pi R_o T_o L_o = 2\pi R_o(1+\in_{cp}) T_p L_o(1+\in_{LP})$$

or

$$T_p = \frac{T_o}{(1+\in_{cp})(1+\in_{LP})}.$$

Thus, radius and thickness of the main artery at each value of pressure is known. Using these, the stresses at each value of pressure are calculated.

Figure 7.10a shows a plot of stress versus strain for the circumferential and longitudinal directions. The figure indicates that the artery has different properties in the circumferential and longitudinal directions, and this data can be used to describe the arterial material as orthotropic and nonlinear.

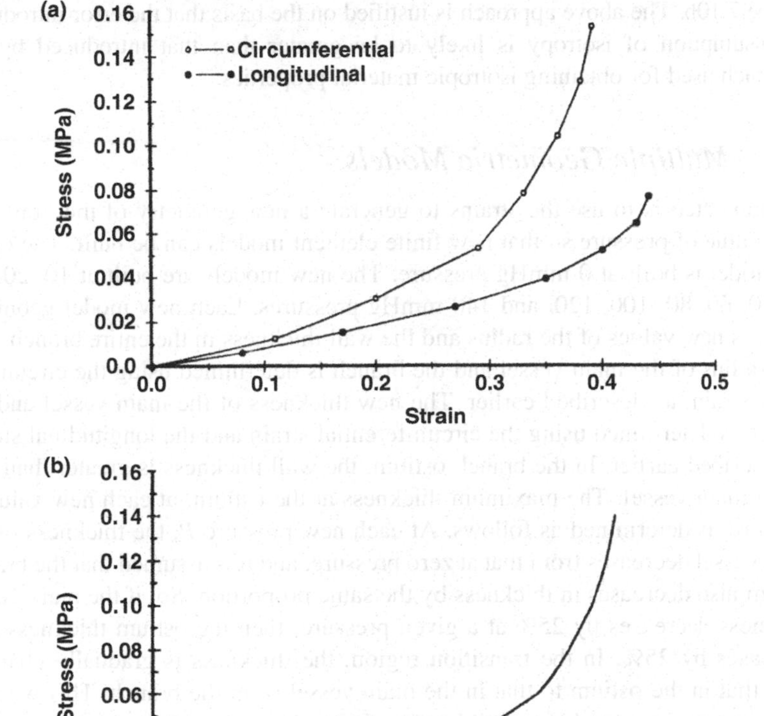

FIGURE 7.10. (a) The stress-strain curves for one artery showing orthotropic properties. (b) The stress-strain curves for the same artery showing isotropic properties.

4.4. *Isotropic Nonlinear Material Properties*

For the analysis, the artery is assumed to be isotropic and nonlinear. Therefore, the first task is to determine isotropic nonlinear material properties from the orthotropic properties shown in Figure 7.10a. This is done as follows. First, the incremental values of orthotropic elastic moduli are obtained, considering the artery to behave linearly over a small pressure range. Then the two moduli are averaged to yield a single value, which represents incremental isotropic modulus. This procedure is repeated in small pressure steps to cover the entire pressure range. The orthotropic relationships for the artery under biaxial loading may be expressed by Eqs. (3)–(6) of Section 3 described earlier. This assumes that the pressure load on the artery is predominantly a biaxial load. These equations are used to determine ε_c, ε_L, ν_c, and ν_L for each pressure step. ε_c and ε_L are then averaged for each pressure step to obtain isotropic modulus, and this approach is continued until the complete experimental range is covered. The resulting isotropic nonlinear stress-strain curve is shown in Figure 7.10b. The above approach is justified on the basis that the error introduced by assumption of isotropy is likely to be greater than that introduced by the approach used for obtaining isotropic material properties.

4.5. *Multiple Geometric Models*

The next step is to use the strains to generate a new geometry of the branch at each value of pressure so that new finite element models can be built. The original model is built at 0 mmHg pressure. The new models are built at 10, 20, 30, 40, 50, 60, 80, 100, 120, and 140 mmHg pressures. Each new model geometry requires new values of the radius and the wall thickness in the entire branch. The new radius of the main vessel and the branch is determined using the circumferential strain, as described earlier. The new thickness of the main vessel and the branch is determined using the circumferential strain and the longitudinal strain, as described earlier. In the branch ostium, the wall thickness is greater than that in the main vessel. The maximum thickness at the ostium, at each new value of pressure, is determined as follows. At each new pressure *P*, the thickness of the main vessel decreases from that at zero pressure, and it is assumed that the branch ostium also decreases in thickness by the same proportion. So, if the main vessel thickness decreases by 25% at a given pressure, then the ostium thickness also decreases by 25%. In the transition region, the thickness is gradually changed from that in the ostium to that in the main vessel or in the branch. This way, all of the parameters needed to build a new finite element model at each new pressure are available.

4.6. *Changing the Modulus of Elasticity*

For each new model at each value of pressure, the complete geometry is known and we need to determine the modulus of elasticity. The modulus of elasticity needs to be considered carefully because upon loading, different regions of the branch develop different stresses and consequently could have a different modulus. In general, the branch region could have a greater modulus than the main vessel. These considerations are implemented as follows.

For the straight region, the E value at each pressure step is determined as follows. The circumferential stress at various pressures, such as 10, 20, 30, 40, 50, 60, 80, 100, 120, and 140 mmHg is plotted on the stress strain curve (Fig. 7.10b) and the incremental modulus is determined for each pressure step. For instance, if the starting geometry of the model is at 40 mmHg and the model is loaded with 10 mmHg, then the E value is that given by the two stress values, one at 40 mmHg and the other at 50 mmHg. In this manner, the modulus in the straight region is known for all of the models.

For the ostial region, first the finite element model geometry at zero pressure is loaded with the 10 mmHg pressure. The same initial value of E is used for the entire geometry. As the pressure increases, the artery develops more stress at the branch. At higher stress, the branch region is stiffer. The branch stress is plotted on the experimental curve (Fig. 7.10b). This is the initial value of the stress for the next load step. Then the final value of the branch stress is approximated for that load step. Using these two values, the incremental modulus is then read from the curve. With some trial and error, it is possible to achieve this with good accuracy. This approach favors the behavior of the artery in the circumferential direction rather than in the longitudinal direction. The values of E for various models for the straight and the branch regions are listed in Table 7.5. Once the modulus at the branch ostium is obtained, then the modulus through the transition region is interpolated (similar to the thickness change) between the value at the ostium and the value in the main vessel. These procedures are quite tedious but they are followed in the current analysis. Thus, each of the models can be analyzed as incremental models where they maintain both the geometric continuity and the material property continuity in a reasonable fashion so that the stress and the strain values can be added to get absolute values at any pressure.

TABLE 7.5. Incremental elastic modulus E (Mpa).

P (mmHg)	$E_{\text{Non Branch}}$	$E_{\text{Distal Ostium}}$	$E_{\text{Prox Ostium}}$
0–10	0.113	0.113	0.113
10–20	0.113	0.113	0.113
20–30	0.113	0.195	0.185
30–40	0.160	0.320	0.195
40–50	0.195	0.850	0.286
50–60	0.222	1.600	0.436
60–80	0.260	2.000	0.836
80–100	0.339	2.000	2.000
100–120	1.000	2.000	2.000
120–140	1.600	2.000	2.000

Non branch = straight.

4.7. *Analytical Results*

A typical finite element mesh, used for the arterial branch at 0 mmHg pressure, is shown in Figure 7.5. The straight segments, the ostial region, and the transition zone are identified in the figure. The other models built at each pressure step had a similar geometry and mesh. The model is loaded with a pressure of 10 mmHg (or 20 mmHg), and the results are obtained in terms of various stresses and strains. In particular, the stresses on the middle surface and the inner surface of the artery are examined. The stress distribution in the branch at various internal pressures is shown in Figures 7.11 and 7.12. The stress on the middle surface increases toward the branch ostium, and it is larger in the distal portion than in the proximal portion of the ostium. On the inner surface also, the stress is highest at the distal region of the ostium, higher at the proximal region of the ostium, and lowest in the straight segment. Overall, as expected, the stress is greater in the ostial region than in the straight segment. Stress contour maps reveal the details of the stress gradient in the ostial region. The absolute values of stress in the branch region (maximum stress) and in the straight segment (nominal stress) for a pressure range of 0–140 mmHg are shown in Figure 7.13a and Table 7.6. This figure shows the key results of the current analysis and *the absolute values of the stresses are now known*. The stress increases more rapidly at the branch than in the straight segment as luminal pressure increases.

The stress-concentration factor (*K*) as a function of pressure is shown in Figure 7.13b and Table 7.6. For the middle surface, *K* reaches maximum value of 4.6 at 100 mmHg pressure. The stress and the stress-concentration factor (*K*) at the proximal region of the ostium is also of interest and is listed in Table 7.6.

To evaluate the accuracy of the analysis in predicting stresses and strains, it is necessary to compare experimental results with the analytical results. An excellent agreement was seen in the circumferential strains obtained analytically and experimentally for the straight segment of the artery. Longitudinal strains on the other hand showed poor agreement between the measured and analytical values. Tables 7.7 and 7.8 show the comparison of experimental and analytical results in the branch ostium. It is interesting to note that in the distal portion, the circumferential strains show a better match rather than the longitudinal strains, whereas in the proximal portion longitudinal strains show a good match.

Because the stresses on the inner surface of the branch are the highest, it is of interest to examine them (Fig. 7.12). The stress contours once again suggest the presence of stress-concentration and stress gradient in the ostial region. The stress-concentration factor (*K*) is larger on the inner surface than on the middle surface because the stress is larger on the inner surface. Overall, due to various assumptions and simplifications made in the analytical approach, the stress on the middle surface provides a better representation of the branch than the stress on the inner surface.

In conclusion, the analysis described above has provided the absolute values of the stresses in the arterial branch at various luminal pressures (Fig. 7.13 and Table 7.6).

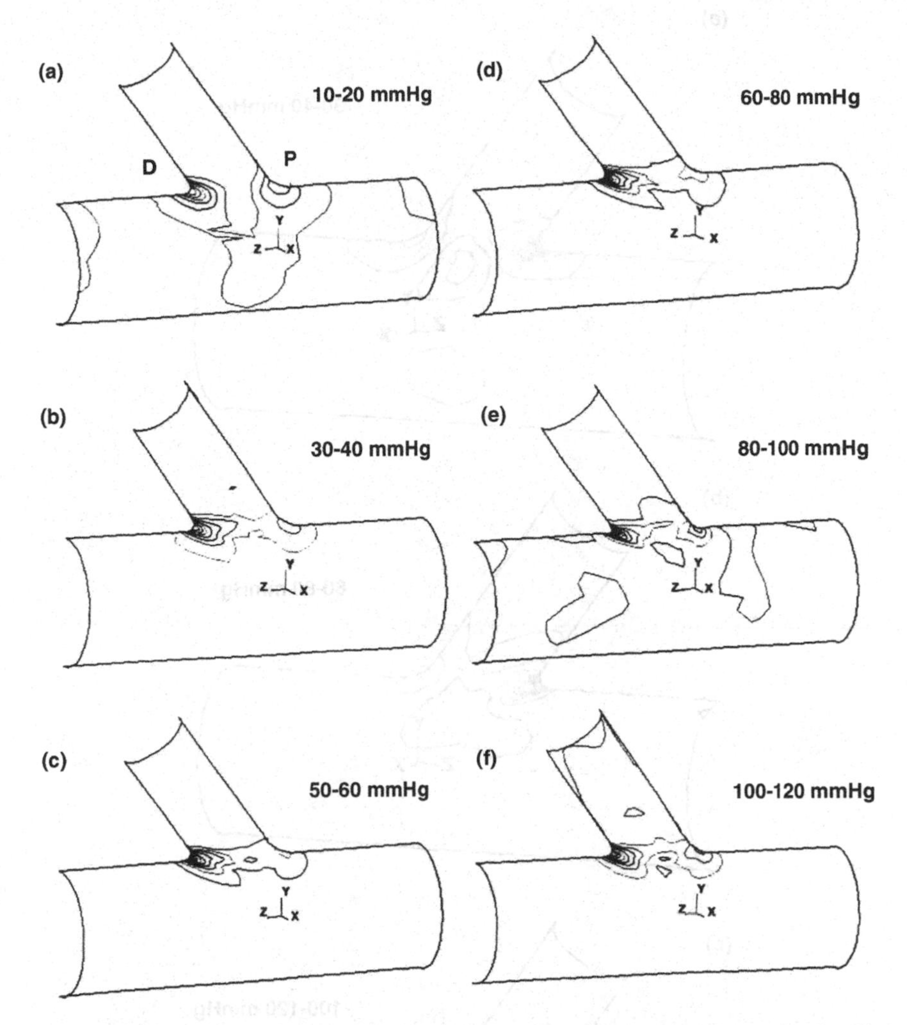

FIGURE 7.11. The first principal stress on the middle surface of the artery for the ilio-femoral branch. The incremental pressure values for which the stresses are determined are listed. The stress-concentration occurs at the ostium throughout the pressure range studied.

At the branch, the maximum stress can be as high as 37.2 N/cm^2 while the stress in the straight segment is only about 8.3 N/cm^2 at a pressure of 120 mmHg. The occurrence of very high stress at the branch ostium must make us stop and think that it is the most likely cause of the development of arterial pathology at that location.

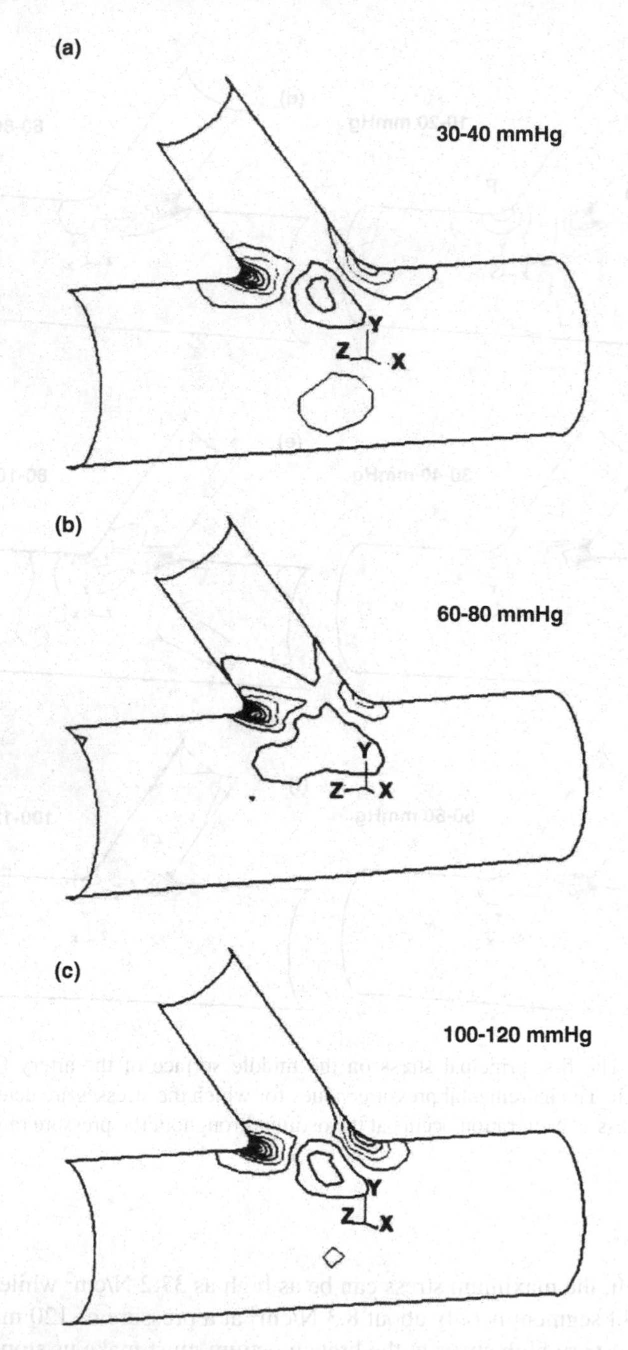

FIGURE 7.12. The first principal stress on the inner surface of the ilio-femoral branch. The incremental pressure values for which the stresses are obtained are listed. The distribution of stress contours indicates the presence of stress-concentration around the ostium.

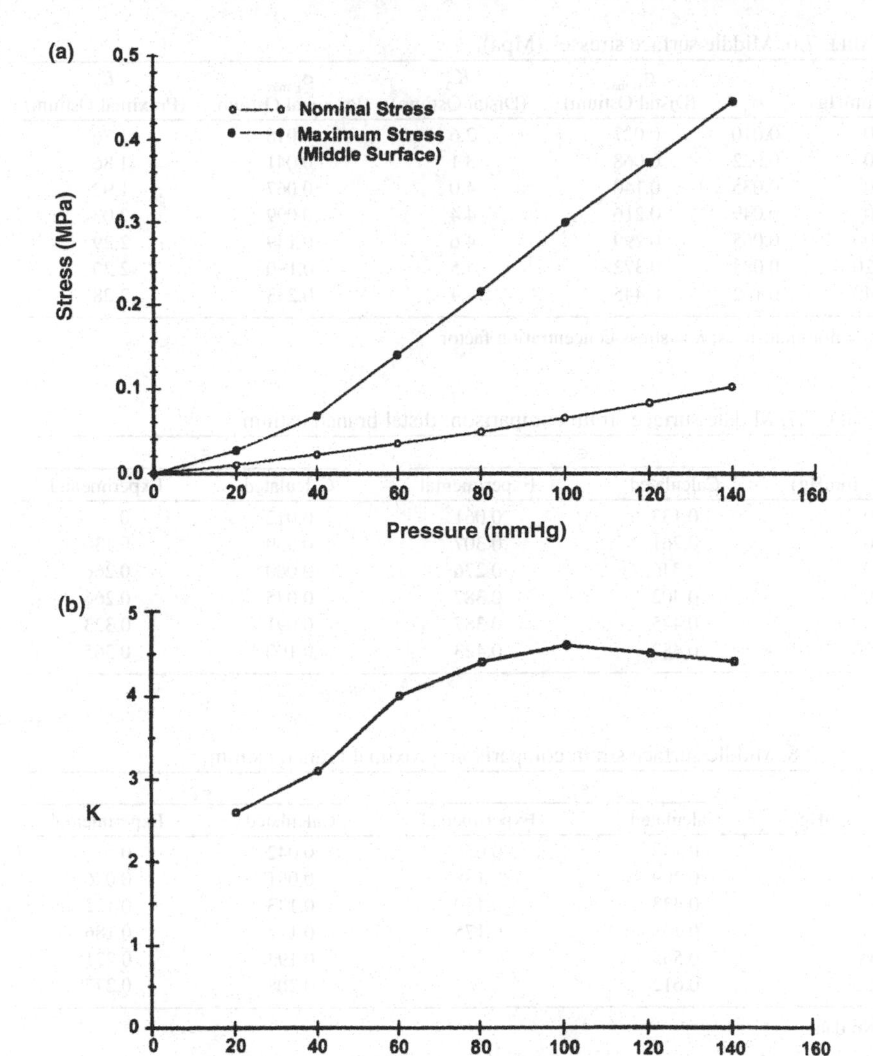

FIGURE 7.13. (a) Maximum and nominal principal stresses on the middle surface of the artery as a function of pressure for the ilio-femoral branch. (b) The stress-concentration factor (K) versus pressure for the ilio-femoral branch.

TABLE 7.6. Middle surface stresses (Mpa).

P (mmHg)	σ_n	$\sigma_{D\,max}$ (Distal Ostium)	K_D (Distal Ostium)	$\sigma_{P\,max}$ (Proximal Ostium)	K_P (Proximal Ostium)
20	0.010	0.027	2.6	0.018	1.76
40	0.022	0.068	3.1	0.041	1.86
60	0.035	0.140	4.0	0.067	1.92
80	0.049	0.216	4.4	0.099	2.03
100	0.065	0.299	4.6	0.149	2.29
120	0.083	0.372	4.5	0.190	2.30
140	0.102	0.445	4.4	0.233	2.28

σ_n = nominal stress; K = stress-concentration factor.

TABLE 7.7. Middle surface strain comparison: distal branch ostium.

	ϵ_C		ϵ_ℓ	
P (mmHg)	Calculated	Experimental	Calculated	Experimental
20	0.133	0.061	0.012	0
40	0.261	0.307	0.039	0.130
60	0.346	0.276	0.060	0.266
80	0.402	0.387	0.075	0.266
100	0.445	0.387	0.091	0.323
120	0.482	0.423	0.100	0.365

TABLE 7.8. Middle surface strain comparison: proximal branch ostium.

	ϵ_C		ϵ_ℓ	
P (mmHg)	Calculated	Experimental	Calculated	Experimental
20	0.148	0.050	0.042	0
40	0.299	0.138	0.091	0.076
60	0.423	0.150	0.143	0.122
80	0.498	0.175	0.177	0.186
100	0.555	*	0.193	0.221
120	0.612	*	0.208	0.277

*No data.

5. Stresses in the Human Aortic Arch: Analysis Using Nonlinear, Hyperelastic, and Isotropic Properties

5.1. Introduction

In this section, we will consider determination of stresses and strains in the human aortic arch and the great vessels originating from the arch. In the past section, to analyze the artery over a pressure range of 0 to 120 mmHg, a tedious approach was used. A new model had to be built at each pressure step. A more desirable approach would be to use a single model of the artery and study it over the entire pressure range. Such an approach is possible with more advanced analytical programs in which the artery is considered as a nonlinear hyperelastic material and very large dimensional changes are possible. We will consider this approach. As always, there are two parts to the study of the artery. The first part deals with the experimental data and the second part with the modeling and analysis. It is desirable to consider the study of the artery in these two parts because the experimental data can be taken at a face value, whereas analytical results are still based on assumptions and simplifications.

5.2. Strains in the Human Aortic Arch

Three samples of the intact human aortic arch with the great vessels—brachiocephalic artery, left common carotid artery, and left subclavian artery—were studied by Rao and Thubrikar (6). The arteries were obtained from three young adults (ages twenties and thirties) by CryoLife, Inc. (Marietta, GA, USA) and provided for the study. The aortic arch was cannulated at the level of the ascending aorta and at about half an inch distal to the left subclavian artery (Fig. 7.14).

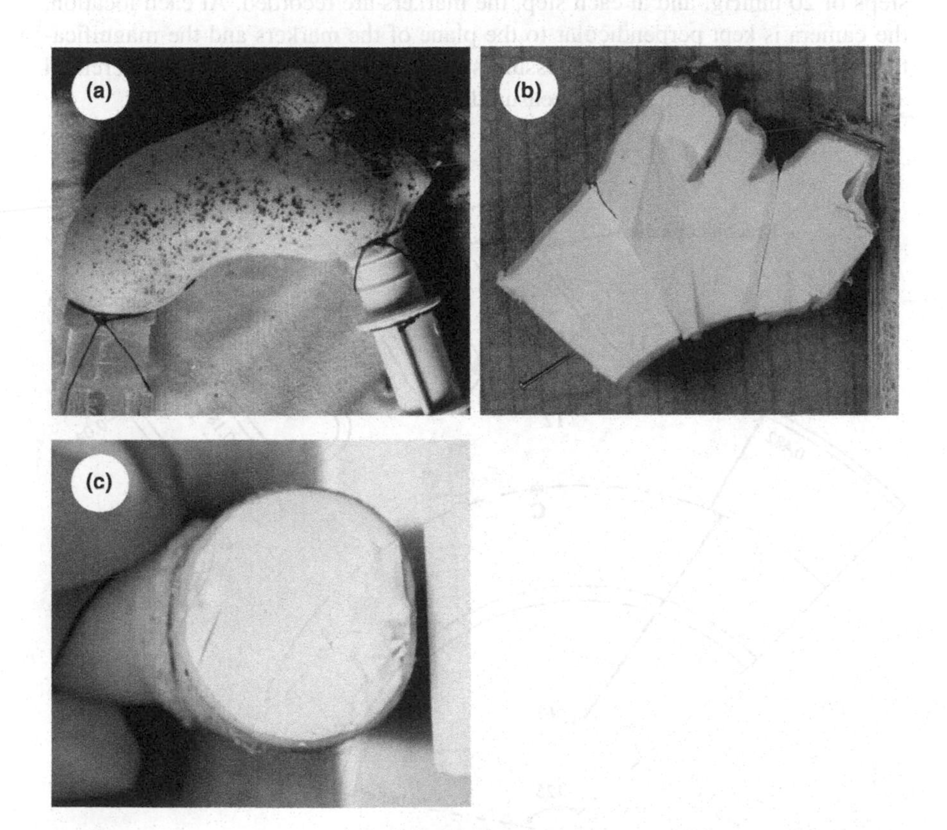

FIGURE 7.14. (a) The human aortic arch and the branches with potassium permanganate markings. (b) Longitudinal section of the aortic arch with silicone rubber on the inside. Various branch angles and the tissue thickness changes can be noted. (c) Cross section of the ascending aorta in the arch region. Variation in the aortic wall thickness can be noted.

The three great vessels were tied at the ends. The arch and several locations around the first branch were marked using potassium permanganate powder sprinkled on the tissue.

The goal is to determine, under pressure, the elongations in the circumferential and longitudinal directions at four locations: in the middle/anterior section of the arch, and in three places around the ostium of the brachiocephalic artery (Fig. 7.15). The setup for the experiment was described earlier in Chapter 6. The distal cannula is closed and the proximal cannula is connected to a saline reservoir. The pressure inside the arch is measured and varied by raising or lowering the reservoir. During the experiment, the proximal cannula is fixed while the distal end of the arch is allowed to elongate.

To measure strains, two markers are chosen in the circumferential direction and two in the longitudinal direction at each of the four locations (Fig. 7.15). The positions of the markers are recorded with a video camera at 23x magnification. During the experiment, the aortic arch is pressurized from 0 to 120 mmHg in steps of 20 mmHg, and at each step, the markers are recorded. At each location, the camera is kept perpendicular to the plane of the markers and the magnification is kept constant. At each pressure step, the distances in the circumferential and longitudinal directions are measured.

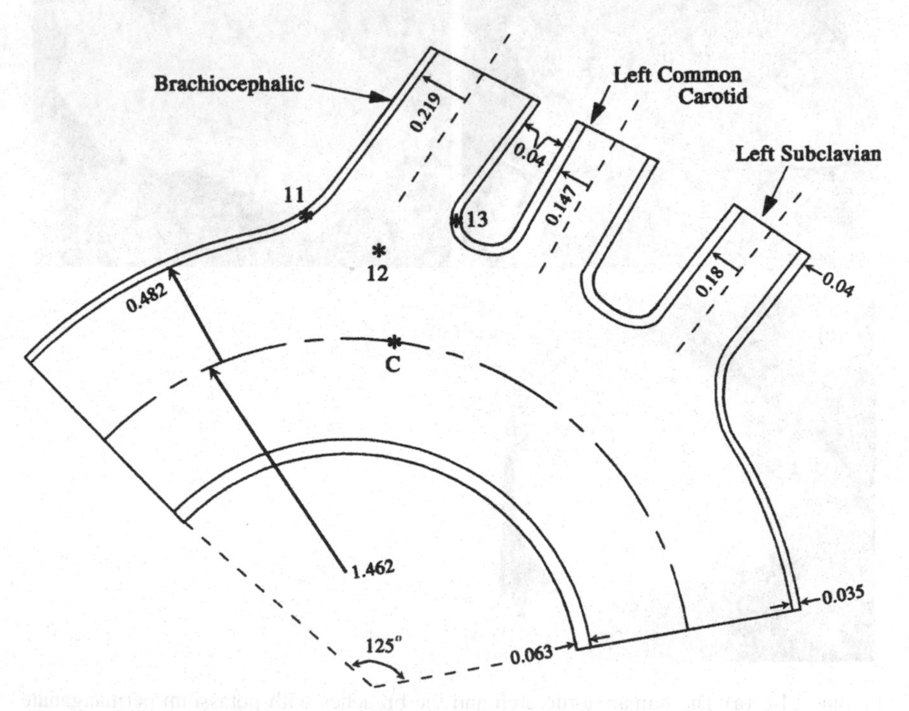

FIGURE 7.15. Details of the geometry of the aortic arch and the branches. Various radii and wall thicknesses (in inches) are reported. Locations 11, 12, and 13 are around the brachiocephalic branch ostium, and location C is on the anterior surface of the arch.

5.2.1. Geometry of the Aortic Arch

To determine the geometry, the aortic arch is filled with liquid silicone rubber compound and maintained at 40 mmHg pressure during the solidification process. The reason for choosing 40 mmHg instead of 0 mmHg for the arch geometry is that the arch often was oval (not circular) under its own weight at 0 mmHg pressure. The basic geometry is better represented by the silicone mold at 40 mmHg pressure. Photographs of the aorta and the mold are taken to determine various geometric parameters. Also, to determine other details, the mold and the aorta are cut along various planes and the geometric parameters measured. The details around the ostium of each of the branches as well as details pertaining to the curvature of the arch are important. The wall thickness of the branches, the arch, and around the ostium as well as branch angles are measured. Figures 7.14 and 7.15 illustrate various cutting planes used and the parameters measured. Of the three samples studied, sample 2 was the most representative and therefore chosen for the modeling and analysis. Its geometric details are reported in Figure 7.15.

5.2.2. Strains in the Aortic Arch and Around the Ostium

For all three human samples, engineering strains were determined from the distances between the markers at four locations shown in Figure 7.15. The engineering strain (ε) is a ratio of change in the length (ΔL) divided by the original length (L_o).

Thus,

$$\varepsilon = \frac{\Delta L}{L_0} = \frac{L - L_0}{L_0}$$

where L is the length at a given pressure and L_0 is the original length at zero pressure. The length is the distance between the two chosen markers in the circumferential or longitudinal direction.

The strains in the middle of the arch on the anterior surface (location C) in the circumferential and longitudinal directions for all three samples are shown in Figure 7.16. The strains increase with pressure indicating that the arch is expanding in both the diameter and the length. Also, at any given pressure, strains in the circumferential direction are greater than those in the longitudinal direction. Overall, the pressure-strain curves indicate that the arch is more compliant in the circumferential than longitudinal direction, particularly up to a pressure of 80 mmHg. These data could be converted into the stress-strain data to understand whether the arch behaves as an isotropic or an orthotropic material. The data in Figure 7.16 are very important because they represent fundamental properties of the aortic arch from young humans and are made available for the first time. The figure also shows that the variation in the samples, from three different humans, is small and thereby suggests that the mechanical properties of the aortic arch in young humans are similar. At 120 mmHg pressure, the circumferential strains are 0.57–0.63 while the longitudinal strains are 0.38–0.44.

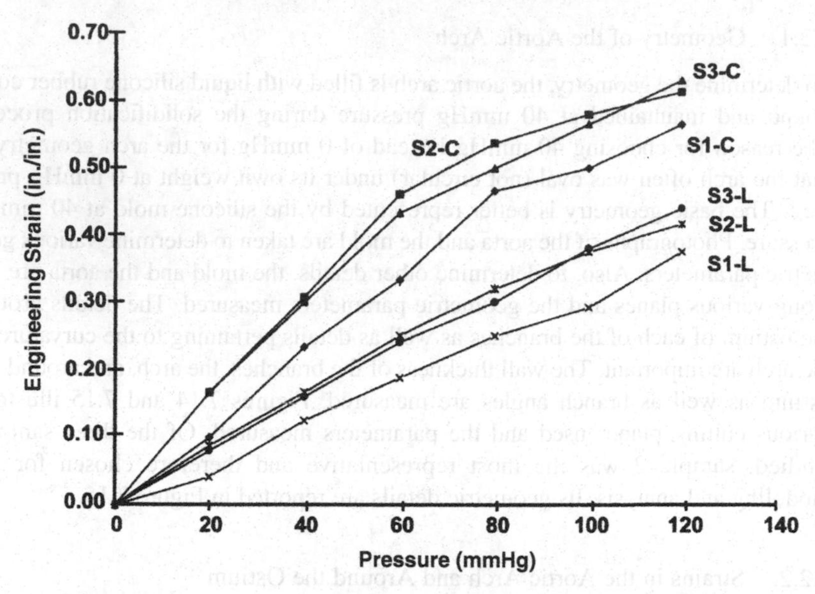

FIGURE 7.16. Pressure versus strain for location C on the aortic arch. S1, S2, and S3 denote samples 1, 2, and 3, respectively. C, circumferential direction; L, longitudinal direction.

The strains around the ostium of the brachiocephalic artery at three locations were also measured. However, the experimental measurements at these locations are less accurate because these locations undergo large and complex changes in their curvature, and because it is difficult to ascertain that the markers lie in a plane and the video camera is perpendicular to this plane. The authors indicate that the measured strains are likely to be an underestimation. Figure 7.17 shows the plots of strains at these locations for sample 2. Overall, both the circumferential and the longitudinal strains at these locations show similar trends as those in the arch (location C). Also, most often, location 11 shows larger strains than locations 12 and 13. Furthermore, all three human samples showed strains around the ostium to be similar, indicating a significant consistency in the mechanical behavior of the ostium in young people.

5.3. *Determination of Stresses in the Human Aortic Arch*

The human aortic arch has been modeled as a nonlinear, hyperelastic, isotropic material for determination of stresses by Rao and Thubrikar (6). Typically, it is necessary to know the material properties of the arch, the geometry of the arch, pressure loads, and boundary conditions. The results are obtained in terms of various stresses and strains and model strains are compared with the experimentally measured strains to determine the accuracy of the model.

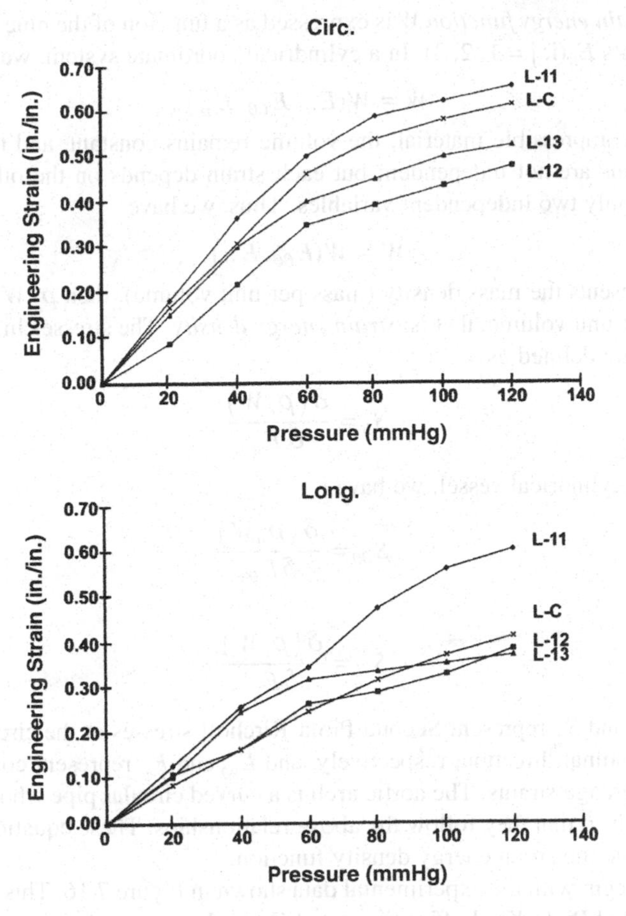

FIGURE 7.17. The strains as a function of pressure at various locations. L-11, L-12, L-13, and L-C denote locations 11, 12, 13, and C on the aortic arch (see Fig. 7.15). Circ., circumferential direction; Long., longitudinal direction.

5.3.1. Determination of Nonlinear, Hyperelastic, Isotropic Material Properties of the Human Aortic Arch

Hyperelasticity refers to materials that can experience a "large" but finite elastic deformation that is completely recoverable. Rubber and many other polymer materials fall in this category, and aortic tissue will be considered in this category. The stresses for these materials are usually derived from *strain energy density* functions. The stress analysis program ANSYS, used by Rao et al. (6), provides the Mooney-Rivlin option, which is a material law suitable for incompressible materials, and the aorta is assumed to be incompressible. Below is the description of the approach.

The *strain energy function* W is expressed as a function of the nine Green strain components E_{ij}(i, j = 1, 2, 3). In a cylindrical coordinate system, we have

$$W = W(E_{rr}, E_{\theta\theta}, E_{zz}).$$

For an incompressible material, the volume remains constant, and therefore the three strains are not independent but each strain depends on the other two. So, there are only two independent variables. Thus, we have

$$W = W(E_{\theta\theta}, E_{zz}).$$

If ρ_o represents the mass density (mass per unit volume), then $\rho_o W$ is the strain energy per unit volume, that is, *strain energy density*. The stresses in the sense of Kirchoff are defined as

$$S_{ij} = \frac{\delta(\rho_o W)}{\delta E_{ij}}.$$

For a thin cylindrical vessel, we have

$$S_{\theta\theta} = \frac{\delta(\rho_o W)}{\delta E_{\theta\theta}} \tag{7}$$

and

$$S_{zz} = \frac{\delta(\rho_o W)}{\delta E_{zz}} \tag{8}$$

where $S_{\theta\theta}$ and S_{zz} represent Second Piola-Kirchoff stresses in the circumferential and longitudinal direction, respectively, and $E_{\theta\theta}$ and E_{zz} represent corresponding Green-Lagrange strains. The aortic arch is a curved circular pipe whose walls are thin enough so that they follow the above relationships. These equations are used to determine the strain energy density function.

Let us begin with the experimental data shown in Figure 7.16. This data is converted to 2nd Piola-Kirchoff stresses and Green-Lagrange strains as follows and then plotted in Figure 7.18. If R_o, z_o are the outer radius of the aorta and longitudinal length, respectively, before pressure loading, and r_o, z are lengths after pressure loading, then the stretch ratios λ_θ, λ_z in the circumferential and longitudinal directions are given as

$$\lambda_\theta = \frac{r_o}{R_o} \text{ and } \lambda_z = \frac{z}{z_o}.$$

The *Green strains* in the circumferential and longitudinal directions are defined as

$$E_{\theta\theta} = \frac{1}{2}\left(\lambda_\theta^2 - 1\right) \text{ and } E_{zz} = \frac{1}{2}\left(\lambda_z^2 - 1\right).$$

Engineering strains (ε) in Figure 7.16 are converted first into stretch ratios (λ) and then into Green strains as follows.

$$\varepsilon = \frac{\Delta L}{L_o} = \frac{L - L_o}{L_o} = \frac{L}{L_o} - 1$$
$$= \lambda - 1$$
$$\text{or } \lambda = 1 + \varepsilon.$$

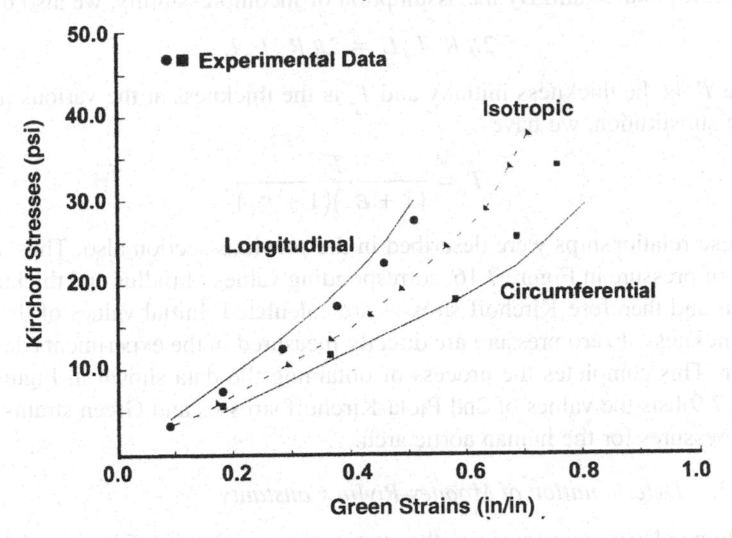

FIGURE 7.18. Material properties of the human aorta. Isotropic properties were derived. (Kirchoff stresses = Second Piola-Kirchoff stresses; Green strains = Green-Lagrange strains.)

Thus for each value of engineering strain, the Green strain is known.

The average values of the circumferential and longitudinal stresses in a thin cylindrical vessel are given as

$$\sigma_{\theta\theta} = \frac{P_i r_i}{t}; \text{ and } \sigma_{zz} = \frac{P_i r_i}{2t} = \frac{\sigma_{\theta\theta}}{2}$$

where P_i is the internal pressure and t is the wall thickness. These stress equations apply to location C in the arch as explained in Chapters 4 and 5.

Second Piola Kirchoff stresses $S_{\theta\theta}$ and S_{zz} in the circumferential and longitudinal directions are defined as:

$$S_{\theta\theta} = \frac{\sigma_{\theta\theta}}{\lambda_\theta^2} \text{ and } S_{zz} = \frac{\sigma_{zz}}{\lambda_z^2}.$$

For each value of pressure in Figure 7.16, first $\sigma_{\theta\theta}$ and σ_{zz} are calculated and then Kirchoff stresses are determined. To calculate $\sigma_{\theta\theta}$ and σ_{zz}, it is necessary to know the radius (r_i) and wall thickness (t) at each value of pressure.

The radius increases with circumferential strain, and we have

$$R_p = R_o(1+\varepsilon_c)$$

where R_o is the initial radius, R_p is the radius at pressure P, and ε_c is the engineering strain. Similarly, the longitudinal length changes with the longitudinal strain and is given by the relation

$$L_p = L_o(1 + \varepsilon_\ell)$$

where L_o is the initial longitudinal length, L_p is the length at pressure P, and ε_ℓ is the longitudinal strain. By the assumption of incompressibility, we also have

$$2\pi R_o T_o L_o = 2\pi R_p T_p L_p$$

where T_o is the thickness initially and T_p is the thickness at the various pressure P. On substitution, we have

$$T_p = \frac{T_o}{(1+\varepsilon_c)(1+\varepsilon_\ell)}.$$

These relationships were described in the previous section also. Thus, at each value of pressure in Figure 7.16, corresponding values of radius and thickness are known and therefore Kirchoff stresses are calculated. Initial values of the radius and thickness at zero pressure are directly measured in the experiments described earlier. This completes the process of obtaining the data shown in Figure 7.18. Table 7.9 lists the values of 2nd Piola-Kirchoff stresses and Green strains at various pressures for the human aortic arch.

5.3.1.1. *Determination of Mooney-Rivlin Constants*

The hyperelastic constants in the strain energy density function determine mechanical response of the material. For a successful hyperelastic analysis, it is necessary to assess the Mooney-Rivlin constants using the experimental stress-strain data. For the description of the hyperelasticity, the following form of Mooney-Rivlin strain energy function (the five parameter option) was used (6).

$$W = A(I_1-3) + B(I_2-3) + C(I_1-3)^2 + D(I_1-3)(I_2-3) + E(I_2-3)^2. \tag{9}$$

The analysis is concerned only with the two directions 1 and 2, or circumferential and longitudinal.

$$I_1 = \lambda_1^2 + \lambda_2^2 + \lambda_3^2$$

$$I_2 = \lambda_1^2 \cdot \lambda_2^2 + \lambda_2^2 \cdot \lambda_3^2 + \lambda_1^2 \cdot \lambda_3^2$$

$$I_3 = \lambda_1 \cdot \lambda_2 \cdot \lambda_3 = 1$$

where λ_1, λ_2, and λ_3 are stretch ratios in the three coordinate directions.

TABLE 7.9. Calculated Second Piola-Kirchoff stresses and Green strains on the outer surface (location C) of sample 2.

P (mmHg)	$S_{\theta\theta}$ (psi)	$E_{\theta\theta}$ (in./in.)	S_{zz} (psi)	E_{zz} (in./in.)
20	5.84	0.18	3.36	0.09
40	11.98	0.36	7.53	0.18
60	18.46	0.57	12.62	0.28
80	26.08	0.68	17.70	0.37
100	34.39	0.75	22.76	0.45
120	42.90	0.80	27.89	0.50

By substitution, we have:

$$I_1 = \lambda_1^2 + \lambda_2^2 + \left(\frac{1}{\lambda_1 \lambda_2}\right)^2$$

and

$$I_2 = \lambda_1^2 \cdot \lambda_2^2 + \left(\frac{1}{\lambda_1}\right)^2 + \left(\frac{1}{\lambda_2}\right)^2$$

noting that

$$E_{\theta\theta} = \frac{1}{2}(\lambda_\theta^2 - 1)$$

or,

$$2E_{\theta\theta} + 1 = \lambda_\theta^2 \quad \text{and}$$

$$2E_{zz} + 1 = \lambda_z^2.$$

By substitution, we have:

$$I_1 = 2E_{\theta\theta} + 1 + 2E_{zz} + 1 + \left[\frac{1}{(2E_{\theta\theta}+1)(2E_{zz}+1)}\right]$$

or,

$$I_1 = 2E_{\theta\theta} + 2E_{zz} + 2 + \frac{1}{(2E_{\theta\theta}+1)(2E_{zz}+1)} \tag{10}$$

and,

$$I_2 = (2E_{\theta\theta}+1)(2E_{zz}+1) + \frac{1}{(2E_{\theta\theta}+1)} + \frac{1}{(2E_{zz}+1)}. \tag{11}$$

Noting the relationships from Eqs. (7) and (8), and using differentiation by parts, we have:

$$S_{\theta\theta} = \frac{\delta(\rho_o W)}{\delta E_{\theta\theta}} = \left(\frac{\delta(\rho_o W)}{\delta I_1}\right)\left(\frac{\delta I_1}{\delta E_{\theta\theta}}\right) + \left(\frac{\delta(\rho_o W)}{\delta I_2}\right)\left(\frac{\delta I_2}{\delta E_{\theta\theta}}\right) \tag{12}$$

and

$$S_{zz} = \frac{\delta(\rho_o W)}{\delta E_{zz}} = \left(\frac{\delta(\rho_o W)}{\delta I_1}\right)\left(\frac{\delta I_1}{\delta E_{zz}}\right) + \left(\frac{\delta(\rho_o W)}{\delta I_2}\right)\left(\frac{\delta I_2}{\delta E_{zz}}\right). \tag{13}$$

From Eq. (9), $\frac{\delta W}{\delta I_1}$ and $\frac{\delta W}{\delta I_2}$ can be determined; from Eq. (10), $\frac{\delta I_1}{\delta E_{\theta\theta}}$ and $\frac{\delta I_1}{\delta E_{zz}}$ can be determined; from Eq. (11), $\frac{\delta I_2}{\delta E_{\theta\theta}}$ and $\frac{\delta I_2}{\delta E_{zz}}$ can be determined, and then Eqs. (12) and (13) can be written in terms of all known quantities except constants A, B, C, D, and E. For the circumferential direction, $S_{\theta\theta}$ and $E_{\theta\theta}$, are known for pressures 20, 40, 60, 80, 100, and 120 mmHg (Table 7.9). Thus, at each pressure, one equation is obtained in terms of all known quantities except constants. Six such equations are obtained at various pressures, and they can be used to determine five unknown constants. Using a SAS software package, the best fit (a least

TABLE 7.10. Mooney-Rivlin constants ($B = 0$).

Mooney-Rivlin constants	Circum. set	Long. set	Weighted
A	3.8650	2.060	2.8398
C	35.001	20.4061	24.1896
D	–42.4779	–20.0688	–24.3642
E	14.3526	5.9571	7.2276

squares fit) values are obtained for the constants. These values are reported in Table 7.10 and the curve corresponding to them is shown in Figure 7.18. For the longitudinal direction, a similar process is carried out. The constants obtained for this direction are also listed in Table 7.10, and the corresponding curve is shown in Figure 7.18.

The two sets of constants are then averaged to compute a single set, which describes the aortic arch as an isotropic material. The weighted averages of Mooney-Rivlin constants are listed in Table 7.10 and the curve corresponding to isotropic elastic properties is shown in Figure 7.18. These constants (the weighted set) are the final descriptors of the isotropic, nonlinear, hyperelastic material properties of the aortic arch. These constants are used in the model for determination of the stresses.

5.3.2. Solid Model and Mesh Generation of the Aortic Arch

A finite element model of one-half of the aortic arch was constructed using ANSYS with dimensions shown in Figure 7.15, taking into account the symmetry. It should be noted that the geometry used is at a pressure of 40 mmHg obtained from the silicone mold of the aortic arch, as described in the previous section. This geometry was assumed to be the zero pressure zero stress geometry for the finite element model because a reliable geometry could not be obtained at zero pressure. The primary errors introduced by this geometry would be an underestimation of the wall thickness and overestimation of the radius and, consequently, overestimation of the stresses. The authors decided to use the geometry as is, because there is no dependable way to correct for the thickness in various regions of the arch because strains vary in different regions. This becomes one of the acceptable assumptions just as the other assumptions (e.g., isotropic properties) involved in the analysis. The model was built as a parametric model with the parameters being the radii of the arch and the three branches and the wall thickness in various regions. This allowed exploration of the effect of various parameters on the results, although this will not be described here.

The element HYPER 58 is suitable for 3D modeling of solid hyperelastic structures. It is applicable to incompressible rubber-like materials with large displacements and strains. The element is defined by eight nodes having three degrees of freedom at each node. The tetrahedral shape option is available by collapsing the nodes to a total of four. The element input data includes the isotropic material properties and the constants defining the Mooney-Rivlin strain energy function. Pressures may be input as surface loads on the element faces. The hyperelastic formulation is nonlinear and requires an iterative solution.

The HYPER 58 element was used and the model was meshed using tetrahedral elements. The smallest element size of 0.125 in. was chosen for generating the mesh near the branches and the element size of 0.2 in. was used elsewhere. The mesh was generated automatically by ANSYS resulting in 2234 elements and 837 nodes. An element plot of the model is shown in Figure 7.19. For the hyperelastic nonlinear material properties, Mooney-Rivlin function constants (Table 7.10) were used. The analysis was done on a SunSparc 10 workstation and on the CRAY supercomputer.

5.3.3. Loading and Boundary Conditions

The ends of the arch as well as the three branches were capped to take into consideration the longitudinal stresses on the wall of the aorta. The inlet end of the arch was constrained completely and the distal end restrained radially. Symmetric boundary conditions were applied, and the model was prevented from rigid body motion by constraining all nodes in the XY plane against moving in the Z-direction. The inner surface of the model was pressurized to impose the loading. The model was loaded to a pressure of 120 mmHg in a single load step.

For this nonlinear analysis, the pressure was loaded as a time variable, and after the solution was obtained, solutions at any other pressure were obtained by setting the ANSYS postprocessor pointer to the corresponding time. The solution

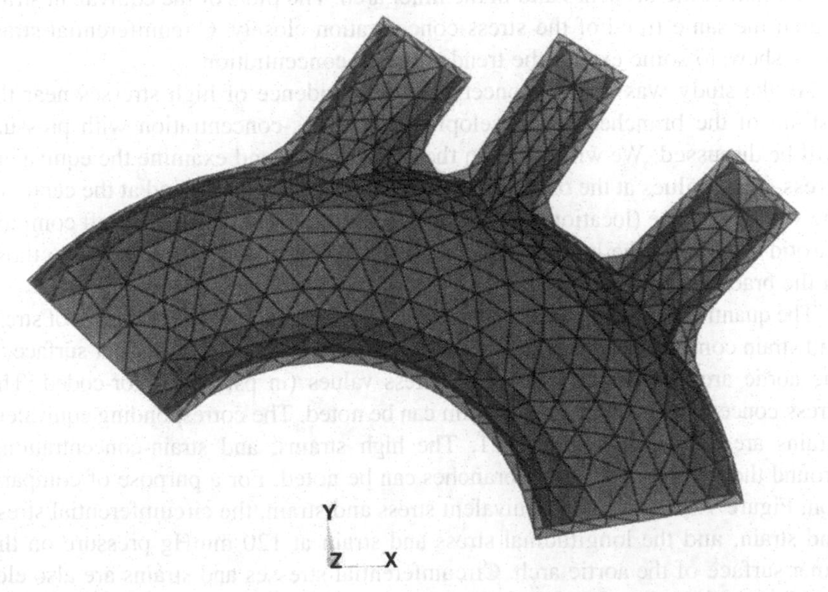

FIGURE 7.19. Element plot of the finite element model of the aortic arch and the branches.

was obtained using the full Newton-Raphson method available in ANSYS 5.0 for solving hyperelastic nonlinear problems. Solutions were obtained for the whole range of 120 mmHg, in steps of 20 mmHg.

5.4. *Results: Finite Element Model Solution*

A host of information is available from the solution, but only the basic information was retrieved. Results for stress and strain in the circumferential, longitudinal, and equivalent sense are obtained for the inner and outer surfaces at 20, 40, 60, 80, 100, 120 mmHg as well as the deformation at these pressures. The stresses are obtained as true (Cauchy) stresses, and the strains are obtained as logarithmic (Hencky) strains. Logarithmic strains (ε_{Ln}) can be converted into engineering strains (ε_{eng}) using the following relationships:

$$\varepsilon_{\ln} = \ln\left(\frac{L}{L_o}\right)$$

or,

$$\varepsilon_{\ln} = \ln\left(1 + \varepsilon_{eng}\right).$$

The contours for the stresses and strains in the circumferential, longitudinal, and equivalent sense are examined carefully for both the inner and the outer surfaces at all pressures. Results at the ostium of the three branches, center of the ventral surface, and the inner arch are of particular interest.

It is observed that all the entities of interest such as stress, strain, and deformation increase rapidly following a nonlinear curve as the pressure increases. The plots of the equivalent stresses show the best incidence of stress-concentrations at the ostium of the branches and in the inner arch. The plots of the equivalent strain match the same trend of the stress-concentration closely. Circumferential strain plots show, to some extent, the trend of stress-concentration.

As the study was mainly concerned with incidence of high stresses near the ostium of the branches, the development of stress-concentration with pressure will be discussed. We will focus on the inner surface and examine the equivalent stress-strain values at the ostium of the brachiocephalic artery and at the center of the ventral surface (location C). The results around the ostium of the left common carotid artery and the left subclavian artery were similar in magnitudes to those in the brachiocephalic artery.

The quantitative results are best understood by examining the color plots of stress and strain contours. Figure 7.20 shows the equivalent stress on the inner surface of the aortic arch at various pressures. Stress values (in psi) are color-coded. The stress-concentration around the ostium can be noted. The corresponding equivalent strains are shown in Figure 7.21. The high strains, and strain-concentrations, around the ostium of all three branches can be noted. For a purpose of comparison, Figure 7.22 shows the equivalent stress and strain, the circumferential stress and strain, and the longitudinal stress and strain at 120 mmHg pressure on the inner surface of the aortic arch. Circumferential stresses and strains are also elevated around the branch ostium. The circumferential stresses at location C for a

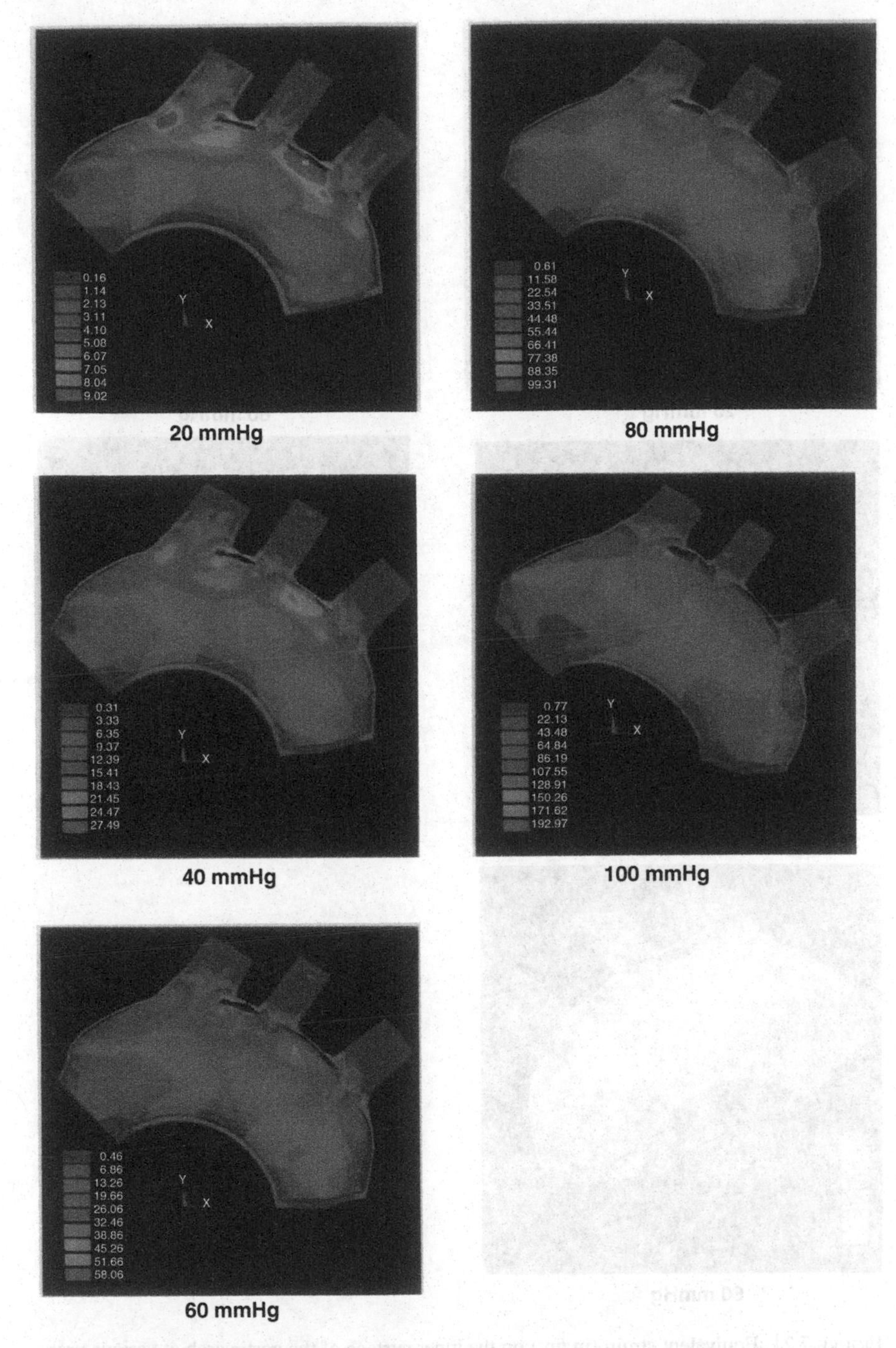

FIGURE 7.20. Equivalent stress (psi) on the inner surface of the aortic arch at various pressures. Stress-concentration is present around the branch ostium. Stress values are color-coded. (*Please see color version on CD-ROM.*)

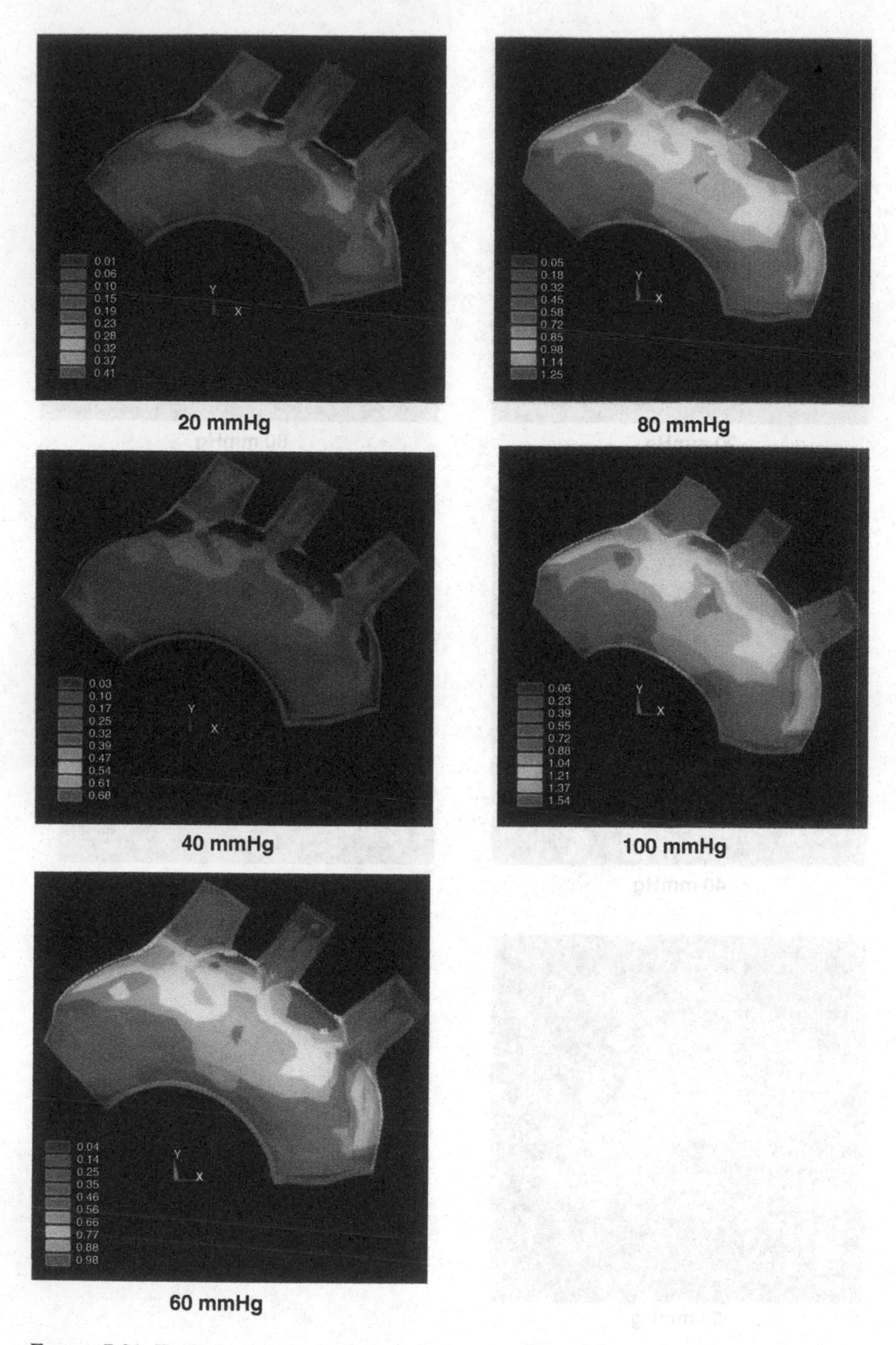

FIGURE 7.21. Equivalent strain (in./in.) on the inner surface of the aortic arch at various pressures. High strains can also be noted around the branch ostium. Strains are color-coded. (*Please see color version on CD-ROM.*)

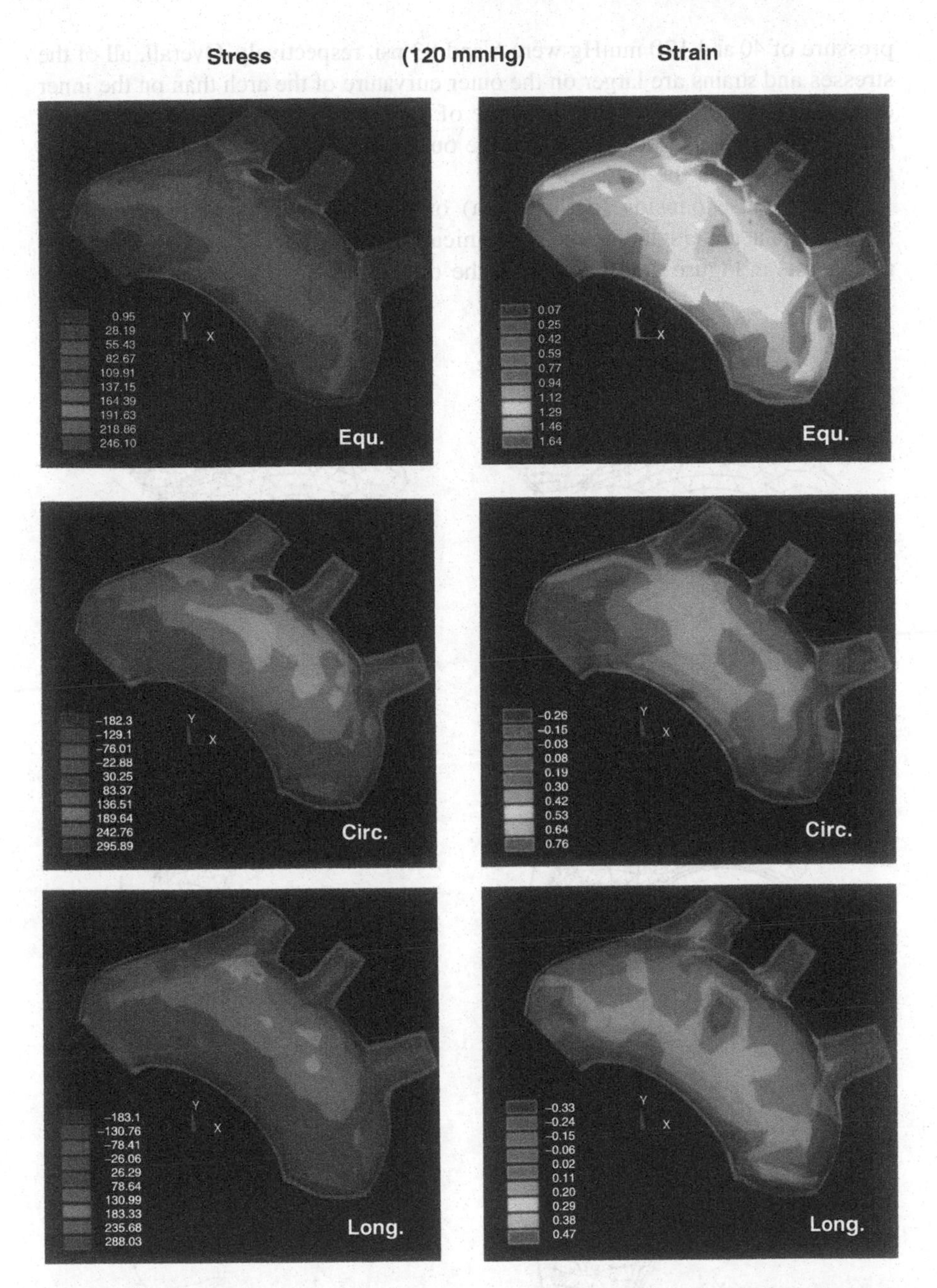

FIGURE 7.22. Stresses and strains on the inner surface of the aortic arch at 120 mmHg luminal pressure. Equ., equivalent; Circ., circumferential; and Long., longitudinal. Stresses are in psi and strains are in in./in. (*Please see color version on CD-ROM.*)

pressure of 40 and 120 mmHg were 6 and 83 psi, respectively. Overall, all of the stresses and strains are larger on the outer curvature of the arch than on the inner curvature primarily due to the presence of the branches. The equivalent stresses and strains were also determined on the outer surface of the arch and they were similar to those on the inner surface.

The overall expansion (deformation) of the arch with pressure is of great importance in understanding the mechanical behavior. This is shown as displacement plots in Figure 7.23 along with the original geometry, so that the relative

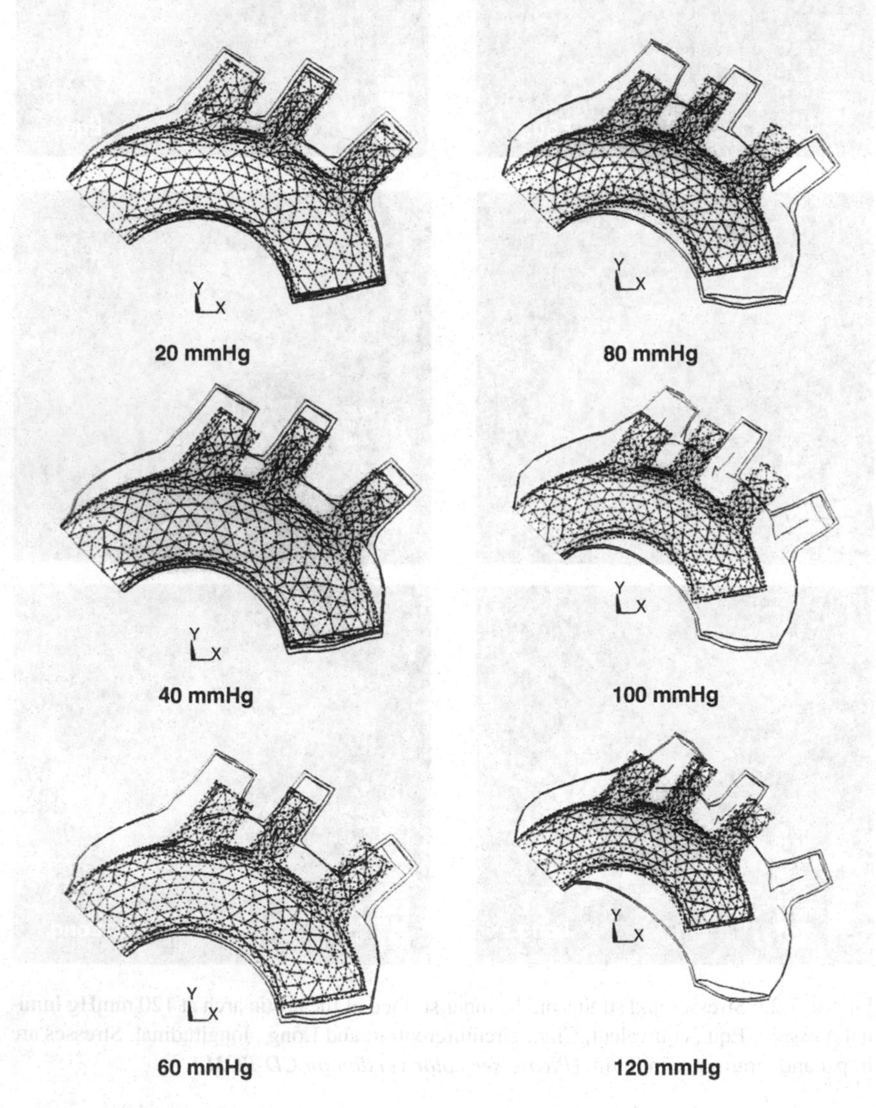

FIGURE 7.23. Deformation in the aortic arch at various pressures. Overall, there is a greater circumferential expansion from 0 to 60 mmHg pressure and greater longitudinal expansion from 60 to 120 mmHg pressure.

expansion can be visualized. There is a significant increase in both the diameter and the length of the arch as the pressure increases from 0 to 120 mmHg.

The quantitative values of the equivalent stresses and strains at locations C, 11, 12, and 13 (see Figure 7.15) are reported in Table 7.11. Location C is on the anterior surface of the arch and it is expected to behave as a part of a cylinder for stress considerations. Locations 11, 12, and 13 are around the ostium of the first branch (brachiocephalic artery) and are influenced by stress-concentration. As expected then, both the stresses and strains are lowest at location C, they are higher at location 12, they are yet higher at location 11, and they are the highest at location 13. In other words, the stresses and strains increase in the order of the location C, 12, 11, and 13. This is true at any given pressure. Also, all of the values increase with increasing pressure. The stresses and strains are higher at location 13 than at location 11 because the branch makes an acute angle at this location and consequently experiences a greater stress-concentration.

The stress values reported in Table 7.11 appear quite high for the aortic arch to sustain. It is of interest to examine the high stress values such as 246 psi and relate it to a possible scenario in vivo. For example, arteries are known to burst when the internal pressure is raised to 1200–1600 mmHg in the laboratory setting. Using pr/t as a first-order approximation for circumferential stress, and using a r/t ratio of 10, which is close to that in vivo at a pressure of 100 mmHg, the stress at 100 mmHg is $1.935 \times \frac{10}{1} = 19.35$ psi (100 mmHg = 1.935 psi). The bursting stress at 1200 mmHg pressure will be $12 \times 19.35 = 232$ psi. Thus, the stress of 246 psi (Table 7.11) cannot be sustained by the arch.

The reported high values of the stress must therefore be understood as significant overestimations and can be explained as follows. It was mentioned earlier that a lower value for the wall thickness was used in the starting geometry. Also, the radius of the aorta would be less at 0 mmHg than at 40 mmHg used in the model. The combined effect of the higher thickness and the lower radius, which would represent the true initial geometry, would be to reduce the values of the stresses by a factor of 2 to 4. In other words, all of the values reported in Table 7.11 may truly be less by a factor of 2 to 4, and in that case, the stresses could be sustained by the aortic arch without bursting. The initial geometry and the assumption of isotropy are the two main areas for future improvement in the stress analysis. Nevertheless, the importance of the study can be seen in establishing the relative differences in the stresses and strains at the four locations of interest as well as throughout the

TABLE 7.11. Equivalent stresses and strains at locations C, 11, 12, and 13.

	Equivalent stress (psi)				Equivalent strain (in./in.)			
P (mmHg)	Loc. C	Loc. 11	Loc. 12	Loc. 13	Loc. C	Loc. 11	Loc. 12	Loc. 13
20	3.11	7.05	5.08	9.02	0.18	0.36	0.32	0.40
40	6.35	21.44	15.41	27.49	0.31	0.61	0.53	0.68
60	19.66	38.86	32.46	58.05	0.45	0.87	0.76	0.97
80	33.51	66.41	44.48	99.31	0.58	1.11	0.98	1.24
100	43.48	107.55	80.00	192.97	0.71	1.37	1.04	1.53
120	82.60	137.10	109.90	246.10	0.76	1.46	1.28	1.63

aortic arch. The stress-concentration factors predicted by the study are in a similar range as those reported in other studies described earlier in Chapter 6. From Table 7.11, the stress-concentration factor at the proximal region of the ostium is 1.65 ($K = 137.10 / 82.6 = 1.65$) and at the distal region it is 3.0 ($K = 246.1/82.6 = 3.0$) at 120 mmHg pressure.

5.5. *Verification of the Finite Element Model: Strain Comparison*

The only verification procedure for a hyperelastic analysis is to correlate the strains obtained from the model with those obtained experimentally. Such a comparison at location C and experimental strains at locations around the brachiocephalic artery ostium are shown in Figure 7.24.

While comparing these strains, it may be recalled that the experimentally measured strains around the ostium were underestimates due to difficulty in measurements. For the model, it may be remembered that all of the analytical results were overestimates as explained earlier. From Figure 7.24, it may be seen that at location C and at other three locations where analytical strains are not

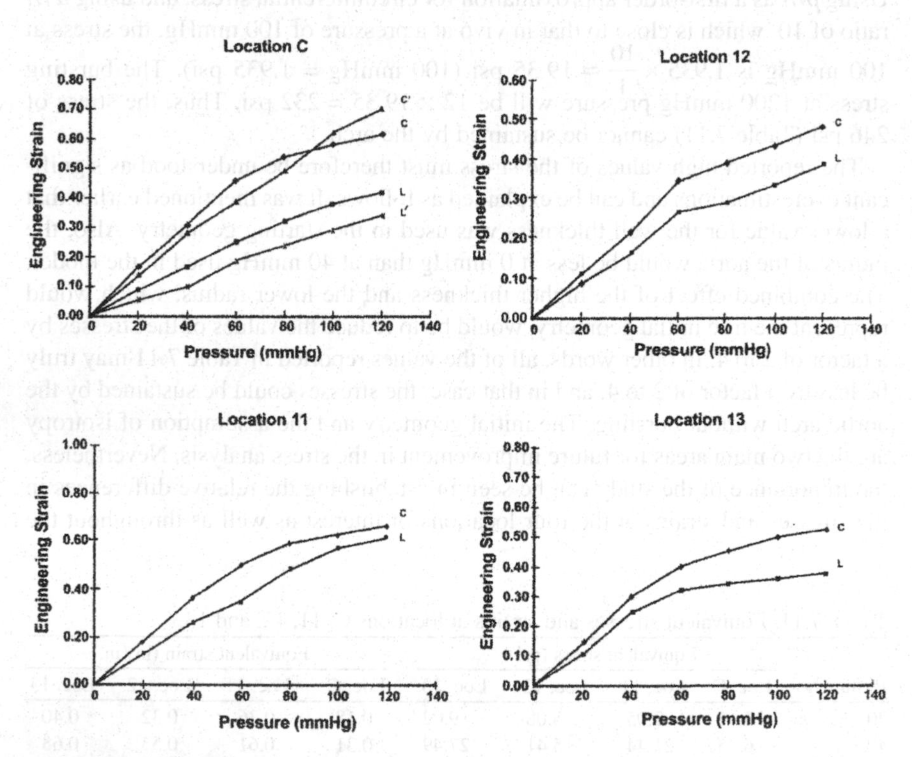

FIGURE 7.24. Experimentally measured strains at locations C, 11, 12, and 13 as a function of pressure in the aortic arch. C, circumferential; L, longitudinal; C′ and L′ represent finite element results.

shown, the circumferential strains show a reasonable match between the model and the experiment. The longitudinal strains, however, show a reasonable match at location 12. At the other three locations, the model values are lower than the experimental values. The reason for this could be that the isotropic material properties favor those in the circumferential direction more than those in the longitudinal direction. The model strains increase with pressure at all locations just as the experimental strains do.

5.6. *Conclusions and Comments*

This study is a beginning toward the determination of intramural stresses in the human aortic arch. Regions of stress-concentration are identified in the study. Even though the aortic arch is modeled from the data obtained from the experiments done in vitro, the results obtained are qualitatively true in vivo. The equivalent stress and strain plots show an indication of bunching of their contours near the ostium of the three branches. The intramural stresses show regions of concentration at both the proximal and distal parts of the ostium of the three branches. The measured strains at the branch areas show the same trend as the stresses.

The region of highest stress-concentration is observed near the distal portion of the ostium of the brachiocephalic artery. The stresses are around 3 times greater than those on the ventral surface. The stress near the ostium of the other branches are also about 2 to 3 times that at the ventral surface. The distal part of the subclavian artery shows significantly lower stresses than the other two arteries.

Aside from the assumptions about material and geometry in the model, some other distinctions are obvious between experimental conditions and in vivo conditions. The experiments were conducted in vitro with the ends of the aorta unconstrained. The aortic arch, however, is held in position by tethering forces in vivo. The arch was analyzed under static pressure loading, whereas in the body it is under dynamic pressure, which varies through the arch. The contractile properties and the viscoelastic properties of the aortic tissue are not included because the experiments are conducted in vitro. In spite of these differences, it is clear that the results give reasonable approximation of the magnitude of stresses present in vivo.

6. Circumferential Stress in the Artery: Comparing Different Studies

We have considered several studies in Chapters 4, 6, and 7 where we derive circumferential stress for the artery. For the reader, it is confusing to see that different studies report different values. Therefore, it is important to compare different studies in order to understand why the reported values are similar or different, as the case may be, and what might then be the true value of the circumferential stress.

Table 7.12 shows the values of circumferential stress and corresponding pressure load reported in the different studies. In Chapter 4, the stress values are

TABLE 7.12. Circumferential stress in the artery.

Studies	Comments	Pressure load (mmHg)	Circumferential stress (N/cm²)
Chapter 4	Any artery, equilibrium of forces	105	12.6
Chapter 6			
i) Bovine coronary artery	Incremental pressure – geometry at 80 mmHg	40	6.0
ii) Human carotid artery	Incremental pressure – geometry from cadaver specimen	40	2.0
iii) Human carotid artery with residual stress	Absolute value	120	7.5
Chapter 7			
i) Canine Ilio-femoral branch, in vivo	Incremental pressure – geometry at 80 mmHg	40	4.3–6.0
ii) Canine Ilio-femoral branch, in vitro	Absolute value	40 120	2.2 8.3
iii) Human aortic arch in vitro	Absolute value Stress overestimated due to *R/t* at 40 mmHg	40 120	4.1 57.2

derived on the basis of equilibrium of forces and the R/t ratio was taken to be constant at 9. Let us use this stress value as a reference.

In Chapter 6, for the bovine coronary artery the stress is derived for 40 mmHg pressure. The stress value appears only slightly higher than the reference value for the comparable pressure load. In the case of the carotid artery, the stress is lower for a pressure load of 40 mmHg than we may expect from the reference value. Because in this case the geometry is taken to be that present in a cadaver, the artery is likely to be collapsed, and the R/t ratio will be lower. In the case of the carotid artery with residual stress, the stress value seems lower than expected. The reason for this is that the authors used a different analytical approach and their R/t ratio will also be different than that used in the case of the reference value.

In Chapter 7, in the case of the ilio-femoral branch in vivo, the stress value reported is comparable with that in the bovine coronary artery. In the case of the ilio-femoral branch in vitro, the absolute values reported are similar to those in the case of the carotid artery with residual stress. For the aortic arch, the value for stress seems reasonable for a pressure of 40 mmHg but quite high for a pressure of 120 mmHg. The reasons for this are the high R/t ratio in the starting geometry; the reported stress is on the inner surface rather than on the middle surface of the artery; and the model uses isotropic properties.

In summary, a careful examination of the procedures followed in different studies allows us to understand why there may be differences in the reported stress values. Furthermore, all of the studies show the results that are qualitatively in agreement with each other, and all of the studies show the presence of stress-concentration at the branch ostium.

References

1. Macfarlane TWR, Canham PB, Roach MR: Shape changes at the apex of isolated human cerebral bifurcations with changes in transmural pressure. Stroke 1982;14:70-76.
2. Hassler O: Morphological studies on the large cerebral arteries, with reference to the aetiology of subarachnoid haemorrhage. Acta Psychiat Neuro Scand 1961;36(Suppl 154):141-145.
3. Campbell GJ, Roach MR: Fenestrations in the internal elastic lamina at bifurcations of human cerebral arteries. Stroke 1981;12:489-496.
4. Manuel L: A study of the stress concentration at the branch points of arteries in vivo by the finite element method. MS thesis, University of Virginia, 1986.
5. Sharretts JM: Intramural stress analysis at the branching regions of canine arteries. MS thesis, University of Virginia, 1990.
6. Rao S: Stress analysis of the human aortic arch with great vessels. MS thesis, The University of North Carolina at Charlotte, 1994.

8
Endothelial Cells and Low-Density Lipoproteins at the Branch

1. Introduction

We have seen in previous chapters how the stress in the artery is unevenly distributed and in particular how it is concentrated in the branch region. One of the fundamental principles of cell biology is that the cells, and therefore tissues, respond to physical and chemical forces or stimuli. It is therefore expected that the cells and the tissues in the arterial branch region will be different and have different properties. In the body, several different types of forces and stimuli are present simultaneously, and therefore it becomes difficult to correlate the cellular structure or the tissue properties to a particular force or stimulus. For instance, we have elaborated in previous chapters how the wall stress may be concentrated in the branch region. Another obvious force in the branch region occurs due to fluid shear, which is also unevenly distributed in the region. There are also forces due to tethering and those that may occur due to physical motion such as motion of the heart or lungs. Of course, we must also consider forces due to chemical stimuli, the obvious of which can be noted as vasoconstriction or vasodilation. In spite of multiplicity of forces/stimuli, it is important to examine the cells and tissue properties at the branch and compare them with those at the nonbranch region for two reasons: i) an insight can be gained into the possible causes of such a difference; and ii) to have a better understanding of arterial pathology that may be associated with the branch region.

We also expect different forces to have different effects on cells and tissues, and the influence may not be singular but multiple. In this chapter, we will focus on morphology of the endothelial cells and on permeability of the artery in the branch and nonbranch regions and examine these properties under various conditions such as hypertension, constriction of the artery, and mechanical stretch. We will also examine how these properties occur in the aortic valve and how they relate to cell replication. The chart below shows how the various parameters are known to be associated with each other. The associations shown are minimum, and many more associations are likely to exist. Using the chart to define the scope of the subject, we are now ready to examine some of the parameters.

The Chart

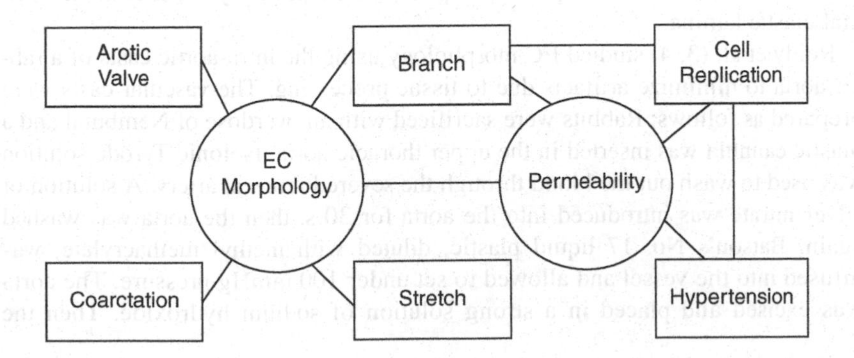

2. Endothelial Cell Morphology at the Branch and Nonbranch Regions

Endothelial cell (EC) morphology has been described in terms of cell shape, size, orientation, and cellular integrity. Endothelial cell morphology has been studied using multiple techniques and in different animals. When we consider the data from various groups, we find that all of them report that the ECs at the branch region are different from those at the nonbranch region. However, some groups report the differences only in shape and orientation, whereas others report differences also in the cellular integrity.

Garity et al. (1) studied the swine aortas that were perfusion fixed. The endothelial surface was studied using scanning electron microscopy (SEM) and the intima was studied using transmission electron microscopy (TEM). They reported that the arterial endothelium and intima of the ostial (branch) region was different from those of the nonbranch (straight) region (Fig. 8.1). In the nonbranch region, the ECs are elongated, flattened, and oriented in the direction of blood flow (Figs. 8.1a and 8.1b). In contrast, the ECs in the ostial region showed almost a cuboidal structure in both surface and sectional view (Figs. 8.1c and 8.1d). The ECs in the branch region are larger, more rounded, and have a larger surface area than those in the nonbranch region.

Zarins et al. (2) studied the EC morphology in perfusion-fixed rabbit aortas. They believed that it was important to maintain the curvature of the branch region while examining the endothelial surface by scanning electron microscopy. They reported that the ECs were well preserved at all positions about the ostia. The luminal surface was not ridged or folded, and nuclei and cell bodies were oriented in the direction of the flow. Groups of bulging cells were frequent only at or near cut edges of the specimen or where perfusion had been poor as evidenced by longitudinal endothelial ridges. According to them, when the ostial region had been removed from the aorta, opened by a single cut and flattened to some extent, then the evidence of EC disruption was seen in the ostial region. Such preparations showed the fusiform or spindle-shaped cells, and linear ridges were also noted. However, these authors also reported that they observed a small amplitude of buckling or wrinkling of the innermost layers under endothelial cells, and in these instances the cells become spindle-shaped but remain firmly attached to the intimal elastic lamina.

Reidy et al. (3, 4) studied EC morphology using the intra-aortic casts of a rabbit aorta to minimize artifacts due to tissue processing. The vascular casts were prepared as follows: Rabbits were sacrificed with an overdose of Nembutal and a plastic cannula was inserted in the upper thoracic aorta. Isotonic Tyrode solution was used to wash out the blood through the severed femoral artery. A solution of silver nitrate was introduced into the aorta for 30 s, then the aorta was washed again. Batson's No. 17 liquid plastic, diluted with methyl methacrylate, was infused into the vessel and allowed to set under 100 mmHg pressure. The aorta was excised and placed in a strong solution of sodium hydroxide. Then the

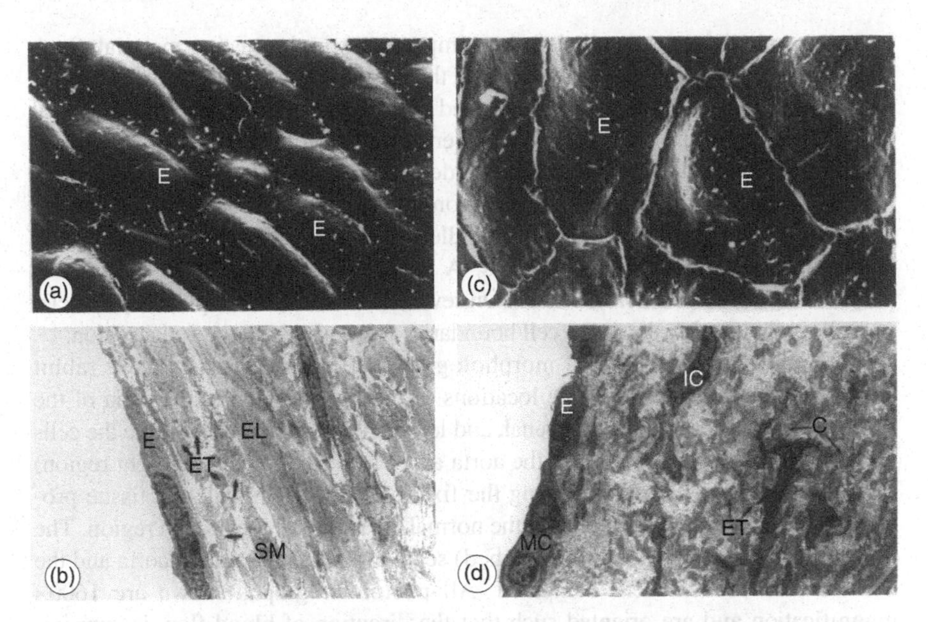

FIGURE 8.1. (a) Scanning electron micrograph of endothelial surface from a nonbranch area of normal swine aorta. Endothelial cells (E) are elongated and aligned in direction of blood flow. (b) Transmission electron micrograph of section through endothelium (E) and intima of nonbranch area. Intima is delineated by internal elastic lamina (EL) and contains fragments of elastic tissue (ET), collagen (arrows), and portions of smooth muscle cells (SM). Endothelial cells are flat and elongated. (c) Scanning electron micrograph of endothelial surface from a branch area of normal swine aorta. Endothelial cells (E) are large, cuboidal, and not aligned in direction of blood flow. (d) Transmission electron micrograph of section through endothelium (E) and greatly expanded intima of branch area. Internal elastic lamina is not visible. Intima contains elastic tissue fragments (ET), collagen (C), elongated undifferentiated intimal cells (IC), and occasional mononuclear cells (MC). Endothelial cells are cuboidal with simple end-to-end junctions. (Reproduced from Glagov S, Newman III WP, Schaffer SA (eds), Pathobiology of the Human Atherosclerotic Plaque. Springer-Verlag, New York, 1990:28, 29, with permission.)

aorta was digested, the cast was washed with distilled water, and sections of the cast were prepared for SEM.

In the vascular cast, the impression of the cell nuclei and cell boundaries were clearly seen. The normal endothelial cells were observed in the straight region of the aorta. The cells are oriented with their long axis parallel to the direction of flow. The entire surface of the cast is covered with such cells except around the aortic ostia. At the junction of an intercostal branch with the aorta, the cells were examined at the flow divider (at the distal lip of the ostium). The flow divider is the ridge separating the lower aortic surface from the upper intercostal branch. The impression of normal silver-stained endothelial cells in the plastic was visible on both the surfaces. Here the normal pattern of the endothelium was disrupted, and curled up spindle-shaped cells left deep impressions in the casting material. Such abnormal-shaped cells were only found at the center of the flow divider. The number of spindle cells at the branch increased with the size of the ostia. At the renal ostium, many spindle endothelial cells were present on both the aortic and renal side of the flow divider. Also, a cell-denuded area was visible at the center of this region. These cells extended as a band from the flow divider up to the distal face of the branch for several millimeters. Spindle endothelial cells were also seen on the proximal face of most aortic branches. A wide band of these cells was seen at the proximal face of the celiac branch. On several occasions, the displaced cell nuclei as well as obvious breaks in the cell boundaries were visible in the ostial region.

We have examined the EC morphology by SEM in perfusion-fixed rabbit aortas (5). Figure 8.2 shows the locations of the distal region in the ostia of the celiac, superior mesenteric, right renal, and left renal artery branches where the cells were studied. Also, a location in the aorta away from the branch (straight region) was studied for comparison. During the fixation, tissue removal, and tissue processing, care was taken to maintain the normal curvature of the branch region. The morphology of the straight (nonbranched) segment of the abdominal aorta and the aorta-arterial branch were compared. All photomicrographs shown are 1600× magnification and are oriented such that the direction of blood flow is approximately from top to bottom. All photomicrographs of the branch areas show the aortic luminal surface just distal to the central portion of the flow divider.

2.1. *Straight Segment*

The intimal surface of a straight segment of the abdominal aorta is shown in Figure 8.3. SEM of this area demonstrates a topographic pattern of endothelial cells having uniform size and a preferential orientation along the length of the aorta in the direction of blood flow. The endothelial cells are flat and the cell borders appear intact. No spindle-shaped cells are seen.

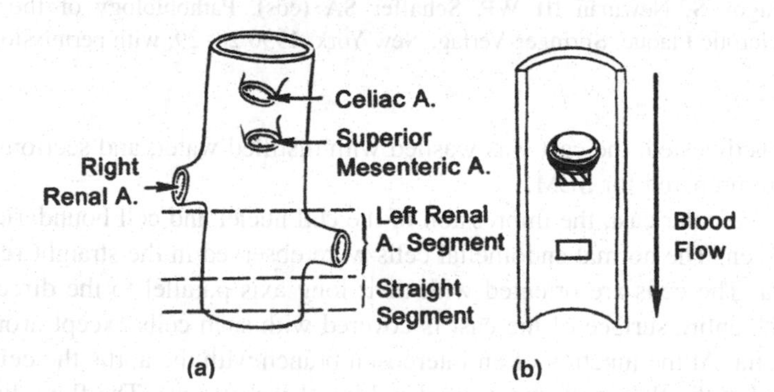

FIGURE 8.2. (a) Schematic presentation of the abdominal aorta showing four major arterial branches. (b) Schematic presentation of the opened aorta showing the ostium of a branch. SEM photographs were taken at the distal lip of the ostium (⛀) and at a straight region of the aorta (□).

FIGURE 8.3. SEM photomicrographs of the luminal surface of a straight portion of the abdominal aorta. The intimal surface is smooth. Endothelial cells have uniform size and they are flat and elongated. Cell borders are intact and the cells are oriented in the direction of blood flow (from top to bottom, approximately).

2.2. *Branch Region*

The intimal surface of the aorta-celiac artery branch is shown in Figure 8.4. When compared with the straight portion, the overall topographic pattern is not uniform. In some areas, the endothelial cells appear oriented with the long axis in the direction of blood flow, but in other areas they are oriented almost 45 degrees to the direction of flow (Fig. 8.4a). The cells are less uniform in size when compared with those of the straight segment. Some endothelial cells are thickened and spindle-shaped. Cell borders no longer appear uniformly intact, especially in the areas occupied by spindle-shaped cells. The intima has a convoluted surface. At another celiac branch site, there is a different morphologic pattern. The endothelial cells are rounded and disoriented with respect to blood flow (Fig. 8.4b). The central portions of the cells are prominent and protrude into the lumen of the artery and have a cobblestone appearance. Several endothelial cells appear to have nuclei that bulge into the lumen. Similar patterns occur at the branch sites of the intercostal, superior mesenteric, right renal, and left renal arteries. Overall, the endothelial cell morphology at the branch sites display a morphologic spectrum ranging from spindle-shaped cells with disrupted cell borders to cobblestone-shaped cells with loss of cell orientation.

Svendson et al. (6) studied the rabbit aorta by light, scanning, and transmission electron microscopies to compare cellular morphology in the areas near intercostal artery branches, at the inner curvature of the aortic arch, and in the nonbranch region. The aortas were perfusion fixed. They reported that in the nonbranch region, a regular pattern of intimal folds was seen where the endothelial surface was

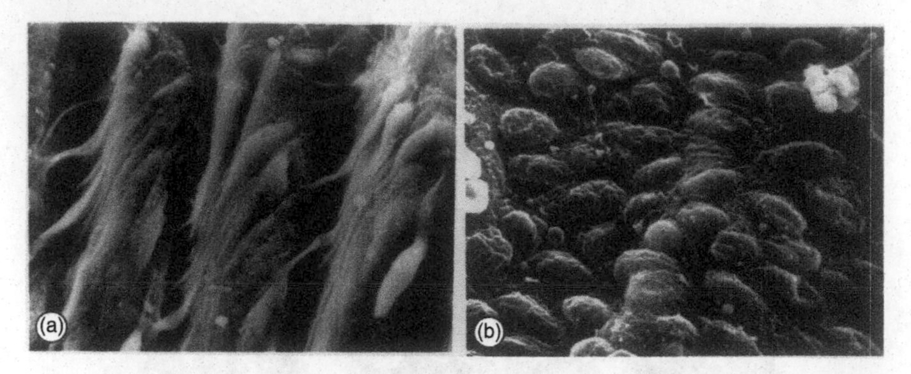

FIGURE 8.4. SEM photomicrographs of the luminal surface of the aorta at the celiac artery ostium of rabbits. (a) An intimal surface that is convoluted and has frequent spindle-shaped endothelial cells with separated cell borders. (b) Cobblestone-shaped endothelial cells. These cells are not oriented in the direction of bulk blood flow. Many cells have bulging surfaces that protrude into the lumen.

smooth. TEM also showed unaltered ECs in this area. In the areas of inner curvature of the arch, protruding elongated cells were found. In other branch regions, there were signs of cellular injury such as ruptured plasma membranes. The cell nucleus was of irregular shape and often protruded into the lumen. Often the cell appeared bulging although it was still attached to the neighboring cells at intercellular junctions. Loosening of the cell from the basement membrane and twisting of the cell was also noted. Overall, they concluded that in comparison with the nonbranch regions, the branch regions presented endothelial cells, which depicted features of cellular injury.

3. Endothelial Cell Morphology at the Site of High Permeability

After examining the EC morphology at the branch and nonbranch regions in the aorta, the study of the same at the site of high permeability is of interest. It so happens, as will be described later in this chapter, that the sites of high permeability are predominately the branch (ostial) regions. Expectedly then, most of the observations presented below will be along similar lines as those presented earlier. The emphasis, however, will be on the EC morphology in the region of high permeability, and this region may extend beyond the branch region.

Gerrity et al. (7) examined EC morphology in a pig aortic arch using transmission electron microscopy and compared the morphology in the areas of high versus low permeability to Evans blue dye. Evans blue was prepared as an 0.5% solution in sterile 0.85% saline, and 50 mL of this solution was injected into the jugular vein in young pigs. Three hours later, the perfusion fixation was begun, using carotid and femoral arteries. The animal was allowed to bleed through the femoral artery, while through the carotid 500 mL of Krebs-Ringer-bicarbonate solution was perfused at systolic pressure. This was followed by perfusion of 1 L of fixative at systolic pressure and 37°C temperature. Different fixatives were tried initially and 2% glutaraldehyde in 0.1 M sodium cacodylate buffer, pH 7.35, was chosen for subsequent experiments. After perfusion the aortas were removed, opened longitudinally, and examined for the blue areas.

Evans blue dye is an azo dye that binds to albumin and accumulates in the intima in the areas where aortic endothelium is permeable to albumin. The aortic surface appears blue in these areas. The blue and white areas were mapped in the aorta as shown in Figure 8.5, and 1 mm^3 blocks of tissue were excised from these areas. All tissue blocks were fixed further for 6 h. Ultrathin sections were made from these tissues and further processed for examination by transmission electron microscope. In addition, silver nitrate–stained endothelial Häutchen preparation was produced for light microscopy. The details of these techniques can be found in Ref. 7.

The authors report that blue and white areas were clearly seen and are shown in Figure 8.5. The ultrastructural details were well preserved. The endothelial

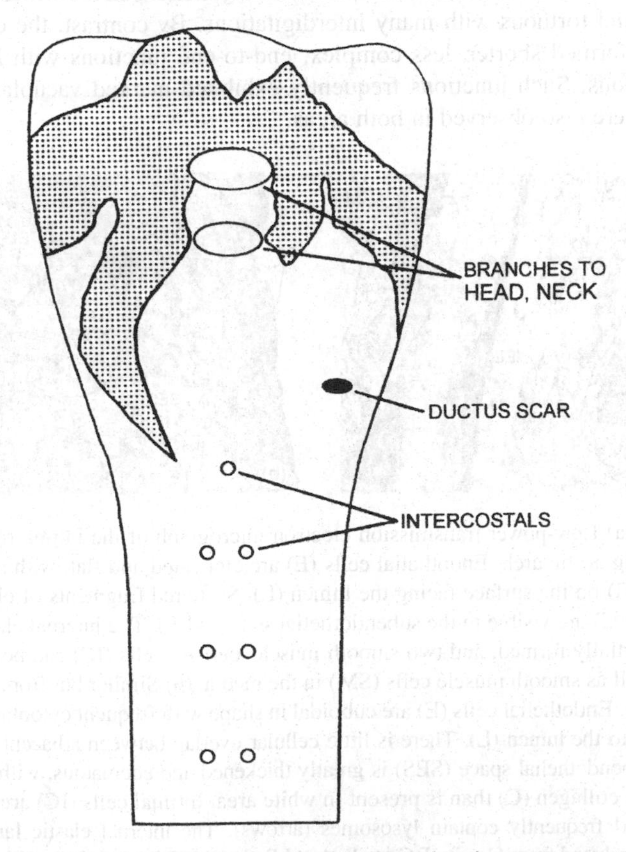

FIGURE 8.5. Schematic diagram of pig aorta arch. Dotted region represents typical area of Evans blue uptake. (Reproduced from Gerrity RG et al., Am J Pathol 1977;89:313-334, with permission from the American Society for Investigative Pathology.)

cells in the white areas appear flattened and elongated with considerable overlap of adjacent cells (Fig. 8.6). In contrast, the cells in the blue areas are more cuboidal with little or no overlap of adjacent cells. Endothelial cytoplasmic flaps or projections into the lumen were present more frequently in the blue areas than in the white areas. The projections are often of considerable length and complexity and are commonly located at junctional complexes.

When endothelial cell organelles were compared between the blue and white areas, a number of these such as nucleoli, mitochondria, and pinocytic vesicles appeared similar on a qualitative basis. In contrast, definite differences in the prominence and development of rough endoplasmic reticulum (RER), Golgi apparatus, and lysosomal bodies were noted between the blue and white areas. The RER was more prominent in the cells of the blue areas. The Golgi apparatus was well developed in the cells of the white areas, whereas it was rarely seen in cells of the blue areas. A variety of inclusions were seen in cells of both areas, but they occurred in greater numbers in the blue areas. No potent channels or chains of vesicles extending across the endothelium were seen in the blue or white areas.

When cellular junctions were compared, most junctions in the white areas were elongated and tortuous with many interdigitations. By contrast, the cells in the blue areas formed shorter, less complex, end-to-end junctions with little or no interdigitations. Such junctions frequently exhibited marked vacuolation. Tight junctions were also observed in both areas.

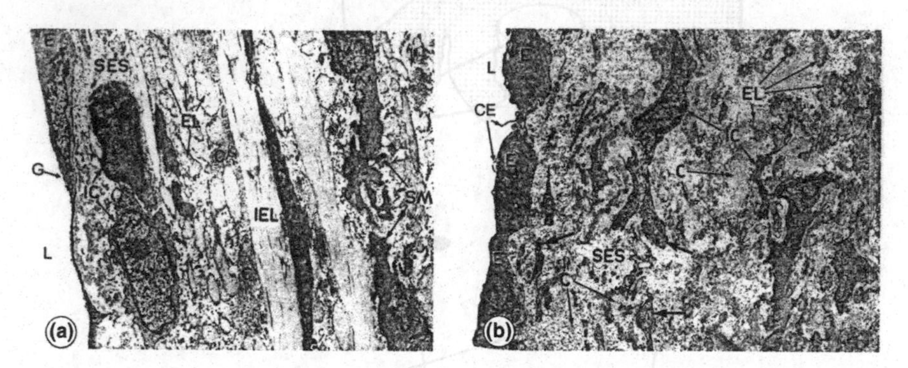

FIGURE 8.6. (a) Low-power transmission electron micrograph of the intima from a white area in the pig aortic arch. Endothelial cells (E) are elongated and flat, with a prominent glycocalyx (G) on the surface facing the lumen (L). Scattered fragments of elastica (EL) and collagen (C) are visible in the subendothelial space (SES). The internal elastic lamina (IEL) has partially formed, and two smooth muscle–derived cells (IC) can be seen in the intima, as well as smooth muscle cells (SM) in the media. (b) Similar but from a blue area of dye uptake. Endothelial cells (E) are cuboidal in shape with frequent cytoplasmic extensions (CE) into the lumen (L). There is little cellular overlap between adjacent endothelial cells. The subendothelial space (SES) is greatly thickened and edematous, with more elastica (EL) and collagen (C) than is present in white area. Intimal cells (IC) are numerous, elongated, and frequently contain lysosomes (arrows). The internal elastic lamina is not visible. (Reproduced from Gerrity RG et al., Am J Pathol 1977;89:313-334, with permission from the American Society for Investigative Pathology.)

Subendothelial space was consistently and strikingly thicker in the blue than in the white areas (Fig. 8.6). In the white areas, internal elastic lamina is evident at about 7–8 μm below the endothelium, whereas in the blue areas even at 30–35 μm depth it is not visible. In the blue areas, the subendothelial space appeared markedly edematous, with collagen, elastic tissue, and microfibrils scattered in the region. The intimal cells in the blue areas were usually not in close proximity to the endothelium. In the white areas, endothelium was close to the internal elastic lamina or to the intimal cells.

Endothelial cell injury as exhibited by the spectrum of changes—ranging from minor morphologic alterations to overt cell death—was consistently observed in the blue areas but rarely in the white areas. Changes noted include disruption of mitochondrial cristae, mitochondrial swelling, and a general loss of cytoplasmic organelles, as well as prominent cytoplasmic and junctional vacuolation and ultimately cellular death and sloughing. Highly vacuolated cells were also observed in these regions. In silver nitrate—stained Häutchen preparations of endothelium, injured or dead cells were found to occur with a significantly greater frequency in the blue (2.91%), than in the white (0.71%) areas. The authors (7) also speculate that focal endothelial cell injury or death observed might result in a number of physical defects in the endothelium, which could function as "ultralarge pores" and thus contribute to the enhanced permeability of blue areas.

The surface-associated mucopolysaccharide coat, or glycocalyx, was threefold thicker over the endothelium of the white areas than the blue areas. This surface layer may influence processes such as permeability, filtration, binding, ion exchange, and cell adhesion.

The authors speculate that the differences described above in glycocalyx and endothelial cells in the blue and white areas could be the result of focal differences in the hemodynamic stress in these two regions. For example, stress concentration and increase of strain in the blue areas compared with those in the white areas could be implicated in causing those differences. A more detailed description and discussion of the above observations can be found in Ref. 7.

Bjorkerud et al. (8) studied endothelial cells using a technique of dye exclusion in which cellular viability is decided on the basis of ability of cells to exclude certain anionic dyes. They studied rabbits, rats, and guinea-pigs using Evans blue solution or Nigrosin solution, which is a microscopical stain. In the animals, they perfused the aorta for about 90 s with Ringer's solution, and after 1 min the perfusion pressure was gradually decreased from 100 mm to 10 mm of mercury. Either Evans blue or Nigrosine was then introduced in the aorta through the perfusion catheter at a pressure of 10 mmHg. After 3–6 min of perfusion with the dye, the perfusion continued with Ringers solution to wash off the excess stain. The perfusion pressure was then increased to 100 and the perfusion changed to buffered formaldehyde or formaldehyde-glutaraldehyde solution. More details on the technique can be found in Ref. 8.

Evans blue as a marker for albumin was mentioned earlier. Nigrosin has been used as a dye exclusion marker for cell injury in tissue cultures. It seems to stain injured or dead endothelial cells selectively. In contrast, Evans blue stains slightly or markedly changed cells. Thus, Nigrosin stains are confined to areas that show heavy stains of Evans blue. The staining to explore injured endothelial cells must be done at low pressure because at high pressure dye penetration due to filtration would occur and obscure the dye present in the injured cells. The authors also emphasize that in the experiments, the animals should be anesthetized when the perfusion is started. If the endothelium from dead animals is investigated, several artifacts may appear.

It has been well recognized that Evans blue and some other dyes accumulate in certain regions of the aorta after intravenous injection. The ascending arch of the aorta, the minor curvature of the arch, and the regions around the orifices of major branches show such accumulation. Very rapidly after the injection, Evans blue is bound to serum proteins, primarily albumin, and it penetrates the arterial wall as a component of an Evans blue–protein complex. This is consistent with the observations that there is an increased uptake of labeled plasma proteins in areas of in vivo Evans blue uptake.

It has been suggested that the increased permeability in areas of Evans blue uptake in vivo was due to the presence of young endothelial cells (9). This idea would fit with the data, which suggests increased endothelial cell turnover in these areas. However, injured endothelial cells are also greater in number in these areas and would contribute to increased permeability. We may also note that, after entering the tissue, the dye-protein complex is mainly confined to the intercellular space, whereas the uncomplexed dye occupies mainly the intracellular space in injured cells.

Scanning electron microscopy in relation to the dye exclusion suggested the following. In areas with no dye uptake, there was little variation in the endothelial surface structure. Individual cells were polygonal with cell borders delineated by microvillous structures or interdigitating cell processes. The cell surface was generally flat and the position of the nucleus was indicated by a gentle bulging in the center of the cells. In areas with dye uptake, the surface topography was quite irregular. Cell shrinkage and formation of fissures were seen, however, the authors considered these to be due to preparation artifacts. Some cells were large and distended and appeared injured. There were also spindle-shaped cells with an orientation different from the long axis of the vessel. Some spindle-shaped cells were partially detached from the underlying structure. When these results are combined with other similar observations, where cells did not exclude Nigrosin or Evans blue (i.e., injured cells), the distended cells observed with SEM represent markedly injured or dead endothelial cells.

Overall, the areas with Evans blue or Nigrosin stains show much more variable cell structure. The cells are more bulging, with larger oblate to round nuclei, and occasional cytoplasmic or nuclear vacuoles. They show character of premortal cells seen in cell cultures such as cytoplasmic edema, intercellular vacuolization, nuclear edema, and so forth. Thus, spindle-shaped cells may also represent injured or dead cells.

3.1. *Quantitation of Endothelial Injury*

Bjorkerud et al. (8) also report the effect of hypertension on the integrity of rat aortic endothelium. The degree of injury was evaluated from the areas stained with Evans blue as well as the degree of staining. The rats were made hypertensive using a Goldblatt clip on one renal artery. Systemic blood pressure was measured in control and in hypertensive rats. After Evans blue injection intravenously, the thoracic aorta was studied for endothelial injury. Densitometry was used to assess the degree of staining and planimetry to assess the stained area. The integrated number was taken as a measure of *degree of injury*. Figure 8.7 shows a plot of degree of injury versus systemic blood pressure. It may be noted that the level of blood pressure was significantly related to the degree of endothelial injury. The injury increases as the pressure increases. This result is important because it shows that increased wall stress (hypertension), increases endothelial injury in the nonbranch (straight) segments of the artery also.

4. Changes in Endothelial Cell Shape Due to Imposed Coarctation

It is fascinating to examine how cells may change in response to mechanical forces. In this section, we will consider how the shape of the endothelial cell changes when coarctation is imposed on the aorta. Levesque et al. (10) performed a very nice study of the endothelial cell shape in dogs where they deliberately

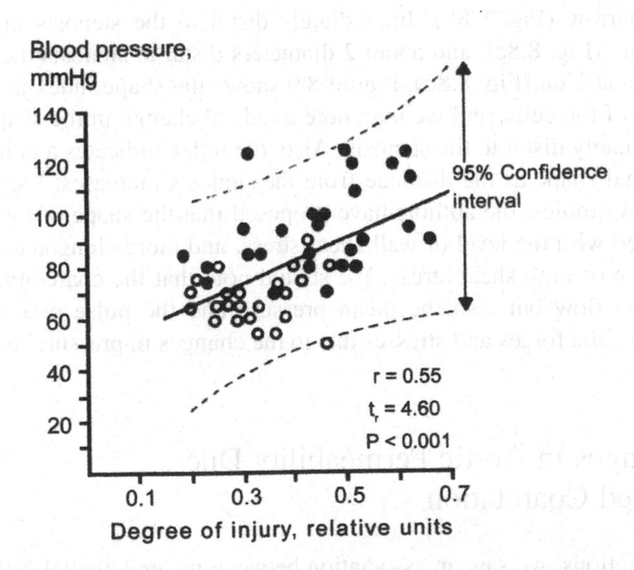

FIGURE 8.7. Systemic blood pressure and the degree of endothelial injury in the thoracic aorta as evaluated by combined densitometry and planimetry in hypertensive rats and unoperated controls (open circles). The degree of endothelial injury was highly significantly related to the level of blood pressure. (Reproduced from Bjorkerud S et al., Studies of arterial endothelial integrity with the dye exclusion test – A review. In: Schwartz CJ, Werthessen NT, Wolf S (eds), Structure and Function of the Circulation, vol. 3. Plenum Press, New York, 1981, with permission from Springer Science and Business Media.)

created a coarctation (stenosis) in the descending thoracic aorta. The surgical procedure was to anesthetize the dog and open the thorax at the level of the first set of intercostal arteries through a long incision between the ribs. Through this opening, a cotton band, 0.3 mm wide, was placed around the aorta. The band was tightened until the presence of a thrill or a bruit was felt distal to the band. This usually produced a 71 ± 3% decrease in the cross-sectional area of the aorta. The authors also observed small poststenotic dilation in the aorta. The animals recovered well and were sacrificed after 12 weeks. The aortas were prepared for a vascular casting procedure, as described earlier (3, 4), and the casts were used for the measurement of cross-sectional areas. Also, the ventral aspects of the casts were studied by light microscopy to determine endothelial cell morphology. From the micrographics of the endothelial cells, they measured the area, perimeter, length, width, and calculated the shape index. The shape index of the endothelial cell was defined as 4π area/perimeter^2. By definition, the shape index is 1.0 for a circle and 0.0 for a straight line. The smaller the shape index, the more elongated is the cell. The authors also carried out flow studies on the model, which they made from one of the vascular casts of the dog aorta with coarctation.

Figure 8.8 shows the morphology of the endothelial cells at and near the region of coarctation. As we see, the shape of the cells has changed dramatically with the location. For the normal morphology, we consider Figure 8.8a, which is at a location 3 diameters away from the coarctation. Here we can see the normal elongated shape of the cells. At the throat of the stenosis, the cells have become long and narrow (Fig. 8.8b). Immediately distal to the stenosis the cells are almost round (Fig. 8.8c), and about 2 diameters distal to stenosis the cells have a mixed orientation (Fig. 8.8d). Figure 8.9 shows the shape index as a function of location of the cells, and we may note a radical change in the shape index at and immediately distal to the stenosis. Also, the index indicates a gradual return to the normal shape as the distance from the stenosis increases. Using the data of their flow studies, the authors have proposed that the shape of the cells may be correlated with the level of wall shear stress, and more elongated cells occur in the region of high shear stress. We should note that the coarctation changes not only the flow but also the mean pressure and the pulse pressure in that region. Thus, the forces and stresses due to the changes in pressure must be considered.

5. Changes in Aortic Permeability Due to Imposed Coarctation

In earlier sections, we saw an association between the endothelial cell morphology and the aortic wall permeability. In the previous section, we considered the effect of coarctation on the shape of the endothelial cells. The logical next step is to consider the effect of coarctation on the permeability of the aortic wall.

Somer et al. (11) carried out experiments in pigs, 6–12 weeks old and weighing 30–50 lb, in which they created a coarctation in the descending thoracic aorta

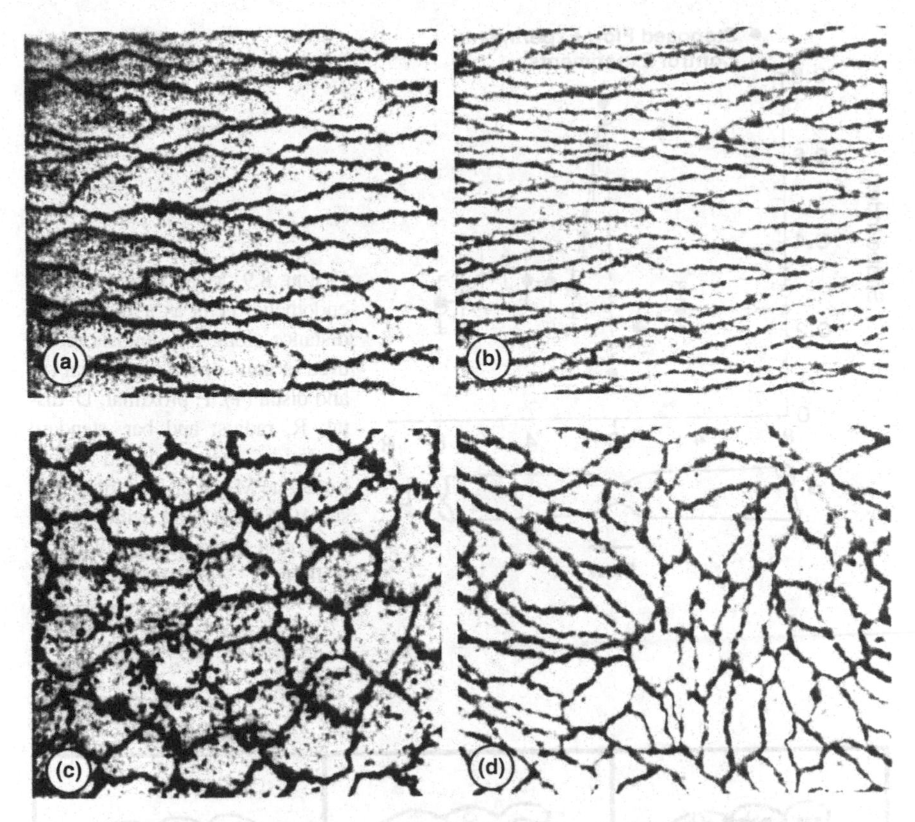

FIGURE 8.8. Light micrographs of endothelial cell patterns obtained from stenosed aortas at selected locations: (a) 3 diameters proximal to stenosis, (b) throat of the stenosis, (c) immediately distal to stenosis, (d) 2 diameters distal to stenosis. Flow is from left to right. (Reproduced from Levesque MJ et al., Arteriosclerosis 1986;6:220-229, with permission from Lippincott, Williams & Wilkins.)

and studied the permeability of the aorta to Evans blue dye. In pigs, they performed thoracotomy through the 4th intercostal space and approached the aorta at about 1 cm below the level of the obliterated ductus arteriosus. A plastic band was placed around the aorta at this level and the band tightened until they felt the evidence of a jet as well as reduced pressure below. They created stenosis where the cross section of the aorta was reduced by 90% (tight) or somewhat less (moderate to loose). The animals were sacrificed 28–30 days later, and 3 h prior to sacrifice, 45 mL of Evans blue was injected intravenously. Evans blue was prepared as a 0.5% solution (weight/volume) in 0.85% saline. The areas of Evans blue accumulation were mapped and the extent of aortic narrowing was recorded.

Figure 8.10 shows the typical patterns of Evans blue uptake in the aorta in case of normal conditions (control), tight coarctation, and loose coarctation. In the normal aorta, the areas of Evans blue uptake were in the dorsal section of the ascending arch, proximal to and on both sides of the two main branches to the forelimbs, head,

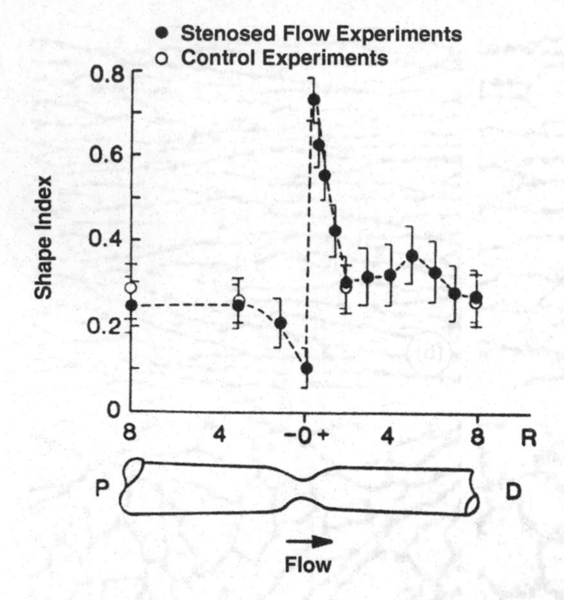

FIGURE 8.9. Mean shape index of endothelial cells as a function of distance along the stenosis measured in units of radii proximal (–) and distal (+). P, proximal; D, distal; R, radius; and bar, standard deviation. (Reproduced from Levesque MJ, et al., Arteriosclerosis, 1986;6:220-229, with permission from Lippincott, Williams & Wilkins.)

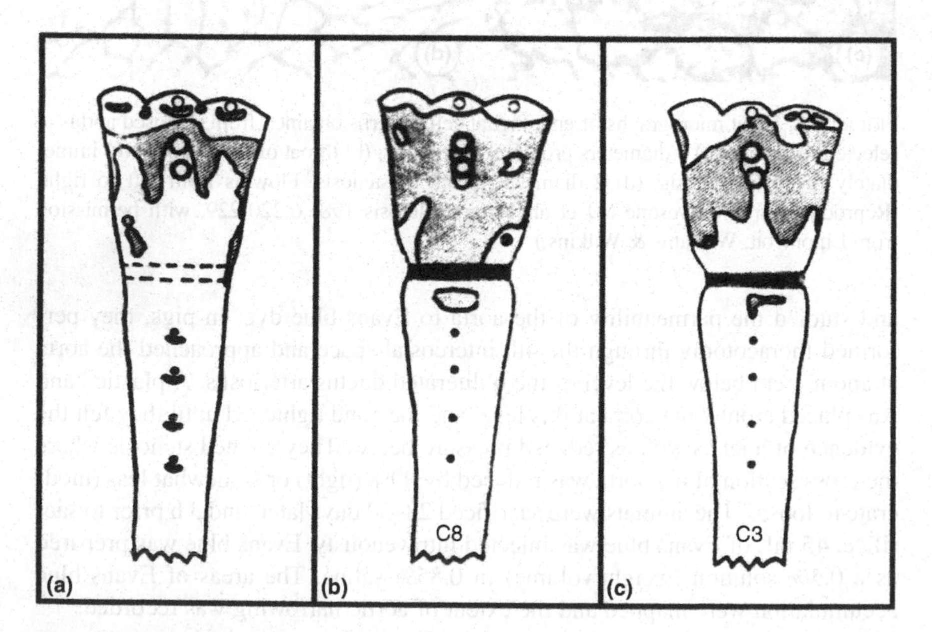

FIGURE 8.10. Typical spontaneous patterns of aortic Evans blue uptake (cross-hatching) in the young pigs: (a) unoperated, (b) tight aortic coarctation (solid line), and (c) moderate to loose aortic coarctation (solid line). (Reproduced from Somer JB et al., Atherosclerosis 1972;16:127-133, with permission from Elsevier.)

and neck. This blue descended distally as two extensions, usually to a point just below the lower branch. Patches of bluing were also noted around the ostia of intercostal arteries. A tight coarctation had a significant effect on the pattern of Evans blue uptake. The blue area proximal to the coarctation was greatly enhanced. Obviously, a significant amount of area that had no dye uptake previous to coarctation had now become blue. In the animals, which had loose coarctation, the pattern of dye uptake was not significantly different from that in the normal aortas.

In the aorta distal to the coarctation, the dye uptake was reduced in the case of tight coarctation compared with control. Also, new areas of dye uptake emerged perhaps as a result of jet striking the aortic wall. In the case of loose coarctation, the dye uptake in the distal region was usually reduced.

In an attempt to correlate the changes in dye uptake with the changes in hemodynamic parameters, the authors make the following observations (11). The hemodynamic changes associated with aortic coarctation are complex, and therefore it is not possible to identify any one component as being responsible for changes in Evans blue uptake. The aorta proximal to the coarctation is subjected to large and sudden pressure fluctuations. The mean pressure, the pulse pressure, and aortic stretching are increased in this region and influence the Evans blue uptake. Distal to the coarctation there is reduced pressure and lower pulse pressure, which could explain reduction in Evans blue uptake. New areas of Evans blue uptake could likewise be a result of jet impingement on the wall. Such areas could reflect an early stage in the evolution of so-called jet lesions customarily seen at this site in less acute situations.

If we compare these observations with those described previously, it becomes obvious that changes in the endothelial shape do not always go hand in hand with those in the arterial permeability. For instance, in coarctation, the most change in the shape of the endothelial cells occurred at the throat and immediately distal to it (Fig. 8.8) whereas most change in permeability occurred proximal to the coarctation (Fig. 8.10). These observations indicate that besides the function of the endothelial cells, there are other factors (such as pressure) that influence the arterial permeability.

6. Endothelial Cell Orientation in the Aortic Valve

The studies of endothelial cells in the arteries have had an overwhelming influence on our thinking, in that we believe that the endothelial cells are aligned with their long axis in the direction of flow. This thinking is so prevalent that often flow is thought to be the *only* factor responsible for the orientation of these cells. In this section, however, we will consider the endothelial cell morphology in the aortic valve, particularly to emphasize that in the valve the cells are *not* oriented in the direction of flow. This observation is of vital importance because it points to *factors other than flow* that are responsible for the orientation of the cells. This puts us in a better position to understand not one but multiple factors that may influence the cell morphology and function.

Deck (12) studied the endothelial cells on the aortic valve leaflets in dogs. The aortic valve leaflets were fixed in the closed position under diastolic pressure with 2% buffered glutaraldehyde. Some of the leaflets were also fixed with the valve in the open position using continuous perfusion at systolic pressure. The leaflets were prepared for SEM. The details of the technique are described in Ref. 12.

Deck observed that the endothelial cell surface on the valve leaflets was readily seen with SEM, however, the shape and arrangements of individual cells were less obvious because cell boundaries were not clearly visible. Nevertheless, cell patterns could be inferred from the shapes and orientations of cell nuclei, which routinely appear as a raised mound on the endothelial surface. Round nuclei reflect symmetrical cells, and long oval nuclei imply cells with elongated shapes. In the leaflets, fixed in the *open* position, on the aortic surface of the leaflet in the central portion, the cells with long oval nuclei were seen. These cells as well as those on the rest of the surface are aligned with the internal fibrous cords of the leaflet in the circumferential direction. Thus, the aortic surface is characterized by cells oriented circumferentially. On the aortic surface of the leaflets fixed in the *closed* position, endothelial cells are again aligned with the leaflet free edge, that is, in the circumferential direction (Fig. 8.11a). The endothelium on the redundant flap of the leaflet near its attachment to the aortic wall also shows the same arrangement. The tendinous cords beneath the epithelium especially stand out in the redundant region, appearing as gross ridges on the surface. Overall, the endothelial cells on the aortic surface of the valve leaflets are oriented parallel to the leaflet free edge, whether the leaflet is preserved in the open or closed position.

The ventricular surface of the valve leaflets, unlike the aortic surface, is directly exposed to the thrust of the systolic blood flow and therefore to the flow force that may tend to align the endothelial cells. The surface itself is smooth and lacks tendon-like cords. Nonetheless, in the leaflets fixed open, the cells are elongated circumferentially whether the cells are in the middle or at the base of the leaflet (Fig. 8.11b). Likewise, in the leaflets fixed closed, the cells are similarly oriented both at the center and near the free edge (i.e., the cells are aligned circumferentially and parallel to the free edge). Often, the cells on the ventricular surface are longer and spindle-shaped compared with those on the aortic surface, but like the aortic surface cells, they are almost always aligned with the free edge of the leaflet. Obviously, this orientation is *perpendicular* to the direction of blood flow.

As mentioned earlier, it has usually been assumed that the pulsating blood flow exerts a shearing stress on the directly exposed endothelial cells and brings about their elongation in the direction of flow. The evidence presented above, however, indicates that cells on both sides of the leaflets are aligned perpendicular to the flow rather than parallel to it. It seems to follow, therefore, that endothelial cell orientation on the leaflets is governed by forces other than blood flow.

To understand the forces that may govern the orientation of endothelial cells in the leaflet, let us consider the following. In case of the leaflets, the hydrostatic pressure of blood appears to be the force that ultimately determines the orientation

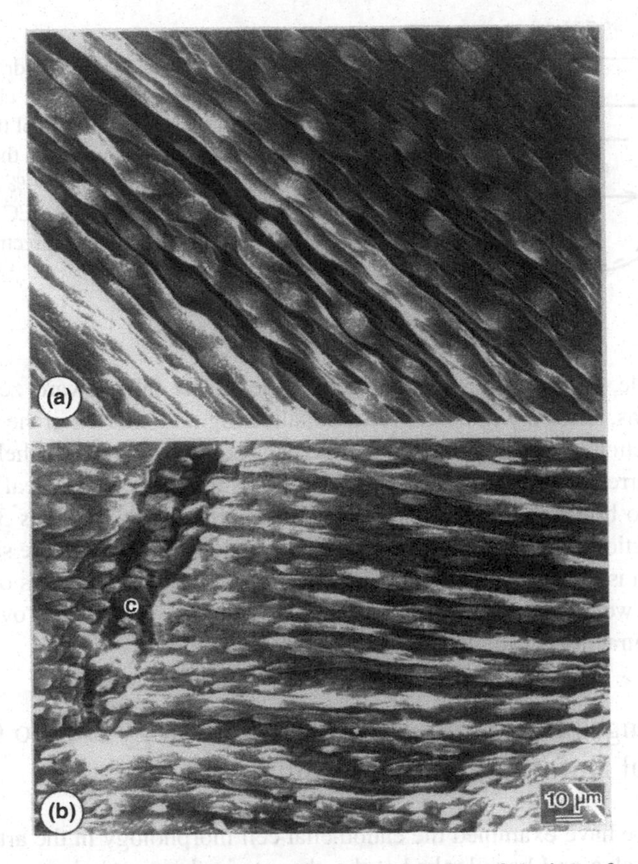

FIGURE 8.11. (a) Electron micrograph showing central region of aortic surface of a closed leaflet. Endothelial cells, elongated from upper left to lower right, parallel the leaflet free edge. (b) Electron micrograph showing ventricular surface view of endothelial cells in the central region of an opened leaflet. Cell nuclei are aligned (left to right) with the free edge of the leaflet and stand up in raised rows. The vertical cleft, c, is a wrinkle caused by drying the tissue. (Reproduced from Deck JD, Cardiovasc Res 1986;20:760-767.)

of the leaflet tissues. In diastole, the pressure gradient across the closed leaflets acts to distend the leaflets fully, in both the circumferential and the radial directions (Fig. 8.12). Stress analyses of the leaflets, however, indicate that principal loading occurs in the circumferential direction (13). This kind of loading presumably helps during development to produce and subsequently maintain the pronounced circumferential alignment of the leaflet's principal collagenous tissues. These tissues are particularly aggregated just beneath the aortic surface and are collected into tendinous cords. It seems that the endothelial cells covering this aortic surface, exposed to the same tensions of diastole, simply follow the orientation of the collagenous tissues.

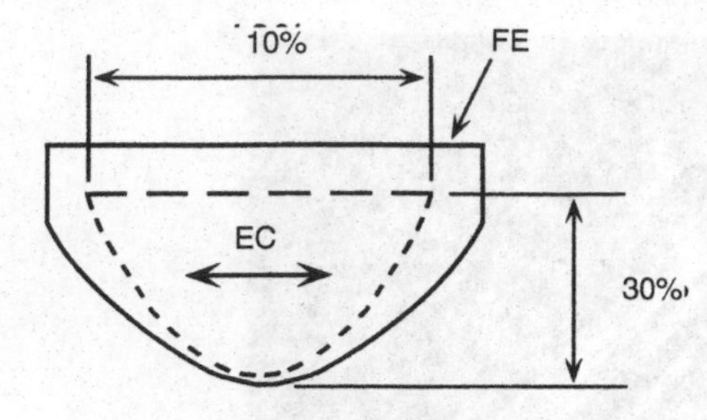

FIGURE 8.12. Schematic drawing of an aortic valve leaflet. From closed (solid) to open (dotted) position of the valve, the leaflet shortens by 10% in the circumferential direction and by 30% in the radial direction. FE, free edge; EC, endothelial cells oriented in the direction (arrows) parallel to the FE.

In systole, the pressure gradient across the leaflet falls toward zero. As the leaflet opens, tension within it is relieved, allowing it to shorten in the circumferential direction by about 10% (13) and perhaps permitting the endothelial cells to shorten correspondingly. Radially directed elastic fibers in the leaflet appear, however, to bring about an even greater contraction of as much as 30% in the radial direction (13), tending to compress the endothelial cells in the same direction, which is their short axis. The net effect on the endothelial cells of any such alterations would favor the circumferential orientation of the cells owing to the greater contraction of the leaflet in the radial direction.

7. Changes in Endothelial Cell Morphology Due to Cyclic Tensional Deformation (Cell Culture Studies)

Thus far, we have examined the endothelial cell morphology in the artery and in the aortic valve and have looked at the changes in the morphology produced by coarctation. We also have considered various forces that could influence the morphology and orientation. In an attempt to isolate a certain factor, for example blood flow, and to establish how that particular factor may influence EC orientation, there have been studies performed on the endothelial cell culture. Typically, a culture of endothelial cells is developed and then the flow is imposed as the cells are continuing to grow, and then the orientation of cells is examined with respect to the imposed flow (14). Such studies have shown that the cells in the culture do orient with their long axis in the direction of flow. Because, in vivo, the cells are subjected to both, the blood flow and the mechanical stresses and strains resulting from pulsating blood pressure (and other forces), it is obvious that the cells are influenced by multiple forces. Some studies, therefore, have been performed where the cells in the culture are subjected to both the flow and cyclic mechanical strain. The scope of such studies is vast, and the readers are better served by going to those reports. A portion of such studies will be considered in this section to correlate the effect of cyclic mechanical strain with the EC morphology. Both the mechanical forces and the biological processes in the body are quite complex, and therefore it is important to limit the scope to describing the relevant feature of the processes in the context of vascular mechanics.

Sumpio et al. (15) studied the EC morphology under the conditions of cyclic tensional deformation. Thoracic calf aortas were removed septically and transferred on ice to the tissue culture facility. The vessels were opened and the endothelial cells detached by gentle scraping of the intimal surface. The cells were dispersed into culture dishes containing Dulbecco's modified Eagle's medium supplemented with 10% heat-inactivated fetal calf serum and grown to confluence at 37°C in an atmosphere of 5% CO_2 in a humidified incubator. Endothelial cells were identified by their typical light microscopic morphology, by the growth of polygonal closely opposed cells forming confluent monolayers, and by other properties.

Cells were cultured on a flexible-bottomed silicone rubber culture plate. When a precise vacuum level was applied to the system, the culture plate bottoms were deformed to a known percent elongation that was transferred to the cultured cells. When the vacuum was released, the plate bottoms returned to their original conformation. In these experiments, the vascular endothelial cells in the culture were subjected to repetitive cycles of 10 s of 24% maximum elongation and 10 s of relaxation (three cycles per minute) for 5 days.

The degree of actin filament organization in the bovine aortic endothelial cell was evaluated with a fluorescent F-actin probe, rhodamine phalloidin. Endothelial cells, grown for 5 days under either static conditions or tension/relaxation cycles, were used for comparative morphologic studies. In parallel experiments, the level of protein synthesis in these cells was also determined and compared. In these experiments, bovine aortic endothelial cells were seeded in 6-well plates at 200,000 cells per well. After attachment overnight, the plates were either placed on the vacuum-operated stress-providing instrument or in a static environment in the same incubator. Cells were incubated overnight in 1 mL of growth medium containing 500 μCi of ^{35}S-methionine before harvesting. Gel electrophoresis of the samples was performed. The details of the technique for determination of ^{35}S-methionine–labeled endothelial cell proteins can be found in Ref. 15.

The authors report that the cells not subjected to mechanical deformation (controls) displayed a more rounded configuration with a more diffuse distribution of actin. The stress fibers were absent. In contrast, the ECs subjected to tension were larger and appeared in one of two distinct forms: either elongated with prominent pseudopods and actin stress cables *or* as large polygonal-shaped cells with marked peripheral vacuolization. When the cells in the periphery subjected to mechanical stretch were examined, they were found to not align themselves in the direction of maximum flow shear. They were oriented randomly. The protein distribution of the mechanically deformed cells was found to be significantly different from that of the static controls. The authors report characteristics of proteins whose synthesis increased significantly after mechanical stretching. Also, they found other proteins whose synthesis was depressed by the stretching. Overall, the authors conclude that ECs subjected to cyclic tension had a more polygonal shape and demonstrated pseudopods and actin stress fibers, whereas ECs cultured under static conditions were more rounded and did not express actin stress cables.

Thus, the ECs subjected to cyclic mechanical deformation respond by altering their cellular morphology and protein synthesis.

Ives et al. (16) studied the endothelial cell shape under conditions of cyclic deformation. Human umbilical vein endothelial cells (HUVECs) were harvested from umbilical cords obtained from a labor and delivery department. To remove the endothelial cells, the veins were cannulated, rinsed with 50 mL of phosphate-buffered saline (PBS), and then filled with 0.03% collagenase in Medium 199 and incubated for 20 min. After incubation, the enzyme solution was flushed through the cord with 40 mL of PBS; the effluent was collected and centrifuged for 10 min. After centrifugation, the cell pellet was resuspended in Medium 199, supplemented with 20% fetal bovine serum, 100 U/mL penicillin, and 100 μg/mL streptomycin. Between 7.4×10^3 and 1.5×10^4 cells/cm^2 were seeded onto the substrate for experiments.

The apparatus for stretching the cells consisted of a polycarbonate chamber divided into two compartments. In the experimental compartment, the substrate was mounted with clamps so that it was positioned close to the floor of the chamber. In the control compartment, the membrane was fixed to the bottom of the chamber. Cell suspensions (approximately 1×10^6 cells) were seeded onto the membrane (polyetherurethane urea membrane; Mitrathane) in each compartment. Then, 13 mL of medium was added to each compartment and the chamber placed in the incubator. After 3 to 4 days in stationary culture, when the cells had reached confluence, the chamber was removed from the incubator and mounted on a frame attached to the stage of an inverted phase microscope. The movable clamps were connected to a motor-driven camshaft. Variation of the cam eccentricity controlled the stretch amplitude. In the experimental chamber, the movement of the clamp produced a cyclic stretching and relaxation of the membrane together with viscous drag-induced movement of the medium. In the control chamber, the movement of the clamp affected only the medium. Fluid motion due to clamp motion was nearly identical in both chambers.

For the cyclic stretch experiments, the motor speed and cam eccentricity were adjusted to subject the experimental membrane to cyclic deformation of 10% of its unstretched length at 1 Hz.

To determine the effect of mechanical forces on the microtubule components of the cytoskeleton, control and experimental cell monolayers, grown on either glass or polyurethane, were processed to visualize their microtubule networks using antibody binding. The cell monolayers were fixed in formalin and then lysed using Triton X-100. The cytoskeletal elements were incubated with a buffer solution of antibody to tubulin. After incubation, a second anti-I_gG antibody with a fluorescein isothio-cyanate (FITC) tag was used to label the bound antibody. The FITC label was visualized using a Nikon epifluorescent microscope. The details of the technique can be found in Ref. 16.

Endothelial cells from human umbilical veins aligned perpendicular to the direction of the membrane deformation in which the cells were cyclically stretched for 24 and 48 h (Fig. 8.13). The alignment response was quantitated by morphometric analysis (Fig. 8.14), and it was observed that the EC had a

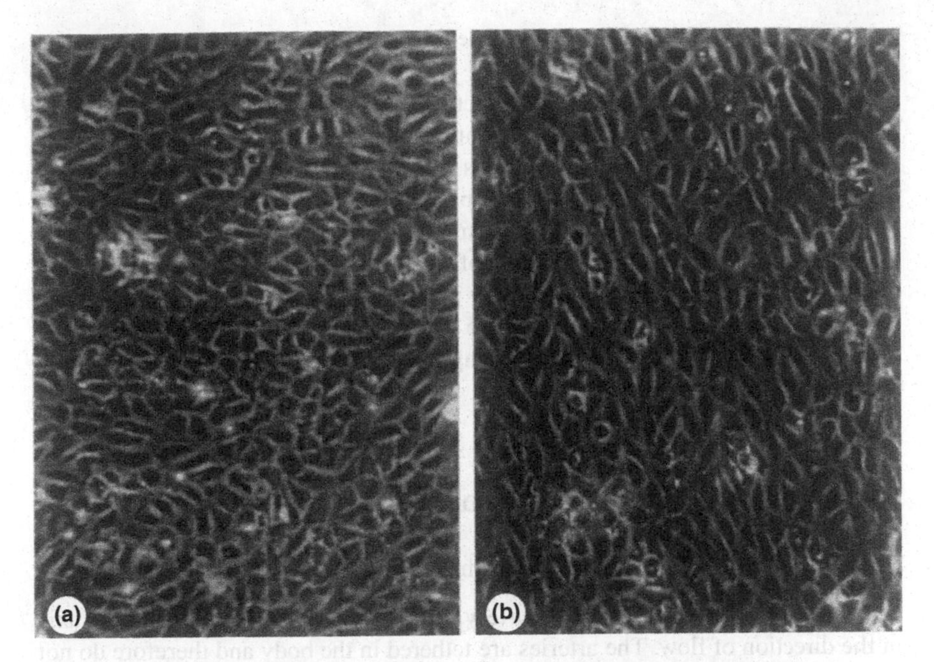

FIGURE 8.13. Human umbilical vein endothelial cells (primary) cultured on Mitrathane for 3 days then subjected to 48 h of (a) movement of the medium over the cells and (b) 10% cyclic stretching of the Mitrathane at 1 Hz. Axis of stretch is horizontal. (Reproduced from Ives CL et al., In Vitro Cell Dev Biol 1986;22:500-507, with permission from the Society for Invitro Biology.)

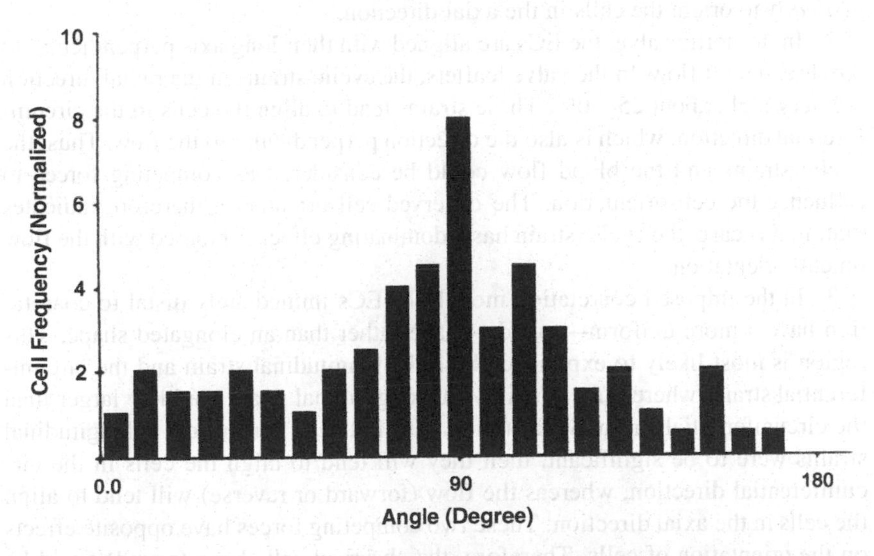

FIGURE 8.14. Histogram of cells in Figure 8.13b shows distribution of angles made by the cells' maximum dimensions with the axis of stretch, defined as 0 degrees to 180 degrees. The cells aligned most frequently 90 degrees away from the direction of stretch. (Reproduced from Ives CL et al., In Vitro Cell Dev Biol 1986;22:500-507, with permission from the Society for Invitro Biology.)

pronounced tendency to align in the direction 90 degrees to the stretch. The degree of confluence did influence the alignment rate, and cells in less confluent portions aligned more rapidly. Figure 8.13 shows that the movement of the medium over the cells, without cyclic stretch, did not produce alignment of the cells.

Using the antitubulin antibody, the microtubule complexes were visualized. Consistent with the elongated nature of cyclically stretched cells, there was a greater density of microtubules running parallel to the major axis of the cell. In summary, the endothelial cells elongate and orient perpendicular to the axis of cyclic deformation, and these shape changes are also reflected in the alignment of the microtubule network.

8. Endothelial Cell Orientation and Mechanical Forces

Let us summarize certain salient points that emerge out of the description thus far.

1. In the straight segment of the artery, the ECs are aligned with their long axis in the direction of flow. The arteries are tethered in the body and therefore do not experience much change in the axial direction from pulsating pressure. However, in the circumferential direction, they undergo cyclic strain with each pressure pulse. In this case, the cyclic strain will tend to orient the cells in the axial direction. Also, the blood flow will tend to orient the cells in the axial direction. Thus, the two mechanical forces, the blood flow and the pulsating pressure, act *synergistically* to orient the cells in the axial direction.

2. In the aortic valve, the ECs are aligned with their long axis perpendicular to the direction of flow. In the valve leaflets, the cyclic strains in the radial direction are very high, about 25–30%. These strains tend to align the cells in the circumferential direction, which is also the direction perpendicular to the flow. Thus, the cyclic strains and the blood flow could be considered as competing forces to influence the cell orientation. The observed cell orientation therefore indicates that, in this case, the cyclic strain has a dominating effect compared with the flow on cell orientation.

3. In the imposed coarctation model, the ECs immediately distal to coarctation have a more uniform—round—shape rather than an elongated shape. This region is most likely to experience both the longitudinal strain and the circumferential strain, where quite possibly the longitudinal strain could be larger than the circumferential strain. Such studies have not been performed. If longitudinal strains were to be significant, then they will tend to align the cells in the circumferential direction, whereas the flow (forward or reverse) will tend to align the cells in the axial direction. These two competing forces have opposite effects on the orientation of cells. Therefore, the observed cell shape (round) could be taken to indicate that neither of the two forces have a dominating effect on the cell orientation.

4. In the branching region of the artery also, the ECs have a more uniform—round—shape. In this region, the presence of stress concentration and increased

strain should be considered as forces competing with the direction of flow for determining the cell shape and orientation.

9. Distribution of Low-Density Lipoprotein in the Aorta

9.1. *Introduction*

We saw earlier (Chapters 1 and 2) that atherosclerosis is one of the major diseases of the artery. Atherosclerosis is enhanced by some well-identified risk factors such as the cholesterol level in the blood, blood pressure, and diabetes. Within the family of cholesterol, low-density lipoprotein (LDL) is known to be the major factor to enhance the risk of atherosclerotic disease. In the following sections, we will consider how LDL is present in the artery, how its distribution may be influenced by the branch and nonbranch regions, and how the pressure and the heart rate may influence its distribution. As we have seen earlier (Chapters 6 and 7), the wall stress in the artery changes with location and pressure, and its effects can change with the heart rate. Therefore, the study of LDL under various conditions can allow us to formulate a link between the arterial mechanics and pathology.

9.2. *Distribution of LDL in the Rabbit Thoracic Aorta*

Bratzler et al. (17) studied the distribution of human LDL in the thoracic aorta of rabbits in vivo. LDL (density 1.025–1.05 g/mL) was isolated from human blood plasma by ultracentrifugation. There are several other steps involved in the process of LDL preparation, for the details of which the reader may go to Ref. 17. The LDL was then made radioactive by binding it with radioactive iodine, which produces iodinated LDL, ^{125}I-LDL. For this binding reaction, the iodine monochloride method of McFarlane (18), as modified by Bilheimer (19), was used with 1:1 molar ratio of iodine to LDL protein. After more processing, the final product obtained had a specific activity of 0.4 ± 0.2 mCi/mg LDL apoprotein. By chloroform-methanol extraction, it was estimated that less than 4% of the precipitable radioactivity was associated with lipid.

The radiolabeled LDL (1.5–7 mL of saline containing 0.4–4 mCi and 1–8 mg LDL apoprotein/mL) was then injected intravenously into normal conscious New Zealand white rabbits (4.5–5 kg). Blood samples from rabbits were taken at various intervals to determine plasma radioactivity and total cholesterol. The rabbits were sacrificed after 10 min, 20 min, 4 h, 24 h, or 67 h, and the descending thoracic aorta was removed and frozen. The samples of frozen aorta were sectioned (frozen sections, 20 μm thick) parallel to the intimal surface. The sectioning proceeded from the adventitial side. The volume of each slice was estimated from the thickness and the cross-sectional area. Each tissue slice, the injection solution, and the plasma samples were extracted with TCA to remove non-protein-bound activity before the radio assay in a gamma counter. All slices and samples were counted for 10 min. We will consider some of their findings.

Figure 8.15 shows the average profiles of I-LDL distribution (TCA-precipitable radioactivity) from the intima, through the thickness of the aorta, to the adventitia. The profiles indicate how the LDL is distributed after different intervals of circulation ranging from 10 min to 67 h. The X-axis represents normalized depth of the section where the distance from the intima to the middle of the section has been divided by the wall thickness (L). The Y-axis represents the relative tissue concentration, which is the ratio of the tissue count C_T to the initial

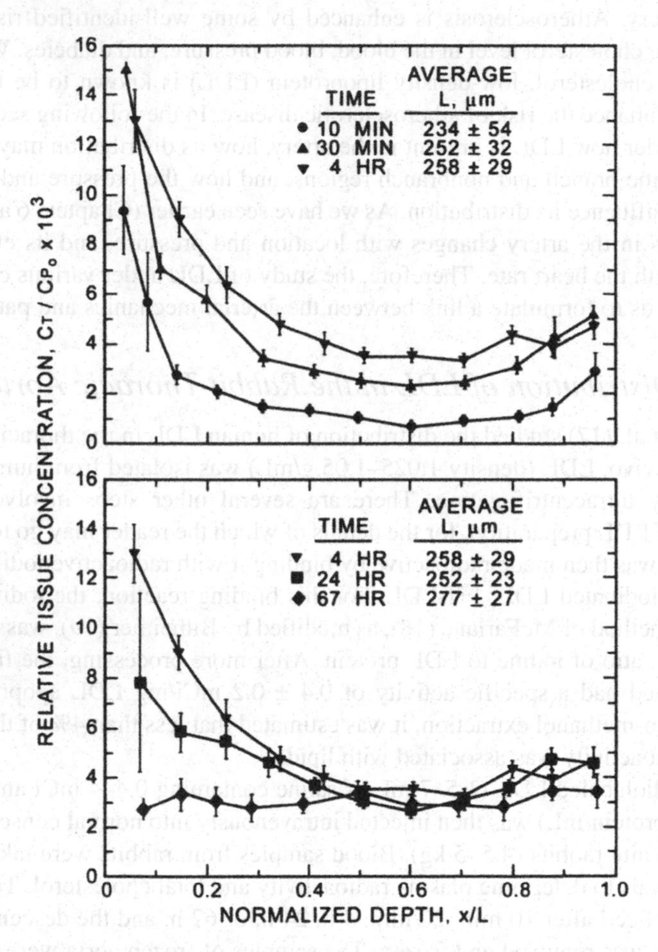

FIGURE 8.15. Grand average profiles of LDL distribution in the aortic wall at various intervals after radioactive LDL injection. C_T/C_{PO}, relative tissue concentration of TCA – precipitable radioactivity. C_T, tissue radioactivity (counts min^{-1} cm^{-3} wet tissue). C_{PO}, plasma radioactivity (counts min^{-1} cm^{-3} plasma) 3–9 min after injection. X, distance from the initimal surface to the midpoint of the tissue slice. L, distance between intimal surface and medial-adventitial border. (Reproduced from Bratzler RL et al., Arteriosclerosis 1977;28:289-307, with permission from Lippincott, Williams & Wilkins.)

plasma count C_{po}. The tissue count (C_T) represents the TCA-precipitable tissue radioactivity (counts min^{-1} cm^{-3}) and the initial plasma count (C_{po}) is the number of counts min^{-1} cm^{-3} of plasma.

The initial I-LDL concentration in plasma ranged from 1 to 3×10^7 counts min^{-1} cm^{-3}. The plasma concentration dropped to 91% of its initial value after 30 min, 63% after 4 h, 27% after 24 h, and 7% after 67 h. In the 10-min experiments, relative tissue concentration ranged from about 0.0005 to 0.002 near the middle of the media and was about 5- to 10-fold higher near the intimal surface. It was at an intermediate level near the medial-adventitial border. Thus, a typical distribution profile of LDL was the highest LDL concentration at the intima, a rapid drop over the next few sections, a relatively flat profile through most of the media, and a rise again near the adventitial surface.

The grand average profile suggests that at 10 min, C_T/C_{po} was about 0.009 at the intimal surface and 0.001 near the middle of the media. Similar qualitative distribution of I-LDL was observed at 30 min and at 4 h where, in addition, the medial concentration was greater than that at 10 min. At 24 h, the I-LDL distribution was comparable with that at 4 h with the exception that the concentration was reduced at the intimal surface. At 67 h, all gradients had virtually disappeared, and the LDL profile was flat through the entire aortic wall.

The authors state that radiolabeled human LDL rapidly enters the rabbit aortic wall from both the luminal and adventitial sides, although the LDL concentration gradients are much higher near the intimal surface. The entry from the adventitial side is related to the presence of vasa vasorum, which carries the blood supply to the adventitia and to part of the media. This study eloquently describes how the LDL is distributed and how it may accumulate in the aortic wall with time. It also suggests that after a certain period, the accumulation does not rise. This suggests that a degradation process may now be active along with the accumulation process to produce the net concentration observed. The distribution profiles of LDL seen above suggest that the penetration of LDL in the artery wall may be driven by the process of diffusion based on the concentration gradient.

10. Distribution of LDL in the Branch and Nonbranch Regions of the Aorta

In the previous section, we considered the distribution of LDL in the straight segment of the aorta. However, the branch region is the more interesting location because of its propensity to develop atherosclerosis and because the mechanics becomes more significant there. Therefore, we will now consider the LDL distribution in the arterial branch region and compare it with the distribution in the straight segment.

Thubrikar et al. (20) studied the distribution of LDL in the branch and nonbranch regions using a similar approach as described in the previous section. They studied the distribution of rabbit LDL (^{125}I labeled) in the rabbit aorta in vitro.

In brief, LDL was prepared as follows. LDL donor rabbits were fed a 1.0% cholesterol and 2% corn oil diet for 7 days. Fresh whole blood in EDTA was obtained from them. LDL (*d* 1.019–1.063) was separated from the plasma by sequential ultracentrifugation and gel filtration and then iodinated using Bilheimer's modification (19) of the iodine monochloride method of MacFarlane (18). Contamination with free iodine was assessed by precipitation with 10% w/v trichloroacetic acid (TCA). Free iodine contributed <1.5% of the final radioactivity. Less than 10% of the radioactivity in the labeled LDL was associated with lipid, as estimated by chloroform-methanol extraction.

The aorta was harvested and used as follows. The rabbit was sedated with intramuscular ketamine and acepromazine, given 3000 units of heparin sodium intravenously, euthanized with pentobarbital, and the aorta exposed. Thirty milliliters of vascular smooth muscle (VSM) nutrient media (Medium 199) was injected immediately into the proximal descending thoracic aorta. The aorta was separated from the surrounding tissues and all branches were ligated. The abdominal aorta was harvested from the celiac to the left renal artery, with 3-cm lengths proximal and distal to these vessels. The cannulae were placed into the proximal and distal ends of the segment. The vessel was filled with VSM nutrient media. The cannulae were attached to two 3-mL syringes suspended in a rack. ^{125}I-labeled LDL (1.5 mL) was instilled into the vessel through the proximal cannula and mixed with the media within the vessel. This resulted in the LDL concentration in the intraluminal fluids of approximately 0.3 mg LDL protein/mL. A fluid column of 3 cm was maintained to assure a small but uniform intraluminal pressure. The rack was placed in a bath of VSM nutrient media. The entire apparatus was incubated in a shaker water bath at 37°C for 1 h. Samples of luminal LDL and nutrient media from the bath were removed for counting radioactivity. The aorta was then removed from the nutrient bath and rinsed with saline.

The aorta was sectioned transversely above and below the celiac, superior mesenteric, and left renal arteries. Each segment was opened longitudinally opposite the branch and immediately frozen between two glass slides. From the frozen segment, a small disk was cut from the aortic wall, adjacent but distal to the ostium of each branch, using a 1.8-mm-diameter punch. This is the branch sample. The second disk of each pair was cut from the aortic wall opposite the ostium. This is the nonbranch sample. Each sample was placed between two glass slides, after recording intimal orientation, then frozen and stored at –80°C until cryosectioning.

Figure 8.16 shows the locations of the tissue samples. Each sample disk was sectioned parallel to the intimal surface into 16-µm sections. Two 16-µm sections were placed in each vial and counted for 10 min using the Beckman Biogamma counter. Counts were recorded as counts/min per 32-µm section.

The distribution of ^{125}I-LDL through the aortic wall was plotted for each branch and nonbranch pair of disks. The summary distribution profile compiled from all branch samples was compared with that compiled from all nonbranch samples. Radioactivity reflecting precipitable ^{125}I-LDL accumulations was calculated using the TCA precipitation assay. As we note, there are many technical

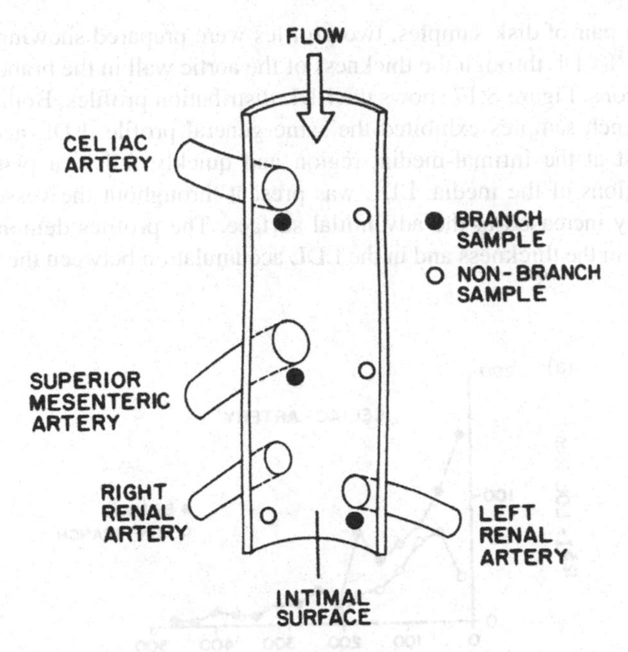

FIGURE 8.16. A drawing of the opened aorta and major arterial branches. Disks were removed from the aortic wall in branch and nonbranch regions. Branch samples were taken immediately distal to the ostia and the nonbranch samples from opposite the ostia.

details associated with the experiments above, and for those details readers may refer to Ref. 20.

We will now consider some of the important observations. The aortic wall was significantly thicker in the branch than nonbranch region. The average wall thickness was 257 ± 86 μm in the branch versus 154 ± 86 μm in the nonbranch region.

The LDL accumulation in each pair of sections at the same depth from the luminal surface was compared between the branch and nonbranch samples. The sections from the branch sample had a significantly greater LDL content. The average radioactivity per section was 194.34 counts/min in the branch versus 122.82 counts/min in the nonbranch region. The intima is approximately 4 to 8 μm thick. They evaluated LDL accumulation in the first two sections to compare LDL accumulation in the intimal-medial junction. The intimal-medial sections at the branch had significantly more LDL accumulation (280 counts/min) than those at the nonbranch region (149 counts/min). The total LDL accumulation in the branch region (1621 counts/min) was significantly greater than that in the nonbranch region (739 counts/min).

In a typical experiment, the average intraluminal LDL counts were 33,737 counts/min per μL, whereas the average tissue counts in the first two intimal-medial sections were 280 counts/min per 0.1 μL. Hence, LDL concentration in the intimal-medial region was approximately 8% of that in the intraluminal fluid.

For each pair of disk samples, two profiles were prepared showing the distribution of ^{125}I-LDL through the thickness of the aortic wall in the branch and nonbranch regions. Figure 8.17 shows the LDL distribution profiles. Both the branch and nonbranch samples exhibited the same general profile. LDL accumulation was highest at the intimal-medial region and quickly fell to a plateau in the deeper sections of the media. LDL was present throughout the vessel wall and occasionally increased at the adventitial surface. The profiles demonstrated the differences in the thickness and in the LDL accumulation between the branch and

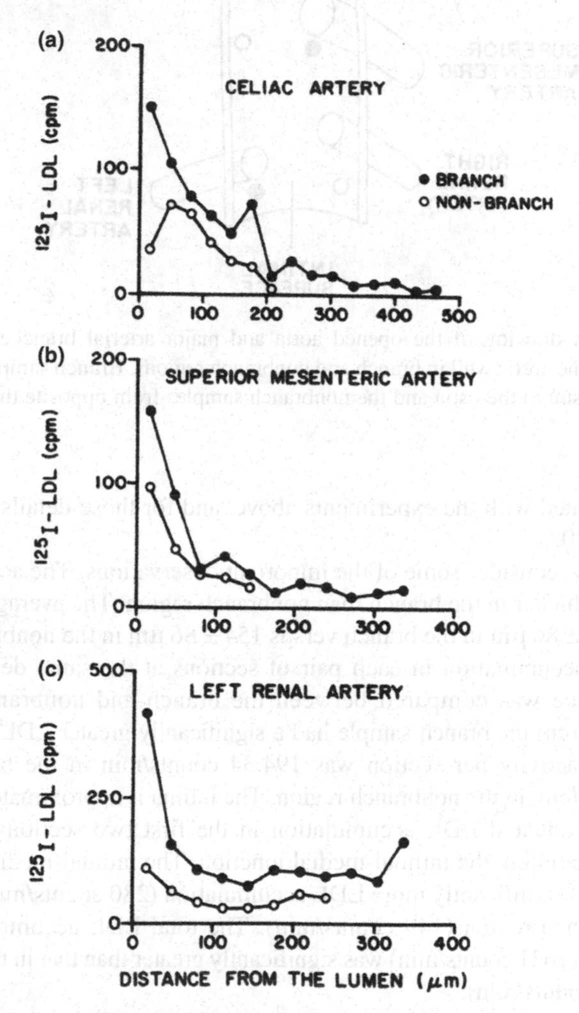

FIGURE 8.17. Profiles of LDL distribution through the thickness of the aortic wall in the branch and nonbranch regions. Accumulation of LDL is represented by counts per minute (cpm). Each symbol represents LDL accumulation at the midpoint of a 32-μm-thick tissue slice. LDL accumulation was higher in the branch than nonbranch region at any given depth.

nonbranch samples. The LDL distribution profile for the branch region was always higher than that for the nonbranch region, indicating that LDL accumulation in each section was greater in the branch region. Also, the profile was longer in the branch sample, indicating a greater wall thickness.

The TCA precipitation analysis was carried out to assess the amount of protein-bound radioactivity. The analysis of tissue samples revealed that 86% of the total radioactivity was in a precipitable form, bound to LDL apoprotein, and this did not change significantly through the thickness of the aortic wall. Overall, LDL accumulation at a given depth was greater in the branch region than in the nonbranch region. This is important in understanding the transport mechanism for LDL. This observation and the increased wall thickness suggests that there is more LDL and it stays longer in the artery wall in the branch region than in the nonbranch region. Concentration of LDL in the tissue ranged from 2% to 8% of that in the luminal fluid. The low tissue concentration implies that endothelial cells are still the dominant barrier to the permeation of LDL particles. Greater permeability of the artery to LDL at the branch implies that endothelial cell barriers may be less effective in that region.

11. Replication of Endothelial Cells and Its Preferential Localization

Cellular replication is needed for growth and repair of the tissue. Obviously then, a pattern of cell replication can be taken to suggest the pattern of growth or a pattern of cellular damage and repair. We will consider cell replication and its occurrence at preferred locations in this section. In the previous sections, we saw that the endothelial cells have different morphology in branch and nonbranch regions of the artery. We also found that the permeability of the artery to LDL is different in the branch and nonbranch regions. Therefore, it is of interest to examine whether the replication pattern of the endothelial cells is also different in the branch and nonbranch regions.

Wright (21) studied the pattern of endothelial cell replication in the branch and nonbranch regions, among other conditions, in the artery of guinea pigs. We will consider some of his findings. He used Häutchen (en face) preparations of endothelium to examine the cell replication. Normal male guinea-pigs of 250 to 750 g weight were used for the experiments. The animals were injected intracardially with tritiated thymidine (0.05 mCi/g of body wt) 16 and 24 h before they were sacrificed.

A carotid cannula was inserted under sodium pentobarbitone anesthesia and the animal was sacrificed. The aorta was washed free of blood by perfusion with saline and partially fixed by perfusion for 1 h with 4% formaldehyde solution at an arterial pressure (100 mmHg). Häutchen (en face) preparations of endothelium were made by a modification (22) of the techniques of Poole et al. (23) and of Warren (24). The layer of single cells was covered with Kodak AR 10 stripping film and exposed for autoradiography for 6 weeks. After developing, the slides were stained with Ehrlich's hematoxylin or neutral fast red to show the nuclei. The nuclei in which tritiated thymidine was incorporated during the phase of DNA synthesis of mitosis were readily seen by the localized reduction of the silver in the film. The labeled nuclei were counted in various areas and expressed as a percentage of the total number of nuclei present. About 1000 nuclei were counted in each area.

The thoracic aorta was divided into four equal lengths between the arch and the diaphragm; the abdominal portion was similarly divided. Figure 8.18 shows the average numbers of labeled nuclei in the various regions of the aorta. The highest mitotic index was found in the arch. The numbers of labeled nuclei decreased somewhat down the length of the aorta though they were frequent in the region of the diaphragm, just above the renal arteries. The area proximal to the bifurcation also showed considerable mitosis, but the upper parts of the iliac were low.

11.1. Areas Around the Mouths of Branches

The labeled endothelial cells were counted in areas immediately around the mouths of branches and in areas at the same level but away from the branch.

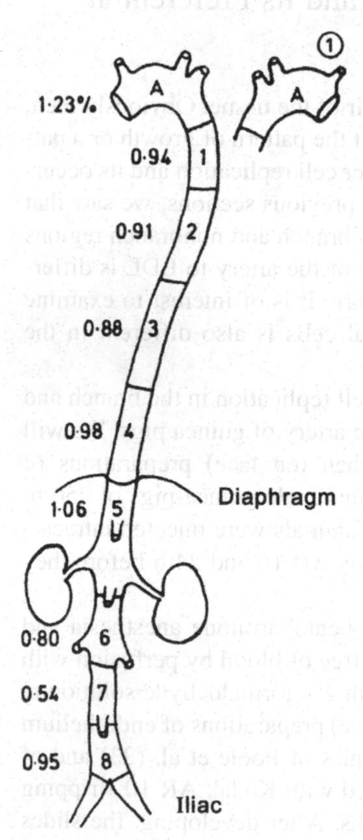

FIGURE 8.18. Diagram of aorta showing the percentage of mitoses at various levels. (Reproduced from Wright HP, Atherosclerosis 1972;15:93-100, with permission from Elsevier.)

The numbers of nuclei showing [^{3}H]thymidine uptake were greater at the mouths of branches (1.3%) than in the areas away from the branch (0.78%). The numbers of labeled nuclei were more than half as numerous around the branches than in the nonbranch areas.

11.2. The Effect of Constriction

The thoracic aorta was constricted by placing a U-shaped clip, made from silver wire 0.31 mm thick and 3.18 mm wide. This procedure produced significant constriction without completely occluding the aorta. In these animals, cell replication was studied.

The average percentage of labeled nuclei in the endothelium above the constriction, in the constricted area and below the constriction, was compared with normal controls. The highest incidence of [^{3}H]thymidine uptake was seen in the constricted area (3.31% vs. 0.86% in the normal). The percentage of labeled nuclei (1.13%) above the clip was also considerably higher than in the controls (0.9%). The aorta below the constriction, however, showed no difference from the controls (0.9%).

Overall, the mitosis is most frequent in the aortic arch, gradually declines down its length, and considerably more frequent around the mouths of branches.

The authors speculate that if mitosis represents the reparative replacement of endothelial cells, then the observations may be explained as follows. The constant bombardment of the blood in this region may cause damage to the endothelial cells, which are then replaced by the repair processes. Similar conditions are present at the diaphragmatic and renal level where there are numerous branches likely to produce eddies in the flow of blood. Just above the bifurcation of the aorta, streaming across the fork also produces turbulence and shearing that may injure the cells; in this region also mitosis is increased. The observation that mitosis is more frequent around the mouths of branches is consistent with the process of replacement of damaged cells (25).

In the partially constricted thoracic aorta, which increased pressure proximally, the rate of endothelial mitosis was increased above the constriction. Where blood flow-rate was increased in the narrowed lumen, the numbers of labeled nuclei were very high. In the artery distal to the clip where pressure was low, the mitosis was not significantly different from the control.

We may also interpret these findings in a slightly different way. The arch undergoes more displacement and lengthening due to the downward pull exerted on it by the beating heart than the remaining aorta. Thus, more displacement in the arch could be correlated with increased cell replication there. The area around the mouths of branches is the region of high stress and strain and that can be correlated with a high labeling index. At the constriction itself, there are increased strains and stresses. Thus, more labeled cells at the constriction can be correlated with the increased stresses and strains besides the altered flow.

12. Endothelial Cell Replication at the Site of Increased Permeability

We can already relate to a phenomenon that appears to suggest a link between arterial mechanics and cellular processes. For instance, at the branch region we have noted that there is a stress-concentration and an increase in strain. Also, in this region the endothelial cell morphology is altered, the arterial permeability is greater, and the cell replication is increased. Furthermore, the stresses as well as cellular processes are influenced by the systemic pressure. In this section, we will describe some observations that further relate cell replication to the permeability, and in the following sections we will consider the effect of hypertension.

Caplan et al. (26) studied the endothelial cell turnover in the areas of Evans blue uptake in pigs using Häutchen preparation and the tritiated thymidine technique described in the past section. Male Yorkshire pigs weighing 30–40 lb were used for the experiments.

Pigs were anesthetized by sodium pentobarbital and 45 mL of Evans blue solution was injected through a jugular vein. Evans blue was prepared as an 0.5% solution (w/v) in 0.85% saline. Three hours later, the pigs were sacrificed, the aortas removed intact from the aortic valve to the trifurcation and immersed in ice-cold Krebs-Ringer bicarbonate (KRB). The aorta was opened longitudinally. Areas of Evans blue dye accumulation on the intimal surface were mapped, and full-thickness segments with an area of 1.0–1.5 cm^2 were excised from both blue and white areas in the aortic arch and also from white areas in the upper abdominal aorta. The typical pattern of Evans blue uptake observed is shown in Fig. 8.10a.

Excised aortic segments used for autoradiography were first pre-incubated in oxygenated KRB at 37°C for 20 min and then transferred to oxygenated KRB containing 5 μCi/mL of thymidine-6 [^{3}H]. Tritiated thymidine had a specific activity of 23.3 Ci/mmole. After incubation with shaking in a water bath for 1 h at 37°C, the segments were rinsed. "En face" endothelial preparations were made from [^{3}H]thymidine incubated segments. Aortic segments were dried and the endothelial surface covered by cellulose acetate strips. These strips were held firmly and uniformly on the endothelium with the aid of glass slides and weights, until the cellulose acetate paper had dried (usually 15 min), after which the aortic wall was removed with fine forceps, leaving the endothelium adherent to the cellulose acetate.

The Häutchen preparations from blue and white aortic segments were coated with Kodak Nuclear Emulsion NTB-2 in a dark room at a temperature of 18–22°C. Slides were dipped in emulsion for 2 s and allowed to dry for 20–30 min. After coating, the cellulose acetate strips attached to glass microscopic slides were placed in black slide boxes that were sealed and placed in an X-ray film storage box. Exposure time was 3–5 days at 4°C. Autoradiographic preparations were stained with hematoxylin and eosin.

12.1. Endothelial Cell Labeling

The [^{3}H]thymidine labeling of endothelial cells derived from blue and white areas in the aortic arch showed interesting results. The percentage of labeled cells (labeling index) was greater in areas of Evans blue accumulation than in contiguous white areas. Of a total of 26,330 endothelial cells counted from white areas, 0.63% were labeled, as compared with 21,929 cells counted from blue areas of which 0.86% were labeled. Figure 8.19 illustrates the labeling patterns. Thus, endothelium from areas of aortic Evans blue accumulation was found to show a significantly greater [^{3}H]thymidine labeling index than white areas, a finding that can be interpreted as indicating a more rapid spontaneous endothelial cell turnover in blue areas. In other words, the areas of the aorta that accumulate Evans blue dye are probably subject to continuous endothelial injury, which is reflected in enhanced endothelial regeneration.

The authors note that the permeability is most often increased around the branch regions and therefore cell turnover in those regions is understandably more, as was seen in the previous section. However, they state that the cell turnover is higher in the regions where the permeability is higher and that such regions are not limited to the branch sites.

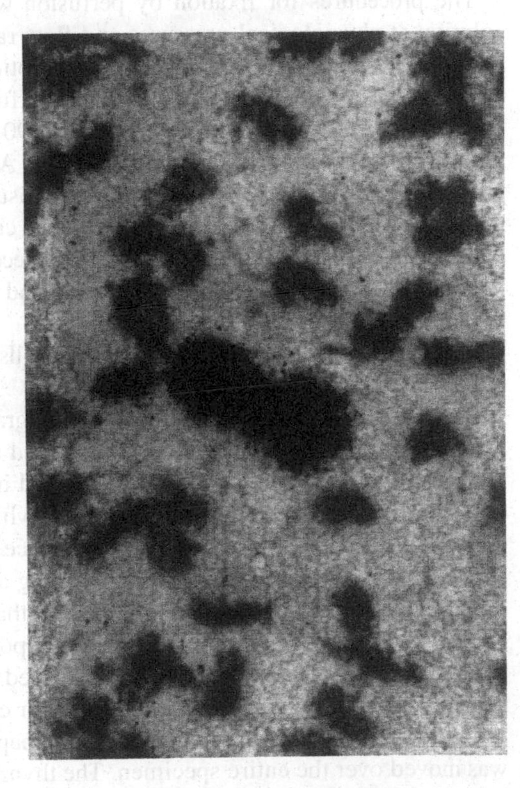

FIGURE 8.19. Autoradiograph of Häutchen preparation from an area of Evans blue uptake in the thoracic arch. Two labeled nuclei in close juxtaposition can be seen. (Reproduced from Caplan BA et al., Atherosclerosis 1973;17:401-417, with permission from Elsevier.)

13. Endothelial Cell Replication Under Hypertension

Increase in the blood pressure increases the load on the arterial wall and thus increases tension (hypertension) in the wall. The artery begins to respond to the increased blood pressure by increasing the wall thickness. In chronic hypertension, the artery wall continues to thicken until it reaches a certain value at which point the wall stress becomes similar to that before hypertension. As we saw in Chapters 6 and 7, in chronic hypertension the mean wall stress may be the same as that in the control, however, the stress on the inner surface and the stress gradient through the wall must increase. With this observation in the background, it is of interest to examine how the endothelial cell replication changes in hypertension.

Schwartz et al. (27) studied the effect of hypertension on the endothelial cell replication in rats. They used female Sprague-Dawley rats 5–6 months old. Tritium-labeled thymidine was injected intraperitoneally (50 μCi/100 gm) in the rats. The experiments were designed to determine the effect of hypertension on the labeling index. Three doses of tritiated thymidine were administered at 8-h intervals. This schedule was chosen to minimize diurnal variation and to provide large numbers of cells to permit accurate counting. This schedule provides an estimate of the number of cells that replicate each day.

The procedures for fixation by perfusion with formaldehyde-glutaraldehyde solution at physiological pressures and flow rates were as follows. Under ether anesthesia, a catheter was inserted in the carotid artery. A flowmeter was used to estimate the pressure drop between the perfusion bottle and the catheter tip. Vessels were fixed at an aortic pressure of 90–100 mmHg and at flow rates of 20–40 mL/min over an interval of 5–10 min. After removal of the vessel, all aortas were divided into five pieces. For the measurement of the aortic surface area, the aortas were not divided, instead they were cut just below the subclavian artery and just above the trifurcation. This single piece was pinned out flat without distending the vessel. Measurements of length and width were made at several points to determine the surface area.

Cell density (CD; the average number of cells per unit area), was obtained from en face preparations described below.

Preparation of endothelium for autoradiography consisted of three steps. 1) Removal of endothelium: The aorta is opened and pinned out flat. The tissue is dehydrated and the endothelium is embedded in a layer of collodion. The collodion film is stripped away from the aorta, which then removes the endothelial layer as a sheet. 2) Exposure of luminal surface of endothelial cells: The exposed subendothelium is fixed to a glass slide using a layer of gelatin. Then the collodion is removed by dissolving it in 1:1 ether/ethanol. 3) Autoradiography: A sheet of endothelium, with the luminal surface exposed, is coated with emulsion for autoradiography. The preparations are exposed for 2 weeks and then developed, stained with hematoxylin, and mounted under coverslips.

Cell counts were made using a square-shaped reticule and the square field was moved over the entire specimen. The thymidine index was calculated as the fraction of cells labeled divided by the total number of cells.

13.1. Creation of Hypertensive Rats

The surgical procedure used intraperitoneal anesthesia with pentobarbital. The renal artery was exposed. A figure-8 suture was placed on the vessel snugly but without totally occluding flow. Rats were defined as hypertensive once the pressure remained over 130 mmHg for a minimum of 6 weeks. For further details of the techniques, readers may refer to Ref. 27. We will now consider some of the important findings.

Cell replications represent the sum of the rate of cell proliferation (i.e., rate of population increase) and the rate of cell turnover (i.e., rate of replacement of spontaneously lost cells). In most young animals, growth is very rapid, and the portion of the dividing cell population involved in expansion of cell number constitutes a large portion of the number of replicating cells. The contribution of proliferation to the thymidine index therefore should reach a minimum value coincident with cessation of growth.

13.2. Effects of Hypertension

The rats were sacrificed within 2 weeks after their blood pressure exceeded 130 mmHg. As can be seen in Figure 8.20, hypertensive rats showed a 5- to 10-fold increase in replication rate compared with the controls. The hypertensive rats showed an elevated rate of replication in all segments examined.

The principal effect of hypertension is a 10-fold increase in the *replication* rate of endothelium. This increase could be interpreted to indicate cell damage due to hypertension as reflected in the increased rate of cell *turnover*. It could also be interpreted to indicate increased cell *proliferation* as a response to increased stretching of the aortic wall resulting in increased surface area. The authors state that the increase in cell replication in hypertension represents, at least in part, a proliferative response to cover the expanded luminal surface of the dilated vessel.

14. Association Between Endothelial Cell Replication, Arterial Permeability, and Hypertension

We saw earlier that there is an association between various factors (see The Chart, above) such as endothelial cell function, arterial permeability, hypertension, and so forth. Let us therefore consider an interesting study made by Wu et al. (28) that explores the three parameters mentioned above.

The study was done in spontaneously hypertensive rats (SHR), age 3–4 months and weighing 257–333 g. Their blood pressures ranged from 155 to 250 mmHg. These hypertensive rats were inbred descendants of the Wistar strain originally developed by Okamoto et al. (29). Male, age-matched, normotensive (blood pressure 111 ± 8 mmHg) Wistar-Kyoto (WKY) rats, weighing 287–405 g, were the control group.

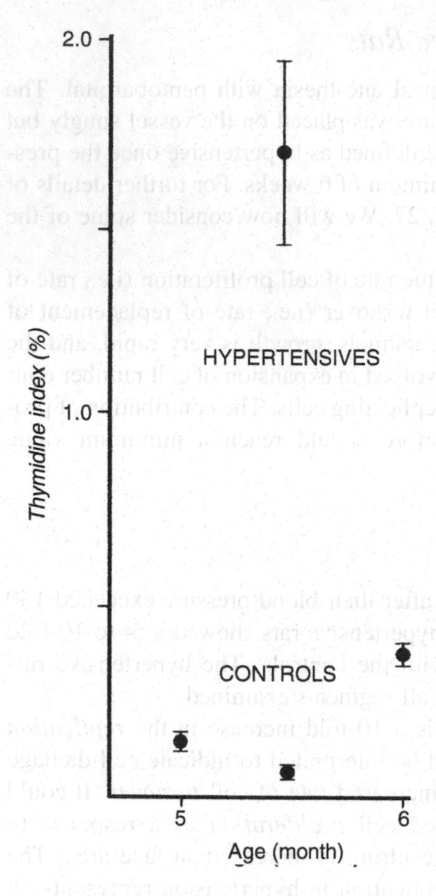

FIGURE 8.20. Comparison of the endothelium of hypertensive rats with endothelium of normal rats of similar age. The mean replication rate has increased by 10-fold. (Reproduced from Schwartz SM et al., Circ Res 1970;41:248-255, with permission from Lippincott, Williams & Wilkins.)

Evans blue–albumin (EBA) conjugate was prepared by adding 0.5 g bovine serum albumin and 0.1 g Evans blue to 50 mL normal saline. The experiments were performed while the rats were under pentobarbital anesthesia. Blood pressures were measured through the carotid artery. The catheter placed in the femoral vein was used for intravenous injection of a macromolecular tracer; the catheters that were placed in the femoral artery and the femoral vein both served as the egress routes during perfusion. A total of 5 mL EBA solution was slowly injected into the left femoral vein. At 3 minutes after EBA injection, an overdose of pentobarbital was given. The carotid artery was perfused immediately with a heparinized saline solution at a pressure of 110 mmHg. The perfusate was then switched to 10% formaldehyde, which was perfused at the same pressure and a flow of approximately 20 mL/min for 10 min. After perfusion fixation, the aorta was excised between the aortic root and the diaphragm and immersed in 10% formaldehyde for 1 h.

14.1. Methods for Detecting Dying or Dead Endothelial Cells

The thoracic aorta was cut open longitudinally. The specimen was dissected into five pieces. For en face preparation, each piece was pinned with the endothelial side up. Four aortic pieces were incubated with rabbit anti-rat IgG. One aortic piece was used as control. All five pieces were incubated for 90 min with horseradish peroxidase–conjugated goat anti-rabbit IgG. Horseradish peroxidase was visualized by a modification of the method described by Grahan et al. (30).

14.2. Quantification of Immunoglobulin G–Containing Dying or Dead Endothelial Cells

After overnight immersion in 10% formaldehyde, the aortic specimens were stained with hematoxylin. IgG-containing endothelial cells (as revealed by horseradish peroxidase staining) and mitotic endothelial cells (as revealed by hematoxylin staining of mitotic nuclei) were identified and counted on en face preparations of the thoracic aorta under a fluorescence microscope. The distribution of IgG-containing endothelial cells and mitotic endothelial cells was mapped as a function of topographic locations in the aorta and correlated with the distribution of macromolecular tracer (EBA).

14.3. Fluorescence Microscopy

A Nikon epifluorescence microscope was used for en face observation. All mitotic and dying (IgG-positive) cells, as well as EBA leaky foci, were registered. The entire endothelial surface including the branching region was scanned. The numbers of EBA leaky foci, mitotic endothelial cells, and IgG-containing dying or dead endothelial cells were counted. EBA fluorescence was detected. Some of their important findings are as follows.

The fluorescent leaky spots correlated well with endothelial cell mitosis (Fig. 8.21). A comparison of the findings of SHR and WKY rats showed EBA leaky foci per unit endothelial surface area 2.8-fold larger in SHR (7.46/mm^2) than in WKY (2.70/mm^2) rats (Fig. 8.22). This increase of leakage could be attributed at least partially to a higher endothelial cell turnover rate in SHR, whose average mitotic frequency was 2.9 times and cell death frequency was 3.2 times the corresponding values in WKY rats.

Comparison between branching and nonbranching regions showed significant increases of EBA leakage and endothelial cell turnover (mitosis and death) in the branching regions. These findings suggest that hypertension correlates with increased endothelial cell turnover, which may produce more leaky endothelial junctions and thus lead to an enhanced endothelial permeability to macromolecules. The authors also state that clinical and postmortem studies indicate that it is often the regions of arterial branching and sharp curvature that have the greatest predilection for the development of atherosclerosis. These are also the regions

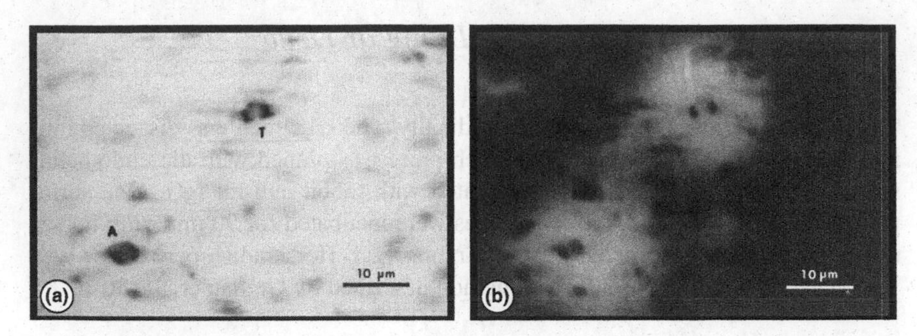

FIGURE 8.21. (a) Mitotic cells in anaphase (A) and telophase (T), observed en face under the light microscope. (b) Fluorescence photomicrograph showing the same field as (a). Note endothelial leakage to Evans blue–albumin that corresponds with the two mitotic cells. (Reproduced from Wu CH et al., Hypertension 1990;16:154-161, with permission from Lippincott, Williams & Wilkins.)

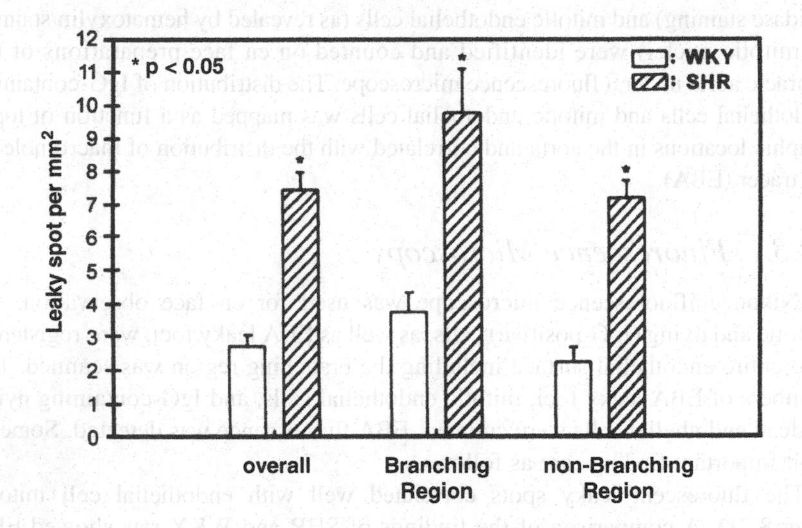

FIGURE 8.22. Leaky spots (endothelial leakage to Evans blue–albumin) per unit endothelial surface (en face preparation) in spontaneously hypertensive rats (SHR) and Wistar-Kyoto rats (WKY). Data are presented for entire thoracic aorta (overall) and also separately for branching and nonbranching regions. Leaky spots density is significantly higher in SHR. Also, permeability in branching region is higher than that in nonbranching region for both SHR and WKY. (Reproduced from Wu CH et al., Hypertension 1990;16:154-161, with permission from Lippincott, Williams & Wilkins.)

that correspond with the "blue area" in animal experiments. This study also showed a significant increase of EBA leaky foci in the branching regions of the aorta, where higher frequencies of endothelial cell mitosis and death were simultaneously noted. This finding could serve as a pathogenic basis for the higher propensity for atheromatous lesions in the branching areas.

References

1. Gerrity RG: Arterial endothelial structure and permeability as it relates to susceptibility to atherogenesis. In: Glagov S, Newman WP III, Schaffer SA (eds). Pathobiology of the Human Atherosclerotic Plaque. Springer-Verlag, New York, 1986:13-45.
2. Zarins CK, Taylor KE, Bomberger RA, Glagov S: Endothelial integrity at aortic ostial flow dividers. Scan Electron Microsc 1980;(3):249-254.
3. Reidy MA, Levesque MJ: A scanning electron microscopic study of arterial endothelial cells using vascular casts. Atherosclerosis 1977;28:463-470.
4. Reidy MA: Arterial endothelium around rabbit aortic ostia: A SEM study using vascular casts. Exp Mol Pathol 1979;30:327-336.
5. Baker JW, Thubrikar MJ, Parekh JS, Forbes MS, Nolan SP: Change in endothelial cell morphology at arterial branch sites caused by a reduction of intramural stress. Atherosclerosis 1991;89:209-221.
6. Svendsen E, Jorgensen L: Focal "spontaneous" alterations and loss of endothelial cells in rabbit aorta. Acta Pathol Microbiol Scand 1978;86(1):1-13.
7. Gerrity RG, Richardson M, Somer JB, Bell FP, Schwartz CJ: Endothelial cell morphology in areas of *in vivo* Evans blue uptake in the aorta of young pigs. Am J Pathol 1977;89(2):313-334.
8. Björkerud S, Bondjers G, Bylock A, Hansson G: Studies of arterial endothelial integrity with the dye exclusion test – a review. In: Schwartz CJ, Werthessen NT, Wolf S (eds). Structure and Function of the Circulation. Plenum Press, New York, 1981:211-238.
9. Friedman M, Byers SO: Excess lipid leakage: A property of very young vascular endothelium. Br J Exp Pathol 1963;43:363.
10. Levesque MJ, Liepsch D, Moravec S, Nerem RM: Correlation of endothelial cell shape and wall shear stress in a stenosed dog aorta. Arteriosclerosis 1986;6:220-229.
11. Somer JB, Evans G, Schwartz CJ: Influence of experimental aortic coarctation on the pattern of aortic Evans blue uptake *in vivo*. Atherosclerosis 1972;16(1):127-133.
12. Deck JD: Endothelial cell orientation on aortic valve leaflets. Cardiovasc Res 1986;20(10):760-767.
13. Thubrikar MJ: The aortic valve. CRC Press, Boca Raton, 1990:97-127.
14. Dewey CF, Bussolari SR, Gimbrone MA, et al. The dynamic response of vascular endothelial cells to fluid shear stress. J Biomech Eng 1981;103:177-185.
15. Sumpio BE, Banes AJ, Buckley M, Johnson G: Alterations in aortic endothelial cell morphology and cytoskeletal protein synthesis during cyclic tensional deformation. J Vasc Surg 1988;7:130-138.
16. Ives CL, Eskin SG, McIntire LV: Mechanical effects on endothelial cell morphology: In vitro assessment. In Vitro Cell Dev Biol 1986;22(9):500-507.
17. Bratzler RL, Chisolm GM, Colton CK, Smith KA, Lees RS: The distribution of labeled low-density lipoproteins across the rabbit thoracic aorta in vivo. Arteriosclerosis 1977;28:289-307.
18. McFarlane AS: Efficient trace labeling of proteins with iodine. Nature (London) 1958;182:53.
19. Bilheimer DW, Eisenberg S, Levy RI: The metabolism of very low density lipoprotein proteins. I. Preliminary in vitro and in vivo observations. Biochim Biophys Acta 1972;260:212.
20. Thubrikar MJ, Keller AC, Holloway PW, Nolan SP: Distribution of low density lipoprotein in the branch and non-branch regions of the aorta. Atherosclerosis 1992;97:1-9.

21. Wright HP: Mitosis patterns in aortic endothelium. Atherosclerosis 1972;15:93-100.
22. Obaze DER, Wright HP: A modified technique for producing "en face" (Häutchen) preparations of endothelium for autoradiography. J Atheroscler Res 1968;8:681.
23. Poole JCF, Saunders AG, Florey HM: The regeneration of aortic endothelium. J Pathol Bacteriol 1958;75:133.
24. Warren BA: A method for the production of "en face" preparations one cell in thickness. J Roy Microsc Soc 1965;24:407.
25. Wright HP: Endothelial mitosis around aortic branches in normal guinea-pigs. Nature (London) 1968;220:78.
26. Caplan BA, Schwartz CJ: Increased endothelial cell turnover in areas of *in vivo* Evans blue uptake in the pig aorta. Atherosclerosis 1973;17(3):401-417.
27. Schwartz SM, Benditt EP: Aortic endothelial cell replication. I. Effects of age and hypertension in the rat. Circ Res 1977;41(2):248-255.
28. Wu CH, Chi JC, Jerng JS, Lin SJ, Jan KM, Wang DL, Chien S: Transendothelial macromolecular transport in the aorta of spontaneously hypertensive rats. Hypertension 1990;16:154-161.
29. Okamoto K, Aoki K: Development of a strain of spontaneously hypertensive rats. Jpn Circ J 1963;27:282-293.
30. Grahan RC, Karnovsky MJ: The early stages of absorption of injected horseradish peroxidase in the proximal tubules of mouse kidney: Ultrastructural cytochemistry by a new technique. J Histochem Cytochem 1966;14:291-302.

9 Smooth Muscle Cells and Stretch

1. Introduction

Smooth muscle cells (SMCs) are the essential part of the arterial structure. They are also the essential part of both the normal arterial function and the arterial pathology. In the structure of the artery, the medial layer makes up the majority of the wall thickness in the larger elastic arteries and an even greater proportion of the wall thickness in the smaller contractile arteries. The media is comprised primarily of the smooth muscle cells and the connective tissue surrounded by space containing mucopolysaccharides (see Chapter 3). The SMCs are generally of two phenotypes: i) synthetic and ii) contractile. In the elastic arteries, the

SMCs are primarily involved in their synthetic function whereby they synthesize the connective tissue required for the maintenance of the elasticity of the artery. In the smaller arteries, SMCs are primarily involved in their contractile function, and through contraction they change the vascular diameter and consequently the vascular resistance, thereby regulating the blood flow.

In atherosclerosis, aneurysm, balloon angioplasty, and hypertension, the role of SMCs is of prime importance. In atherosclerosis for instance, the bulk of the plaque involves SMCs in at least three obvious ways: i) SMCs proliferate and therefore there are more of them, ii) SMCs take up lipid and therefore occupy more space, and iii) more SMCs synthesize more connective tissue, which also occupies more space (i.e., larger plaque). In clinically significant atherosclerosis, the tendency of SMCs to proliferate could be the critical step, and therefore, we will look at the proliferation of SMCs in relation to vascular mechanics; in particular, stretch in the artery.

In the formation of aneurysm, almost the opposite of atherosclerosis occurs. There is a loss of SMCs and a thinning of the media. It is of great interest there to consider the following possibility: Could it be that a certain amount of stretch stimulates SMCs to proliferate, whereas overstretching may actually rupture and destroy the SMCs thereby causing the loss of cells? We will therefore examine the relationship between the quantity of stretch and the response of SMCs.

Balloon angioplasty has become a routine practice in cardiology, and it is known that in a significant number of cases, the region of angioplasty develops restenosis. In the balloon angioplasty procedure, the artery diameter is forcefully expanded by the balloon (i.e., the artery wall is stretched significantly in the circumferential direction), and the development of restenosis after angioplasty involves, once again, a major proliferation of SMCs. It is not known at this time what degree of stretch may produce restenosis or what may be a "safe" stretch for opening up the blocked artery.

In case of hypertension, it is known that the artery wall becomes thicker and that there is an increase in the mass of SMCs. Also, hypertension (increased wall stress) is a well-documented risk factor for atherosclerosis. Thus, it is important to examine the relationship between vascular mechanics and the response of SMCs to gain an insight into the development of vascular pathology.

2. SMC Proliferation and Hypertension

Bevan (1) studied SMC proliferation using an autoradiographic technique in rabbit arteries and correlated it with hypertension. Male white New Zealand rabbits, 2.4–3.0 kg, were used. The rabbits were anesthetized, and a direct recording of the blood pressure through the femoral artery was made. A silk ligature was placed around the abdominal aorta just proximal to the superior mesenteric artery and tightened to reduce the mean blood pressure distal to the ligature by 50%. Sham-operated animals with a loose ligature around the aorta were used as controls. At 3, 6, 14, 21, and 28 days postoperatively, the animals were given intravenous [^{3}H]methyl thymidine 2 μCi/g body weight, specific activity 3.0 Ci/mmol. After 2 h, arterial blood pressure recordings were made under light Nembutal anesthesia and the animals were sacrificed. Heart and kidneys were weighed, and portions of the aorta, common carotid, ear, brachial, superior mesenteric, renal, and femoral arteries, gall bladder, pancreas, mesentery, and kidney were fixed in Carnoy's solution, dehydrated, and embedded in paraffin. Five-micrometer serial sections of tissues were processed for light microscope autoradiography or stained with the following stains to define components of the arterial wall: Verhoeff's elastin, Mallory's phosphotungstic acid hematoxylin and Gomori trichrome for connective tissue, Fraser-Lendrum method for fibrin, and PAS–Alcian blue for mucopolysaccharides.

Autoradiograms were prepared by a dipping technique using Kodak NTB_2 emulsion and exposed for 2 and 4 weeks. They were developed and poststained with hematoxylin and eosin. Labeled nuclei were counted. Additional details of the technique can be found in Ref. 1.

In control rabbits, the mean blood pressure was 84 ± 1.7 mmHg. In the rabbits with partial constriction, the pressure in the upper abdominal aorta increased progressively and reached a hypertensive level (above 120 mmHg) at about 6 days and then remained stable. Figure 9.1 shows typical examples of [^{3}H]thymidine labeling indices of SMCs proximal and distal to the ligature.

There is a marked parallel between the development of hypertension and the increase of [^{3}H]thymidine-labeled SMCs in arteries subject to the rise in blood pressure. The normotensive vessels below the constriction did not differ from controls. The number of mitotic smooth muscle cells increased over the same

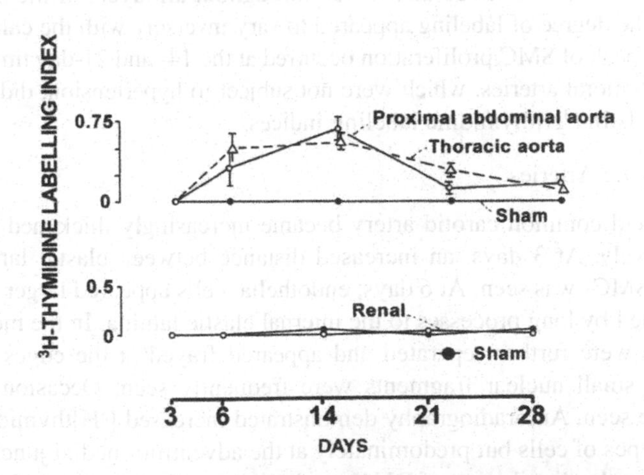

FIGURE 9.1. Time course of [^{3}H]thymidine labeling indices of vascular smooth muscle cells in the media of various arteries proximal to the aortic constriction (top) and distal to the constriction (bottom). The labeling index is increased significantly proximal to the constriction whereas it remains unchanged distal to the constriction. (Reproduced from Bevan RD, Blood Vessels 1976;13:100-128, with permission from S. Karger AG, Basel.)

time course in hypertensive arteries. After the blood pressure became relatively stable, between 14 and 21 days, both the [^{3}H]thymidine and mitotic indices fell but remained above control levels.

At 3 days postoperatively, the arterial pressure had risen above preoperative levels but was not hypertensive. The proximal aorta and the heart were dilated and venous congestion was present. At 6 days, during the phase of rapidly rising blood pressure, evidence of congestive heart failure had increased and varying degrees of pleural, pericardial, and peritoneal effusions were present. At 14 days, after blood pressure had stabilized, congestive heart failure was diminished and the left ventricle wall was further thickened. At 21 and 28 days, signs of heart failure were few or absent and further cardiac hypertrophy had occurred. Kidneys of experimental animals appeared normal macro- and microscopically. No evidence of renal arterial disease was seen and no difference was observed between [^{3}H]thymidine labeling of renal tissue or arteries from sham-operated and control animals.

2.1. *Changes in Arteries Proximal to Ligature*

The time course of pathological and proliferative changes and distribution of [^{3}H]thymidine-labeled cells in the artery wall differed according to the histological structure of vessels, whether they were elastic or muscular.

2.1.1. Medium-sized Muscular Arteries

The ear, brachial, gastric, and splenic arteries were studied. By 21 days, the media appeared thicker in hypertensive animals (Fig. 9.2). Autoradiography demonstrated proliferation of SMCs distributed throughout all layers of the media (see Fig. 9.7). The degree of labeling appeared to vary inversely with the caliber of the vessel. The peak of SMC proliferation occurred at the 14- and 21-day time periods. Renal and femoral arteries, which were not subject to hypertension, did not differ from controls in [^{3}H]thymidine labeling indices.

2.1.2. Elastic Arteries

The aorta and common carotid artery became increasingly thickened with time postoperatively. At 3 days, an increased distance between elastic laminae and individual SMCs was seen. At 6 days, endothelial cells appeared larger and many were attached by long processes to the internal elastic lamina. In the media, elastic laminae were further separated and appeared frayed at the edges. Necrotic SMCs and small nuclear fragments were frequently seen. Occasional mitotic SMCs were seen. Autoradiography demonstrated increased [^{3}H]thymidine labeling of all types of cells but predominately at the adventitio-medial junction zone. In the media, the distribution of DNA-synthesizing cells was mainly in the outer third and the inner three layers of cells adjacent to the internal elastic lamina.

By 14 days, the subendothelial space had increased in cellularity. Endothelial cells remained larger and in some areas had separated completely from the internal elastic lamina. The subendothelial space contained a few SMCs, orien-

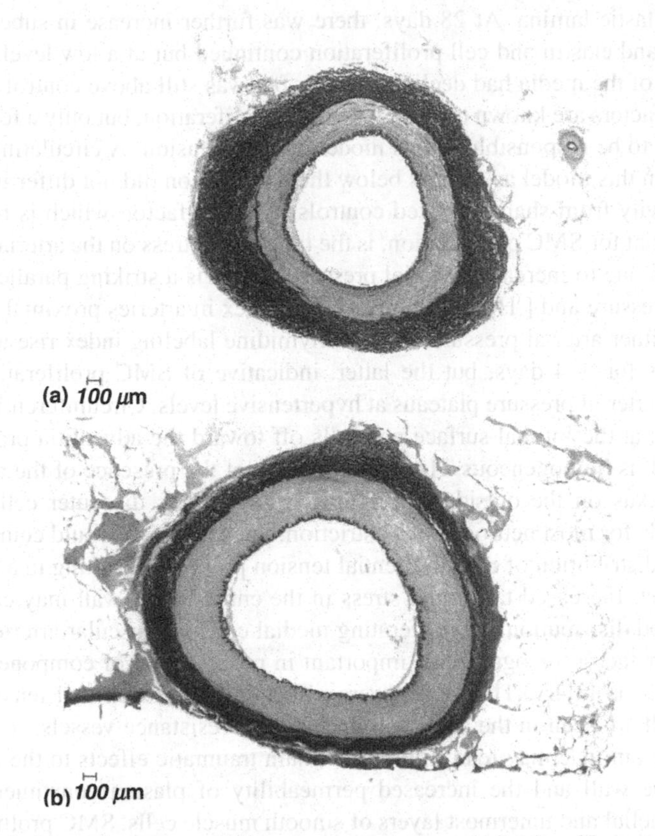

FIGURE 9.2. Left gastric artery at 21 days in normotensive rabbit (a) and hypertensive rabbit (b). The medial thickness has increased significantly in the hypertensive rabbit. Verhoeffs elastic stain. (Reproduced from Bevan RD, Blood Vessels 1976;13:100-128, with permission from S. Karger AG, Basel.)

tated as in the media. [^{3}H]thymidine labeling indicated increasing proliferation in this layer, and occasionally a labeled SMC was seen on both sides of the internal elastic lamina. Subendothelial connective tissue, such as elastin, collagen, and mucopolysaccharide, had increased, and fibrin plaques were present adjacent to the internal elastic lamina. In the media, further separation of elastic laminae had occurred and they frequently appeared discontinuous in the inner layers. Autoradiography demonstrated a similar distribution of labeled cells, but adventitial cells synthesizing DNA had diminished while SMCs labeling with [^{3}H]thymidine had increased.

At 21 days, endothelial labeling had further diminished although the cells were still more prominent. The subendothelial space contained increased numbers of SMCs. Subendothelial cell labeling had slightly diminished from 14 days. In the media, the distribution of labeled SMCs differed in that they were generalized throughout the vessel wall with a slight predominance in the layers adjacent to the internal elastic lamina. At 28 days, there was further increase in subendothelial collagen and elastin and cell proliferation continued but at a low level. Labeling of SMCs of the media had declined further but was still above control levels.

Many factors are known to stimulate SMC proliferation, but only a few of these are likely to be responsible in this model of hypertension. A circulating factor is unlikely in this model as arteries below the constriction did not differ in proliferative activity from sham-operated controls. Another factor, which is recognized as important for SMC proliferation, is the tangential stress on the arterial wall and distension due to increased arterial pressure. There is a striking parallel between arterial pressure and [^{3}H]thymidine labeling index in arteries proximal to the ligature. Neither arterial pressure nor [^{3}H]thymidine labeling index rise above normal levels for 3–4 days, but the latter, indicative of SMC proliferation, peaks when the arterial pressure plateaus at hypertensive levels. Circumferential tension is greatest at the intimal surface and falls off toward the adventitia provided the artery wall is homogeneous. However, because of the presence of the adrenergic neural plexus on the outside of the muscle layer, it is the outer cells that are responsible for most neurogenic constriction. Such an effect would counteract the expected distribution of circumferential tension probably resulting in a more uniform effect. Increased tangential stress in the entire artery wall may explain the generalized distribution of proliferating medial cells in muscular arteries.

Another factor, recognized as important in proliferation of components of the artery wall, is injury. The most severe effects of distension and tension in the artery wall are seen in the aorta and in the small resistance vessels. In the aorta, the elastic laminae may tend to limit the main traumatic effects to the innermost part of the wall and the increased permeability of plasma constituents to the subendothelial and innermost layers of smooth muscle cells. SMC proliferation is predominately adjacent to both sides of the internal elastic lamina. The inflammatory response was associated with acute injury and evidence of necrosis of the vessel wall. This was observed in the phase of acute rise of blood pressure and accompanied by proliferation of all components of the vessel wall and the formation of a subendothelial layer in the affected arterioles and the aorta.

3. Response of SMCs to Cyclic Stretch: A Cell Culture Study

In the artery, the environment of SMCs is different than the environment of endothelial cells (ECs). SMCs reside in the media, which is a three-dimensional structure of considerable thickness, and they are responsible for structured integrity of the vessel wall as well as regulation of the vasomotor tone. This regulatory function is accomplished through their circumferential orientation and their contractile properties. In the artery, SMCs are continuously subjected to the pulsatile pressure, which produces circumferential stretching of the wall with each heartbeat. SMCs are not directly exposed to the blood flow, which is in contrast with the ECs. Therefore, the periodic pulsatile mechanical strain may be the most influential for SMC behavior in the artery. For these reasons, the study of

SMCs in cell cultures, which are two-dimensional preparations or three-dimensional preparations, may show different results. For ECs, the 2D cell culture studies could indicate their behavior in the artery because they form a single cell layer in both the artery and the cell culture. For SMCs, however, we could distinguish between the 2D and 3D cell culture studies and take 3D cell culture studies as more representative of the environment in the artery.

3.1. 2D Cell Culture Studies

Sumpio et al. (2) studied the SMCs in a 2D cell culture subjected to cyclic tensional deformation to determine, in particular, the growth rate and orientation of the cells. Aortas were removed from young pigs, and the endothelial cell layer and inner portion of the media were scraped off. Two-millimeter punch biopsies of the remaining media of the vessel were obtained and placed in a culture dish with medium consisting of Dulbecco's modified Eagle's medium supplemented with 10% (v/v) heat-inactivated fetal calf serum, antibiotics, and 0.2 M L-glutamine. The medial wall biopsies were left at 37°C in a 5% CO_2 incubator until SMC explants were detected. A uniform population of SMCs was obtained, which at confluency displayed a "hill and valley" pattern and on ultrastructural examination demonstrated prominent cytoplasmic filaments. The cells were subcultured using 0.01% trypsin and used.

3.1.1. Application of Stress to Cultured Cells

The stress unit consists of a vacuum unit controlled for the duration and frequency of the applied strain or relaxation. Cells were cultured on flexible-bottom culture plates with a hydrophilic surface that could be deformed by vacuum. The degree of deformation was regulated by controlling the vacuum level. Because the cells are attached to the surface of the culture dish, they experience the same deformation as that applied to the plate. The SMCs in culture were subjected to cycles of 10 s of a maximum 24% elongation (average elongation = 10%) and 10 s of relaxation for 7 days. Proliferation of cells was assessed by cell counts and by measurement of [^{3}H]thymidine incorporation into DNA. The SMCs were plated (approximately 200,000 cells per well) onto the 6-well plates with flexible bottoms. After an overnight attachment period, the culture media and detached cells were aspirated and the number of adherent cells in a representative 6-well plate was counted. Some plates were maintained in a static environment (control group).

SMCs were analyzed for cell number and DNA synthesis on days 1, 3, 5, and 7. The cells were harvested with 0.25% trypsin and the cell number determined by means of a Coulter cell counter. Five percent trichloroacetic acid–0.125% tannic acid (TCA-TA) was added to the remaining cell suspension, and the samples were allowed to precipitate on ice and then were sedimented at 1000 × *g*. The supernatant fluids were discarded, and the pellets were resuspended in ice-cold TCA-TA and repeatedly washed. The final sample pellets were solubilized in scintillation fluid, and the amount of radioactivity was determined.

SMCs were examined from 1 h to 7 days under phase-contrast microscopy. Additional details of the technique can be found in Ref. 2.

The cells grown in a static environment showed a typical proliferative response, increasing in number 3.2-fold from day 1 to day 7 (Fig. 9.3). In contrast, the cells grown on plates, subjected to 3 cycles/min tensional deformation, had a significantly increased generation time, increasing in number only 2.1-fold between days 1 and 7. The retardation in growth rate occurred by the first day the cells were subjected to cyclic tension. The deceleration in growth rate was maximal between days 1 and 3. After that interval, the growth rate was similar to that of the cells in the control culture.

DNA synthesis, assessed by the incorporation of [^{3}H]thymidine, paralleled the rate of cell division for the growth curve for both the control cells and those subjected to cyclic tension. The orientation of SMCs subjected to mechanical deformation for 3 days was as follows: The cells were uniformly aligned perpendicular to the direction of the strain vector, in an annular fashion, at the periphery of the culture dish. A tendency toward alignment at the periphery was seen by 24 h of stretching.

Overall, there was a decrease in the rate of cell proliferation and DNA synthesis of SMCs subjected to repetitive mechanical deformation in culture. Also, the cells showed propensity to align perpendicular to the strain vector. This orientation was more evident in the well periphery where the strain was 24% while the cells were more randomly oriented in the center where the strain approached zero.

Kanda et al. (3) also studied the orientation of SMCs in cell cultures where the cells were subjected to cyclic stretching. Smooth muscle cells harvested from bovine aortas were seeded onto a substrate at a density of 2.5×10^5 cells per cm^2 and cultured for 12 h. After the adherent cells were well spread out, the substrates

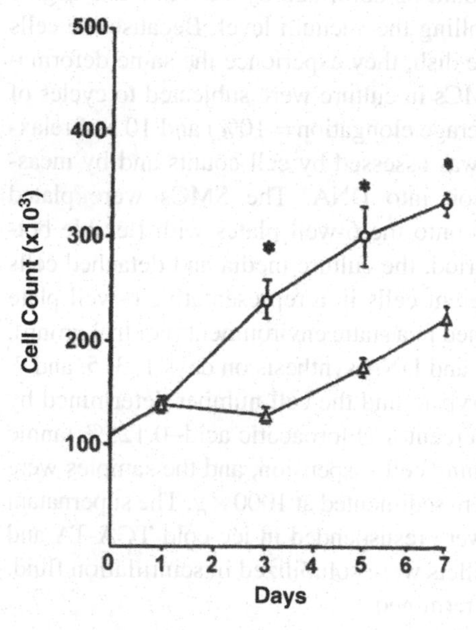

FIGURE 9.3. Effect of cyclic 10 s of 24% elongation and 10 s of relaxation on smooth muscle cell growth in culture. Cell number was plotted against the number of days with (triangle) and without (circle) constant tension/relaxation cycles. Each point represents the mean ± SD of six determinations. *Represents $p < 0.05$ compared with static controls. (Reproduced from Sumpio BE et. al., J Surg Res 1988; 44:696-701, with permission from Elsevier.)

were subjected to mechanical stresses. To evaluate cellular orientation and morphology under a phase-contrast microscope, a translucent, extensible film was used as a substrate. Polyurethane film was prepared by casting a tetrahydrofuran solution. The substrate was 20 μm thick and 10 × 20 mm in size.

3.1.2. Stress Chamber and Apparatus for Stretching

A plastic flask was handcrafted to fix the films. The equipment for periodic and directional mechanical stress loading on the adherent cells consisted of a DC motor, steering eccentric disk, piston rod with a linear bearing, positional transducer, and stress chamber. Periodic reciprocal movement was generated with a steering eccentric disk designed to provide amplitudes of 5%, 10%, and 20% of the unstretched length of the membrane. The eccentric disks were driven by a DC motor controlled to produce frequencies of 15 to 120 revolutions per minute (RPM). Membranes were exposed to cyclic directional stretch and relaxation at various amplitudes and frequencies for up to 24 h. Smooth muscle cells on unstretched membranes placed in the same chamber served as controls.

Smooth muscle cells subjected to the stress on extensible membranes were examined at time intervals of 0, 1, 3, 5, 12, and 24 h under a phase-contrast microscope. The long axis of each cell was marked to measure the cellular longitudinal length and orientation angle to the direction of stretching. Additional details can be found in Ref. 3.

Quantitative analysis of cellular orientation is shown in Figure 9.4. Smooth muscle cells showed a polygonal shape and random orientation before stress loading. The SMCs on the stressed membranes aligned perpendicular to the direction of stretching within a few hours, whereas those on the nonstressed membranes remained randomly oriented. The average orientation angles of the stressed cells increased to higher angles depending upon variables such as amplitude, frequency, and stress loading time. Higher amplitude and frequency of stress, and longer stress loading times, resulted in more profound increases in the average orientation angles.

3.2. Cyclic Stretch and 3D Cultured SMCs

Kanda et al. (4) studied SMCs for their orientation and phenotypic modulation in response to cyclic stretch in three-dimensional cell culture. They designed an experimental model of stress-loaded hybrid tissue in which SMCs are cultured in 3D collagen fibrous networks and subjected to periodic stretch-relaxation.

Complete Dulbecco's modified Eagle's medium supplemented with 10% heat-inactivated fetal calf serum was used for cell culture. SMCs were harvested from bovine aortic media. A bovine dermal Type I collagen solution, complete medium, and SMC suspension were mixed at 4°C. The final concentration of collagen was 2.5 mg/mL, and the cell density was 1.0×10^6 cells/mL. The mixture was poured into a ring-shaped capsule (outer and inner diameters were 53 and 39 mm, respectively) and kept at 37°C for 15 min in an incubator to form a gel.

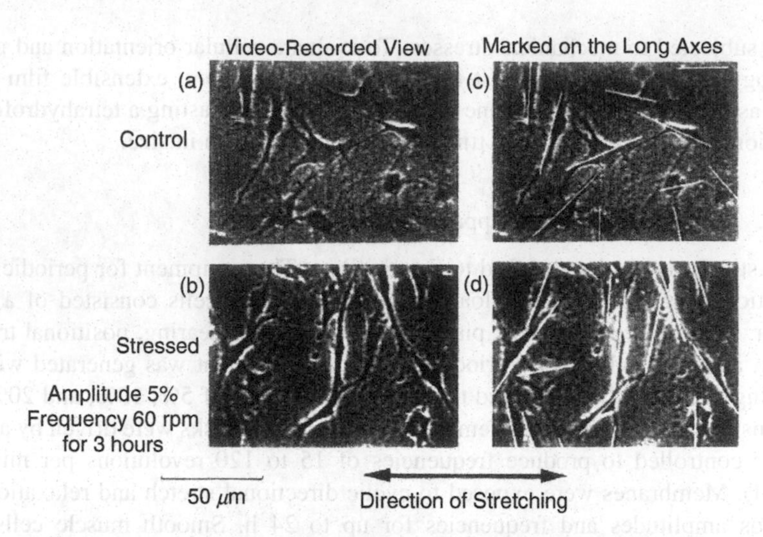

FIGURE 9.4. Phase-contrast micrographs. (a) Nonstressed SMCs show a random pattern. (b) Stressed SMCs align perpendicular to the direction of stretch. (c, d) Cellular axes were marked by a digitizer. (Reproduced from Kanda K et. al., ASAIO J 1992;38:M382-M385, with permission from Lippincott, Williams & Wilkins.)

The ring-shaped gel was then allowed to float in the complete medium for 3 days. The spontaneously shrinking gel, prior to stress loading, was approximately 15 mm in inner diameter.

The cell-incorporated collagen gels were subjected to three different modes of mechanical stimulation: isotonic strain, isometric strain, and periodic strain in a specially designed stress chamber. A plastic flask was hand crafted to install three ring gels in the flask as follows: 1) A ring gel was floated in the medium and allowed to shrink (isotonic control). 2) A gel was stretched into an isometric position (static stress) by use of two glass rods fixed on the bottom of the flask. 3) A gel was suspended at one end by a fixed glass rod. The opposite end of the gel was periodically stretched and relaxed (dynamic stress) with a glass rod driven by an apparatus. A DC motor drove a steering eccentric disk at 60 RPM, which moved the glass rod back and forth, which caused periodic stretch relaxation of the dynamically stressed ring gel with 10% amplitude of the relaxed gel length. The gels were subjected to mechanical stimulation for up to 4 weeks.

After stress loading, specimens were fixed for 30 min. Specimens were processed for transmission electron microscopy. Specimens for light microscopy were embedded in paraffin. The 3-μm-thick sections, stained with hematoxylin and eosin for evaluation of the cell shape and with azan for collagen fiber bundles, were examined. The gels (39 mm in inner diameter) were quite fragile as prepared. After a 3-day incubation, the gels spontaneously shrank to form compact gels (15 mm in inner diameter) that were stiff enough to be subjected to mechanical stress loading.

3.2.1. Cellular Orientation, Morphology, and Structure

SMCs in statically and dynamically stressed gels exhibited an elongated bipolar spindle shape and were regularly oriented parallel to the direction of stretch, whereas those in control gels were polygonal or spherically shaped and randomly oriented. On the surface of stress-loaded gels, spindle-shaped SMCs overlapped to form a multilayered structure with orientation parallel to the direction of stretch. Azan-stained sections revealed that, regardless of static or dynamic stress loading, the dense collagen fiber bundles in stressed gels were oriented parallel to the direction of stretch. On the other hand, coarse collagen fiber bundles in control gels were randomly oriented to form a loose network. Further culturing for up to 4 weeks did not cause significant changes in the morphology of cells and collagen compared with those from 2-week cultures.

Spherically shaped SMCs in the control collagen lattices had representative features of the synthetic phenotype (Fig. 9.5). They were filled with synthetic organelles, such as rough endoplasmic reticulum (RER), free ribosomes, mitochondria, and Golgi complexes, and had a significantly low density of myofilaments. In stress-loaded gels (Fig. 9.5) or in dynamic stress loading, bipolar spindle-shaped SMCs were oriented parallel to the direction of stretch and were abundant in synthetic organelles as in the control gels. However, after 4 weeks of stress loading, high-power electron micrographs showed that some of the SMCs in dynamic stress-loaded gels had increased their contractile apparatus with things such as myofilament bundles (MFs), dense bodies (DBs), and extracellular filamentous materials (EFMs) similar to basal lamina, all of which were rarely detected in either control or statically stressed gels. Thus, dynamic stress loading induced phenotypic modulation of SMCs from *the synthetic to the contractile*

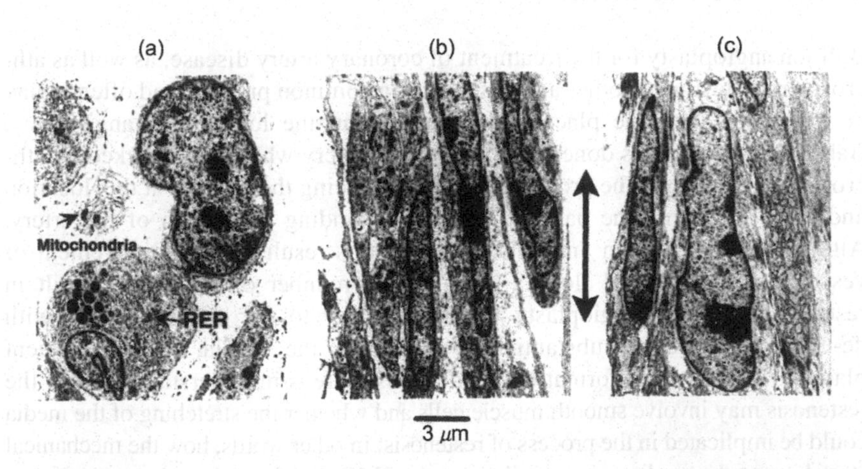

FIGURE 9.5. Transmission electron micrographs of the hybrid tissues after stress loading for 2 weeks. (a) Control, (b) statically stressed, (c) dynamically stressed. Direction of stretch is marked by the arrow. (Reproduced from Kanda K et. al., ASAIO J 1993;39: M686-M690, with permission from Lippincott, Williams & Wilkins.)

state. Collagen fibers adjacent to SMC membranes were oriented parallel to the long axes of the cells in both static and dynamic stress-loaded gels, which implies that the interaction between collagen fibers and SMCs played an important role in their orientation response.

The authors state that interaction of cells with their environment plays a vital role in such cellular behavior as adhesion, migration, orientation, proliferation, differentiation, and metabolism. To reconstitute a vital functional cell in vitro, it is essential to provide an appropriate environment. SMCs are the most predominant cell type in the arterial media. Their major function is to actively maintain tension in the arterial wall via contraction and relaxation. This function is reflected by the morphologic features of the cells, such as regular circumferential orientation and intracellular ultrastructure of the contractile phenotype, which responds to hormonal contraction-relaxation control. SMCs in native arterial media are embedded in 3D extracellular matrices. Moreover, they are continuously exposed to the periodic strain induced by pulsatile blood pressure. Therefore, both 3D environment and periodic strain are required to build a functional hybrid medial tissue.

The majority of SMCs cultured in collagen gels had the synthetic phenotype characteristics of intracellular ultrastructure. However, with an increase in stress loading time, SMCs in dynamic stress-loaded gels had increased proportions of contractile apparatus. Hence, dynamic stress loading plays a role in phenotypic modulation of SMCs from the synthetic to the contractile state.

4. Artery Response to Balloon Injury and Cholesterol Diet: Contribution of SMCs

Balloon angioplasty for the treatment of coronary artery disease, as well as atherosclerotic lesions in other arteries, is now a common practice and often it may be combined with the placement of a stent in the location of angioplasty. Balloon angioplasty is done to "open up" the artery where it is blocked by atherosclerotic lesions. The procedure involves placing the balloon at this location and then expanding the balloon, thereby expanding the lumen of the artery. Although many balloon angioplasty procedures result in the improvement of vessel lumen for blood flow, a considerable number of these also result in restenosis. Balloon angioplasty produces injury to the arterial vessel, with de-endothelialization, substantial stretching of the media, and subsequent platelet-rich thrombus formation. Our interest here is in understanding how the restenosis may involve smooth muscle cells and whether the stretching of the media could be implicated in the process of restenosis; in other words, how the mechanical stretching of the media may contribute to the SMC proliferation and restenosis.

Stadius et al. (5) studied the subject of restenosis and involvement of SMCs in cholesterol-fed rabbits that underwent balloon injury. They studied the temporal response of the artery to balloon injury followed by cholesterol feeding in rabbits. The temporal response is assessed using angiographic, histologic, morphometric, and immunocytochemical techniques. Quantitative angiographic measurements document the dimensions of the iliac arteries as the stenotic lesions develop, histologic morphometric measurements document the changes in the intimal and medial cross-sectional areas, and immunocytochemical techniques allow identification and localization of smooth muscle and macrophage cellular components.

New Zealand white rabbits were anesthetized, and the areas over the iliac arteries were prepared. An arteriotomy was performed, and a 2F Fogarty balloon catheter was inserted, advanced to the distal ascending aorta where the balloon was inflated, and then pulled back. The iliac arteries were then ligated, and the animals were allowed to recover and placed on a 2% cholesterol diet.

At various intervals after the initial injury, the animals were returned for angiographic study and sacrifice. Under general anesthesia, a carotid artery cutdown was performed, and a catheter was introduced into the descending thoracic aorta and positioned for bilateral iliac artery angiogram. Angiography demonstrated that stenotic lesions developed between the proximal and distal deep femoral branches in the iliac arteries; therefore, this segment was identified and divided into three equal segments. The segments were fixed in formalin solution for routine histologic analyses or Carnoy's solution for immunocytochemical analysis.

Arterial sections submitted for immunocytochemical analysis were stained with RAM-11, a monoclonal antibody that identifies macrophages, and HHF-35, a monoclonal antibody that identifies muscle actin and reacts only with vascular smooth muscle cells. Additional technical details can be found in Ref. 5. The serum cholesterol level was measured in animals at the time they were sacrificed and it ranged from 214 mg/dL at 3 days to a mean value of 1560 mg/dL at 5–6 weeks.

Angiographic assessment showed that there was a progressive decrease in lumen diameter as the time interval between injury and angiography increased. There was a progressive increase in the intimal cross-sectional area. The first intimal cellular response was detected in 2 of 10 specimens 3–4 days after injury; by 7–9 days after injury, all eight artery specimens showed evidence of an intimal cellular infiltrate. Medial cross-sectional area changes were less prominent than those in the intima.

Neointima was present in 2 of 10 arteries assessed 3 days after injury, and 85% of this neointimal area was immunostained for macrophages. By 7 days, all arteries demonstrated neointima. HHF-35 immunostaining of the neointima did not appear until 7 days. Five of eight arteries assessed between 7 and 9 days after injury demonstrated HHF-35 immunostaining in intimal cells. All arteries 11 days or more demonstrated HHF-35 immunostaining of the intima. The relative contribution of RAM-11– and HHF-35–immunostained intima to the total intimal area appears to change in the first 2 weeks after injury; at 3 days, RAM-11 immunostaining is predominant in the neointima, whereas at days 11–15, HHF-35 immunostaining predominates. Between days 17 and 41, there is little apparent change in the ratio of neointimal area immunostained by RAM-11 and HHF-35.

A pattern of spatial orientation of RAM-11 and HHF-35 immunostaining within the developing intima was present in 75% of arteries 11 or more days after injury. Intimal cells that were immunostained by RAM-11 predominated at the inner portion of the intima adjacent to the internal elastic lamina, whereas those that were HHF-35 immunostained predominated at the luminal edge (Fig. 9.6).

The rabbit iliac artery responds to balloon injury and subsequent cholesterol feeding with the development of a progressively enlarging intima. The temporal development of the neointima appears similar to the temporal development of intimal hyperplasia in normal rat carotid arteries (6, 7) and porcine carotid arteries (8) subjected to balloon injury. The rat carotid artery subjected to balloon injury responds with progressive intimal hyperplasia during the first 8 weeks. The porcine carotid artery subjected to balloon dilatation responds with intimal hyperplasia that is progressive during the first 2 weeks. The progressive intimal cellular response seen in these three disparate models occurs despite differences in the

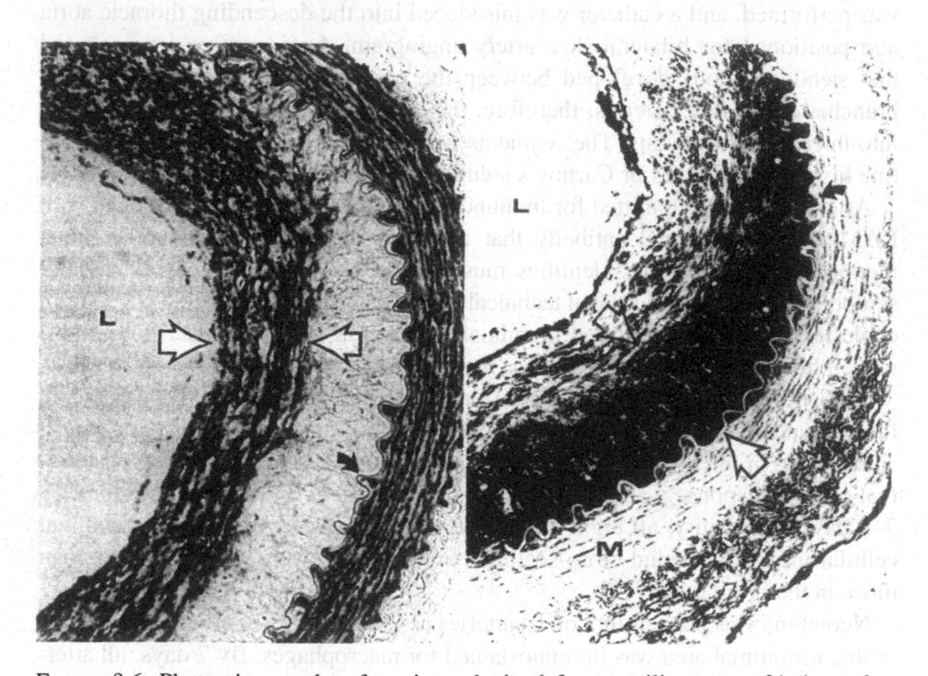

FIGURE 9.6. Photomicrographs of sections obtained from an iliac artery 21 days after injury. (Left) Immunostained with HHF-35 monoclonal antibody, indicating the pressure of SMCs. L, vessel lumen; closed arrow, internal elastic lamina; M, media. Two regions of immunoreactivity are present. The first region is in the intima (between two open arrows) and the second region is the media. (Right) A section adjacent to that in the left panel. This is immunostained with RAM-11 monoclonal antibody, indicating the location of macrophages. Predominant immunoreactive region in the intima is next to the internal elastic lamina. In the intimal hyperplasia in the balloon-injured aorta, the SMCs occur on the luminal side while the macrophages occur deeper into the intima. (Reproduced from Stadius ML et. al., Arterioscler Thromb 1992;12:1267-1273, with permission from Lippincott, Williams & Wilkins.)

elastic and muscular characteristics of these vessels and differences in the cholesterol status of the animals after injury.

At 7–9 days after injury, all arteries demonstrated an intimal cellular infiltrate, and there was now evidence of both RAM-11 and HHF-35 immunostaining within the intima. The RAM-11–immunostained cells persisted and increased in the area along the internal elastic laminal aspect of the developing intima. The HHF-35–immunostained cells predominated in the portion of the intimal lesion adjacent to the lumen of the vessel. Migration of smooth muscle cells from the media to the intima has been identified as an important contributing factor to the developing neointimal lesion in the balloon-injured rat carotid artery (9,10). The spatial relation of HHF-35 and RAM-11 immunostaining identified in this study strongly suggests that smooth muscle cell migration also plays an important role in the development of this intimal lesion.

5. SMC Proliferation in Relation to Aortic Stretch

In the previous section, we considered how balloon angioplasty results in the intimal hyperplasia, which can sometimes advance to produce restenosis of the vessel lumen. The lesions that produce restenosis and the lesions from primary atherosclerosis have many features in common. One feature that is common to both is the proliferation of smooth muscle cells. This feature is very important because it is responsible for the bulk of the mass in both. Naturally, we would like to examine which parameters could have produced SMC proliferation in balloon angioplasty and whether those parameters are present also in primary atherosclerosis. Specifically, we are interested in the parameters of vascular mechanics.

In balloon angioplasty, we saw that the process results in severe endothelial injury, formation of platelet-rich thrombus, and stretching of the artery wall. From the mechanics point of view, we are interested in how the stretching of the artery wall could be related to the proliferation of SMCs. In the branching regions of the artery, we saw in Chapters 6 and 7 that both the stresses and the strains are considerably increased and that the region has predilection to atherosclerosis. If we could now establish a correlation between aortic stretch (strain) and SMC proliferation, then we would have identified aortic stretch as one of the common parameters involved in both restenosis and primary atherosclerosis.

Thubrikar et al. (11) studied the SMC proliferation in the rabbit model where they produced two different degrees of wall stretch during balloon angioplasty. We will consider their studies. NZW male rabbits, weighing 2.5–3 kg, were anesthetized and subjected to balloon injury of the aorta with either 2-French or 3-French Fogarty catheters (2F-inflated diameter 4 mm and 3F-inflated diameter 5 mm). The catheter was inserted through the femoral artery, advanced into the upper abdominal aorta, inflated, and gently withdrawn up to the aortic bifurcation. The rabbits were allowed to recover and 48 h later were anesthetized and injected intravenously with tritiated thymidine (0.6 mCi/kg). After the isotope had circulated for 1 h, the rabbits were euthanized. The abdominal aorta was removed and immersed in buffered glutaraldehyde. The aorta was dissected into six 5-mm-long segments at the following locations: at the celiac, below the celiac, at the superior mesenteric, at the right renal, at the left renal, and below the left renal artery.

5.1. *Autoradiography*

The aortic segments were embedded in methacrylate plastic and cut into 2-μm-thick longitudinal sections. Every 20th section was saved and mounted on a slide serially. The slides were dipped in photographic emulsion for autoradiography. Tritium-labeled thymidine gets incorporated into the DNA of dividing cells, and the beta emission from tritium exposes the photographic emulsion present over the section. The sections were exposed to isotope emission for 3 weeks. The slides were then developed and stained with hematoxylin and eosin for microscopic examination.

The total number of labeled cells in each section was counted, and the depth at which the labeled cells were located was noted. The labeled cells present on the luminal surface were classified as endothelial cells (ECs) and those that lay deeper were classified as SMCs (Fig. 9.7). Because no specific technique was used for cell identification, it was understood that the cells on the surface, classified as ECs, could have been different cells.

5.2. *Luminal Surface Area, Volume, and Weight of the Artery Segments*

They also investigated the effect of taper of the aorta on the degree of cell proliferation. They grouped the three proximal segments (at celiac, below celiac, and at superior mesenteric) as the proximal aorta and the three distal segments (at right renal, at left renal, and below left renal) as the distal aorta. In the proximal and the distal segments, the number of labeled ECs was normalized to the luminal surface area, whereas the number of labeled SMCs was normalized to the volume and weight of that segment.

They also prepared silicone rubber molds of the rabbit aorta at 90 mmHg pressure. From these molds, they measured the diameter of the abdominal aorta at the mid-plane of each of the six segments. From the diameters, they determined the relative proportion of the surface area and volume of the proximal and distal segments as follows: surface area of an aortic segment = $\pi D \times L$, where D is the inner diameter and L is the length of the segment. The ratio of the surface area of the proximal segment to that of the distal segment, therefore, is D_p/D_d, where D_p is the average of the diameters of the three proximal segments and D_d is the average of the diameters of the three distal segments. The volume of the aortic segment is given by $\pi D \times L \times T$, where T is the thickness of the aortic wall. For the aorta, thickness is proportional to the diameter, and therefore, the ratio of the volume of the proximal segment to that of the distal segment is equal to $(D_p/D_d)^2$. The weight of each of the six segments was also measured.

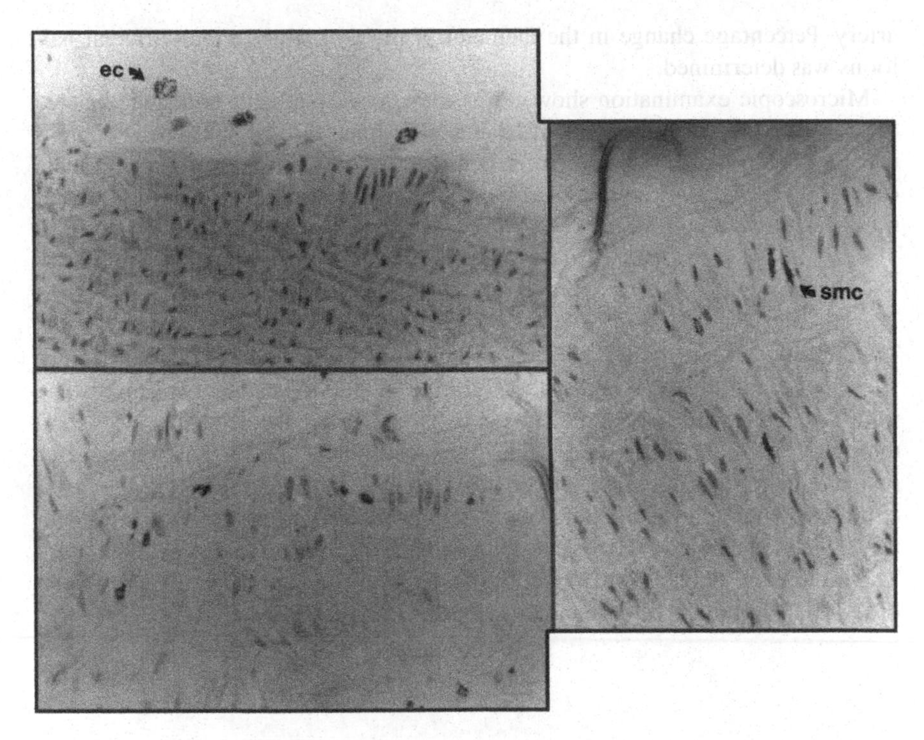

FIGURE 9.7. Three typical autoradiographs of the abdominal aorta from balloon-injured rabbits. Dark grains represent tritium label in the nucleus. Labeled cells on the luminal surface were classified as endothelial cells (ec) and those that lay deeper as smooth muscle cells (smc). Many labeled smooth muscle cells were close to the intimal-medial junction.

5.3. Change in the Aortic Diameter Due to Balloon Catheters

To establish that the amount of stretch imposed by the 2F balloon catheter was less than that imposed by the 3F balloon catheter, the diameter of the aorta and of the balloons was measured as follows: The rabbits were anesthetized, and a 5F dye injection catheter was inserted through the femoral artery into the abdominal aorta. During fluoroscopy, a radiopaque dye was injected in the aorta and the image was videotaped. The catheter was removed, and a 2F balloon catheter was inserted through the femoral artery into the abdominal aorta. The balloon was inflated using a radiopaque dye as a filling fluid so as to make the balloon diameter appear during fluoroscopy. The balloon was then pulled through the aorta and was videotaped. The catheter was then removed, and a 3F balloon catheter inserted in its place and the procedure repeated. The diameter of the aorta and of the 2F and 3F balloons were measured at the following levels: below the diaphragm, at the right renal artery, at the left renal artery, and below the left renal artery. Percentage change in the diameter of the aorta due to two different balloons was determined.

Microscopic examination showed that the arterial segments contained labeled cells. Relatively small numbers of labeled cells lined the arterial lumen, whereas the vast majority of labeled cells occurred in the intima and inner media (Fig. 9.7). The adventitia contained significant numbers of labeled cells, however, they were not counted.

5.4. *Effect of the Balloon Size*

The proliferative response of medial smooth muscle cells to balloon injury varied, depending upon whether the 2F or the 3F balloon catheter was used. The injury caused by the 2F balloon catheter resulted in more labeled ECs (410 ± 324) than SMCs (38 ± 23), whereas the injury caused by the 3F balloon catheter resulted in many more labeled SMCs (1750 ± 634) than EC (294 ± 168) (Fig. 9.8). The number of labeled endothelial cells was significantly greater with the 2F balloon catheter (410 ± 324) than with the 3F balloon catheter (294 ± 168). The number of labeled SMCs with the 3F balloon catheter (1750 ± 634) was an order of magnitude greater than that with the 2F balloon catheter (38 ± 23).

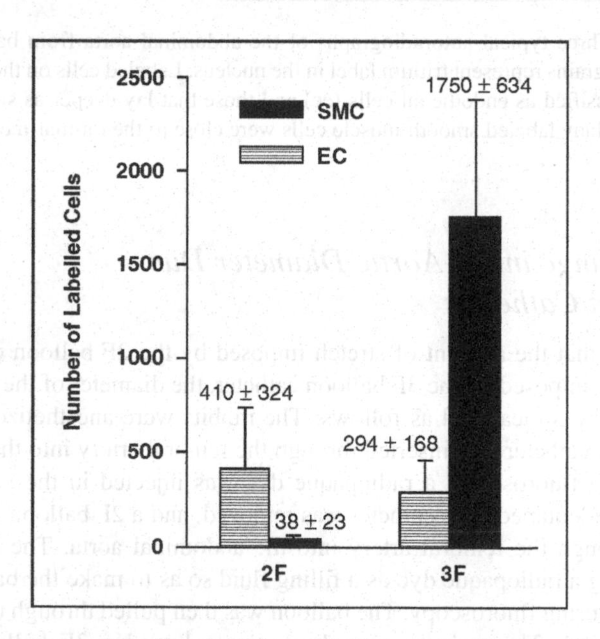

FIGURE 9.8. Average number of labeled ECs and SMCs per 5-mm-long aortic segment (mean ± SEM) determined by counting every 20th section, 2 μm each. The number of labeled ECs was similar in the 2F and 3F balloon experiments, whereas the number of SMCs was an order of magnitude greater in the 3F balloon experiments.

5.5. Effect of Tapering of the Aorta

There was a distinct increase in both the labeled ECs and SMCs as the aorta tapered distally. With the 2-French balloon catheter, labeled ECs were more in the distal aorta (575 ± 432) than in the proximal aorta (245 ± 438), although this difference was not statistically significant. Labeled SMCs, however, were significantly more in the distal aorta (58 ± 54) than in the proximal aorta (17 ± 26). With the 3-French balloon catheter, labeled ECs were the same in the proximal and the distal aorta, however, labeled SMCs were slightly more in the distal aorta (1974 ± 1012) than in the proximal aorta (1525 ± 835).

In rabbits, the ratio of the diameter of the proximal to distal aorta (D_p/D_d) was 1.33. This represents the ratio of the luminal surface area of the two segments and indicates that the surface area of the proximal segment is 33% greater than that of the distal segment. Because the surface area reflects the total number of endothelial cells, the number of labeled ECs in the distal segment was multiplied by 1.33. Upon normalization, it was observed that there were more labeled ECs in the distal aorta than in the proximal aorta, in both the 2F and the 3F balloon catheter experiments.

Similarly, the ratio of the volume of the proximal to distal segment was 1.77, which indicates that the proximal segment was 77% larger in the tissue volume than the distal segment. Because the tissue volume reflects the total number of SMCs in the segment, the number of labeled SMCs in the distal segment was multiplied by 1.77. The ratio of the weight of the proximal to distal segment was determined to be 1.56. This weight ratio was also used to normalize the labeled SMCs in the two segments. When the labeled SMCs were normalized for the volume or weight of the aortic segment, there were many more labeled SMCs in the distal aorta than in the proximal aorta in both the 2F and the 3F balloon catheter experiments (Table 9.1).

5.6. Location of the Labeled SMCs

In a vast majority of cases, the labeled SMCs were not distributed uniformly through the media but were concentrated toward the intimal-medial junction (Fig. 9.7). Also, through the entire length of the artery, the labeled cells were not spread uniformly but appeared in patches or clusters. These observations indicate a strong tendency toward clustering of replicating SMCs.

TABLE 9.1. Average number of labeled SMCs per 5-mm aortic segment (mean ± SD).

	Proximal aorta	Distal aorta Normalized for volume (× 1.77)	Distal aorta Normalized for the weight (× 1.56)
2F	17 ± 26	103 ± 96	90 ± 84
3F	1525 ± 835	3494 ± 1791	3079 ± 1579

5.7. Aorta and Balloon Diameter

Table 9.2. shows the percentage increase in the diameter of the aorta due to 2F and 3F balloon catheters. In case of the 2F balloon catheter, the aortic diameter either did not change or increased only a small amount (0–11%). With the 3F balloon catheter, however, the aortic diameter increased in all cases and the increase was much greater (in the range 18–35%). Furthermore, due to the tapering of the aorta, the increase in the diameter distally was greater than that proximally. Hence, the 2F balloon appears to cause injury mostly to the luminal surface, without stretching the aortic wall significantly, whereas the 3F balloon appears to cause injury to both the surface and the aortic wall by stretching the wall significantly.

5.8. Balloon Injury and Cell Proliferation

These results indicate that the degree of endothelial injury caused by 2F or 3F balloon catheters is similar because the number of labeled ECs is similar in both cases (Fig. 9.8). The degree of medial injury caused by the two balloon catheters, however, is not the same because the number of labeled SMCs with the 3F balloon catheter far exceeds that with the 2F balloon catheter (Fig. 9.8). This appears consistent with the finding that the 2F balloon may or may not stretch the aorta, whereas the 3F balloon definitely and substantially stretches the aorta (Table 9.2). Because the endothelial injury is similar with the 2F and the 3F balloon catheters, an overwhelmingly large number of labeled SMCs seen with the 3F balloon catheter appears to be in response to the aortic stretch.

5.9. Effect of Tapering of the Aorta

The aorta has a natural taper distally. The distal aorta has a smaller diameter and less wall thickness than the proximal aorta. Consequently, the total number of ECs and SMCs are less in the distal aorta than in the proximal aorta.

The labeled ECs and SMCs, before normalization, were equal or slightly more in the distal segment compared with those in the proximal segment in both the 2F and the 3F catheter experiments. However, when normalized to the surface area, the volume, and the weight of the tissue, the results changed dramatically. After normalization, both the labeled ECs and SMCs were significantly greater in the distal segment than in the proximal segment in both the 2F and the 3F balloon catheter experiments. Because the aorta tapers distally, it is expected that the

TABLE 9.2. Percent increase in the aortic diameter due to balloons.

	2F (%)	3F (%)
Below diaphragm	0	18
Right renal	0	26
Left renal	7	27
Below left renal	11	35

balloon will stretch the aorta by a greater amount distally. It may also be noted that below the left renal artery, the aorta does not always expand to the diameter of the balloon because the balloon is too large; instead, the balloon may deform to the shape of the aorta. Hence, a much larger number of labeled SMCs observed in the distal segment (3079 vs. 1525) appears to correlate with the increased aortic stretch distally and not with the EC injury, which is similar in both the proximal and the distal segments.

6. Quantitative Relationship Between SMC Proliferation and Aortic Stretch

In the previous section, we saw that in the balloon angioplasty model in the rabbit aorta, SMC proliferation was an order of magnitude more with a 3F balloon than with a 2F balloon. We also saw that the 3F balloon stretched the aorta by 18–35%, whereas the 2F balloon stretched it by only 0–11%. Thus, greater aortic stretch produced greater SMC proliferation. Because the correlation between aortic stretch and SMC proliferation is very important, it is of interest to establish a quantitative relationship between these two parameters. The quantitative relationship is important also for other reasons stated below:

i) The strains in the arterial branch regions are known, and we can evaluate whether they are sufficient to cause a significant increase in SMC proliferation.
ii) In balloon angioplasty, very high strains (18–35%) can be produced, and such high strains are not sustainable by the artery wall without significant damage; and consequently.
iii) such high strains could have caused disintegration of connective tissue and cell death in the media, and these details can be very important in the fate of the artery.

Here we consider studies by Thubrikar et al. (12) where they attempt to establish a quantitative correlation between SMC proliferation and aortic stretch. They studied the quantitative relationship between the change in the aortic diameter and cellular proliferation in the rabbit model. They used the balloon angioplasty technique to produce dilation of the aorta and tritiated thymidine as a radioactive tracer to detect proliferating cells. In brief, the rabbits were anesthetized, and an angiographic catheter was inserted and positioned into the aorta. A radiopaque dye was injected during fluoroscopy so that the aortic internal diameter could be measured at two locations described below. Between the left renal artery and the aortic bifurcation, four small arterial branches, almost equally spaced, can be identified (Fig. 9.9). The aortic diameter is measured between the first and the second branch (location 1) and between the third and the fourth branch (location 2). Then the angiographic catheter was withdrawn, and a balloon angioplasty catheter was inserted and positioned at location 1. The balloon was then inflated by filling it with radiopaque dye so as to be able to measure its diameter during fluoroscopy. The balloon was inflated to produce a

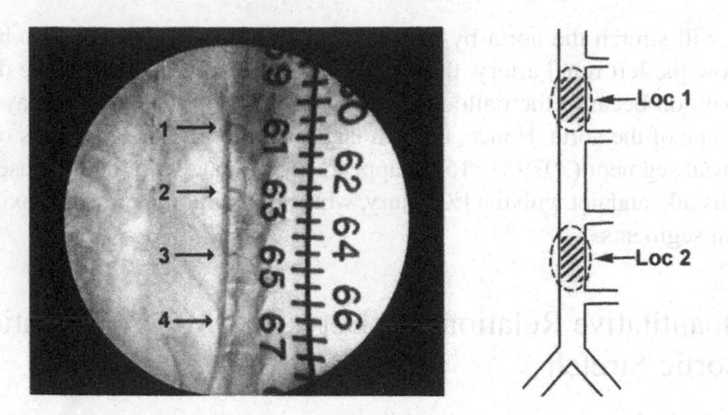

FIGURE 9.9. (Left) Angiogram of the rabbit aorta. The aortic region shown is from below the left renal to the aortic bifurcation. The four side branches (numbered) form the identification marks. (Right) Schematic of the same aortic region. For ballooning, locations 1 and 2 were used. The four branches identifying the locations are also shown.

predetermined diameter increase in the aorta and held in this position for 15 s. The balloon was then deflated and re-inflated two more times. At location 2, a similar procedure was repeated to produce a predetermined but different amount of dilation. Two days later, tritiated thymidine (0.6 mCi/kg) was injected in the rabbit intravenously and allowed to circulate for 1 h. The rabbits were then sacrificed, and the aortic segments under study were removed and processed for autoradiography. The aortic segments were embedded in paraffin and serial sections, 4 μm thick, were cut in the circumferential direction. These sections were coated with photographic emulsion and exposed for 21 days and then developed and stained with hematoxylin and eosin. The labeled cells were counted in every 10th section over a 2-mm length of the aorta (2 mm = 2000 μm = 500 sections of 4 μm each). Every 10th section gives a total of 50 sections for counting.

During a histological examination, it was observed that there were areas devoid of cells, that is, there were regions in the media that contained fragmented, dead, or no cells (Fig. 9.10). Therefore, the number of *dead cells* was also estimated because they represent injury to the artery. To estimate the number of dead cells, the area containing fragmented, dead, or missing cells was measured by planimetry and the number of cells calculated after determining the cell density (number of cells per unit area) in nearby locations. Even though this technique has its limitations, the replicating and dead cells within a given aorta can be compared at two locations, subjected to two different dilations.

Figure 9.11 shows the number of replicating cells in a typical aorta at the proximal and distal locations. Only the smooth muscle cells in the media were counted. Replicating cells in the adventitia and on the luminal surface (e.g., ECs) were not counted. In a given aorta, it was observed that the greater the aortic dilation, the greater was the number of replicating cells.

It was observed that up to a certain dilation (about 25%), the number of replicating SMCs was small and almost constant (Fig. 9.12). However, beyond this

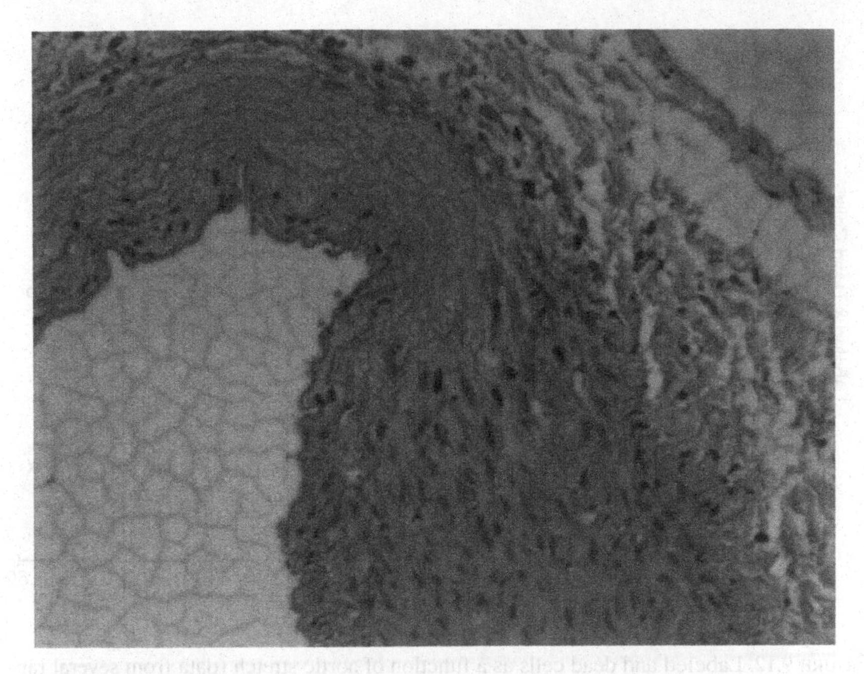

FIGURE 9.10. Autoradiograph of the ballooned aorta showing the region without any cells near the top (dead cell region) and the region with labeled nuclei halfway to the bottom. From such sections, the number of dead cells and labeled cells were estimated.

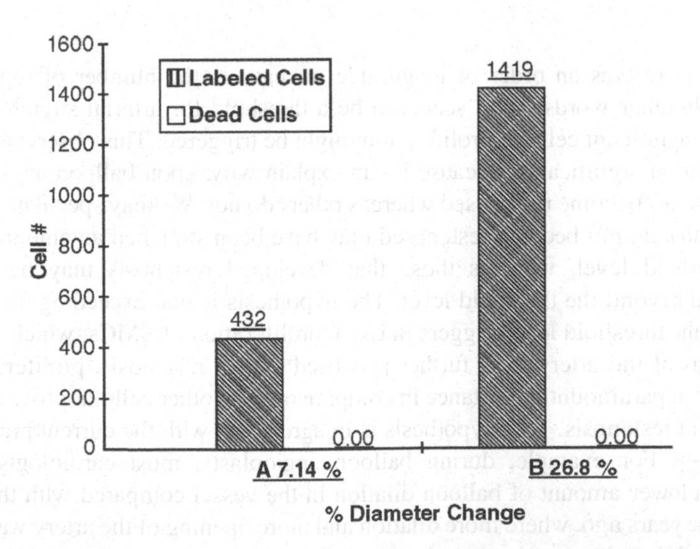

FIGURE 9.11. The number of dead cells and labeled cells in a rabbit aorta after balloon injury in response to two different degrees of stretch. Larger stretch (B) produces more labeled cells than the smaller stretch (A).

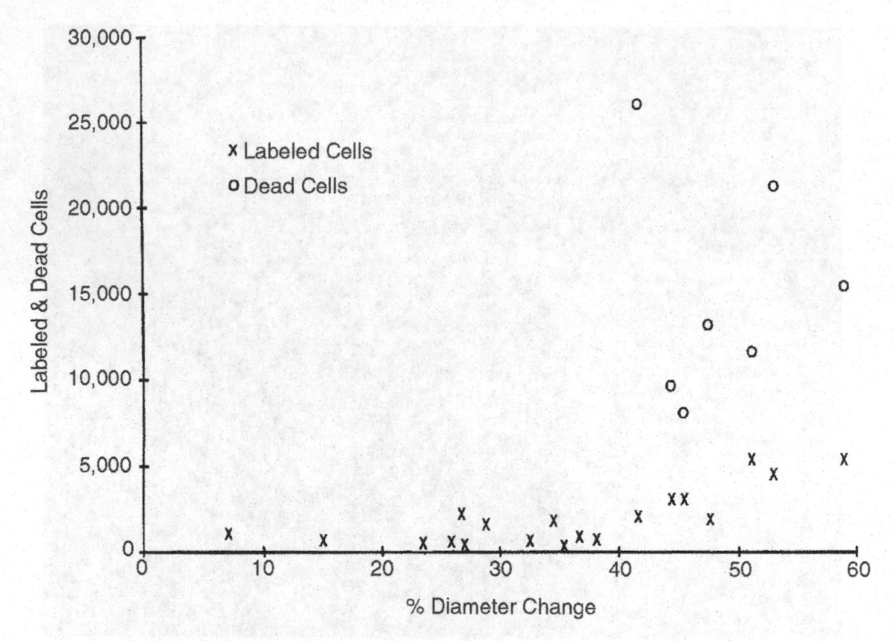

FIGURE 9.12. Labeled and dead cells as a function of aortic stretch (data from several rabbits). There is a tendency toward a larger number of labeled cells being associated with a greater amount of stretch. Furthermore, excessive stretch causes cell death. There appears to be a threshold (at about 25% stretch) beyond which cell replication increases significantly. There may also be another threshold (at about 40% stretch) beyond which cell death becomes dominant.

dilation there was an order of magnitude increase in the number of replicating SMCs. In other words, there seems to be a threshold for arterial stretch beyond which a significant cellular proliferation might be triggered. This observation is of great clinical significance because it can explain why, upon balloon angioplasty, some arteries become restenosed whereas others do not. We may speculate that the arteries that do not become restenosed may have been stretched an amount below the threshold level, whereas those that developed restenosis may have been stretched beyond the threshold level. The hypothesis is that exceeding the stretch beyond the threshold level triggers massive proliferation of SMCs, which leads to restenosis of the artery. It is further proposed that in restenosis, proliferation of SMCs is of paramount importance in comparison with other cellular processes that are part of restenosis. This hypothesis is in agreement with the current practice in cardiology. For example, during balloon angioplasty, most cardiologists now induce a lower amount of balloon dilation in the vessel compared with the practice some years ago, where more dilation and more opening of the artery was being performed. In other words, there is an acknowledgment in the clinical setting that a larger amount of artery dilation may produce undesirable results (restenosis).

Figure 9.12 also shows the number of dead and replicative SMCs as a function of aortic dilation. The dead cells reported in the figure do not represent all of the dead cells present in the artery because only a 2-mm-long segment was examined. The figure shows that the dead cells start to appear only beyond a certain amount of stretch (about 43%). This implies that there is a threshold stretch for cellular apoptosis also.

The stretch as a causative force for apoptosis is an important concept because this forms a link between the stretch and the aneurysm of the blood vessel. As we continue to examine the concept of vascular mechanics in pathogenesis, this is a good place to make an observation that the stretch in the artery has a potential to do both, to cause significant cellular proliferation and to produce cellular apoptosis. From this, it naturally follows that the stretch can lead to either atherosclerosis or aneurysm depending upon whether the cells are proliferating or undergoing apoptosis. The concept may also be described as follows: "the high strains which normally are present at the branch are always stimulating SMCs to respond by proliferation. When the SMCs are able to respond by proliferation the process results in atherosclerosis and when they are not, then they undergo apoptosis and that leads to aneurysm." This concept is of significant interest to us because as we see in Chapters 1, 2, 13, and 14, both atherosclerotic lesions and aneurysms do form in the branch region. In fact, the aneurysms of the cerebral arteries are often called VBA (vessel branching aneurysms) because they are located at the branch points. It is worth repeating that the enhanced mechanical strain challenges the SMCs to respond, and when SMCs are able to respond by proliferation, we have atherosclerosis, and when they are not, they undergo apoptosis, and we have an aneurysm. This hypothesis, though it remains to be proved, is the most logical hypothesis that can explain the occurrence of both atherosclerosis and aneurysm at branch locations.

References

1. Bevan RD: An autoradiographic and pathological study of cellular proliferation in rabbit arteries correlated with an increase in arterial pressure. Blood Vessels 1976;13: 100-128.
2. Sumpio BE, Banes AJ: Respone of porcine aortic smooth muscle cells to cyclic tensional deformation in culture. J Surg Res 1988;44(6):696-701.
3. Kanda K, Matsuda T, Oka T: Two-dimensional orientational response of smooth muscle cells to cyclic stretching. ASAIO J 1992;38(3):M382-M385.
4. Kanda K, Matsuda T, Oka T: Mechanical stress induced cellular orientation and phenotypic modulation of 3-D cultured smooth muscle cells. ASAIO J 1993;39:M686-M690.
5. Stadius ML, Rowan R, Fleischhauer JF, Kernoff R, Billingham M, Gown AM: Time course and cellular characteristics of the iliac artery response to acute balloon injury. Arterioscler Thromb 1992;12(11):1267-1273.
6. Clowes AW, Reidy MA, Clowes MM: Mechanisms of stenosis after arterial injury. Lab Invest 1983;49:208-215.

7. Clowes AW, Reidy MA, Clowes MM: Kinetics of cellular proliferation after arterial injury: I. Smooth muscle cell growth in the absence of endothelium. Lab Invest 1983;49:327-333.
8. Steele PM, Chesebro JH, Stenson AW, Holmes DR Jr, Dewanjee MK, Badimon L, Fuster V: Balloon angioplasty: Natural history of the pathophysiological response to injury in a pig model. Circ Res 1985;57:105-112.
9. Clowes AW, Schwartz SM: Significance of quiescent smooth muscle cell migration in the injured rat carotid artery. Circ Res 1985;56:139-145.
10. Schwartz SM, Reidy MR, Clowes AW: Kinetics of atherosclerosis, a stem cell model. Ann N Y Acad Sci 1985;454:292-304.
11. Thubrikar MJ, Moorthy RR, Deck JD, Nolan SP: Smooth muscle cell proliferation: Is it due to endothelial injury or aortic stretch? Eur J Lab Med 1996;4(1):65-71.
12. Thubrikar MJ, Cribbs M: Laboratory report on work done at the Heineman Medical Research Center, Charlotte, NC, 1995–1997, unpublished.

10
Stress Reduction and Atherosclerosis Reduction

1. Introduction

This is the most significant chapter where the principles that have been used in pressure vessel technology to prevent rupture are applied to the artery to prevent disease, particularly atherosclerosis. We expect that if the wall stress is at the center of causing atherosclerosis, then the reduction of wall stress should lead to reduction or prevention of atherosclerosis. This reasoning can also suggest new modalities for treatment of the disease, based on the principles of vascular mechanics. Here, the vascular mechanics becomes not only most relevant in clinical medicine but also quite enjoyable to learn about. We will examine how stress reduction is used in the experiments to prevent atherosclerosis, how stress reduction occurs naturally in our body and prevents atherosclerosis, how treatment with drugs (antihypertensive drugs) can reduce stress and atherosclerosis, and how treatment with beta-blockers can reduce artery fatigue and atherosclerosis. The stress here refers to the mechanical stress induced by pressure load. Quite a bit of work on this subject was done by Thubrikar et al (1–4), and we will consider their findings.

2. Inhibition of Atherosclerosis by Reduction of Arterial Wall Stress

2.1. *Reduction of Arterial Wall Stress*

The principle of reinforcement of the arterial wall by external support was used for reducing the stress in the arterial wall (1). This principle is illustrated in Figure 10.1. At an internal pressure (P), the artery develops a circumferential stress (σ_c) and a longitudinal stress (σ_L) (panel I). Along line AB, circumferential stress remains nearly constant (panels II and III). Along line CD, circumferential stress begins to increase as one approaches the branch and it becomes several times greater at points E and F compared with that along AB (panel III). Although the exact magnitude and distribution of the stress depends upon the geometry and properties of the branch area, the stress around the branch is always higher because the branch eliminates the arterial wall and creates a "hole" in the main artery, which leaves less arterial wall to sustain the pressure load. This stress concentration near the arterial branch has been described in detail in Chapters 4, 5, and 6.

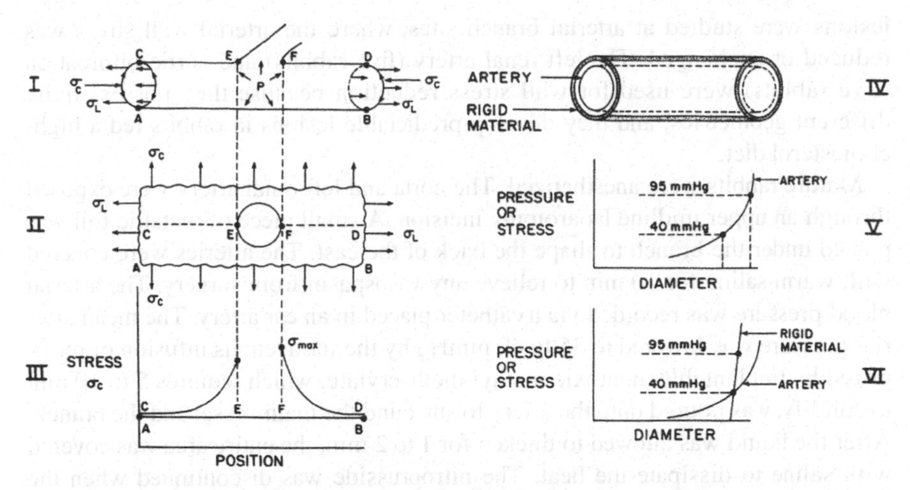

FIGURE 10.1. Schematic presentation of an artery with the arterial branch (I, II, and III) and without the branch but surrounded by rigid material (IV, V, and VI). Panel I: P, intraluminal pressure; σ_L and σ_C, intramural arterial stress in the longitudinal and circumferential directions, respectively; A, B, C, D, E, and F are the points on the artery. Panel II: An artery opened longitudinally along the line AB. σ_C and σ_L are indicated. Panel III: Qualitative variation of circumferential stress σ_C along the lines AB and CD. The stress increases toward the ostium and reaches maximum (σ_{max}) at branch points E and F. Panel IV: Artery surrounded by rigid material. Panel V: Schematic presentation of the variation of internal pressure or stress with the arterial diameter. Panel VI: After the artery has been surrounded by rigid material at 40 mmHg intraluminal pressure, the rise in pressure up to 95 mmHg does not produce the same increase in the arterial diameter that it would have without the rigid material. The stress in the artery, therefore, remains at a low value in spite of high pressure.

A typical pressure-diameter relationship for an artery is shown in Figure 10.1 (panel V). When the luminal pressure is decreased from 95 to 40 mmHg, the diameter may decrease by 4–8% (panel V). If the artery is encased in rigid material at 40 mmHg (panel IV) and the pressure is then increased to 95 mmHg, the stress from the increased pressure will be transmitted to the rigid material, leaving the artery at a lower circumferential stress (panel VI). Although the actual stress distribution in the artery and surrounding material is complicated, the arterial stress will be less than that without the surrounding rigid material. This principle of reinforcement for maintaining low arterial wall stress in spite of normal systemic pressure is used in the study.

2.2. *Rabbit Model for Atherosclerosis*

Atherosclerotic lesions in humans commonly develop at arterial branch sites. Rabbits fed high-cholesterol diets also develop lesions at arterial branch sites, and these animals have been used for the study of atherosclerosis. Atherosclerotic lesions were studied at arterial branch sites, where the arterial wall stress was reduced or unchanged. The left renal artery (five rabbits) and aortic bifurcation (five rabbits) were used for wall stress reduction because they possess quite different geometries, and they develop predictable lesions in rabbits fed a high-cholesterol diet.

Mature rabbits were anesthetized. The aorta and left renal artery were exposed through an upper midline laparotomy incision. A small piece of metallic foil was placed under the branch to shape the back of the cast. The arteries were covered with warm saline for 10 min to relieve any vasospasm from surgery. The arterial blood pressure was recorded via a catheter placed in an ear artery. The mean arterial pressure was lowered to 35 to 40 mmHg by the intravenous infusion of nitroprusside. Implantable, nontoxic methyl methacrylate, which requires 5 to 10 min to solidify, was poured onto the artery to surround the main vessel and the branch. After the liquid was allowed to thicken for 1 to 2 min, the entire area was covered with saline to dissipate the heat. The nitroprusside was discontinued when the cast was set, and the laparotomy was closed. For the aortic bifurcation cast placement, a lower midline laparotomy incision was made and a similar procedure was followed. Figure 10.2 shows the casts at the left renal branch and the aortic bifurcation. To test the model, silicone rubber, another rapid-setting casting compound, was also used for cast placement on the left renal arterial branch and on the aortic bifurcation. Methyl methacrylate casts were placed in four rabbits at a mean pressure of 95 mmHg, which was achieved by IV infusion of angiotensin. Casting at this pressure should not reduce wall stress, as it is similar to the control pressure of conscious rabbits (110 mmHg systolic/80 mmHg diastolic). Over

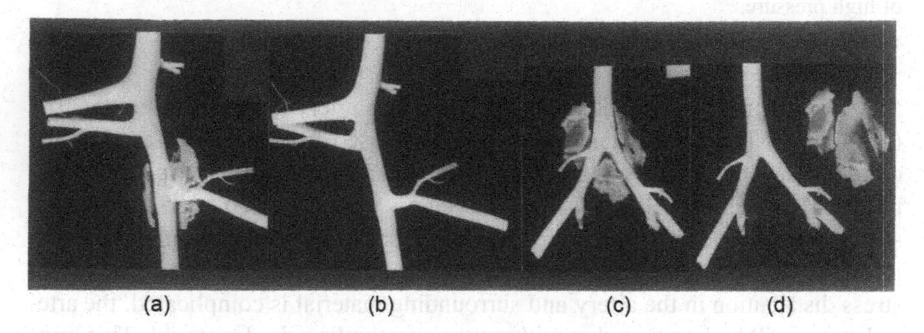

FIGURE 10.2. Photographs of casts placed on the left renal arterial branch (a) and the aortic bifurcation (c, d). (a) and (b) also show the silicone rubber form of the inside geometry of the abdominal aorta, superior mesenteric artery, right renal artery, and left renal artery. (c) and (d) show the silicone rubber form of the inside geometry of the aortic bifurcation. Externally placed cast of methyl methacrylate (a) and (c) did not produce noticeable change in the inside geometry of the artery at either the left renal branch area or the aortic bifurcation. Silicone rubber forms were made 21 days after external casting. In (a), the superior mesenteric artery (top) and right renal artery (bottom) can be seen on the left side.

a period of time, the arteries should respond to decreased wall stress in a manner opposite to that induced by hypertension, which increases wall stress. A response time of 2 weeks was allowed for the artery to adapt to the conditions of locally decreased wall stress. Therefore, the rabbits were fed a regular diet for the first 2 weeks and then switched to a 2% cholesterol diet.

The renal artery-casted rabbits were fed a 2% cholesterol diet for 7–10 weeks, and the aortic bifurcation-casted rabbits were on the diet for 8–11 weeks. Total plasma cholesterol was determined from venous blood samples. At the end of the diet period, all of the rabbits were anesthetized, and the carotid artery and femoral arteries were cannulated. The rabbits were then euthanized. Physiologic saline at 90 mmHg pressure was flushed from the carotid artery through the femoral artery. The arteries were flushed with 70% ethyl alcohol, followed by a Sudan IV staining solution. After the staining solution emerged from the femoral artery, the artery was clamped and staining continued for 5 min at 90 mmHg pressure. The arteries were then flushed with 70% alcohol, fixed with 4% buffered formaldehyde at 90 mmHg pressure for 2 h, dissected free, and were incised for visual examination of the atherosclerotic lesions. The arteries were also prepared for histological examination by thin sectioning and staining with hematoxylin and eosin.

The overall geometry of the arterial branch is altered only minimally by cast placement. Because increased stress and strain occur locally, geometry change is restricted to the small area around the branch. The stress is decreased by reduced pressure, and even small local geometry changes are preserved by a rigid cast. To determine if the extravascular cast produced constriction of the arterial lumen, silicone rubber forms of the lumen were made as follows. The rabbits were anesthetized and sacrificed. One carotid artery and two femoral arteries were exposed and cannulated. Silicone rubber was introduced through the carotid artery and was allowed to emerge from the femoral artery and solidify for 18 h under a static pressure of 90 mmHg. The left renal arterial branch and the aortic bifurcation were excised, and the silicone rubber form of the arterial lumen was removed. The forms were examined for luminal narrowing, and the arterial diameters were measured at various locations (Fig. 10.2).

To evaluate the effect of the extravascular cast on the arterial tissue, the thickness of the arterial wall under the cast was measured and compared with that in the same uncasted area. The details of the experiments can be found in Ref. 1.

2.3. *Atherosclerosis in Control and Stress-reduced Areas*

The rabbits had a mean arterial pressure of 70–80 mmHg before surgery. Because these pressures were measured under anesthesia, they were below the reported pressure (110/80 mmHg) in conscious rabbits (5). Because normal pressure in conscious rabbits is higher than the casting pressure, the wall stress at the casted sites would have been higher without the cast. The unoperated (control) rabbits on a high-cholesterol diet developed atherosclerotic lesions at the origins of the intercostal arteries, celiac artery, superior mesenteric artery, both renal arteries, and at the aortic bifurcation. The typical lesions were masses raised from the surface of the arterial lumen and stained bright red with Sudan IV. They had a characteristic "V" shape at the arterial branch and increased in size with the duration of the cholesterol diet (Fig. 10.3).

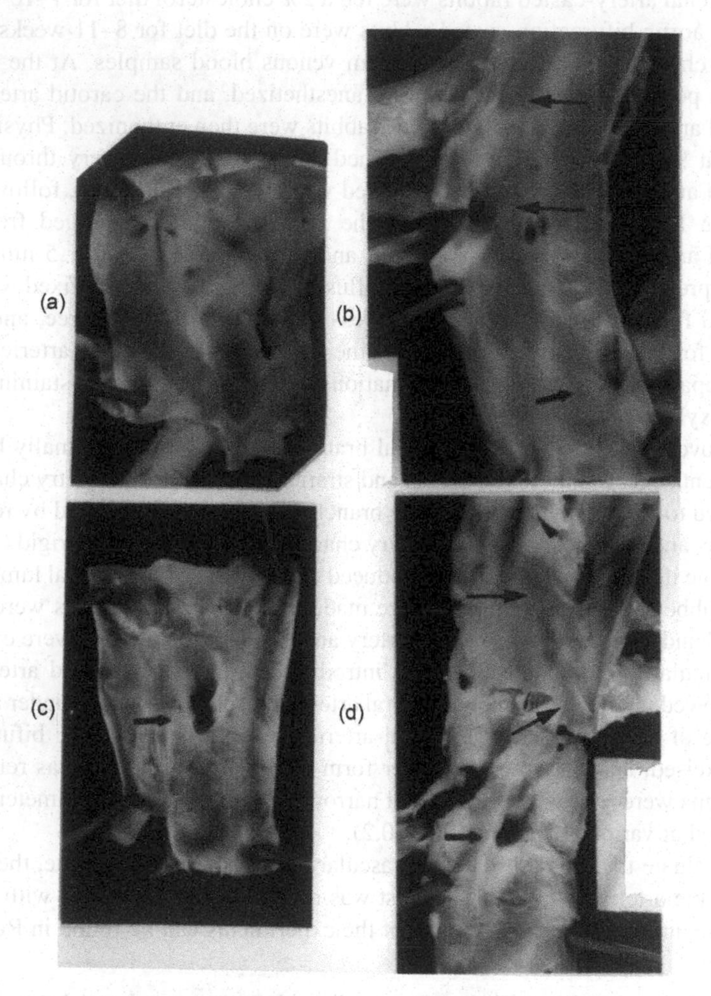

FIGURE 10.3. Photographs showing red atherosclerotic lesions at the arterial branches in three rabbits. Direction of blood flow is from top to bottom. In a rabbit on a high-cholesterol diet for 10 weeks, lesions developed on the superior mesenteric artery (a, arrow), whereas the casted left renal artery (c) was completely free of lesions (arrow). (b) In a rabbit on a high-cholesterol diet for 11 weeks, lesions developed on the left renal (bottom arrow), the right renal (middle arrow), and the superior mesenteric (top arrow) arteries. (d) In a rabbit on a high-cholesterol diet for 10 weeks, no lesions developed on the casted left renal artery (bottom arrow), whereas they did develop on the right renal (middle arrow) and superior mesenteric (top arrow) arteries. (*Please see color version on CD-ROM.*)

Casts were placed on the left renal arterial branch of five rabbits, and atherosclerotic lesions did not develop at this site (Fig. 10.3). However, these rabbits did develop lesions at other arterial branches, including the right renal artery and aortic bifurcation. When casts were placed on the aortic bifurcation of five rabbits, atherosclerotic lesions did not develop at that site. However, these rabbits developed lesions at other locations, including both renal arterial branch sites. In rabbits where a different casting compound (silicone rubber) was used, the lesions did not develop at the casted sites, indicating that the lesions can be prevented with casts of other inert materials. The rabbits, with casts placed at 95 mmHg mean arterial pressure, did develop lesions at the casted sites, indicating that the cast did not inhibit lesions when it did not reduce wall stress.

To determine whether the extravascular cast caused narrowing of the arterial lumen, they measured the diameter of the aorta from intraluminal forms. The aortic diameter in the left renal casted area (2.8 and 3.5 mm) was similar to that in the same uncasted area (2.8–3.4 mm) (Fig. 10.2). This was also true at the aortic bifurcation. Also, the thickness of the aorta in the casted left renal area (0.33 ± 0.16 mm) was similar to that in the uncasted left renal area (0.25 ± 0.06 mm). The thickness in the casted aortic bifurcation (0.32 ± 0.1 mm) was also similar to that in the uncasted aortic bifurcation (0.29 ± 0.08 mm). The thickness of the aorta varied with location and with the animal.

Upon visual examination, the casted artery appeared free of external loose tissue. Histological examinations revealed the atherosclerotic lesions to be in the intima (Fig. 10.4) and to contain lipid-laden foam cells. The casted sites showed the typical three layers of the arterial wall (intima, media, and adventitia) without lesions (Fig. 10.4). The casted artery did not show cell necrosis or inflammatory reaction, and in some sections, vasa vasorum could be observed. The plasma cholesterol level in rabbits on the high-cholesterol diet ranged from 1900 to 3000 mg/100 mL, whereas in the rabbits on a normal diet, it was 40 to 48 mg/100 mL.

Overall, the study describes a rabbit model in which cast placement at reduced arterial pressure leads to inhibition of atherosclerosis at the casted site. The cast does not produce significant narrowing of the arterial lumen or thinning of the arterial wall. The cast is expected to 1) maintain the artery at a low intramural stress in spite of the normal systemic pressure, 2) eliminate pulsatile changes in arterial wall dimensions because of cast rigidity, 3) produce a small but finite change in the geometry of the branch site by maintaining the artery at a casting pressure, 4) not eliminate areas of high and low shear resulting from the blood flow, and 5) possibly reduce the filtration of plasma constituents through the arterial wall by producing an impermeable barrier on the outside. Filtration of plasma constituents, however, does not seem to be reduced merely by the presence of the cast, as the lesions did develop in rabbits that were casted under high pressure. If the filtration of plasma constituents is reduced, it would have to be because of the cast's effect on the artery such as reduction of stress, strain, and adaptive changes in the artery.

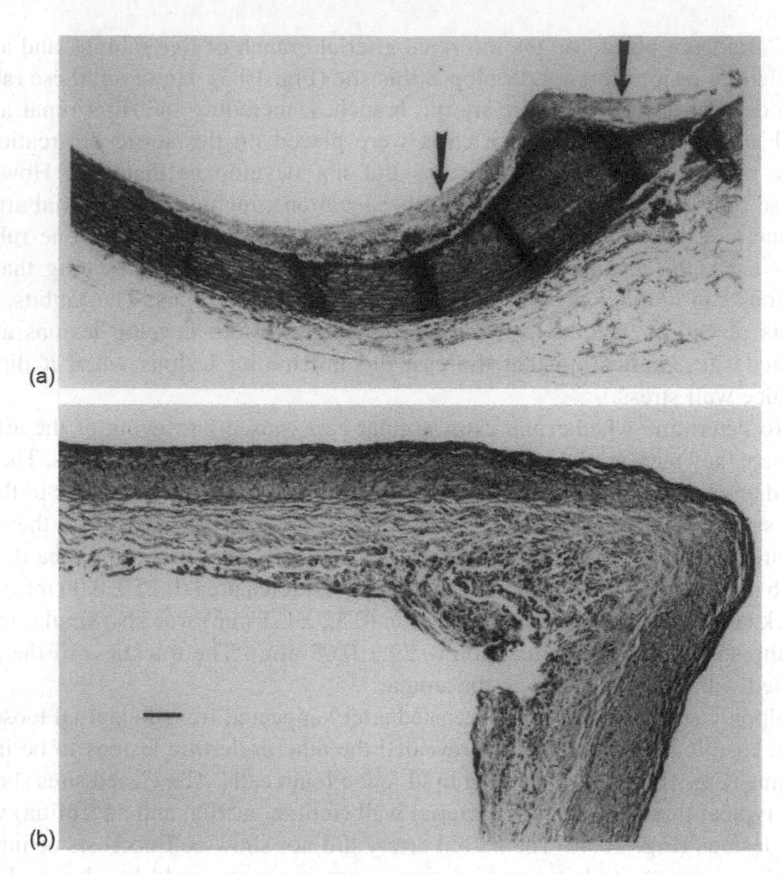

FIGURE 10.4. (a) Circumferential section at the aortic bifurcation from a rabbit on a high-cholesterol diet for 10 weeks. The thickening in the artery represents a flow divider or crotch. Atherosclerotic lesions (arrows) are shown in the crotch area, as well as on both sides of the crotch. (b) Circumferential section at the aortic bifurcation from another rabbit on a high-cholesterol diet for 11 weeks. This casted bifurcation is completely free of atherosclerotic lesions. The crotch in the upper right corner and the arterial wall appears structurally normal. Bar = 0.1 mm for (a) and (b). Hematoxylin and eosin stains.

3. Rhythmic Pattern of Atherosclerosis in Vertebral Arteries

The experiments described in the previous section, where the placement of an external cast on the artery to reduce wall stress is shown to prevent atherosclerotic lesion formation, certainly is a convincing experiment for those who are familiar with vascular mechanics. For others, however, it may raise doubts. For this reason, we should explore if there are natural examples in the body that appear similar to the experiment above. The vertebral artery offers such an example. The vertebral artery passes through multiple bone canals (Fig. 10.5). In the canal region, the artery has an external support of the surrounding bone, and it is unable

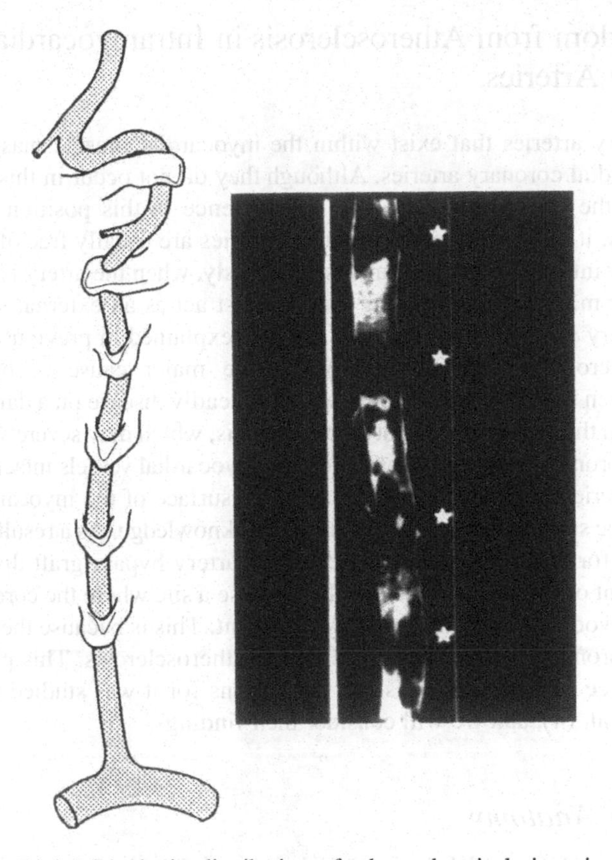

FIGURE 10.5. (Right) Rhythmic distribution of atherosclerotic lesions in the vertebral artery. The star indicates the presence of lipids in the regions where the artery is between the two bone canals. The artery in the bone canal (white) is free of the disease. (Reproduced from Wolf S, Werthessen NT, eds. Dynamics of Arterial Flow. Advances in Experimental Medicine and Biology, Vol. 115. Plenum, New York, 1976:378, with permission from Springer Science and Business Media.) (Left) Schematic presentation of the course of the vertebral artery.

to expand at the arrival of the pressure pulse. This region of the artery is the region of reduced wall stress due to the external support.

The intraosseal portions of the vertebral arteries are known to show an alternating pattern of atherosclerotic changes; the lesions occur in the segments that are free to expand but are absent where the artery is passing through the bone canal and thus is not free to expand with the systolic pulse pressure. The surrounding bone acts as a support and prevents the increase of stress due to systolic stretch in that segment. The portions of the vertebral arteries that lie between the bony canals do not have this protection and therefore are susceptible to the development of high stress, high stretch, and atherosclerosis.

4. Freedom from Atherosclerosis in Intramyocardial Coronary Arteries

The coronary arteries that exist within the myocardial muscle mass are called intramyocardial coronary arteries. Although they do not occur in this anatomical position in the majority of cases, their occurrence in this position is not rare. Interestingly, it is known that the coronary arteries are usually free of atherosclerosis in their intramyocardial segments. Obviously, when the artery is surrounded by a muscle mass, the surrounding muscle must act as an external support, and thus the artery will have reduced wall stress as explained in previous sections.

Atherosclerosis of coronary arteries is the major cause of heart attack. However, even those who are dealing with this deadly disease on a daily basis seldom mention the phenomenon that atherosclerosis, which may severely involve the epicardial coronary arteries, leaves major intramyocardial vessels intact. Epicardial coronary arteries are those that lie on the outer surface of the myocardial muscle mass. Cardiac surgeons are acutely aware of this knowledge. As a result, when they are looking for a place to attach the coronary artery bypass graft distally to the open segment of the coronary vessel, they choose a site where the coronary artery enters the myocardium, if this anatomy is present. This is because they know that the distal coronary artery is open and free of atherosclerosis. This phenomenon of the absence of atherosclerosis and the reasons for it was studied in detail by Robicsek et al. (4), and we will consider their findings.

4.1. The Anatomy

The primary and most of the secondary branches of the human coronary arterial system run on the surface of the heart and are covered by the epicardium and by various amounts of subepicardial fat. From this network rises a number of vertical tributaries, which penetrate the myocardium and link up with a subendocardial small arteries. The intramyocardial artery is a major coronary branch, which after a short epicardial course dives into the myocardial tissue, does not resurface, but continues entirely through an intramuscular course. Interestingly, in the study of coronary arteriosclerosis in the past, the attention has been focused mostly on the epicardial coronary arteries, and the condition of major vessels, which occasionally run within the myocardium, has been largely disregarded. Tissue bridges overlying portions of the left anterior descending coronary artery have been described by Geiringer (6) and Polacek (7). It has been noted that the walls of these "bridged" arteries differ from the rest of the coronary vessels. The epicardial portion of the left anterior descending artery tends to have a considerably thicker intima than the corresponding distal "bridged" segment. Ishii, who studied the phenomenon of bridging extensively, drew a "reverse" conclusion: Not that the intima of the tunneled portion was thinner than the proximal epicardial segment, but that "intimal thickening and macroscopic raised lesions were increased just before the bridge" (8).

Robicsek et al. found that intramyocardial arteries exist in about 10% of their operated cases and that these arteries are almost exclusively the proximal major obtuse marginal branches of the circumflex coronary artery (Fig. 10.6). In the past, there have been controversies about whether or not the intramural position of the major coronary arteries would protect it from atherosclerosis. More recently, needed attention has been given by Roberts and Buja to intramural coronary arteries and to the fact that in contrast with epicardial vessels, they remain free of arteriosclerotic changes; Roberts and Buja wrote: "The intramural coronary arteries in the left ventricular free walls and ventricular septums were free of significant narrowing . . . foam cells, cholesterol clefts, calcific deposits or fibrin platelet thrombi were not observed in any intramural coronary artery" (9).

Atherosclerosis is affected by several factors including increased levels of serum cholesterol, hypertension, diabetes mellitus, endothelial and smooth muscle cell function, platelets, monocytes, and arterial permeability. It is also recognized that certain mechanical forces such as blood flow–generated shear stress and blood pressure–generated arterial wall stress (mural stress) play an important role in the occurrence and localization of atherosclerotic lesions. More recent investigations have suggested that atherosclerotic changes are likely to occur predominately in areas where mural stress in the arterial wall is the most severe. Ongoing studies on this subject are concerned with how and where this stress exerts its effect on the vascular wall, and the mechanism by which it may change the normal adaptive response of the coronary endothelium into a self-perpetuating process of lipid deposition and cell proliferation.

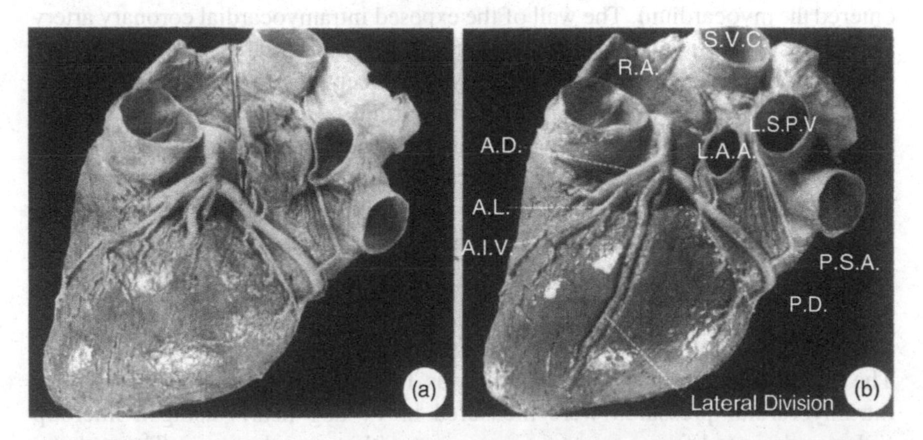

FIGURE 10.6. Photographs of the human heart showing the relationship of coronary arteries to the myocardium. The lateral division of the circumflex coronary artery appears completely hidden inside the myocardium before dissection (a) and becomes visible only after dissection of the myocardium (b). (Reproduced from McAlpine, Heart and Coronary Arteries, Springer-Verlag, New York, 1992, with permission from Springer Science and Business Media.)

4.2. Clinical Investigation

Robicsek (4) studied the phenomenon that intramyocardial coronary arteries remain free of atherosclerosis even if the epicardial arteries do not in 250 patients who underwent routine coronary bypass grafting for severe arteriosclerotic coronary disease. After the patients had been placed on cardiopulmonary bypass, the posterior surface of their heart was exposed, and the coronary vasculature was inspected for the presence of intramyocardial arteries and for the location of atherosclerotic lesions. If such arteries were found, and they were in a need of bypass grafting, a 1- to 2-cm segment was bared from their epicardial coverage just before they entered the myocardium. By incising the overlying myocardium for an additional length (usually about 2 cm), the artery was exposed to prepare it for grafting. The vessel was then incised placing the cut about one-third on the epicardial and two-thirds over the intramyocardial portion, and the entire exposed exterior and the interior portions of the artery were inspected. The operation then proceeded to the completion of the distal anastomosis.

Of the 250 patients, 26 had major primary tributaries of the circumflex coronary artery, which, after an initial short epicardial run, dove into the myocardium and continued on an intramyocardial course and had angiographically proved proximal occlusive arteriosclerotic involvement severe enough to require bypass grafting. Twenty-four of these had critical narrowing, and two complete occlusion. In the latter cases, the vessels were seen on angiogram as a "ghost" filling through collaterals.

They found the epicardial segment of all of the arteries severely atherosclerotic with the changes consisting of diffuse yellow discoloration of the intima, elevated and nonelevated plaques, calcification, and atherosclerotic ulcers (Fig. 10.7). In a rather abrupt manner, all of these changes ceased at the point where the vessels entered the myocardium. The wall of the exposed intramyocardial coronary artery was consistently found to be of pearl-gray color, thin, pliable, and lacking any signs of atherosclerosis. There were only two exceptions to this rule; one of the 26 coronary arteries with a very thin myocardial cover had several islands of calcification, and the one that was totally occluded proximally also had severe distal changes, probably secondary to recanalization of previous total thrombosis. Thus, 24 of 26 intramyocardial arteries were completely free of atherosclerotic lesions, whereas their proximal epicardial segments had severe atherosclerosis.

4.3. Studies in Canine Hearts

The mechanism of this phenomenon was investigated in canine hearts. The intramyocardial pressures were measured in 14 dog hearts using a microtip catheter, designed for pressure measurement within muscle tissue. The catheter contained a pressure-sensing recessed diaphragm of about 1 mm × 1 mm size, located immediately proximal to the beveled needle tip. The length of the needle shaft (3F, 19 gauge, 1 mm O.D.) allowed insertion of the catheter into various depths of the muscle tissue. The intramyocardial pressures were measured in the

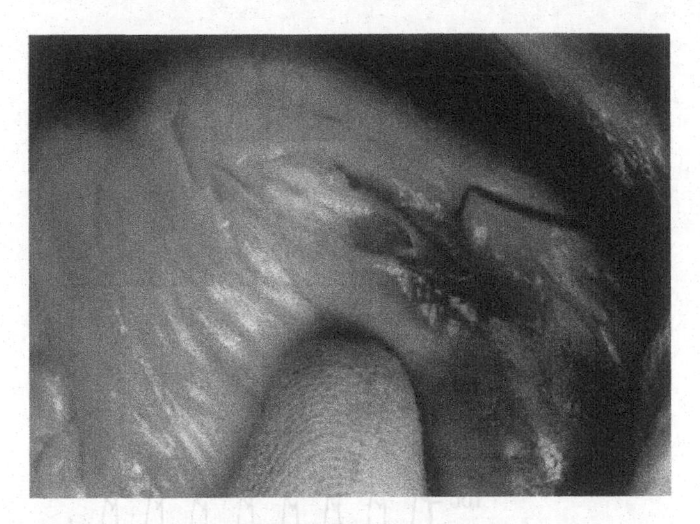

FIGURE 10.7. Photograph of the coronary artery in a patient taken intraoperatively. The coronary artery can be seen to go in the intramyocardial space. The incision in the myocardium exposes the place of entry of the coronary artery into the myocardium. The artery has developed severely occlusive atherosclerotic plaque in the epicardial segment but is completely normal and nonocclusive in the intramyocardial segment. The atherosclerotic lesion abruptly stops at the entry of the artery into the myocardium. (*Please see color version on CD-ROM.*)

lateral wall of the left ventricle (LV) at two locations, 1 cm apart, and also at the apex of the LV. At each site, three depths were chosen for the measurements: 1) close to the epicardial surface (subepicardial), 2) in the center of the LV wall (midwall), and 3) near the endocardium (subendocardial). The pressure measurements were carried out continuously while the catheter was rotated through 360 degrees. Typically, the needle tip catheter was first inserted in the lateral wall of the left ventricle till the pressure sensor was just beneath the epicardial surface and the pressure was recorded. The catheter was then turned slowly and the pressure recording continued. Then it was advanced to the next depth, the midwall region, and the measurements repeated. Following this, the catheter was advanced further to the subendocardial region and the pressure measurements repeated again. The pressure in the LV chamber was measured using a fluid-filled 7F catheter. The details of the technique can be found in Ref. 4.

From the intramyocardial pressure measurements, it was obvious that significant pressures developed within the muscle mass at all times in the cardiac cycle. Even in diastole, the myocardium was not totally relaxed but remained under low pressure. In systole, the pressure close to the epicardial surface was the lowest, in the midwall it was higher, and in the subendocardial region it was the highest (Fig. 10.8). In the 14 dog hearts, the LV cavity systolic pressures ranged from 90 mmHg to 160 mmHg with an average of 116 mmHg. At location 1 on the lateral wall, subepicardial systolic pressures averaged at 61 ± 5 mmHg, midwall systolic pressures at 92 ± 13 mmHg, and subendocardial systolic pressures at

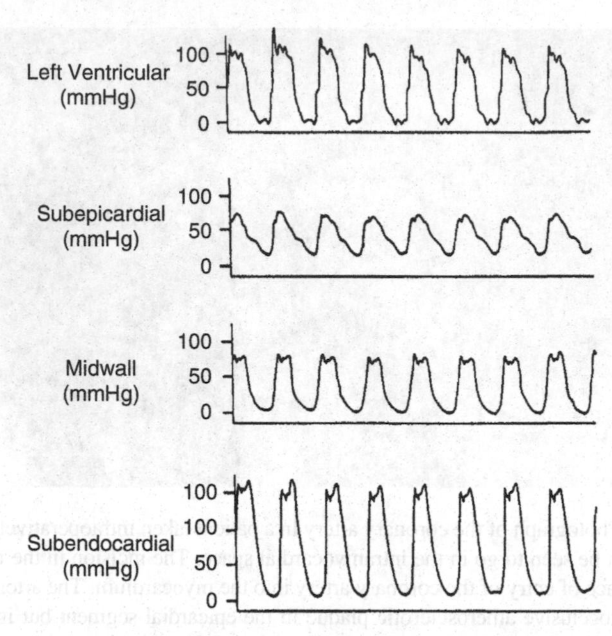

FIGURE 10.8. Typical recordings of the LV chamber pressure and the intramyocardial pressure in subepicardial region, midwall region, and subendocardial region in canine hearts.

124 ± 12 mmHg. At location 2 on the lateral wall, the subendocardial, midwall, and subepicardial systolic pressures were, respectively, 74 ± 6 mmHg, 82 ± 10 mmHg, and 115 ± 8 mmHg. At the apex of the LV, subepicardial, midwall, and subendocardial systolic pressures were, respectively, 80 ± 5 mmHg, 102 ± 8 mmHg, and 122 ± 5 mmHg. The subepicardial systolic pressures were less than the midwall systolic pressures and the midwall systolic pressures were less than the subendocardial systolic pressures at all three locations. Table 10.1 shows the intramyocardial systolic pressures, at the three depths at each of the three locations, in terms of the percentage of LV cavity pressure. Once again, it can be noted that the intramyocardial systolic pressure in the subendocardial region was the highest (104% of the LV pressure), in the midwall it was lower (79%), and in the epicardial region it was the lowest (63%). It is important to note that the systolic pressure in the subendocardial region was greater than the LV cavity pressure. Table 10.2 shows intramyocardial pressures in diastole at the three depths at each of the three locations. The diastolic pressures were almost the same at all depths at all three locations and they were in the range of 4 to 7 mmHg with an average value of 5.5 mmHg. In comparison, LV cavity mid-diastolic pressures were approximately 1 mmHg.

Several other characteristics of the intramyocardial pressure were also observed. The waveforms of the intramyocardial pressure in the midwall and in the subendocardial regions were similar but not identical to the waveform of the

TABLE 10.1. Intramyocardial pressure (% of LV pressure) in systole.

	Subepicardial	Midwall	Subendocardial
Location 1	53 ± 5	79 ± 10	105 ± 7
Location 2	65 ± 7	69 ± 7	100 ± 7
Apex	71 ± 6	89 ± 7	106 ± 4
Average	63 ± 6	79 ± 7	104 ± 7

Grand average ± std. error.

TABLE 10.2. Intramyocardial pressure in diastole (mmHg).

	Subepicardial	Midwall	Subendocardial
Location 1	5 ± 3	5 ± 2	7 ± 2
Location 2	4 ± 3	5 ± 2	7 ± 3
Apex	6 ± 3	7 ± 3	4 ± 2
Average	5 ± 3	6 ± 2	6 ± 2

Grand average ± std. error.

LV pressure, whereas the pressure waveform in the subepicardial region was different (Fig. 10.8). Another significant observation was that for most of the premature ventricular contractions (abnormal beats), the intramyocardial pressures increased more than did the LV cavity pressure. This reflected accentuation of the regional contractility.

4.4. *Wall Stress and Coronary Atherosclerosis*

Let us recall that both intramyocardial pressure and LV pressure are generated by myocardial contraction. Therefore, intramyocardial pressure and myocardial stress are related and should show some similarity. Mirsky (10), in his review of LV wall stresses, reported that the circumferential stress (dominant stress) in the myocardium is highest in the endocardial region and decreases toward the epicardium. This type of distribution of stress through the wall is typical for a thick-wall cylinder geometry, and it has been described in Chapter 4. The decrease of stress through the myocardium is similar to the decrease of intramyocardial pressure (i.e., the intramyocardial pressure is highest in the subendocardial region and decreases toward the epicardial region). Different intramyocardial pressures at different depths probably reflect different contractile states of the myocardial fibers. Intramyocardial pressure has been believed to be external to the blood vessels and thus determines the transmural pressure and may also affect the vessel caliber (11).

From the vascular mechanics point of view, we would like to consider that the phenomenon of freedom of intramyocardial coronary arteries from atherosclerosis is due to the reduction of wall stress, and it occurs in the same way as that seen in the earlier examples. The only difference being, in the earlier examples the artery was supported externally by bone or by the casting compound, whereas the intramyocardial coronary arteries are supported by the surrounding myocardium.

The stress reduction in the intramyocardial arteries could be conceptualized as follows: For a thin cylinder (e.g., artery) we have:

$$\sigma_c = \frac{PR}{T} \text{ and } \sigma_l = \frac{PR}{2T}$$

where σ_c and σ_l, respectively, represent stress in the circumferential and longitudinal directions of the artery, P is the pressure gradient across the artery wall, and R and T are mean radius and thickness of the artery, respectively.

The pressure gradient (P) across the artery wall is the difference between the intravascular blood pressure and the pressure in the tissues surrounding the vessel. The intravascular pressure may indeed be substituted for the pressure gradient when surrounding tissue pressure, in most parts of the body, is small and therefore negligible. This is not the case in areas where significant support is given to the vessel by surrounding tissues. The surrounding contractile myocardium produces tissue pressures, which approach and even exceed the intravascular blood pressure in systole. This reduces wall stress in the artery significantly, particularly in systole, and the amount of stress reduction is directly dependent upon how deep the artery is located in the myocardium. For the artery in the subepicardial region, the wall stress in systole will be reduced by about 63%, if the artery is in the midwall region the stress will be reduced by about 80%, and if the artery is in the subendocardial region the stress may in fact be zero or reversed (Table 10.1). In diastole also the wall stress will be reduced by about 6% (Table 10.2: the gradient will change from 80 mmHg to 75 mmHg). This reduction in the pressure gradient and therefore in the wall stress appears to correlate with the absence of atherosclerosis in the intramyocardial coronary arteries.

It is also probable that in intramyocardial arteries, besides the decrease of the transmural pressure gradient, the absolute amount of wall stress is further decreased by the elimination of the systolic increment of the vessel radius (R). While the epicardial vessels tend to reach their maximum caliber at the height of systole, the intramural arteries being subjected to systolic compression will decrease their diameter compared with what they possess during diastole. In other words, the myocardium surrounding the intramyocardial coronary artery ensures not only a lifelong maintenance of low or no systolic pressure gradient across the artery wall but also prevents systolic increase in caliber.

5. Change in Endothelial Cell Morphology by Reduction of Arterial Wall Stress

In previous sections, we saw excellent examples of inhibition or prevention of atherosclerosis when the arterial wall stress was reduced by external support. We also noted that the artery must adapt to the condition of reduced wall stress and that these adaptive changes must mediate the process of inhibition of the disease. One such adaptive change occurs in the morphology of endothelial cells. Baker et al. (3) explored the endothelial cell morphology after reducing the wall stress at the arterial branch region by placement of an external cast, and we will examine their findings.

Endothelial cell morphology was studied with scanning electron microscopy. Intramural stress in the arterial branch area was reduced by placement of a periarterial cast. New Zealand white rabbits were anesthetized; intravenous and intra-arterial catheters were placed in ear vessels for intravenous infusion and arterial blood pressure measurement. A midline laparotomy was performed, and a segment of the abdominal aorta and proximal left renal artery was isolated. A small piece of metallic foil was placed posterior to the branch area to form the back of the cast. The aorta–left renal artery branch was covered with warm saline to resolve vasospasm. The mean arterial blood pressure was then lowered to 40 mmHg and maintained by the intravenous infusion of nitroprusside. During controlled hypotension, liquid dental acrylic compound was poured around the isolated segment of the aorta and left renal artery. The cast was allowed to harden and nitroprusside was discontinued. The rabbits underwent standard closure of the laparotomy and were allowed to recover for 4–8 weeks. Figure 10.9 shows the cast and its relationship to the arterial geometry. After the period of standard diet, the rabbits were anesthetized. The femoral and carotid arteries were exposed, and 16-gauge catheters were placed into femoral arteries and one carotid artery. The rabbits were sacrificed by intravenous injection of concentrated pentobarbital. The aorta was immediately flushed and perfused with 2.5% glutaraldehyde in a Krebs-Henselyte buffer solution through the carotid catheter. After all intravascular blood had been flushed, these vessels were clamped, and a hydrostatic pressure of 100 mmHg was maintained for 2 h. The aorta was excised and the cast carefully removed. The aortic segments were further fixed in buffered glutaraldehyde for 24–48 h.

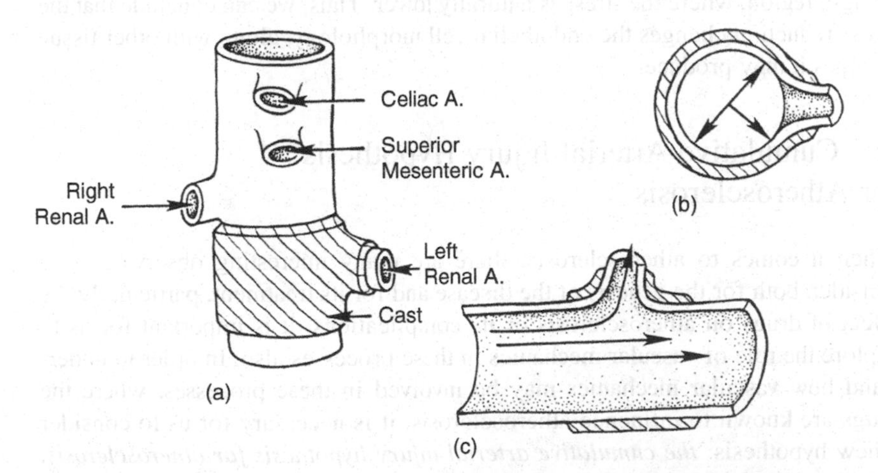

FIGURE 10.9. (a) Diagram of a rabbit abdominal aorta and major branches with a cast around the aorta; left renal artery segment. (b) Perpendicular forces are buttressed by the rigid cast, and intramural stress is reduced. (c) Parallel forces related to blood flow are not significantly altered by the cast.

The aortic segments were processed for SEM of the luminal surface. Specimens studied included the unbranched (straight) abdominal aorta, the control aorta–left renal artery branch segment, and the casted aorta-left renal artery branch segment. During processing, care was taken to maintain the normal aortic wall curvature and to minimize any artifact due to flattening of the specimen. The specimens were stained and fixed in 1% osmium tetroxide. The specimens were mounted on SEM stubs and sputter-coated with gold-palladium. The luminal surface of each specimen was examined with a scanning electron microscope.

Their findings on the endothelial cell morphology in the straight segment and in the control branch segment of the aorta was described earlier in Chapter 8. The morphology of the endothelial cell in the stress reduced (casted) branch region is shown in Figure 10.10. The overall topographic pattern was similar to that of the straight segment of the aorta. The intimal surface was smooth. The endothelial cells were of uniform size and were oriented predominantly in the direction of blood flow. Individual cells were flat, the cell borders were intact, and there were no spindle-shaped cells. The morphology of endothelial cells was the same when the cast was placed for 4, 7, or 8 weeks.

Overall, in the straight segments of the aorta, the endothelial cells (ECs) were flat with intact cell borders. The control branch sites had many morphologically altered endothelial cells. The endothelial cell morphology at these sites displayed a morphologic spectrum ranging from spindle-shaped cells with disrupted cell borders to cobblestone-shaped cells with loss of cell orientation (see Chapter 8). Once the stress in the branch was reduced, the morphology of the endothelial cells changed. There were no endothelial cells with altered morphology, instead the morphology of the cells in the branch region had become similar to that in the straight region, where the stress is naturally lower. Thus, we can conclude that the stress reduction changes the endothelial cell morphology, along with other tissue changes it may produce.

6. Cumulative Arterial Injury Hypothesis for Atherosclerosis

When it comes to atherosclerosis, there are many interesting observations to consider, both for the genesis of the disease and for its treatment, particularly the effect of drugs on atherosclerosis or its complications. It is important for us to explore the role of vascular mechanics in these processes also. In order to understand how vascular mechanics may be involved in these processes, where the drugs are known to influence atherosclerosis, it is necessary for us to consider a new hypothesis: *the cumulative arterial injury hypothesis for atherosclerosis.* This hypothesis will include wall stress, cumulative arterial injury, endothelial injury, as well as injury to the media so that we can develop an understanding of the comprehensive mechanism of atherosclerosis. Then, in the light of this new hypothesis, we will be able to consider observations that involve treatment of atherosclerosis with drugs.

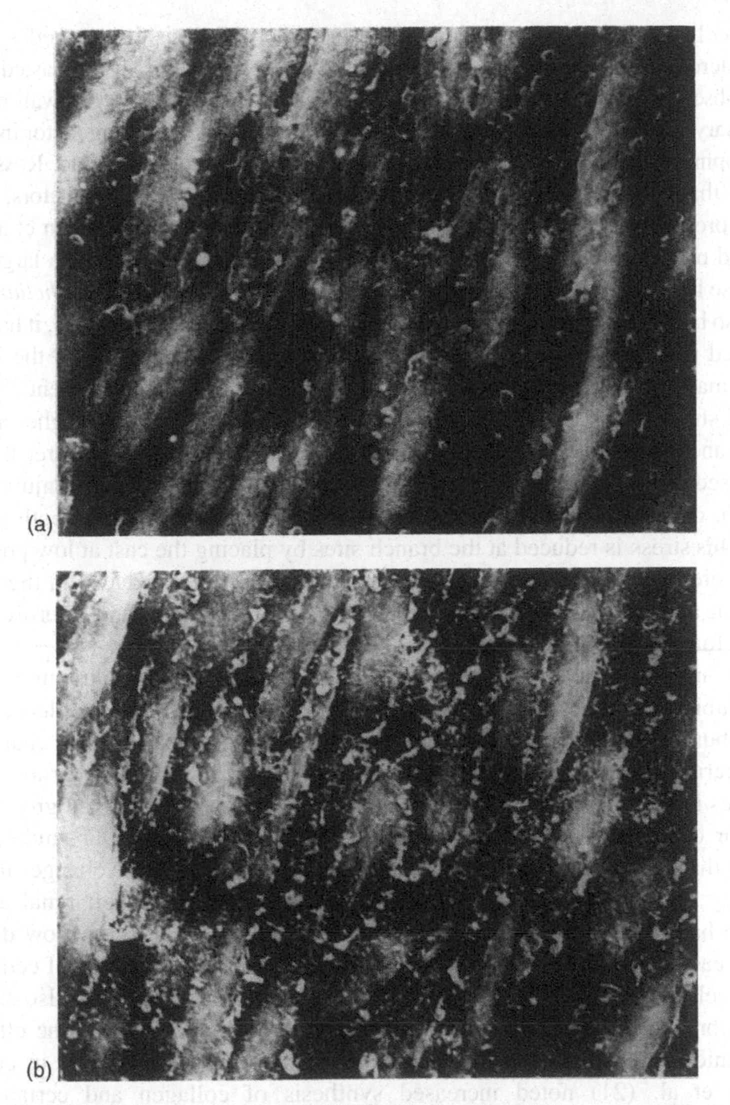

FIGURE 10.10. SEM photomicrographs of the luminal surface of the aorta at the casted left renal artery ostium. The intimal surface is smooth with uniform-appearing, flat, elongated endothelial cells. Cell borders are intact, and the cells are oriented in the direction of bulk blood flow. Cast duration 4 weeks (a) and 7 weeks (b).

6.1. *Wall Stress and Atherosclerosis*

Current theories suggest two possible pathways for the development of atherosclerosis (12). One is endothelial injury, and the other is the lipid hypothesis (i.e, an increase in the serum cholesterol level alone can produce atherosclerotic lesions). In his recent update on the pathogenesis of atherosclerosis, Ross (13) questioned whether hypercholesterolemia alone can cause atherosclerosis. In our model, hypercholesterolemia alone did not produce atherosclerotic lesions in the casted areas. This observation suggests that other functional changes in the arterial wall may be necessary before the plasma cholesterol level becomes a determining factor in lesion development. The endothelial injury hypothesis seems more plausible. Ross mentioned that endothelial injury could result in the release of growth factors, which could provide a mitogenic stimulus to smooth muscle cells. Stemerman et al. (14) pointed out that gross endothelial damage is not necessary to mediate a large local increase in the uptake of low-density lipoprotein (LDL). The term *endothelial injury* has also been used to indicate endothelial cell dysfunction (15). Recently, it has been reported that a different morphology of endothelial cells, observed in the branch areas, may not indicate injury but may be normal to the local environment (16, 17).

Our studies (18–20) have shown that intramural stress is much higher at arterial branch sites than in straight segments. It seems likely, therefore, that the increased stress could have an adverse effect on the artery and cause injury, stimulation, or dysfunction of the endothelial cell, as well as of the smooth muscle cell. This stress is reduced at the branch sites by placing the cast at low pressure. Therefore, it appears likely that with the intramural stress reduced in the casted area, the adverse effect of the stress on the artery is eliminated and atherosclerotic lesion formation is inhibited.

Although atherosclerosis is inhibited in the areas of reduced intramural stress, the inhibition may not necessarily be caused by the direct action of reduced stress itself, but by its effect on the arterial tissue. Such effects may include changes in the arterial permeability to LDL, changes in the endothelial cell turnover rate, changes in the release of endothelial or smooth muscle cell–derived growth factors, or changes in the smooth muscle cell replication rate. Our studies have shown that reduction of intramural stress by the cast did produce changes in morphology of the endothelial cells (3). In normal rabbits, the left renal arterial branch had frequent spindle-shaped endothelial cells distal to the flow divider. When casted, the same areas showed a smooth surface of endothelial cells with intact cell borders and no spindle-shaped cells. These casted areas also showed inhibition of atherosclerotic lesions in cholesterol-fed rabbits (1). The effect of mechanical stress has also been observed on smooth muscle cells in culture. Leung et al. (21) noted increased synthesis of collagen and certain acid mucopolysaccharides by rabbit aortic smooth muscle cells in response to cyclic stretching of the cell substrate. This leads us to understand that the stress, cyclic stress, cyclic stretch, and number of cycles involved would all have an influence on the arterial tissue and thus on atherosclerosis.

6.2. Cumulative Arterial Injury (Artery Fatigue) and Atherosclerosis

It has been reported that a decrease, either of the mean blood pressure or of the pulse pressure, reduces atherosclerotic lesions in rabbits (22, 23), sheep (24), and monkeys (25), and a decrease in the heart rate reduces atherosclerotic lesions in rabbits (26) and monkeys (27, 28) (Table 10.3). In humans, a decrease in mean blood pressure (29) and a decrease in the heart rate (30, 31) is reported to reduce mortality and morbidity originating from atherosclerotic disease. To understand these observations, we must consider a new concept of cumulative arterial injury due to artery fatigue.

As mentioned in Chapters 4 and 5, fatigue is one of the phenomena that occur in all vessels under pressure, particularly when the pressure is pulsatile (32). The phenomenon of fatigue failure is well understood in many nonbiological materials. It is said that, "In pressure vessels virtually all failures are a result of fatigue—fatigue in areas of high localized stress" (32). This type of fatigue damage has been considered by Born and Richardson (33) in relation to rupture of atherosclerotic plaques. We propose that this phenomenon occurs in the artery wall itself because the artery has both areas of high localized stress and pulsatile blood pressure.

The concept of *cumulative arterial injury from fatigue* may be explained as follows: Consider the arterial branch area under systemic pressure of 120/80 mmHg (Figure 10.11). It is logical to postulate that a certain degree of "cumulative arterial injury" occurs at a site of stress-concentration after a certain number of cardiac cycles. In the event that the mean blood pressure increases while the pulse pressure remains unchanged (e.g., 170/130 mmHg), then the same degree of cumulative arterial injury occurs after a lesser number of cardiac cycles. Similarly, when the pulse pressure increases while the mean pressure remains unchanged (e.g., 140/60 mmHg), then a similar amount of cumulative arterial injury occurs after a lesser number of cardiac cycles. If one further assumes that cumulative arterial injury reflects atherosclerotic disease, then increase in either the mean blood pressure or the pulse pressure will increase the injury therefore increase atherosclerosis. If the heart rate is reduced with beta-blockers, then it takes a longer time to reach the total number of cardiac cycles required for a given degree of cumulative arterial injury. Conversely, in a given period, the injury and consequently atherosclerosis is less when the heart rate is reduced. This mechanism of

TABLE 10.3. Effect of mean blood pressure, pulse pressure, and heart rate on atherosclerosis.

Study	Model	Mean BP (mmHg)	Pulse pressure (mmHg)	Heart rate (beats/min)
Snyder and Campbell (23)	Rabbits	107 78[a] ↓	34 12 ↓	
Magarey et al. (24)	Sheep	130 a	32 19 ↓	
Lyon et al. (25)	Monkeys	100 a	44 26 ↓	
Spence et al. (22)	Rabbits	92 79[a] ↓	— —	— —
		102 86[a] ↓	— —	222 185 ↓
Beere et al. (28)	Monkeys	100 a	50	110–150 90–110 ↓

[a] Represents a decrease in atherosclerotic lesions.

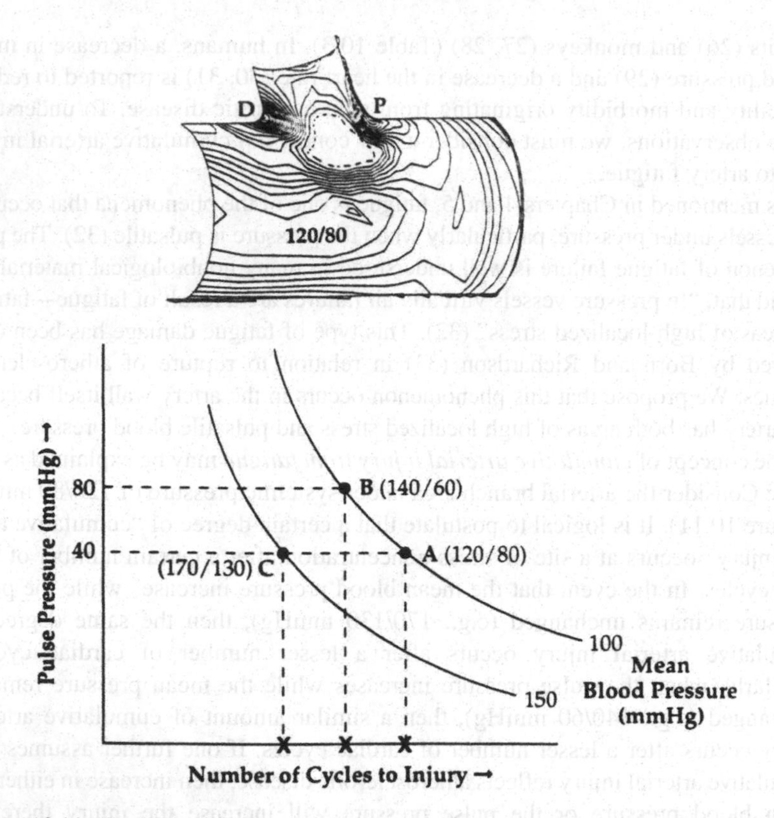

FIGURE 10.11. (Top) Distribution of stress contours in the arterial branch indicating that at both the distal lip (D) and the proximal lip (P) of the ostium, the stresses are high and localized. Fatigue failure is expected to occur at either of these two locations mainly as a result of the pulsatile pressure experienced by the artery. (Bottom) Typical fatigue behavior of a nonbiological material. We propose that this also qualitatively represents the fatigue behavior of an artery. Point A indicates that for the systemic pressure of 120/80 mmHg, a certain number of cardiac cycles are required to produce fatigue damage in the artery. Point B indicates that for a greater amount of pulse pressure, but for the same mean pressure, the number of cardiac cycles required to produce fatigue damage is less. Point C indicates that for the same pulse pressure, but a higher mean blood pressure, the fatigue damage could occur at a lower number of cardiac cycles. In qualitative terms, increasing either the mean blood pressure or the pulse pressure would produce the fatigue damage in fewer cardiac cycles (i.e., atherosclerotic disease will occur sooner). Similarly, for a given period, decreasing the heart rate decreases the fatigue damage (i.e., decreases cumulative arterial injury and hence atherosclerosis).

cumulative arterial injury due to artery fatigue is probably at work in vivo because: 1) the artery has areas of stress-concentration; 2) the artery has pulsatile blood pressure, and 3) this mechanism predicts that a reduction in the mean blood pressure, the pulse pressure, or the heart rate will reduce atherosclerosis, and these effects have already been observed (Table 10.3).

6.3. *Endothelial Injury and Atherosclerosis*

When the arterial wall is denuded of endothelium, platelets and leukocytes adhere to the subendothelial tissue and they can be the trigger for SMC proliferation. In a rabbit model of balloon injury, platelet deposition begins almost immediately after injury, ends within 24 h and is followed by the appearance and proliferation of SMCs within the media (34). In clinical trials, however, platelet inhibitors have failed to consistently show an improvement in rates of restenosis (25). In our study (Chapter 9), balloon denudation in the absence of the artery stretch, which will still cause platelet adhesion, failed to produce a massive SMC proliferation, suggesting that when the artery wall is stretched, additional stimuli are created that are even more important in producing a massive cellular proliferation. Several others (35, 36) also noted that injured ECs, accumulated platelets, and factors in plasma do not fully account for the proliferation of SMCs, and that additional factors must be involved. These additional stimuli could originate from injured SMCs, damaged extracellular matrix, or even damaged cell-to-cell junctions. Thus, the injury to the media could well be more important in atherosclerosis and restenosis than the injury to the endothelium.

6.4. *Medial Injury and Atherosclerosis*

Our previous studies have shown that the arterial wall stress due to blood pressure is very high at the sites where atherosclerotic lesions develop (1, 4). Also, the wall stretch and cyclic changes of both the stress and the stretch are highest where the lesions form. This stress and stretch is primarily borne by smooth muscle cells, connective tissue of the media, and connective tissue of the adventitia. We propose that at these selected sites, the media is injured from the stress and the stretch, which result in SMC proliferation and lesion formation. This hypothesis is supported by our study (Chapter 9), which showed that the EC injury was not sufficient to produce significant SMC proliferation, whereas the aortic stretch provoked a massive proliferative response, which increased even more with the increasing stretch.

Thus, to understand atherosclerosis, we may have to reexamine our concepts of the role of endothelial injury and focus on the role of smooth muscle cells.

6.5. *Comprehensive Mechanism for Atherosclerosis*

Let us now consider a *comprehensive mechanism*, which could explain both the formation of atherosclerotic lesions as well as inhibition/prevention of atherosclerosis. The various steps in this mechanism could be listed as follows:

1. Pressure-induced wall stress is significantly increased (stress-concentration) at the arterial branch sites.
2. The cyclic stresses and the cyclic strains are significantly increased at the branch sites, and they are proportional to the pulsatile pressure in the artery.

3. These stresses and strains are the cause of fatigue injury (damage) to the artery and most dominantly to smooth muscle cells.
4. Fatigue injury to smooth muscle cells is directly proportional to the mean blood pressure, the pulse pressure, and the number of cycles (H.R. × duration) imposed on them.
5. Under normal conditions, the damage is small and repaired, and the artery does not change significantly.
6. This equilibrium is distributed by other changes such as excessive stretch (e.g., in balloon angioplasty), increased blood pressure, increased pulse pressure or increased heart rate, which then leads to increased proliferation of smooth muscle cells.
7. The increased proliferation of smooth muscle cells is the essential step that leads to the development of atherosclerotic lesions.
8. Conditions such as high cholesterol level, smoking, or diabetes are proposed to be responsible for lowering the threshold required for proliferation of smooth muscle cells. In other words, SMCs could proliferate at a higher rate when the LDL level in the blood is high or the plasma sugar level is high, even though the blood pressure is normal.

The central step in this mechanism is that the localized cyclic stress and strain are the stimuli for SMC proliferation, which leads to atherosclerotic lesion formation. Other factors promote or inhibit this step and lead to the presence or the absence of the disease. The above mechanism, though simple and nonmolecular in concept, indicates in the most direct way how vascular mechanics could play a role in atherosclerosis. We are now ready to examine additional observations, which involve treatment of the disease with drugs.

7. Comparing Fatigue Damage Between Nonbiological (Metallic) and Biological Pressure Vessels

There exists a good body of knowledge in the field of pressure vessels made of metals. Because the artery is also a pressure vessel, it is logical that we look for the phenomenon that occurs in the pressure vessels and explore how it might reveal itself differently in pressure vessels made of two completely different materials, where one is nonliving and the other is living tissue. One fundamental difference we expect is that the pressure vessel made of metal will rupture or break while the artery will not rupture, instead it will develop a disease such as *atherosclerosis or aneurysm* with one exception, acute dissection/rupture of the ascending aorta, which is again related to the same phenomenon.

7.1. Common Cause of Atherosclerosis and Aneurysm

With the statement above, we have essentially made the claim that two different diseases like *atherosclerosis* and *aneurysm*, are caused by the same phenomenon—fatigue damage. We need to explain how that is possible. Once again, we go to smooth muscle cells as a common factor for the two diseases. We postulate that fatigue damage is imposed on the SMCs, and they respond to the damaging stimulus by the process of repair and replacement, which requires proliferation. Excessive proliferation results in *atherosclerosis*. Logically then, the failure of repair or replacement results in loss of cells (apoptosis), which results in *aneurysm*. To put it another way, successful but excessive proliferation of SMCs results in atherosclerosis while failure to proliferate (or repair) results in aneurysm. Therefore, it is not surprising that many locations in the artery, where atherosclerosis or aneurysm occur, are common to both; some of the most important risk factors, such as hypertension, are also common to both (again relates to more stress); and quite frequently aneurysms are considered to be of atherosclerotic origin. Throughout this book we describe the commonality between the two diseases (see Chapter 2).

7.2. *Stress-Concentration and Fatigue*

The fatigue damage involves the cumulative effect of numerous small events taking place over many cycles of stress and strain. The three basic factors necessary to cause fatigue damage are 1) a maximum tensile stress of sufficiently high value, 2) a large enough fluctuation in the stress, and 3) a sufficiently large number of cycles of stress. A "benchmark" of fatigue damage in a pressure vessel is the point of initiation of damage and the progression of damage from this point. In the pressure vessel, virtually all failures are a result of fatigue in the area of high localized stress.

Figure 10.12 shows the fatigue failure of (metallic) the pressure vessel. The typical oyster-shell markings focus on the origin of the damage in the region of high local stress concentration at the inside corner of the nozzle and from here the damage propagated throughout the vessel and the nozzle. Let us recall that at the arterial branch site also a similar thicker wall exists at the junction, a similar stress concentration exists at the junction (see Chapters 6 and 7), and the stresses are the highest on the inside surface as in the case above. If we were to look for the damage in the artery, similar to that in the above pressure vessel, then that damage would have to occur to smooth muscle cells located on the inner media of the branch site. This agrees with the location of initiation of atherosclerosis. Furthermore, at the site of stress concentration, a higher stress gradient is set up from the inside to the outside and a complex triaxial state of stress is produced, which allows the damage to progress outward. In the artery, this would mean recruitment of more media in the progression of the disease.

In case of fatigue, the damage occurs at a much lower stress than that required for damage without fatigue. For example, a typical artery may burst by rupture under a pressure of 1200 to 1600 mmHg, whereas the fatigue damage at a site of stress concentration may occur at a considerably lower mean pressure. One of the common ways of reporting the fatigue damage is shown in Figure 10.11. On the x-axis, we have the total number of cycles to failure for the material or a pressure vessel. For the artery, the total number of cycles to failure is equivalent to the

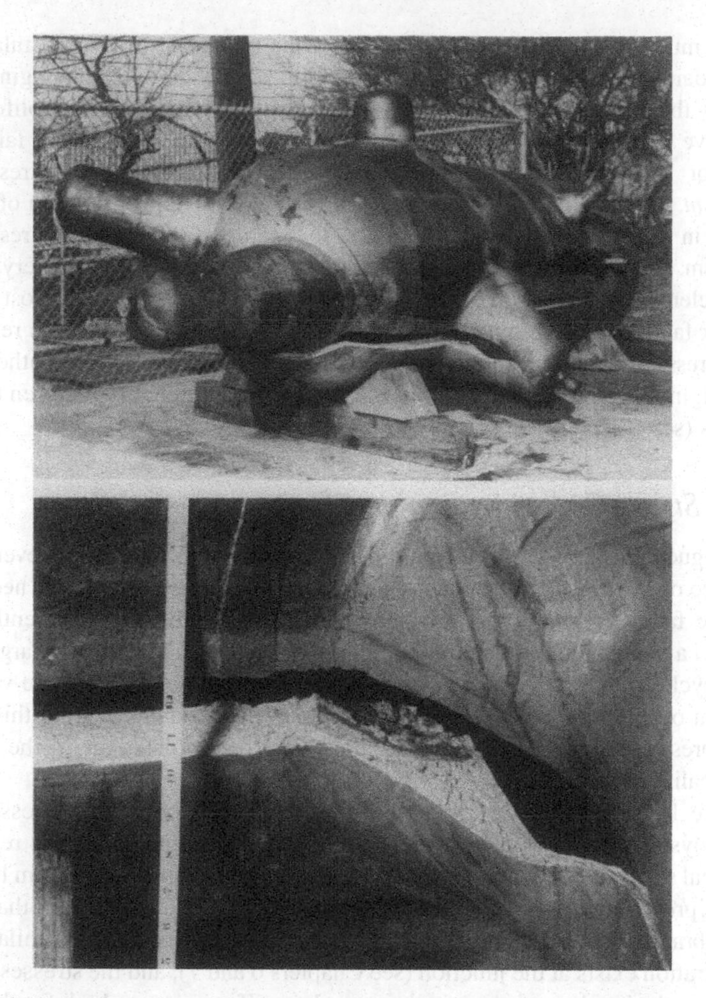

FIGURE 10.12. Fatigue failure of test pressure vessel. (Top) Nature of propagation throughout nozzle and vessel. (Bottom) Origin of fracture at inside corner nozzle. (Reproduced from Harvey JF: Theory and Design of Modern Pressure Vessels, Van Nastrand Reinhold Co, NY, 1974, with permission.)

number of cycles to damage or the number of cycles to disease (e.g., to the development of atherosclerosis). The total number of cycles in the body is simply the heart rate × time in minutes, days, or years. The mean stress is a function of the mean arterial pressure, and the alternating stress is a function of the pulse pressure. Qualitatively then, Figure 10.11 indicates that for a given mean pressure, if the pulse pressure is greater, then the damage occurs in less number of cycles (or in less time). Similarly, for a given pulse pressure, if the mean pressure is greater, then the damage occurs in less number of cycles (or in less time). Also, if the heart rate is reduced, then it will take a longer time to reach the same total cycles (or a longer time for the damage). In other words, reduction in mean blood pressure, pulse pressure, or heart rate will offer a benefit toward less atherosclerosis in a given time. The data that support these interpretations have already been presented in the previous section (Table 10.3).

8. Reduction of Coronary Atherosclerosis by Reduction of Heart Rate in Cynomolgus Monkeys

Conceptually, in the previous section we explored how the reduction of the heart rate could reduce the arterial damage on the basis of fatigue injury in the region of stress-concentration and how that may be seen as a reduction of the disease atherosclerosis. Here we will consider a very important observation related to that. Beere et al. (28) studied the development of coronary atherosclerotic lesions in cynomolgus monkeys and compared the extent of the disease in the two groups: one with a lower heart rate and the other with a higher heart rate.

They lowered heart rates in male cynomolgus monkeys by ablation of the sinoatrial node. This was accomplished in six of nine attempts by electrocautery of the crista terminalis at the entry of the superior vena cava into the right atrium. In eight other monkeys, they performed a sham surgical procedure. One month later, the monkeys were fed an atherogenic diet that included 25% peanut oil and 2% cholesterol. Heart rates during the 24-h test periods were recorded by radiotelemetry at four intervals: preoperatively, postoperatively at 4 weeks, and then twice bimonthly. Serum total and free cholesterol, triglycerides, phospholipids, and cholesterol esters as well as body weights were determined monthly. Blood pressure was measured directly at surgery and before the animals were sacrificed. Six months after the atherogenic diet, each animal was anesthetized, and the heart and great vessels were fixed in situ by pressure perfusion at 100 mmHg with 3% glutaraldehyde. Six samples of the proximal coronary artery tree, including sections of the right, left circumflex, and left anterior descending branches, were removed as complete transverse rings at standard, anatomically defined, sampling sites. Coronary atherosclerosis was evaluated from plastic-embedded semithin (1 μm) histologic sections. The total number of coronary intimal lesions was determined for each monkey, and the lesion cross-sectional area and percent stenosis were determined for each section. The vessels of eight control monkeys not operated on, but maintained on the same atherogenic diet for 6 months, were prepared and studied in the same manner.

The preoperative heart rate for all of the animals treated surgically was 136 ± 22 beats per minute. In the six animals in which sinoatrial node ablation was successful, the heart rate was reduced 31%, from 148 ± 11 to 103 ± 20 beats per minute. Reduced heart rate persisted throughout the experimental period. Five animals had mean heart rates above the preoperative mean (high heart rate group) and 12 had mean heart rates below the preoperative mean (low heart rate group). All six with successful sinoatrial node ablation were in the latter group.

Comparisons between the low and high heart rate groups showed significant differences. In animals with heart rates above the preoperative mean, the number and severity of coronary artery lesions was more than twice that of animals with heart rates below the preoperative mean. The animals in the high heart rate group had an average percent stenosis of 28.1 ± 20.9, whereas those in the low heart rate group had an average percent stenosis of 13.1 ± 12.2; the average percent stenosis for the control group was 23.8 ± 15.6. The average of the lesion areas for the animals with a high heart rate was 0.48 ± 0.47 mm^2, compared with 0.21 ± 0.39 mm^2 for those with a low heart rate; in control animals, the average of the lesion areas was 0.38 ± 0.19 mm^2.

The greatest difference between the groups was observed in the comparison of the maximum stenosis. The most occlusive lesions averaged 55.9 ± 23.1% stenosis in the high heart rate group, 26.1 ± 18.8% stenosis in the low heart rate group, and 49.7 ± 22.2% stenosis in the control group. There were no significant differences between the high heart rate and the low heart rate groups in blood pressure, body weight, or serum lipid levels.

These findings support the hypothesis that a relatively low heart rate for an extended period tends to retard lesion formation in the coronary arteries. Although sinoatrial node ablation was used to produce a sufficient number of animals with a low heart rate, the coronary sparing effect was independent of the surgical procedure. Such a relationship could help to account for the protective effect of regular physical activity against coronary artery disease in man and in experimental animals, in that such conditioning results in lowered average heart rates and therefore in a reduction in total number of heartbeats for extended time periods. Conversely, the association of type A personality traits with coronary heart disease, and the experimental finding that psychosocial stress in monkeys predisposes to coronary plaque formation, could both be the result of frequent stress-related elevations of heart rate that would tend to elevate average heart rate for extended periods.

9. Reduction of Atherosclerosis by Treatment with Beta-Blocker in Cynomolgus Monkeys

From the fatigue damage hypothesis described earlier, it is obvious that the increase of the heart rate should correlate with the increase of the arterial injury and thus with increase of atherosclerosis, just as the decrease of the heart rate should correlate with the decrease of the arterial injury and thus to the decrease of atherosclerosis. Kaplan et al. (37) did nice experiments that addressed both the questions, *effects of increased and decreased heart rate*, in cynomolgus monkeys, and we will consider some of their findings.

In earlier experiments involving cholesterol-fed, socially housed monkeys, they demonstrated that an unstable social environment exacerbates coronary artery atherosclerosis, but only among monkeys that habitually retain a dominant social status; subordinate monkeys develop significantly less atherosclerosis (38). They speculated that the dominant monkeys experience repeated challenges to their social status, resulting in recurrent sympathetic activation, with accompanying increases in heart rate, blood pressure, and catecholamine release, in turn potentiating atherogenesis (38).

If frequent exposure to psychosocial challenge can promote atherogenesis through associated sympathetic mediation, administration of a β-adrenergic blocking agent should inhibit the development of atherosclerosis. They tested this hypothesis in an experiment involving 30 male cynomolgus monkeys fed a moderately atherogenic diet and housed for 2 years in repeatedly reorganized social groupings; half of the monkeys were administered propranolol HCI (at a dose of 0.05 mg J^{-1} day^{-1}) throughout the study, and the social status of each monkey was assessed on a recurrent basis (27). Propranolol HCL is a β-adrenergic blocking agent, and it prevents the heart rate from going up. Measurements of blood pressure, heart rate, serum lipid concentrations, and social behavior were made repeatedly. At the end of the experiment, cross sections of the coronary arteries were prepared after pressure fixation. Atherosclerosis extent (mm^2) was evaluated in 15 cross sections from each monkey.

Chronic administration of propranolol achieved a significant (20%) reduction in heart rate, along with a comparable lowering of blood pressure, relative to untreated controls. Propranolol had no effect on the agonistic behaviors of the treated monkeys, nor were dominance relationships affected by the drug (39).

Evaluation of the coronary arteries revealed that, among untreated monkeys, the more aggressive (dominant) monkeys had [as expected (38)] significantly more atherosclerosis than their subordinate counterparts. In contrast, the atherosclerosis of dominant monkeys treated with propranolol did not differ from that of the subordinate monkeys, treated or untreated. These results are summarized in Figure 10.13 and indicate that the exacerbated atherosclerosis, typically observed among dominant monkeys, is inhibited by treatment with a β-adrenergic blocking agent.

The psychosocial and pharmacological effects on coronary atherogenesis shown in Figure 10.13 were not associated with concomitant variability in serum lipid concentration. Similarly, the generalized effects of propranolol on blood pressure and heart rate, which occurred irrespective of social status, cannot account for the selective protection accorded dominant monkeys in the treated groups. In the latter regard, these heart rate and blood pressure data were obtained under the controlled conditions of laboratory measurements. Such measurements may not reflect the cardiovascular responses experienced during naturally occurring periods of social stress, as the latter physiological reactions are likely to be acute and transient in nature and therefore elude detection in measurements recorded under laboratory conditions (27). To the degree that a substantial increase in heart rate is indicative of sympathetic arousal, propranolol may have exerted an antiatherogenic influence by attenuating sympathetic activation in those monkeys that were behaviorally challenged, that is, the dominant monkeys. This hypothesis suggests, in turn, that treatment with β-adrenergic blocking agents may confer a degree of protection against coronary artery atherosclerosis among people behaviorally predisposed to CHD.

In summary, sympathetic arousal (associated with increased heart rate) increased atherosclerosis, whereas prevention of that (elimination of increase in heart rate) by treatment with propranolol prevented increase in atherosclerosis.

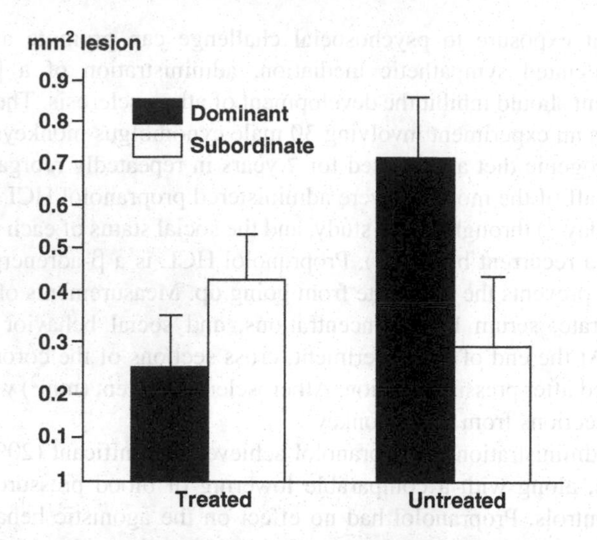

FIGURE 10.13. Bar graph showing extent of coronary artery atherosclerosis (mean ± SEM lesion size) averaged across 15 sections among dominant and subordinate monkeys living in unstable social groups and either treated or not treated with propranolol; n = 6 in each of the four subgroups. (Reproduced from Kaplan JR et al., Circulation 1991;84:VI-23-VI-32, with permission from Lippincott, Williams & Wilkins.)

10. Reduction of Endothelial Cell Injury by Treatment with Beta-Blocker in Cynomolgus Monkeys

The arterial injury due to fatigue damage must reveal itself in terms of parameters such as injury to endothelial cells, smooth muscle cells, extracellular components, or LDL-transport function of the arterial wall in addition to the resulting pathology such as atherosclerosis. In this section, we will consider changes in the endothelial cell injury as a function of heart rate. We expect that any treatment, which reduces atherosclerosis (e.g., treatment with β-blockers) may be expected to reduce endothelial injury. Kaplan et al. (37) explored the effect of β-blocker metoprolol on endothelial cell injury, and we will consider their findings.

They determined whether "naturally occurring" (i.e., behaviorally evoked) sympathetic arousal would be associated with evidence of endothelial injury or dysfunction. They evaluated the effects of a disrupted social environment on the endothelial integrity of different vascular segments in cynomolgus monkeys. They decided to use monkeys subjected to the stress of a disrupted social environment, but under the protection of a β_1-adrenoceptor blocking agent, as treated controls. The principal dependent measure was nondenuding endothelial injury evaluated in two ways: 1) measured directly with IgG immunohistochemistry (40, 41) and 2) measured indirectly via examination of endothelial cell replication (42).

Twenty adult monkeys were used in this experiment. The 10 weeks before the behavioral and pharmacological manipulations comprised a "baseline" period, during which the monkeys were fed a diet containing 0.02 mg cholesterol/J (equivalent to a human intake of ~200 mg cholesterol/day), which resulted in an average total serum cholesterol of approximately 190 mg/dL. They used a social disruption manipulation designed to acutely challenge specific "target" monkeys and thereby *elevate heart rate* and, presumably, blood pressure. The protocol involved exposure of each experimental monkey individually to a social stressor. This individual manipulation involved exposure of each monkey, for a period of 3 days, to a group of four "host" monkeys in a social setting. At the end of 3 days, the experimental monkey was removed and necropsied. The procedure of exposure to the host group was repeated each week with a new experimental monkey until all 20 monkeys in the study were manipulated in this fashion. The heart rates (via radiotelemetry), body weights, and plasma lipid concentrations of the manipulated monkeys were measured during the baseline and experimental periods. In addition, 3 days before the psychosocial procedure, all monkeys were implanted with a subcutaneous osmotic minipump. In half of the monkeys (the metoprolol-treated group), the pump delivered 15 mg kg^{-1} day^{-1} metoprolol (β_1-adrenoceptor blocking agent) from the time of implantation until necropsy; the pump released saline to the rest of the monkeys (the untreated group).

Measurements taken in vivo are shown in Table 10.4. Social disruption substantially increased the heart rate among all of the untreated monkeys; pretreatment with metoprolol resulted in a significant decrease in heart rate. Neither lipids nor body weight was affected by metoprolol treatment.

The postmortem data with respect to aortic endothelial injury as evaluated by the IgG immunoperoxidase technique are depicted in Figure 10.14 for branched and unbranched sites, respectively. Statistical evaluation of these data revealed a significant treatment effect with lower frequencies of injured endothelial cells in metoprolol-treated than in untreated monkeys, but only at the branching sites. Untreated and metoprolol-treated monkeys did not differ in the frequency of IgG-positive cells at unbranched sites. The data reflecting endothelial cell replication in the aorta are shown in Figure 10.15. There was significantly less endothelial cell replication among metoprolol-treated monkeys, but again, only at branching sites; there were no differences between conditions in unbranched areas. Similar to the effect observed in the aorta, in the coronary arteries also, this index of endothelial injury was significantly lower in the metoprolol-treated as compared with untreated monkeys.

The major result of this study was that monkeys subjected to an acute psychosocial stressor showed evidence of endothelial injury when compared with

TABLE 10.4. Measurements taken in vivo during the baseline and postexperimental periods in monkeys.

	Untreated (n = 10)		Metoprolol treated (n = 10)	
	Baseline	Postexperimental	Baseline	Postexperimental
Heart rate (beats/min)	126 ± 19	164 ± 23	146 ± 38	122 ± 14
Total plasma cholesterol (mg/dL)	191 ± 35	150 ± 41	187 ± 25	153 ± 43

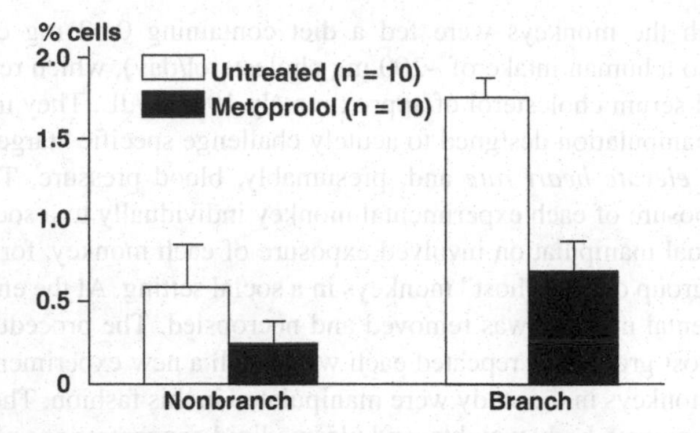

FIGURE 10.14. Bar graph showing percent of endothelial cells with evidence of immunoglobulin G incorporation (injury) at nonbranched and branched sites within the thoracic portion of the aortas of male cynomolgus monkeys (*Macaca fascicularis*). Data are from untreated monkeys and monkeys pretreated with metoprolol. (Reproduced from Kaplan JR et al., Circulation 1991;84:VI-23-VI-32, with permission from Lippincott, Williams & Wilkins.)

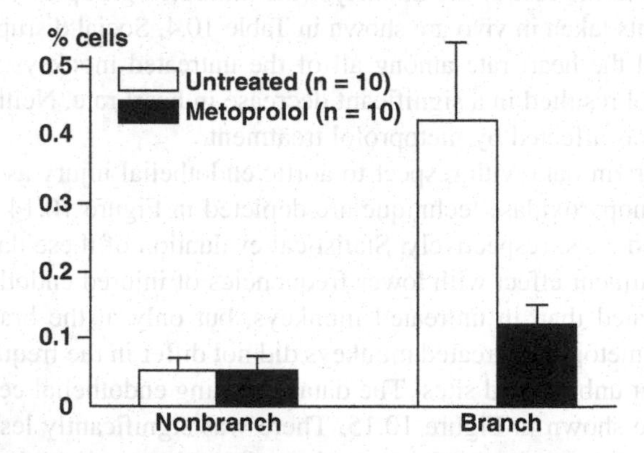

FIGURE 10.15. Bar graph showing percent of endothelial cells with evidence of replication (an indirect index of injury) at nonbranched and branched sites within the thoracic portion of the aorta of male cynomolgus monkeys (*Macaca fascicularis*). Data are from untreated monkeys and monkeys pretreated with metoprolol. (Reproduced from Kaplan JR et al., Circulation 1991;84:VI-23-VI-32, with permission from Lippincott, Williams & Wilkins.)

similarly treated monkeys administered metoprolol. Of note, the presence of injury was indicated by two different techniques (IgG incorporation and endothelial cell replication). Moreover, the observation extends to coronary arteries as well as to the thoracic portion of the aorta, with the effects in the coronary arteries of particular interest because of their potential relevance to human beings. The endothelial effects were paralleled by stress-associated differences in the heart rates of untreated and metoprolol-treated monkeys. A marked elevation of heart rate during the experimental period among untreated monkeys and the concomitant attenuation of heart rate among monkeys treated with metoprolol indicates that the introduction of monkeys to a strange social environment provoked a significant and persistent cardiac response that probably was mediated sympathetically. They conclude that behavioral factors can, through sympathetic mediation, cause damage to the arterial endothelium and thereby, perhaps, initiate atherogenesis in the absence of hypercholesterolemia.

Once again, it is clear that sympathetic arousal (associated with increased heart rate) increased the number of injured endothelial cells, and treatment with metoprolol (associated with prevention of increase in the heart rate) prevented the increase in the number of injured endothelial cells.

11. Reduction of Arterial Permeability to LDL by Treatment with Beta-Blocker in Rabbits

In other chapters (i.e., Chapters 6, 7, 8), we saw that stress-concentration occurs at the arterial branch, EC morphology is altered at the branch, arterial permeability to LDL is high at the branch, and atherosclerosis occurs at the branch. In previous sections, we saw how fatigue damage correlates with the above observations at the branch. We also saw how a reduction of heart rate reduces atherosclerosis and reduces EC injury—as proposed—through reduction in fatigue damage at the branch. Similarly, we may expect that the arterial permeability to LDL at the branch may be changed when the fatigue damage is reduced by the reduction of heart rate, achieved by the treatment with β-blocker.

Thubrikar et al. (43) studied the permeability of the artery to LDL in rabbits and how it might be affected by the treatment with β-blocker metoprolol. We will consider their findings. They used 11 littermate pairs of New Zealand white rabbits (3.5–4 kg). The rabbits were divided into two groups, and Alzeit osmotic pumps with metoprolol (0.4 mg kg^{-1} h^{-1}) were implanted subcutaneously in the experimental group, while saline pumps were implanted in the control group. Heart rates were measured in conscious rabbits once a week. The pumps were replaced after 3 weeks. Blood samples for determination of metoprolol levels were drawn from the ear artery/vein. At the end of 6 weeks, the animals were killed, aortas harvested, and LDL content in the aorta determined.

11.1. LDL Preparation

Rabbit LDL was prepared as follows. Donor rabbits were fed a 1% cholesterol diet for 7 days prior to exsanguination. Fresh whole blood in EDTA was obtained from the carotid artery. LDL (density 1.019–1.063) was separated from the plasma using sequential ultracentrifugation and gel filtration. It was iodinated (^{125}I) by the iodine monochloride method of McFarlane (44). Specific activity of iodinated LDL was 80–200 μCi/mg LDL protein.

11.2. Aorta Preparation and the Experiment

The rabbits were sedated with ketamine and acepromazine. Heparin sodium was injected intravenously, and the rabbits were euthanized. The aorta was exposed through a midline incision. All major branches were ligated 4–5 mm from their origin. The abdominal aorta, starting from a few centimeters proximal to the celiac artery to a few centimeters distal to the left renal artery, was harvested, cannulated, and filled with iodinated LDL solution prepared to give a final concentration of 0.3 mg LDL protein per mL. The intraluminal pressure was maintained between 85 and 90 mmHg using the pressurized air. The aortic segment was immersed in vascular smooth muscle (VSM) nutrient medium in a tray, which was placed in a rack. The entire apparatus was then incubated in a shaking water bath at 37°C for 1 h. After incubation, the aortic segment was emptied and washed.

11.3. Tissue Samples, Cryosectioning, and Counting

The aorta was slit open longitudinally. Two sample disks were cut from the aortic wall at the distal region of the ostium of each branch using a 1.8-mm-diameter metal punch. These disks were called "branch samples." A second pair of disks was cut from the straight areas opposite each ostium and were called "nonbranch samples." One disk from each pair was counted in a gamma counter, while the second disk was saved for cryosectioning. Each disk was sectioned parallel to the intimal surface into 16-μm sections using a cryotome. Two 16-μm sections were placed in each vial and counted, using the Beckman Biogamma II counter for 10 min. The details of the technique can be found in Ref. 43.

11.4. Heart Rate

In the metoprolol-treated group, the average plasma metoprolol concentration over a 6-week period was 320 ± 158 nmol/L. Mean heart rate was lowered by 12% in the metoprolol-treated group (213 ± 3.0 SEM) than in the control group (243 ± 2.0 SEM). In the same group, the heart rate was greater before the drug treatment (246 ± 3.0 SEM) than after the treatment (213 ± 2.0 SEM). The metoprolol treatment lowered the heart rate by 33 BPM (13.4%).

11.5. Aortic Wall Thickness

The aortic wall was significantly thicker in the branch region than in the nonbranch region in the control group (branch 228 ± 38 μm vs. nonbranch 178 ± 19 μm), as well as in the metoprolol group (branch 231 ± 35 μm vs. nonbranch 171 ± 22 μm). There was no difference in the thickness of the aorta between the control and the metoprolol groups, either at the branch region or at the nonbranch region.

11.6. LDL Influx

The whole disks indicated the following. In the control group, LDL influx was substantially higher in the branch region (3061 ± 2784 CPM/disk) than in the nonbranch region (1706 ± 763 CPM/disk). In the metoprolol group, LDL influx was only slightly higher in the branch region (2083 ± 1386 CPM/disk) than in the nonbranch region (1755 ± 1092 CPM/disk). In the branch region, LDL influx was lower by 32% in the metoprolol-treated group than in the control group. In the nonbranch region, LDL influx was almost the same in the two groups. Thus, the treatment with metoprolol reduced LDL influx in the branch regions of the aorta.

The sections indicated the following: In the control group, each section from the branch had a significantly greater LDL influx than the corresponding section from the nonbranch region (Fig. 10.16). The total LDL influx was significantly greater in the branch sample (1457 CPM) than in the nonbranch sample (1226 CPM) (Table 10.5). The intimal LDL influx was also significantly greater in the branch sample (506 CPM) than in the nonbranch sample (417 CPM). The mean LDL influx per section was slightly greater in the branch sample (261 CPM) than in the nonbranch sample (225 CPM).

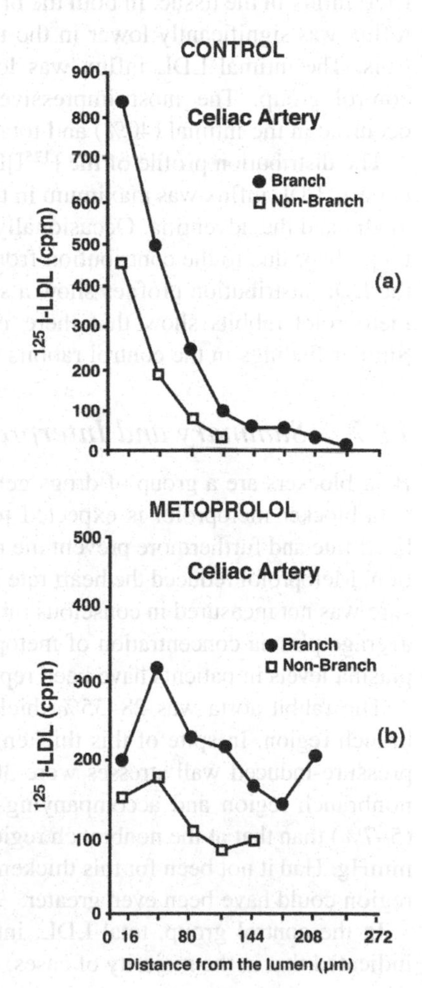

FIGURE 10.16. Typical profiles of LDL distribution through the thickness of the aortic wall in the branch and nonbranch regions of the control (a) and metoprolol-treated (b) rabbits. Influx of LDL is represented by counts per min (CPM). Each symbol represents LDL influx at the midpoint of two 16-μm-thick tissue slices. LDL influx was higher in the branch than in the nonbranch regions. Quantitatively, metoprolol reduced LDL influx in the aorta.

TABLE 10.5. LDL uptake in the aorta in control and metoprolol rabbits.

	Control	Metoprolol	Control – Metoprolol / Control
Branch			
Total	1457 ± 1055	992 ± 625	32%
Intimal	506 ± 342	303 ± 236	40%
Mean	261 ± 183	188 ± 135	28%
Nonbranch			
Total	1226 ± 707	943 ± 631	23%
Intimal	417 ± 270	293 ± 167	30%
Mean	225 ± 125	217 ± 182	3%

Sections (CPM per section; mean ± SD).

The treatment of rabbits with metoprolol had an overall effect of lowering the LDL influx in the tissue. In both the branch and the nonbranch regions, total LDL influx was significantly lower in the metoprolol-treated rabbits than in the controls. The intimal LDL influx was lower in the metoprolol group than in the control group. The most impressive reduction due to metoprolol treatment occurred in the intimal (40%) and total (32%) LDL influx at the branch region.

The distribution profile of the [^{125}I]LDL was in accordance with the following pattern. LDL influx was maximum in the intimal region and decreased toward the media and the adventitia. Occasionally, the LDL influx increased in the adventitia perhaps due to the contribution from vasa vasorum. In the treated rabbits also the LDL distribution profiles show a similar pattern, however, the profiles in the metoprolol rabbits show that there is less LDL influx in the intimal region. Similar findings in the control rabbits have been described in Chapter 8.

11.7. Summary and Interpretation

Beta-blockers are a group of drugs generally used for treating hypertension. The beta-blocker metoprolol is expected to reduce both the blood pressure and the heart rate and furthermore prevent the rise in heart rate upon sympathetic stimulation. Metoprolol reduced the heart rate in rabbits by 33 BPM (13.4%). Blood pressure was not measured in conscious rabbits but could also have been reduced. The average plasma concentration of metoprolol was 320 ± 158 nmol/L. Therapeutic plasma levels in patients have been reported to be 100–300 nmol/L (26).

The rabbit aorta was 28–35% thicker in the branch region than in the nonbranch region. In spite of this thickening, as explained in Chapters 6 and 7, the pressure-induced wall stresses were 300–700% higher at the branch than at the nonbranch region and accompanying stretch was almost double at the branch (5–7%) than that at the nonbranch region (3%) for a pressure change of 80 to 120 mmHg. Had it not been for this thickening, the stress and the stretch at the branch region could have been even greater.

In the control group, total LDL, intimal LDL, and LDL distribution profiles indicated that in the majority of cases, LDL influx was greater at all depths in the branch region than in the nonbranch region. Metoprolol treatment reduced LDL influx in the arterial tissue, but even more significantly, it reduced LDL influx in the branch area by a greater amount. These observations are of particular importance because the reduction of LDL influx by metoprolol treatment seems to be maximum in the locations prone to atherosclerosis.

References

1. Thubrikar MJ, Baker JW, Nolan SP: Inhibition of atherosclerosis associated with reduction of arterial intramural stress in rabbits. Arteriosclerosis 1988;8:410-420.
2. Thubrikar MJ, Robicsek F: Pressure-induced arterial wall stress and atherosclerosis. Ann Thorac Surg 1995;59:1594-1603.
3. Baker JW, Thubrikar MJ, Parekh JS, Forbes MS, Nolan SP: Change in endothelial cell morphology at arterial branch sites caused by a reduction of intramural stress. Atherosclerosis 1991;89:209-221.
4. Robicsek F, Thubrikar MJ: The freedom from atherosclerosis of intramyocardial coronary arteries: Reduction of mural stress – a key factor. Eur J Cardiothorac Surg 1994;8:228-235.
5. Cozma C, Macklin W, Cummins LM, Mauer R: Anatomy, physiology and biochemistry of the rabbit. In: Weisbroth SH, Flatt RE, Kraus AL (eds), The Biology of the Laboratory Rabbit. Academic Press, New York, 1974:57.
6. Geiringer E: The mural coronary. Am Heart J 1951;41:359-368.
7. Polacek P, Kralove H: Relation of myocardial bridges and loops on the coronary arteries to coronary occlusions. Am Heart J 1961;61:44-52.
8. Ishii T, Hosoda Y, Osaka T, Imai T, Shimada H, Takami A, Yamada H: The significance of myocardial bridge on atherosclerosis in the left anterior descending coronary artery. J Pathol 1986;148:279-291.
9. Roberts WC, Buja LM: The frequency and significance of coronary arterial thrombi and other observation in fatal acute myocardial infarction. A study of 107 necropsy specimens. Am J Med 1972;52:425-443.
10. Mirsky I, Ghista DN, Sandler H (eds): Cardiac Mechanics. Wiley, New York, 1974: 381-409.
11. Westerhoff N: Physiological hypotheses – Intramyocardial pressure. A new concept, suggestions for measurements (Editorial). Basic Res Cardiol 1990;85:105-119.
12. Tenth report of the director, National Heart, Lung, and Blood Institute. Heart and vascular diseases, Vol. 2. NIH Publication No. 84-2357, U.S. Department of Health and Human Services, Bethesda, MD, 1982:116.
13. Ross R: The pathogenesis of atherosclerosis – an update. N Engl J Med 1986;314: 488-500.
14. Stemerman MB, Morrel EM, Burke KR, et al.: Local variation in arterial wall permeability to low density lipoprotein in normal rabbit aorta. Arteriosclerosis 1986;6:64-69.
15. Tenth report of the director, National Heart, Lung, and Blood Institute. Heart and vascular diseases, Vol. 2. NIH Publication No. 84-2357, U.S. Department of Health and Human Services, Bethesda, MD, 1982:112.
16. Reidy MA: Biology of disease: A reassessment of endothelial injury and arterial lesion formation. Lab Invest 1985;53:513-520.
17. Reidy MA, Langille BL: The effect of local blood flow patterns on endothelial cell morphology. Exp Mol Pathol 1980;32:276-289.

18. Thubrikar MJ, Eppink RT, Roskelly SK: Finite element stress analysis of coronary arterial branch. Proceedings of the 39th Annual Conference on Engineering in Medicine and Biology. The Alliance for Engineering in Medicine and Biology, Washington, DC, 1986;28:249.
19. Thubrikar MJ, Manual L, Eppink RT: Intramural stress at arterial bifurcation in vivo. Proceedings of the 40th Annual Conference on Engineering in Medicine and Biology. The Alliance for Engineering in Medicine and Biology, Washington, DC, 1987;29:208.
20. Manuel L: A study of the stress concentration at the branch points of arteries in vivo by the finite element method (Thesis). Charlottesville, VA: University of Virginia, 1986.
21. Leung DYM, Glagov S, Matthews MB: Cyclic stretching stimulates synthesis of matrix components by arterial smooth muscle cells in vitro. Science 1976;191:415.
22. Spence JD, Perkins DG, Kline RL, et al.: Hemodynamic modification of aortic atherosclerosis. Effects of propranolol vs. hydralazine in hypertensive hyperlipidemic rabbits. Atherosclerosis 1984;50:325-333.
23. Snyder DD, Campbell GS: Effect of aortic constriction on experimental atherosclerosis in rabbits. Proc Soc Exp Biol Med 1958;99:563-564.
24. Magarey FR, Roser BJ, Stehbins WE, Sharp A: Effects of experimental coarctation of the aorta on atheroma in sheep. J Path Bact 1965;90:129-133.
25. Lyon RT, Runyon-Hass A, Davis HR, et al.: Protection from atherosclerotic lesion formation by reduction of arterial wall motion. J Vasc Surg 1987;5:59-67.
26. Ablad B, Björkman JA, Gustafsson D, Hanson G, Östlund-Lindquist AM, Pettersson K: The role of sympathetic activity in atherogenesis: Effects of betablockade. Am Heart J 1988;116:322-327.
27. Kaplan JR, Manuck SB, Adams MR, Weingard KW, Clarkson TB: Inhibition of coronary atherosclerosis by propranolol in behaviorally predisposed monkeys fed an atherogenic diet. Circulation 1987;76:1364-1372.
28. Beere PA, Glagov S, Zarins CK: Retarding effect of lowered heart rate on coronary atherosclerosis. Science 1984;226:180-182.
29. Roberts WC: The hypertensive diseases. Am J Med 1975;59:523-532.
30. Beta Blocker Heart Attack Trial Research Group: A randomized trial of propranolol in patients with acute myocardial infarction. I. Mortality results. JAMA 1982;247: 1707-1714.
31. Wikstrand J, Warnold I, Olsson G, Thomilehyo J, Elmfeldt D, Berglund G: Primary prevention with metoprolol in patients with hypertension. Mortality results from the MAPHY study. JAMA 1988;259:1976-1982.
32. Harvey JF: Theory and Design of Modern Pressure Vessels. Van Nostrand Reinhold, New York, 1974;251, 316, 338.
33. Born GVR, Richardson PD: Mechanical properties of human atherosclerotic lesions. In: Glagov S, Newman WP III, Schaffer SA (eds), Pathobiology of the Human Atherosclerotic Plaque. Springer-Verlag, New York, 1986;413-423.
34. Steele PM, Chesebro JH, Stanson AW, et al.: Balloon angioplasty: Natural history of the pathophysiological response to injury in a pig model. Circ Res 1985;57:105-112.
35. Inoue K, Nakamura N, Kakio T, Suyama H, Tanaka S, Goto Y, et al.: Serial changes of coronary arteries after percutaneous transluminal coronary angioplasty: Histopathological and immunohistochemical study. J Cardiol 1994;24:279-291.
36. Anderson HV: Restenosis after coronary angioplasty. Disease-A-Month 1993;39: 613-670.

37. Kaplan JR, Pettersson K, Manuck SB, Olsson G: Role of sympathoadrenal medullary activation in the initiation and progression of atherosclerosis. Circulation 1991;84: VI-23-VI-32.
38. Kaplan JR, Manuck SB, Clarkson TB, Lusso FM, Taub DM: Social status, environment and atherosclerosis in cynomolgus monkeys. Arteriosclerosis 1982;2:359-368.
39. Kaplan JR, Manuck SB: The effect of propranolol on social interactions among adult male cynomolgus monkeys (Macaca fascicularis) housed in disrupted social groupings. Psychosom Med 1989;51:449-462.
40. Hansson GK, Bondjers G, Nilsson L-A: Plasma protein accumulation in injured endothelial cells: Immunofluorescent localization of IgG and fibrinogen in the rabbit endothelium. Exp Mol Pathol 1979;30:12-26.
41. Bondjers G, Brattsand R, Bylock A, Hansson GK, Björkerud S: Endothelial integrity and atherogenesis in rabbits with moderate hypercholesterolemia. Artery 1977;3:395-408.
42. Hirsch EZ, Maksem JA, Gagen D: Effects of stress and propranolol on the aortic intima of rats (abstract). Arteriosclerosis 1984;4:526.
43. Thubrikar MJ, Moorthy RR, Holloway PW, Nolan SP: Effect of beta-blocker metoprolol on low density lipoprotein influx in isolated rabbit arteries. J Vasc Invest 1996;2(3):131-140.
44. McFarlane AS: Efficient trace labeling of proteins with iodine. Nature (London)1958; 182:53.

11
The Vein Graft

1. Introduction

When the vein is used as a substitute for a segment of the artery, it is called a vein graft. The vein graft is one of the most elegant examples illustrating the relationship between vascular mechanics and pathology. Vein grafts have a long history of saving lives as they have been used as both coronary artery bypasses and femoro-popliteal bypasses among others. When coronary arteries are blocked due to atherosclerotic plaques, vein grafts are used to bypass the blockages and reestablish the blood flow to the myocardium. Single, double, triple, and quadruple bypasses are common and they have proved to be life-saving treatments. Even though a vast majority of vein grafts are used in the coronary positions, a large number of them are also used in the femoral-popliteal positions to bypass blockages in those arteries.

1.1. Veins from the Lower Leg

The veins are removed from the lower leg, rather than from the thigh, for use as arterial substitutes and the reason for this has to do with vascular mechanics. The pressure in the veins of the lower leg is high, as high as 60 to 80 mmHg, due to the hydrostatic pressure head in the standing position, and therefore these veins are thicker. These veins are more suitable to handle the arterial pressure, because of their thicker walls, even though the arterial pressure is still higher. In contrast, the veins from the thigh are under lower pressure, 20 to 40 mmHg, and therefore have thin walls, and consequently are considered less suitable for the arterial positions.

1.2. Shape of the Vein versus Vein Graft

Veins in the body serve as capacitance vessels as they store a large volume of blood. In the body, the change in the blood volume is accompanied by change in the volume stored in the veins as well as by that stored in other organs, such as the spleen. Change in the volume stored in the vein is accompanied primarily by change in the *shape* of the vein, such as elliptical to circular, rather than by change in the circumference of the vein. In other words, the veins are more suitable for *bending* deformations, which produce a big change in the cross-sectional area with little or no change in the circumference, rather than for *stretching* deformations, which produce change in both the circumference and the cross-sectional area. Veins, therefore, change their shape from elliptical to circular and back to elliptical quite readily. In contrast, *arteries* change their volume by *stretching*, which increases their circumference without changing their circular shape.

This immediately informs us that when the veins are put in the arterial positions, their response to the arterial pressure is going to be of considerable importance. Therefore, it does not come as a surprise that some vein grafts develop atherosclerotic plaques over a period of a few months to several years. We will consider these aspects in this chapter. Besides atherosclerotic plaques, which produce blockages in the vein grafts, there are basic adaptive changes that occur in the vein grafts, and we will consider those also. As we may expect, efforts are also being made to mitigate the atherosclerotic changes that occur in the vein grafts by using the perivenous support of the graft so as to reduce the stretching imposed on it. We will consider these studies also in this chapter.

1.3. *Vein Valves*

There is yet another aspect that is unique to the vein graft: it introduces the valves in the arterial system. There are no valves in the arterial system except for those that separate the arteries from the heart (heart valves). Veins, particularly leg veins, on the other hand have multiple valves. In the venous system, these valves are considered necessary for the forward movement of the blood, which occurs primarily by the compressive action of the muscles on the veins. When the vein is used in the arterial system, these valves can become active and in each arterial pulse, they can prevent the backflow. The backflow is natural in the arterial system, and the use of the vein grafts modifies the flow dynamics. Once again, we may expect that the presence of the valves in the vein grafts can have a significant influence on the long-term performance of the grafts. We will consider this aspect also in this chapter.

1.4. *Reversed, Nonreversed, and in Situ Grafts*

The use of saphenous veins is most common for the aorto-coronary bypasses. The great saphenous vein is present in the entire length of the leg, and usually a significant portion of this vein is removed, preferably from the lower leg, for the bypass (Fig. 11.1). This vein is then "reversed" and cut into the needed number of segments. Usually, one end of the segment is attached to the aorta and the other to the coronary artery, past the point of blockage. The vein must be reversed because the vein has valves in it, and therefore only reversing it will allow the blood to flow through the graft. Reversing the vein means the end that was closer to the heart is now away from the heart and the end that was away from the heart is closer to the heart. The consequence of reversing is that the end that is attached to the aorta is smaller in diameter than the one attached to the coronary artery. Because the coronary artery itself is much smaller in size, there is a significant mismatch at this anastomosis because a much larger cross section of the vein has to be attached to a much smaller cross section of the artery.

To avoid this mismatch in size so as to have a good geometry at the coronary anastomosis, some bypasses are done without reversing the saphenous vein. However, in this case all of the vein valves must be destroyed so that the blood can flow through the graft. These are then "nonreversed" vein grafts in coronary positions. The same options are available also in the femoral artery bypasses, and both reversed and nonreversed grafts are used. In this position, there is even a third category of bypass available known as "in situ" bypass. The anatomy

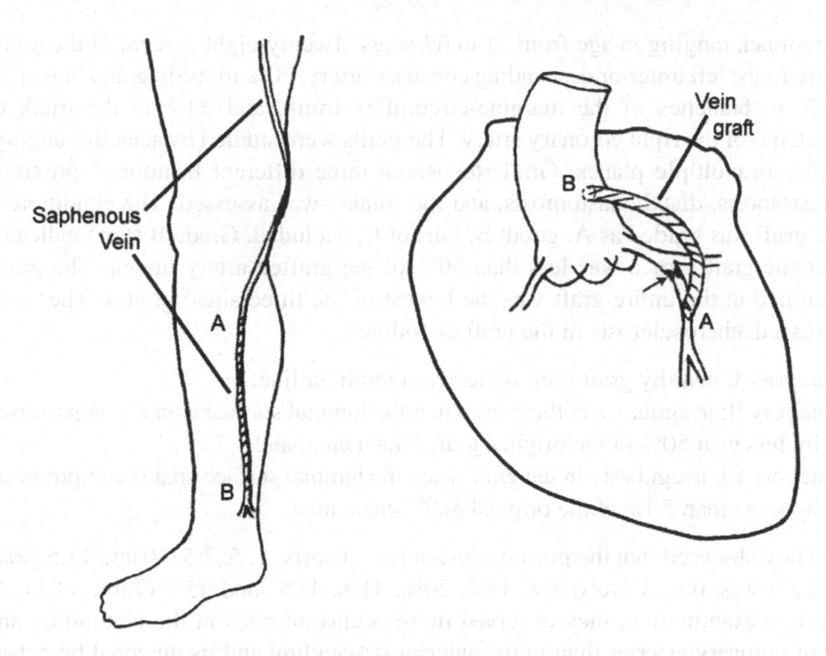

FIGURE 11.1. (Left) Schematic showing the location of saphenous vein. (Right) Vein graft connected to the aortic root at B and coronary artery at A bypassing the blockage (arrow). A segment of the vein AB (left) is reversed and used as a bypass graft BA (right) in this case.

dictates that the artery and the vein run together, and this is the case also for femoral and popliteal arteries. Therefore, to bypass blockages in femoral-popliteal arteries, the adjacent vein is connected to the artery at the two locations, proximal to the blockages and distal to the blockages. This may involve side-to-side or end-to-side anastomosis. The blood flow is now routed through the vein without removing the vein from its place. As must be done, all valves in the veins are then destroyed so that the blood can flow through this in situ graft. In situ grafts are thus nonreversed grafts and we may note that the direction of blood flow in the graft is *opposite* of what it was in the native vein.

Overall, there are several aspects of the vein grafts that offer a very enriching knowledge in the field of vascular mechanics and pathology. We will consider these below, and we will also address several questions such as why do the grafts become diseased and what could be done to prevent it.

2. Vein Graft Atherosclerosis in Coronary Artery Positions

There are several reports on the short- and long-term performance of vein grafts used as aorto-coronary bypasses. FitzGibbon (1) presented observations on 741 saphenous vein coronary bypass grafts in 222 male patients, who were military personnel, ranging in age from 31 to 67 years. Twenty-eight percent of the grafts were to the left anterior descending coronary artery, 20% to its diagonal branches, 28% to branches of the margino-circumflex trunk, and 24% to the trunk or branches of the right coronary artery. The grafts were studied by selective angiography in multiple planes. Graft stenosis at three different locations—proximal anastomosis, distal anastomosis, and the trunk—was assessed. The condition of the graft was graded as A, good; B, fair; or C, occluded. Grade B (fair) indicated that the graft lumen was less than 50% of the grafted artery lumen. The grade assigned to the entire graft was the lowest of the three sites' grades. They also assessed atherosclerosis in the graft as follows:

Category I: healthy graft with perfectly smooth outline;
Category II: irregularity in the graft when the luminal surface area is compromised by less than 50% of the original graft lumen area; and
Category III: irregularity in the graft when the luminal surface area is compromised by more than 50% of the original graft lumen area.

They observed that the graft occlusion rate at early, 1, 5, 7.5, 10 and 11.5 years or later was, respectively, 8%, 13%, 20%, 41%, 41%, and 45% (Table 11.1). At the late examination, they observed more occlusion rates in the circumflex and right coronary arteries than in the anterior descending and its diagonal branches. At the early examination, only a small core of grafts were graded "B" (fair) but 89% of these showed narrowing at the distal anastomosis. After 5 years, more grafts were graded "B" because of stenosis in the trunk. We may note that up to about 5 years, the graft stenosis is dominant at the site of distal anastomosis and this is of great interest from the point of vascular mechanics, as was explained in Chapters 5, 6, and 7. As for atherosclerosis, all grafts were disease-free at early examination, but at 1 year 8% had some irregularity (Table 11.1). At 5 years, 62% of the grafts were healthy and 38% had irregularities. Of these 38%, almost half had a grade III irregularity. At 10 years, 41% of the grafts were occluded and only 25% of the grafts were healthy. At 11.5 years or later, almost 80% of the grafts were diseased.

TABLE 11.1. Patency of saphenous vein coronary bypass grafts.

	Early	1 yr	5 yr	7.5 yr	10 yr	>11.5 yr
No. of grafts	741	741	565	237	403	101
Patency grade						
A	85%	82%	76%	46%	52%	40%
B	7%	5%	4%	13%	7%	15%
O	8%	13%	20%	41%	41%	45%
No. of grafts	683	642	452	140	238	55
Disease category						
I	100%	92%	62%	18%	25%	20%
II and III	0%	8%	38%	82%	75%	80%

A, good; B, fair; O, occluded.
I, healthy; II, less than 50% irregularity; III, more than 50% irregularity.

The graft disease is mainly atherosclerosis involving intimal damage, smooth muscle cell proliferation, locally deposited platelets, and accumulation of lipids in foam cells. Atherosclerotic plaques have different degrees of fibrosis, lipid, thrombus formation, and calcification. In that sense, atherosclerotic plaques, whether they are in a native artery, in a restenosis of the ballooned artery, or in the vein grafts, have the same basic constituents, and this we consider to be of great significance when we examine the process from the point of view of vascular mechanics.

The authors point out that atherosclerotic disease steadily increases in venous grafts and the lesion grows relentlessly, finally ending in total graft occlusion. Freedom from the disease at one stage does not guarantee that the graft will not become diseased later.

3. Etiology of Lesions in the Vein Grafts

The primary failure mode of vein grafts is the development of atherosclerotic lesions, which leads to severe stenosis or total occlusion. Figure 11.2 shows an example of atherosclerotic lesions in the saphenous vein bypass graft removed only 3 months after implantation in the coronary position. A severe narrowing of the lumen from fibrous intimal hyperplasia can be noted. Figure 11.3 shows a similar example of a saphenous vein graft removed 19 months after implantation in the coronary artery position of a 54-year-old man. Once again, a significant intimal lesion made of fibromuscular tissue can be seen.

Davies (2), in review of the subject, describes the process of atherosclerotic lesion development in the vein grafts as follows: Generally, short-term (30 days

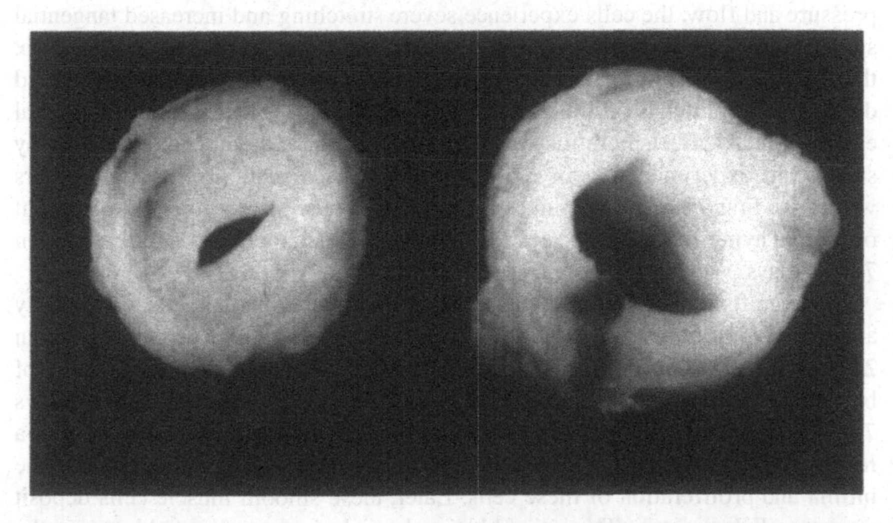

FIGURE 11.2. Appearance of cross sections of saphenous vein bypass graft excised 3 months after implantation in a patient. (Reproduced from Kennedy JH et al., J Thorac Cardiovasc Surg 1974;67:805-813, with permission from Elsevier.)

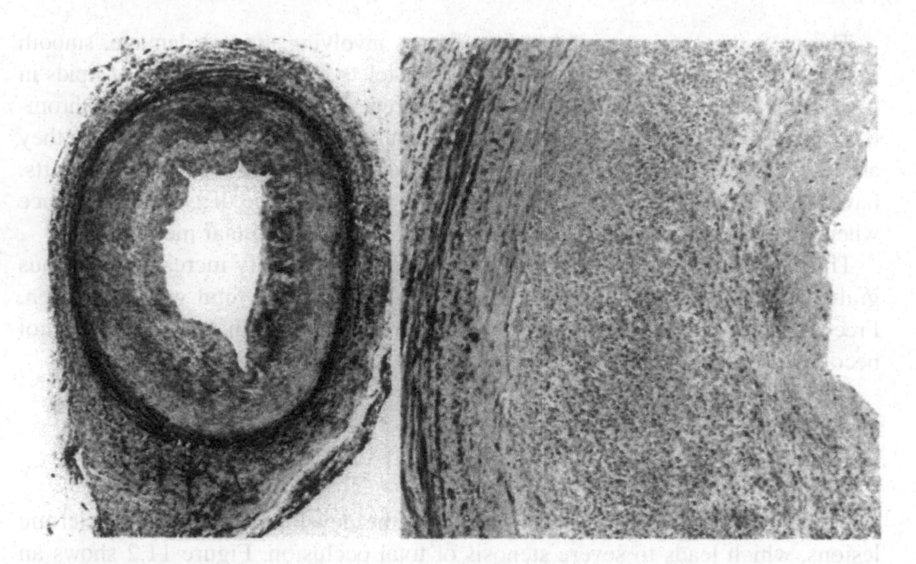

FIGURE 11.3. Saphenous vein graft (left) to left anterior descending coronary artery and close-up of a portion of its wall (right) in a 54-year-old man 19 months after implantation. Luminal narrowing is by fibromuscular tissue (Movat stains). (Reproduced from Kalan JM et al., Am Heart J 1990;119:1164-1184, with permission from Elsevier.)

to 2 years) patency rates range from 80% to 90%, and the failures are due predominately to the development of intimal hyperplasia within the graft. When the vein graft is implanted as the arterial bypass conduit, the endothelial cell layer is preserved after the implantation process. When the graft is exposed to the arterial pressure and flow, the cells experience severe stretching and increased tangential stress, both of which contribute to endothelial cell damage and loss. Within 24 h, there is an infiltration of subendothelial space by polymorphonucleocytes and deposition of platelets on the luminal surface. There is extensive subendothelial edema, which reflects both increased transmural flux and damage to the vein by stretch. In experimental vein grafts, smooth muscle cell proliferation occurs within the first 72 h and continues for at least 7 days. Microscopic development of intimal hyperplasia can be seen in 3 to 5 days, which increases rapidly between 7 to 14 days.

In the grafts retrieved from patients, focal loss of endothelial cells, particularly at the anastomotic areas, and fibrin deposition on the intima was noted within 24 h. In the next 4 days, the deposition of intimal fibrin and the accumulation of blood cell elements on the endothelial surface become more prominent. By days 7 to 14, smooth muscle cells in the intima can be identified. Intimal hyperplasia results from both the migration of smooth muscle cells out of the media into the intima and proliferation of these cells. Later, these smooth muscle cells deposit an extracellular matrix. The intimal hyperplastic lesions are located between the endothelium and the medial smooth muscle cell layer of the vein graft. In general, intimal hyperplasia is a self-limiting process, which does not produce luminal compromise by itself and usually becomes quiescent within 2 years of graft placement. However, in focal areas, the intimal hyperplastic process can proceed to significant stenosis.

The precise initiating stimuli for intimal hyperplasia have not been fully defined but it appears to be the response of the vascular smooth muscle cells to a combination of physical, cellular, and humoral factors accompanied by dysfunctional endothelium. Fibroblast growth factors contribute to the medial proliferation of smooth muscle cells while platelet-derived growth factors promote the migration of smooth muscle cells from the media to the intima. The intimal hyperplastic lesions of vein grafts, retrieved 1 month after aorto-coronary bypass in patients, have been shown to consist of proliferating smooth muscle cells with only scattered macrophages in the subendothelium. Under hyperlipidemic conditions, venous tissue shows an avidity for the uptake of serum lipid. Thus, lipid-laden smooth muscle cells with macrophages in the stage of foam cell formation are also present in the intimal hyperplasia.

Vein grafts retrieved from patients with angiographic evidence of occlusive disease show histologic features of atherosclerosis. The earliest these lesions have been seen is 6 months after implantation. Thus, the occlusions of the vein grafts appear to be due to the development of a rapidly progressive form of atherosclerosis, which has been termed *accelerated atherosclerosis*. One may distinguish it from *spontaneous atherosclerosis* on the basis of its structural morphology. Accelerated atherosclerosis shows lesions that are diffuse, more concentric, and have a greater cellularity with varying degrees of lipid accumulation and mononuclear cell infiltration.

Overall, vein grafts are living, constantly evolving conduits that adapt to the arterial circulation with the development of intimal hyperplasia but subsequently develop accelerated atherosclerosis each of which compromises patency. At present, intimal hyperplasia is the principal impediment to more durable grafts.

4. Vein Grafts in Femoro-Popliteal Positions

Vein grafts have also been in use in femoro-popliteal positions for many years and there are several reports available on their long-term patency. We will consider some of the important findings here. LiCalzi (3) reported their observations on the failure of autogenous reversed saphenous vein grafts used in the femoro-popliteal positions and compared their findings with those of four other centers. Among various centers compared, mean ages of patients varied from 61 to 70 years. Male patients predominated in all series in a ratio of 3:1. Indication for operation was usually disabling claudication or ischemia. The authors state that because of the number of variables involved, all the centers may not have comparable conditions, however, their data may be compared with those from the Massachusetts General Hospital (MGH).

Cumulative life-table patency rates for the grafts through a 5-year follow-up period are listed in Table 11.2 for the various centers. Immediate (0 to 30 days)

TABLE 11.2. Cumulative patency rates.

Interval	Yale I (4)	Yale II (5)	Rochester (6,7)	Henry Ford (8)	MGH (9)	BU-Brigham (10)
0 to 30 days	86.5	90.5	74.0	84.4	94.2	95.3
30 days to 1 yr	75.8	76.5	62.6	70.2	88.4	83.0
1 to 2 yr	71.3	67.6	60.5	65.6	82.4	74.7
2 to 3 yr	69.4	60.6	60.5	61.6	79.1	71.6
3 to 4 yr	64.5	56.1	59.2	59.1	74.2	71.6
4 to 5 yr	64.5	52.3	57.8	55.9	72.3	71.6
5 to 6 yr	64.5	52.3	55.7	51.9	68.2	—

patency rates varied from 74% to 95%. Gradual attrition resulted in a range of patency rates of 52% to 72% at 5 years after surgery. The attrition rates of the grafts (interval difference in cumulative patency) are listed in Table 11.3 for the various centers. It may be seen that the short-term failure rates varied from 4.7% to 26%. In the following period, steady attrition occurred. In MGH for instance, attrition of 4–6% each year was reported. Early (perioperative) graft failures are considered to be primarily due to graft thrombosis as a result of problems with the operative technique, graft kinking, poor runoff, or quality of the vein graft. However, most graft failures occur in the intermediate period (30 days to 3 years), and this may be due to myointimal hyperplasia, which occurs more extensively in the region of anastomosis.

Szilagyi (11) reported their findings on saphenous vein bypasses as follows. They note eight types of morphological alterations on the basis of angiographic data. These are intimal thickening, atherosclerosis, fibrotic valve, fibrotic stenosis, suture stenosis, aneurysmal dilatations, excessively long venous stump at ligated points of tributaries, and loop formation just above the knee.

Intimal thickening was the most common in grafts and appeared as a wavy narrowing of the lumen, sometimes occupying the entire length of the graft. Atherosclerosis was the second most common structural deterioration, which, when advanced, appeared more like typical atheroma in arteries. Fibrotic vein valves could also be observed, which narrowed the lumen at that location. Fibrotic thickening of the graft wall was another reason for narrowed lumen. The ultimate fate of the graft is decided by intimal thickening and atherosclerotic degeneration, both of which are diffuse processes, although they show enhancement in certain regions. The authors state that intimal thickening is manifested by an overgrowth of the subendothelial elements, mainly smooth muscle fibers. Another scenario observed is fibrin layering on the intimal surface.

The atherosclerotic changes are both grossly and microscopically quite similar to those seen in arteries. These grafts examined at 2 or more years showed evidence of atherosclerosis. The authors state that the grafts in the aorto-coronary positions will show similar pathologic changes. They also state that the veins do not develop these pathologic changes in their natural environment, and therefore

TABLE 11.3. Attrition rate*.

Interval	Yale I	Yale II	Rochester	Henry Ford	MGH	BU-Brigham
0 to 30 days	13.5	9.5	26.0	15.6	5.8	4.7
30 days to 1 yr	10.7	14.0	11.4	14.2	5.8	12.3
1 to 2 yr	4.5	8.9	2.1	4.6	6.0	8.3
2 to 3 yr	1.9	7.0	0	5.0	3.3	3.1
3 to 4 yr	4.9	4.5	1.3	2.5	4.9	0
4 to 5 yr	0	3.8	1.4	3.2	1.9	0
5 to 6 yr	0	0	2.1	4.0	4.1	—

* Interval difference in cumulative patency.

the hemodynamic environment of the arterial system must be of cardinal importance in the etiology of graft atherosclerosis.

As mentioned earlier, there are also in situ grafts in use in the femoro-popliteal positions. Moody (12) analyzed the findings on *femoro-popliteal vein grafts*, which were used as either reversed or in situ. They report that there was no significant difference in the long-term patency of grafts in either group. They further report that there was no association between the site of vein graft stenosis and the valve sites, tributaries, clamp sites, or residual valve cusps.

4.1. Additional Comments

Here we consider some of the important features of the vein grafts from the point of view of vascular mechanics. Overall, the arterial conditions (pulsatile pressure and flow) imposed on the vein grafts in the aorto-coronary positions or in the femoro-popliteal positions are similar. Thus, we expect the entire vein to respond to these conditions, and they do so by evoking intimal hyperplasia, which increases the wall thickness. The graft stenosis still occurs in localized regions even though the entire vein is undergoing the thickening process. In reversed veins, the valves are present and they could become the sites of stenosis. However, some or all valves may disappear over days or weeks, and it may be difficult to identify the locations where the valves might have been, particularly if those regions have developed atherosclerosis. In the next section, we will see some of the data that suggests that the valves could be the sites of stenosis. Also, in the reversed vein, there is a greater mismatch of the diameters at the distal anastomosis, and we expect this site to develop stenosis more frequently, which it does.

In nonreversed veins and in veins in situ, all valves must be destroyed. The process of destroying the valves involves inserting a special probe (valvulotome) in the vein and moving it back and forth several times through the entire length of the vein to ensure that all valves have been torn and destroyed. The process obviously scrapes the luminal surface and damages the intima considerably. In this case, we expect the vein grafts to respond to the damaged intima, torn valves, and the arterial conditions. Once again, the massive changes evoked in the process could overwhelm the localized changes, such as that expected at the anastomosis or at the valve sites. In other words, localized atherosclerosis, which is common in the arteries, may not be so common in the vein grafts because the local effects are overshadowed by the generalized response of the vein grafts to the arterial conditions. Hence, from the vascular mechanics point of view, we may expect the following—the dominant response of the vein graft is to the arterial pressure and flow, and the competing response is to the anastomosis and vein valves.

5. Vein Valves and Graft Stenosis

The involvement of venous valves in graft stenosis was reported as early as 1976 (13, 14). Mills and Ochsner (13) reported that when the reversed saphenous veins are used as bypass grafts, the vein valves do not collapse against the wall. The valves may stay in the main stream of the conduit, become a site of turbulent flow, and collect thrombus under the cusps. This is especially true with slower flow and poor runoff. Thrombus under vein valves may lead to perioperative myocardial infarction by embolic phenomena or may cause late graft stenosis. Therefore, they created a special device (later known as Mills Valvulotome) that could be used to destroy the valve cusps.

In femoral, popliteal, and tibial artery positions, using reserved saphenous vein grafts as bypass conduits, Whitney et al. (14) examined 50 late graft failures with a mean occlusion time of 8.14 months. They describe that in 10 patients, 10 lesions in the vein grafts showed web-like stenosis and when the surgical pathology was performed, the lesions were confirmed to be hypertrophic and fibrotic bicuspid vein valves. In 10 additional patients, upon careful follow-up using aortograms, they identified stenotic lesions, which in the majority of cases were considered to be at the site of a normal valve structure. Upon further examination of samples, microscopically when possible, the presence of valvular cusps was confirmed. The presence of a valve could be identified on the aortogram using the following considerations:

1. Because the leaflets opened less than 50% of the vein diameter, the lumen at the valve was considerably narrow, hence the dye appears as a jet.
2. The narrow opening leads to turbulence in the flow downstream, which could be seen in the dye several centimeters beyond the valve.
3. There is an entrapment of dye behind the leaflet—in the sinus cavity—and this dye takes several heartbeats to clear from the cavity.

The procedure they used is to identify the valve sites in the graft and upon follow-up, over several months or years, examine if stenosis has developed at that site. They noticed, 2 to 4 years postoperatively, that the valve leaflets were progressively fixed, more rigid, thickened, and thus readily definable, and there was more turbulence due to increased narrowing. They also noted that when the blood pressure cuff was inflated around the calf—simulating increased peripheral resistance—there was a tendency for the blood to flow backward and the valve was noted to close (i.e., the valve worked as it opened and closed). The authors state that the valves were causing the graft failure because they could observe various stages of graft occlusion, which could be related to the presence of the valves. They propose that with time, arterialized venous valve hypertrophies and undergoes fibrosis, which then leads to fibrotic-hypertrophic valvular occlusion of the graft.

Singh (15) reported the presence of multiple valves as a factor in graft occlusions at the sites associated with the valves in 5 case studies. Each patient had two grafts in place; one staying well preserved and the other deteriorating. The latter had excellent runoff. The presence of two venous valves is the only common factor in the grafts developing occlusion. He observed the turbulence at the valve sites on an angiogram and some years later noted the occlusions at the same sites. He states that flow disturbance at the valve leads to fibrin deposits in the region and could serve as a predisposing factor for early or late graft failure.

The valves are seen to produce turbulence in angiography but besides that, we must consider the possibility that the valves may function, that is open and close during pulsatile flow if the conditions were right to produce reverse flow. If the valves were to open and close, then upon closure, the pressure on the proximal side of the valve will be lower than that on the distal side. Thus, on the basis of a higher pressure on the distal side (after valve closure), we must expect the vein graft to show greater changes on the distal side. This was found to be the case in the study described below. In other words, the vein segment distal to the valve may be influenced by turbulent flow when the valve is open and higher arterial pressure when the valve is closed.

5.1. The Animal Model

Chaux et al. (16) did a well-controlled study in the animal model to explore the role of the valve in the vein graft. In rabbits, they placed a jugular vein graft as a bypass to the carotid artery circulation where one group had veins with valves, whereas the other had veins without valves. All rabbits were fed a 2% cholesterol diet, starting 1 week before the operation and continuing until the end of the experiment. They measured the wall thickness among other parameters at the following three locations: In the middle of the graft without the valves; in the graft with valves—immediately after the valve and immediately before the valve. They observed that at all locations, the wall thickness increased with time (Table 11.4). The wall thickness was larger immediately after the valve than that in the graft with no valve. Also, the wall thickness was larger distal to the valve than proximal to it.

In this elegant model, they also observed that the vein graft reaches maximum thickness at 8 weeks (Fig. 11.4). This thickening is due to proliferation of spindle-shaped cells with (actin positive) features of smooth muscle cells, as well as a matrix composed of collagen, elastic tissue, and acid mucopolysaccharides. At 2 weeks, the intimal and medial thickening was associated with an increase in cell numbers. At 4 weeks, the number of cells started to decrease and the increase in wall thickness was associated with the increase in the extracellular matrix.

TABLE 11.4. Wall thickness (μm).

Weeks	Graft without valve	Graft with valve	
		Distal to valve	Proximal to valve
2	70.3 ± 5.6	96.5 ± 8.3	63.0
4	128.9 ± 14.1	183.0 ± 19.7	147.6
6	165.7 ± 7.1	257.3 ± 11.6	210.1
8	202.5 ± 17.9	266.5 ± 14.4	210.5

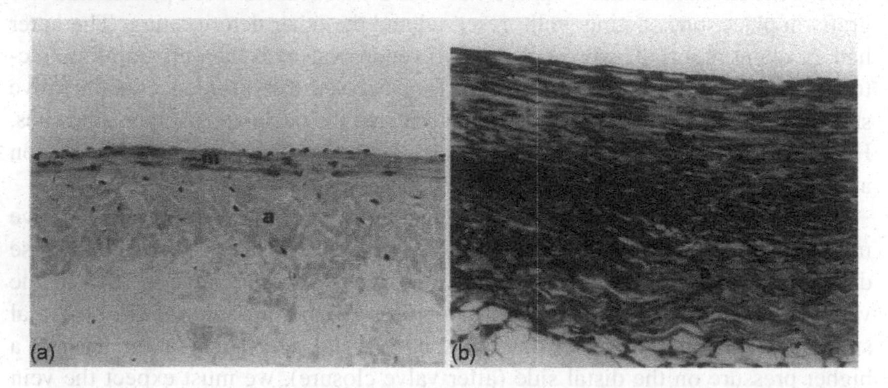

FIGURE 11.4. Photomicrographs showing an increase in medial thickness of vein grafts in rabbits. (a) control, (b) 6 weeks in the arterial position. m, media; a, adventitia. (Reproduced from Chaux A et al., J Thorac Cardiovasc Surg 1995;110:1381-1390, with permission from Elsevier.)

The cell proliferation starts as early as 3 days after the implant, remains active at 2 weeks, is minimal at 4 weeks, and is nonexistent at 6 and 8 weeks. After a longer duration of implantation (6–8 weeks), foam cells were observed. These cells had features of macrophages. In these veins, extracellular lipid was also observed. In the valve itself, thickening and increased cellularity was observed, which was more prominent at the base of the leaflet. Later, the extracellular matrix was present with collagenous tissue being the prominent one. After longer duration, foam cells and extracellular lipid were seen.

Overall, the early changes consisted of smooth muscle cell proliferation, a common response to the vascular wall injury, and later changes included macrophages and extracellular lipid deposition, typical of the atheromatic process.

5.2. *The "Horseshoe" Graft in Patients*

Lajos et al. (17) carried out one of nature's best experiments, to address the role of vein valves in graft occlusion. They connected a saphenous vein with the aorta in a side-to-side anastomosis, in such a way that a single vein produced two limbs of the grafts in a reverse "U" or "horseshoe" configuration, where one limb had valves while the other had no valves. The limb without the valves is in a nonreversed configuration whereas the limb with valves is in a reversed configuration.

Implanting the "horseshoe" graft in the aorto-coronary position is the best way to compare the performance of the grafts with valves and the grafts without valves.

In the vein, the first valve is located at about 12 ± 7 cm above the ankle. First a long segment of the vein is removed, which would have several centimeters of length free of the valves. Then the first valve from the ankle end of the vein is identified. Then a hole is made close to this valve separating the vein in two segments: one with the valves and the other without the valves. At this hole, the vein is attached to the aorta in a side-to-side anastomosis, thereby creating two limbs of the "horseshoe" graft. Each of these limbs is now available to be attached to the coronary artery in one of the two following ways. A limb can be attached to a single artery producing a single graft, or it can be attached to multiple arteries along its length producing a "sequential" graft. In other words, in place of multiple grafts attached to the aorta, the "horseshoe" graft uses only one long vein, with only one attachment to the aorta, and multiple attachments of the limbs to various coronary arteries. Hence, one "horseshoe" graft serves as multiple grafts.

Lajos et al. (17) implanted 1329 grafts in 335 patients in isolated coronary artery bypass graft surgeries. All grafts were saphenous vein to aortic "horseshoe," side-to-side anastomosis, and therefore one horseshoe in a patient produced several grafts. Furthermore, some grafts are single grafts while others are sequential. Thus, we have four different categories: valveless, single or sequential; valvular, single or sequential. Eighty-three patients who had follow-up angiography over 8–12 years, and 252 patients who did not have follow-up angiography, were part of the study. The condition of the graft was classified as patent (0–49% stenosis), diseased (50–99% stenosis), or occluded (100% stenosis). Single grafts were those connected to one coronary artery, and sequential grafts were those connected to two, three, or more coronaries.

They report that the overall graft patency was the same in single grafts (74%) versus sequential grafts (76%). Overall patency of the valveless grafts was slightly better than that of the valvular grafts (78.7% vs. 73.3%). The most significant difference between the valvular and valveless grafts was seen in the category of sequential grafts. The valveless limb had a patency of 88.6% versus the valvular limb of 72%, and the valveless limb also had less disease. The attrition of sequential grafts is shown in Figure 11.5. On an actuarial basis, the attrition of sequential grafts with and without valves is similar up to about 8 years but by 12 years, their patency had become significantly different.

The authors conclude that the valves appear to be responsible for compromising long-term patency of the grafts, and therefore, whenever possible, a valveless segment of the vein should be used for the bypass graft.

6. Vein Valve Motion and Pressure Trap in Vein Grafts

The dynamics of the vein valve in the conditions of arterial pressure and flow is one of the most fascinating examples of vascular mechanics at work in the body. We will examine this, and we will see that the valve is subjected to the mechanics it

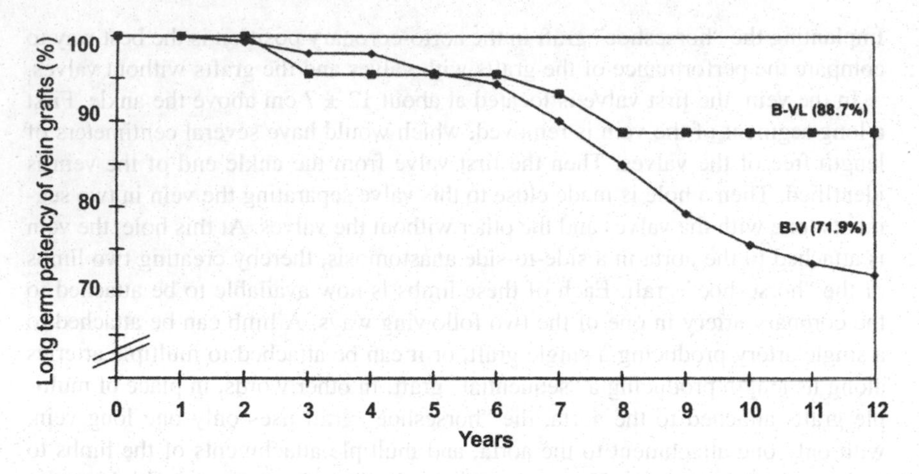

FIGURE 11.5. Long-term patency of vein grafts begins to diverge, starting around 8 years, for the valve-less (VL) grafts and with-valve (V) grafts. B indicates the patient group with sequential grafts, that is, a horseshoe graft where one limb was with valve and another without valve. (Reproduced from Lajos TZ, et al., Eur J Cardio-thorac Surg, 1996;10:846-851, with permission from Elsevier.)

was not designed for. Certainly, the unusual demands placed on the valve in this situation will have its consequence; and it should not come as a surprise then that the graft pathology may be related to the vein valves. There is data showing that the vein valves may not last very long, whereas the graft pathology develops over a much longer period. However, the functional valves in the graft can produce localized changes in the earlier period, which can set the stage for the long-term outcome. We will consider multiple aspects of valves in the vein graft in this section.

6.1. *The Fate of the Vein Valve in the Graft*

Weissenhofer (18) reported that in dogs, in the femoro-popliteal position, the valves in the vein grafts remained functional for 14 days and then degenerated and lost function at 21–28 days. Bond (19) reported that in dogs, 6 of 14 valves remained for 6 months in the femoral veins implanted in arterial positions. Bosher (20) observed that 17 of 25 valves remained intact for a period of 1 week to 6 months in jugular veins implanted in the carotid artery positions in dogs. Phillips (21) reported that *in humans*, in aortocoronary positions, valves in the reversed saphenous vein remained angiographically functional at 1 year after the operation. In patients, therefore, we may expect that some valves may remain functional in the vein grafts while others may degenerate. Furthermore, we may expect that those valves that were not functioning to begin with, or those that lost their function along the way, may degenerate subsequently. In other words, the valves may degenerate when they have no function rather than they degenerate first and lose their function as a consequence. This reasoning may become clearer when we examine the load-bearing function of the valve later in this section.

6.2. Flow Augmentation by the Vein Valve in the Graft

As early as in 1971 (22), it was recognized that the vein valve may prevent the backflow in the arterial position and thereby increase the forward flow. Phillips et al. (22) implanted autologous veins containing valves in the femoral artery positions in dogs. They measured the flow and pressure first only through the artery and then only through the bypass graft distal to the valve. They observed that in the femoral artery, there is a pulsatile flow and pressure, with a clear indication of reversed flow. When the artery is clamped and the flow goes only through the vein graft, then both the flow trace and the pressure trace change their character. There is only forward flow, and the reverse flow is eliminated, and the pressure trace has a profile similar to that of the aortic pressure (with a dicrotic notch) rather than the femoral artery pressure. In this experiment itself, it is evident that the valve closes and has elevated pressure on it (i.e., it is now a load-bearing apparatus in each cardiac cycle). They also reported that prevention of the reversed flow increased total forward flow through the graft. Weissenhofer (18) also made similar measurements in the femoro-popliteal position in dogs and observed that valve-bearing vein grafts showed enhancement of forward flow compared with the grafts without valves. This enhancement disappeared when the valve was degenerated 21–28 days after the implantation.

Baba et al. (23) studied this phenomenon in the aorto-coronary bypass graft in dogs. They implanted a cephalic vein containing a valve as a single bypass graft from the aorta to the left anterior descending (LAD) coronary artery in dogs. They also inserted a wire basket device in the graft, which allowed them to hold the valve open if desired. They reported that the blood flow through the graft was 11% higher when the valve was competent compared with that when the valve was made incompetent. The analysis of flow suggested that the increased forward flow could be attributed to a reduction in the backflow due to the presence of the valve.

Phillips et al. (21) studied this phenomenon in the aorto-coronary bypass grafts in *humans*. In 32 patients, they implanted 39 reversed saphenous vein grafts containing a competent valve. For comparison, they temporarily made the valve incompetent by either pulling on the 7-0 suture that had been placed through the vein wall into each leaflet, or by gentle traction on the adventitia of the vein wall where the valve was located. They reported that a competent valve increased the mean flow in 10 right coronary artery vein grafts and in 12 left coronary artery vein grafts. In fact, they suggested that the vein with valve may be a conduit of choice for the coronary artery bypass graft because it offers enhanced flow.

6.3. Valve Motion and Pressure Trap

Thubrikar et al. (24) studied precise details of the vein valve movement in vitro in the conditions of arterial flow and pressure. We will consider their findings because they help us understand and interpret the observations in vivo. They studied human saphenous veins with valve that were obtained from the operating room when the harvested vein was not used during the bypass surgery for some clinical reason. The vein segments, approximately 3–4 inches long, were cannulated and placed in a pulse duplicator built to simulate the human circulation. The pulse duplicator produced physiologic pressures and pulsatile flows and allowed alterations in the systemic pressure and in the cardiac output while maintaining the physiologic waveforms of pressure and flow. The systemic pressure was regulated by adjusting resistance and compliance on the pump. Cardiac output was regulated by manipulating pump stroke volume and rate. Figure 11.6 shows, schematically, the left heart simulator (pulse duplicator) and the placement of the vein graft in parallel to the arterial circuit. To simulate the distal vascular bed supplied by the vein graft, an adjustable resistance and capacitance was added to the venous outflow.

In vivo, compliance of the distal vascular bed and the vein graft may change under different conditions. For example, different lengths and different elastic properties of the vein will add different compliances. The compliance of the vein will be different at different systemic pressures. Also, the compliance of the distal vascular bed may change with the systemic pressure and the amount of dilation of the vascular bed. In order that the effect of compliance on the dynamics of the vein valve could be studied, a compliance chamber containing either 1 mL or 0 mL of air was added to the distal end of the vein graft (Fig. 11.6). When there is no air in the chamber, the only compliance present is that of the vein segment distal to the valve. When 1 mL air is added to the chamber, the volume of air changes by approximately 0.04 mL when the pressure changes from 80 mmHg to 120 mmHg.

The pulse duplicator was filled with normal saline solution. In the vein, pressures were measured on both sides of the valve. Flow through the vein was also measured. For the higher mean flows (>60 mL/min) the vein valve did not close, and the pressure at the proximal and the distal ends were the same. When the vein valve closed, the pressure at the distal end was higher than the pressure at the proximal end. To visualize the motion of the valve, they used a 4.3F ultrasound intravascular catheter that was introduced in the vein.

These measurements were repeated for the following conditions: 1) venous flow rate ranging from 200 to 0 mL/min; 2) systemic pressures of 80/60, 120/80, and 150/120 mmHg while the cardiac output was held at 3 L/min; 3) cardiac outputs of 3, 4, and 5 L/min while the systemic pressure was kept at 120/80 mmHg; and 4) varying the compliance of the distal vascular bed from 0 to 1 mL air.

They observed that in all conditions, (various systemic pressures and cardiac outputs), the venous valve offered no significant resistance to flow, and the pressures at both ends of the vein were the same but out of phase when the mean venous flow was greater than 60 mL/min. The vein valve remained open during the entire cardiac cycle and the pulsatile flow through the vein was similar to that expected in any nonvalved cylindrical conduit such as an artery. As the resistance of the distal bed was increased, the venous flow decreased and the vein valve

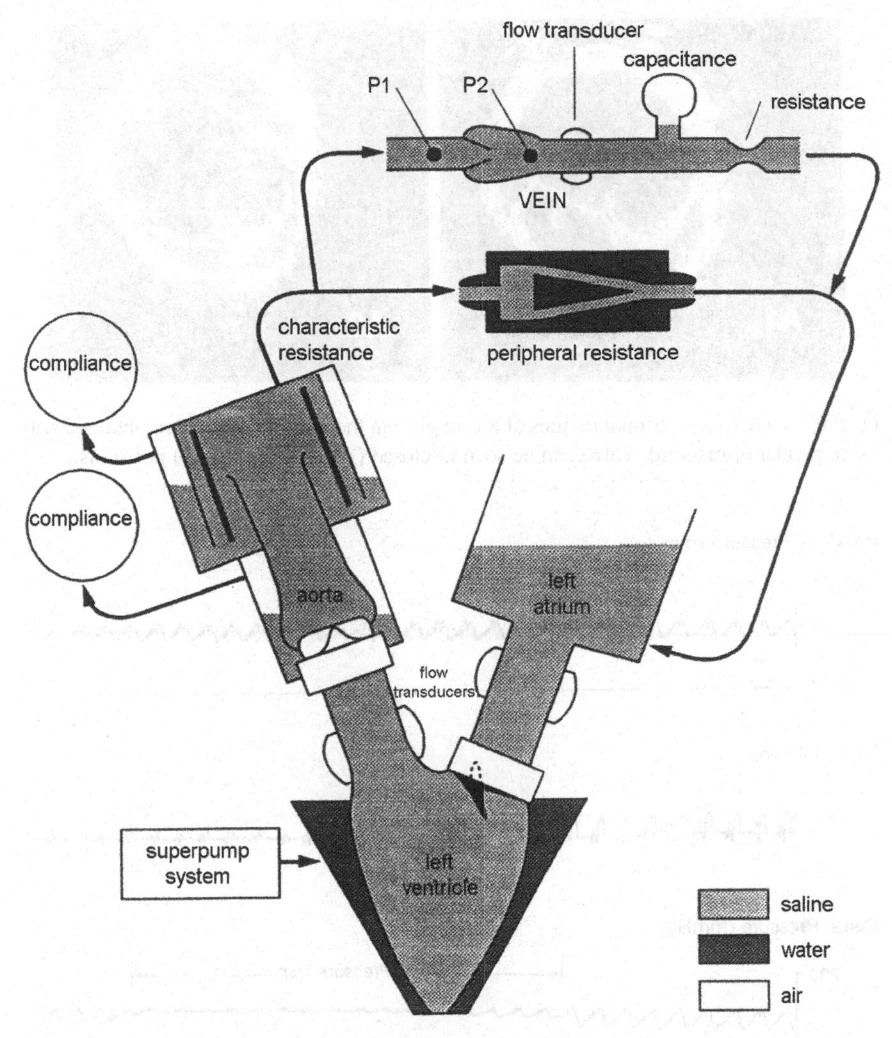

FIGURE 11.6. The left heart simulator to study flow through vein grafts. P_1 and P_2 represent proximal and distal pressures with respect to vein valve. Resistance and capacitance of the distal venous bed are also indicated.

began to open and close in each cardiac cycle (Fig. 11.7). The valve opened and closed rapidly in each cardiac cycle much like the semilunar valves. The orifice of the open venous valve was almost always nearly elliptical, and never circular.

The mean flow rate at which the valve began to close ranged from 32 mL/min to 12 mL/min. When the valve closed, it "trapped" the pressure in the segment distal to the valve. With the proximal pressure of 120/80 mmHg, the systolic pressure in the distal segment was unaltered, but the diastolic pressure rose from 80 to as high as 118 mmHg as the venous flow decreased further. Figure 11.8 shows the

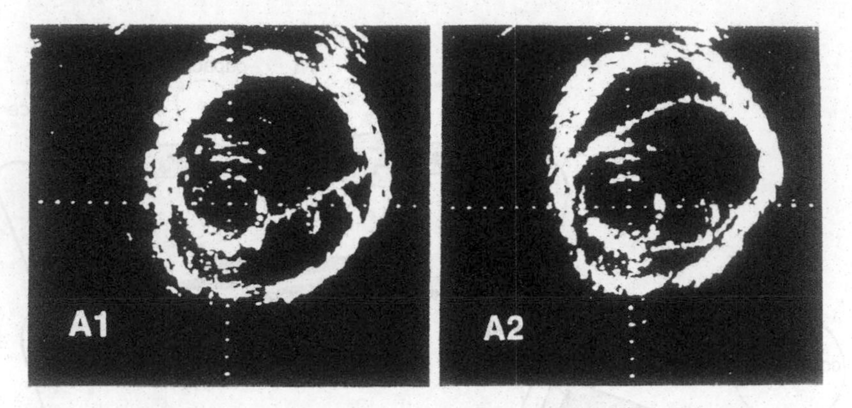

FIGURE 11.7. Cross-sectional images of a vein graft in the region of the valve obtained with intravascular ultrasound. Valve can be seen in closed (A1) and open (A2) positions.

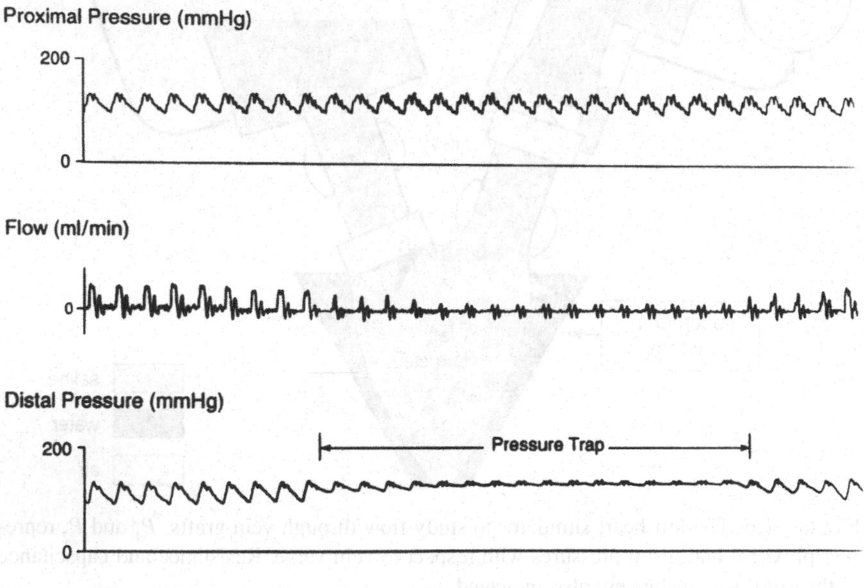

FIGURE 11.8. Actual traces of pressures and flow in the vein graft showing the pressure trap when the flow is reduced and disappearance of the pressure trap when flow is restored.

proximal and distal pressures as they relate to the decrease of the venous flow and demonstrates the pressure trap in the distal segment as an increase in the diastolic pressure.

The phenomenon of vein valve closure in bypass grafts is entirely a function of the relative resistances in the systemic vascular bed and distal (e.g., coronary) vascular bed supplied by the graft. Whenever the systemic vascular resistance is lower than the resistance of the distal vascular bed, the diastolic pressure in the proximal segment of the vein graft will fall below that in the distal segment and thereby produce a tendency for retrograde flow to occur through the graft. The retrograde flow will be stopped abruptly by closure of the vein valve, and the diastolic pressure will be trapped in the distal segment. Thus, the pressure trap occurs because the vein valve closes, and the vein valve closes because there is a tendency for the fluid to drain more rapidly through the proximal than through the distal vascular bed. The functioning valve, therefore, divides the vein graft into two segments: 1) a normotensive proximal segment and 2) a hypertensive distal segment.

The maximum mean flow at which the vein valve starts to close was determined for various compliances, systemic pressures, and cardiac outputs. The compliance of 1 mL air was used in the system. One milliliter of air changes the volume by only 0.04 mL for a pressure change of 80 to 120 mmHg. If one notes that in each pulse the artery may dilate 2.5% circumferentially and about 3% longitudinally, and that the vein may behave somewhat similarly, then the compliance of a 6- to 10-cm-long vein will be equivalent to that produced by 1 mL of air.

The vein valve closed at higher mean flows for a higher compliance (1 mL air) than for a lower compliance (no air). For example, at a systemic pressure of 120/80 mmHg, the valve began to close at a mean flow rate of 20 mL/min for 1 mL air compliance and at 15 mL/min for a 0 mL air compliance. As for the effect of systemic pressure, both, at a high (150/120 mmHg) and at a low pressure (85/60 mmHg), the mean flow at valve closure was lower than that at the normal pressure (120/80 mmHg). The pulse pressure was 30 and 25 mmHg at systemic pressures of 150/120 and 85/60 mmHg, respectively, whereas the pulse pressure was 40 mmHg at a systemic pressure of 120/80 mmHg. The mean flow at valve closure correlated better with the pulse pressure rather than with the systemic pressure. The mean flow rates at valve closure were 20, 12, and 10 mL/min, respectively, when the corresponding pulse pressures were 40, 30, and 25 mmHg. Thus, the mean flow at valve closure decreased as the pulse pressure decreased.

As for the cardiac output, the mean flow rates at the valve closure increased as the cardiac output increased. For example, at 0 mL air compliance, the mean flow at closure increased from 15 to 16.5 to 20 mL/min. as the cardiac output increased from 3 to 4 to 5 L/min.

6.3.1. Pressure Gradient versus Venous Flow

Because flow through the valve occurs in systole, the systolic pressures were the same on both sides of the valve. When the vein valve started to close, the pressure in the distal segment was trapped and the diastolic pressure there became higher than that in the proximal segment (Fig. 11.8). A further decrease in the flow continued to trap more and more pressure distally. Figure 11.9 shows a plot of the diastolic pressure gradient that is the difference between the diastolic pressures in the distal and the proximal segments versus the venous flow. There are three important observations to be made. *First*, as the venous flow decreased, the diastolic pressure gradient increased with the distal diastolic pressure higher all

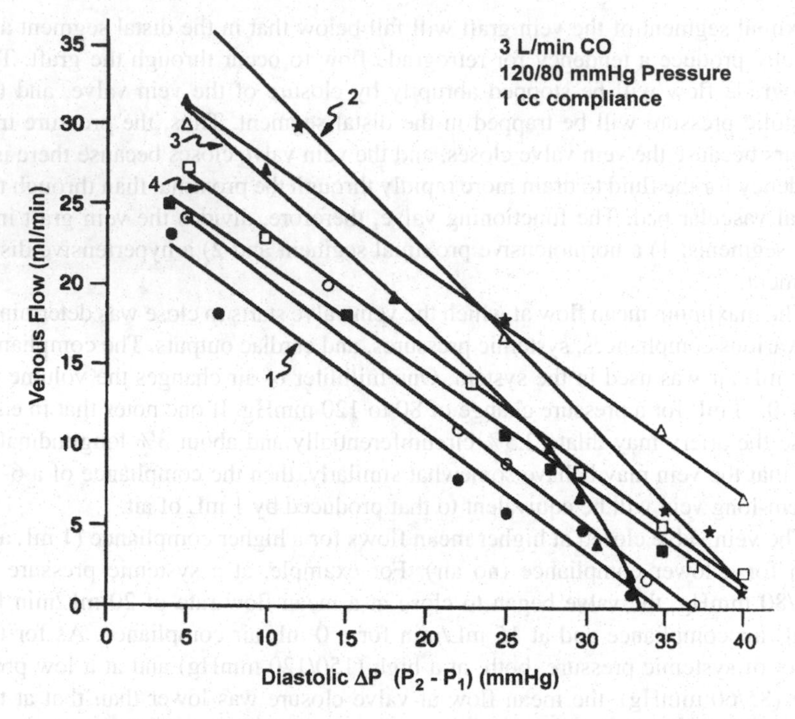

FIGURE 11.9. Difference ($P_2 - P_1$) in diastolic pressures proximal (P_1) and distal (P_2) to the valve versus venous mean flow. Diastolic pressure gradient increased as venous flow decreased. CO, cardiac output.

the time. The maximum gradient attainable equaled the pulse pressure, that is, when there is no distal drainage, the distal segment will have the systolic pressure throughout the cardiac cycle, whereas the proximal segment will have the normal systolic-diastolic pressure fluctuations. *Second*, with increasing compliance, the diastolic pressure gradient will increase at any given venous flow. *Third*, different veins have different flows at the start of the valve closure and different diastolic gradients for a given venous flow. These differences could be due to their different geometry (length, diameter, thickness) and elastic properties that result in different compliances.

This study has brought an important phenomenon in focus by establishing that competent valves in saphenous vein bypass grafts cause hypertension in the segment distal to the valve whenever there is a valve closure. The valve closure is related to the tendency of the flow to reverse through the graft. Once the valve is closed, the pressure distal to the valve remains trapped, and the decay of the pressure is controlled solely by the drainage through the distal bed. As the resistance of the distal bed increases, the flow through the vein graft decreases, and the decay of the distal pressure slows down, which leads to a greater amount of pressure being trapped. Thus, the lower the flow through the vein, the greater the difference in diastolic pressures between the distal and the proximal segments. We saw in the previous section that some investigators have observed that the presence of the vein valve results in augmentation of forward flow. These investigators also suggested that the augmentation occurs because of the increase in forward flow and the prevention of reverse flow. It is obvious that the reported augmentation of flow occurs because of the pressure trap; that is, with the valve closed there is greater pressure in the distal segment to drive the flow through the distal vascular bed and hence to augment the forward flow. These investigators had the pressure trap occurring in their experiments also.

The flow at which the vein valve begins to close (about 25 to 35 mL/min) is in the range of flow rates seen in the vein graft in vivo in both the coronary artery bypass position and the femoro-popliteal bypass position. Graft flows have been reported ranging from 14 to 180 mL/min in the coronary (21, 25, 26) and 13 to 190 mL/min (27, 28) in the femoro-popliteal positions. When the low flow rates occur in patients, then not only will the vein valve close but it will also trap pressure and create a diastolic pressure gradient across the valve of 15–30 mmHg.

6.3.2. The Role of the Pressure Trap in Vein Graft Atherosclerosis

Vein grafts undergo adaptive intimal hyperplegia as a result of their exposure to the arterial pressure and flow. In vivo, the vein valve is expected to open and close and create a pressure trap, which makes the segment of the vein distal to the valve hypertensive. Furthermore, in vivo, forces such as myocardial contraction or compression by the leg muscle may indeed produce reverse flow through the graft, in which case the trapped pressure in the graft may far exceed the systolic pressure. Because the vein is already undergoing adaptive hypertrophic changes, the hypertension in the distal segment may exacerbate these changes. Even if some of the vein valves were to start to disappear at the end of 1–2 years, they will leave the valve remnants behind. The valve remnants could become the sites of endothelial damage, microthrombus formation, and platelet aggregation. The enhanced proliferative changes in the distal segment of the graft could add to the changes at the valve remnant or to the changes in the functioning valve, which could then lead to stenosis at the valve site. This mechanism of graft stenosis could be responsible for reported 20% of the graft failure in the first year of implantation (25, 29, 30) and for the graft stenosis in subsequent years.

6.4. *Pressure Trap in Patients*

It was important to consider the details of the vein valve motion and pressure trap in the previous section because in patients we can get only partial information on the function of the valve. We have to deduce the valve opening and closing on the basis of whether the "pressure trap" is seen or not. Robicsek et al. (31) studied the phenomenon of the pressure trap in reversed saphenous veins used as a femoro-popliteal bypass grafts in patients. Valve function in autogenous reversed saphenous vein grafts was studied in patients undergoing femoro-popliteal bypass surgery. Once the proximal and distal anastomoses were completed and the flow through the graft was established, a 20-gauge angiocath was introduced into the lumen of the graft near the distal end for routine angiography, and this angiocath was used to measure the luminal pressure. The flow through the graft was monitored using an electromagnetic flow probe placed in the midportion of the graft. The systemic arterial pressure was measured either in the radial artery or at the proximal end of the vein graft. In the course of these measurements, resistance to graft flow was increased by manually occluding the artery near the distal anastomosis for brief periods of 2–4 s. They found that increasing the distal resistance decreased the graft flow, which in turn caused the vein valve to open and close. The graft flow and the pressure gradient across the valve during valve closure were measured. When only one pressure in the distal portion of the graft was available, then the diastolic pressure gradient was determined as a difference between the diastolic pressure before occlusion and after occlusion (Fig. 11.10). In case of multiple valves in the graft, the pressure gradient refers to the gradient between the two ends of the graft.

In patients, heart rates ranged from 54 to 84 beats/min and the blood pressures from 170/80 to 110/55 mmHg. Nonmanipulated blood flow in the graft ranged

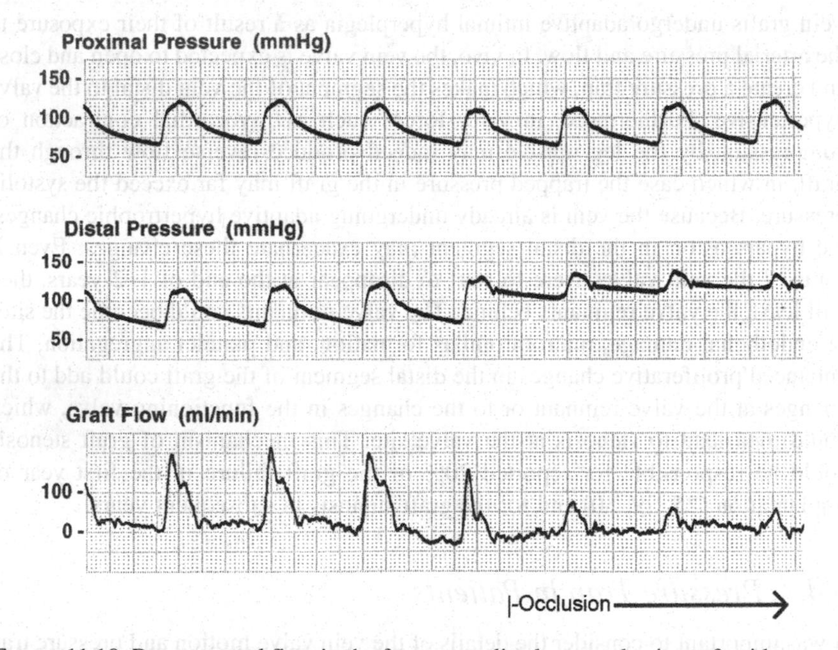

FIGURE 11.10. Pressures and flow in the femoro-popliteal reversed vein graft with competent valves in a patient illustrating the pressure trap. Proximal and distal pressures are at the respective ends of the graft. When the graft flow is reduced by manual occlusion at the distal end, the vein valve starts to open and close cyclically and at closure traps the systolic pressure, which remains trapped during diastole. This raises the distal diastolic pressure and causes a significant reduction in the pulse pressure. The proximal pressure remains unaffected by the occlusion.

from 40 to 180 mL/min with an average of 106 ± 45 mL/min. Deliberate reduction in flow induced valve closure and development of pressure traps in 12 out of 17 grafts. The remaining 5 grafts did not show a pressure trap thereby suggesting indirectly that the grafts either had no valves or had valves that were incompetent. Figure 11.10 shows the pressure and flow in the grafts and how the distal pressure changed when the vein valves started to open and close. Manual occlusion caused the graft flow to drop and the valve to start to open and close. During valve closure, systolic pressure remained trapped even throughout diastole in the distal segment.

The pressure proximal to the valve remained unaffected (i.e., it was the same as the systemic pressure). At partial occlusion of the graft, the distal pressure decayed at a much slower rate than the proximal pressure. At complete occlusion, the column of blood between the valve and the site of occlusion continued to be fed by the energy of systolic pressure, and therefore a small systolic pressure pulse still registered even though there was no blood drainage from this space. With the valve closed, the systolic pulse can diminish in the distal segment to such a degree that it may not be detectable by physical examination. In such cases, the operating surgeon may wrongly conclude that the graft may be occluded, whereas in reality the occlusion could be distal to the distal anastomosis and the valve closure causes the portion of the graft distal to the valve not to pulsate.

Figure 11.11 shows the relationship between the mean graft flow and the degree of the diastolic pressure gradient across the valve (or valves). Overall, the lower the graft flow, the higher the pressure gradient. The development of a pressure trap occurs at the mean graft flow ranging from 20 to 30 mL/min, where the "reverse" diastolic pressure gradient was 5 mmHg or less. At complete occlusion, the "reverse" diastolic pressure gradient was the highest and ranged from

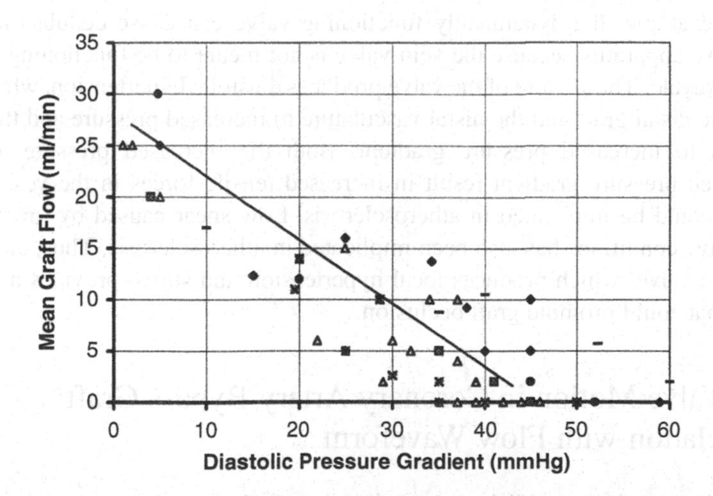

FIGURE 11.11. Mean graft flow versus the diastolic pressure gradient. Each symbol represents the data from a single patient, and the plot includes data from 12 patients having grafts with competent valves. As the mean graft flow decreases, the pressure gradient across the valve (or valves) increases.

35 to 60 mmHg. Combined data from 12 patients revealed a linear relationship between the mean graft flow and the pressure gradient (Fig. 11.11), which was expressed by the following equation:

$$\Delta P = -1.6\,F + 45$$

where ΔP = "reverse" diastolic pressure gradient (mmHg), and F = mean graft flow (mL/min). If the mean graft flow is known, then this equation allows easy determination of the "reverse" diastolic pressure gradient. The *duration of zero flow* is another important parameter to consider. As shown in Figure 11.10, the higher the pressure trapped, the longer the duration of zero flow because the valve remains closed for a longer period.

Thus, in patients the valves in the femoro-popliteal vein grafts are active and do open and close in each cardiac cycle under low flow conditions. When the mean graft flow fell below 30 mL/min, the valves closed. The valve closure produced two important effects: 1) it trapped pressure producing diastolic hypertension in the graft distal to the valve; and 2) it caused stagnation of blood flow in the vicinity of the valve. The lower the mean graft flow, the higher was the pressure trapped, and the longer was the duration of flow stagnation.

It may be good to ask the following question at this time: Could it be that the vein grafts with competent valves are the ones that become occluded in the short term? The graft flow rates of 13–190 mL/min have been reported in the femoro-popliteal positions (27, 28). Patients receiving femoro-popliteal grafts may be wearing elastic stockings or massaging their lower leg regularly. Such circumstances may lower graft flow and cause valve closure for short periods of time. In case of vein grafts with valves, many known pathways for the development of atherosclerosis appear to be present.

For example, the dynamically functioning valve can cause cellular injury to the valve apparatus because the vein valve is not meant to be functioning in each cardiac cycle. The closure of the valve produces diastolic hypertension, which subjects the distal graft and the distal vasculature to increased pressure and the valve leaflets to increased pressure gradient. Both the increased pressure and the increased pressure gradient result in increased tensile forces in the vessel wall, which could be implicated in atherosclerosis. Low shear caused by low-flow or zero-flow conditions has also been implicated in atherosclerosis. Thus, closure of the vein valve, which produces local hypertension and stasis, provides a mechanism that could promote graft occlusion.

7. Valve Motion in Coronary Artery Bypass Graft: Correlation with Flow Waveform

Thubrikar et al. (unpublished data) have studied the valve motion in precise details in reversed saphenous veins used as aorto-coronary bypass grafts in the canine model. Their observations show fascinating valvular dynamics in the coronary position. However, before we consider those, it is good to briefly review a possible role of the valve in the graft stenosis.

Figure 11.12 shows a typical coronary angiogram from a patient where focal graft stenosis can be seen in two places. This particular vein graft was removed from the patient and examined after opening the graft with a longitudinal incision. Figure 11.13 shows the opened graft and three locations where the luminal narrowing was present. All three locations of luminal narrowing are at the site of the vein valves. In all three cases, intimal and medial mass, representing atherosclerotic lesions, is present in the region of the valve leaflet attachment (i.e., near the zone of leaflet flexion). This figure emphasizes the role of the vein valve in the graft stenosis. By this time, the leaflets themselves have disappeared in this patient but the remnant of the valve apparatus, particularly the region of valve attachment, seems to have developed atherosclerotic changes.

Figure 11.14 shows the intact valve in the fresh vein. From this figure, we can appreciate how thin and transparent the leaflets of the vein valves are. In the closed valve, the cotton sponge placed on the leaflets can be seen right through the leaflets. We know that the dynamic motion of the valve must go on in the coronary grafts, but in patients it is difficult to demonstrate. As we saw in the previous section, in patients, we could deduce the valve dynamics on the basis of the pressure trap in the femoro-popliteal position.

In the coronary grafts, Thubrikar et al. studied the valve motion in the canine model. They placed two very tiny radiopaque markers on the two leaflets of the saphenous vein valve after inverting the vein inside out. The vein was then

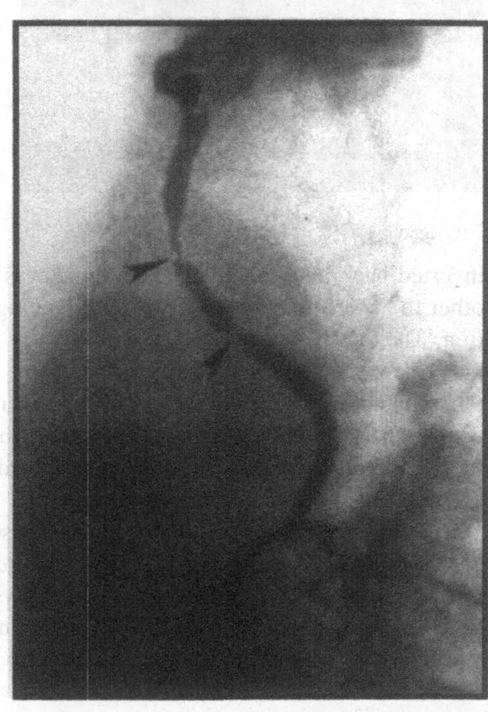

FIGURE 11.12. Aortic root angiogram showing blockages (arrows) in the vein graft in a patient.

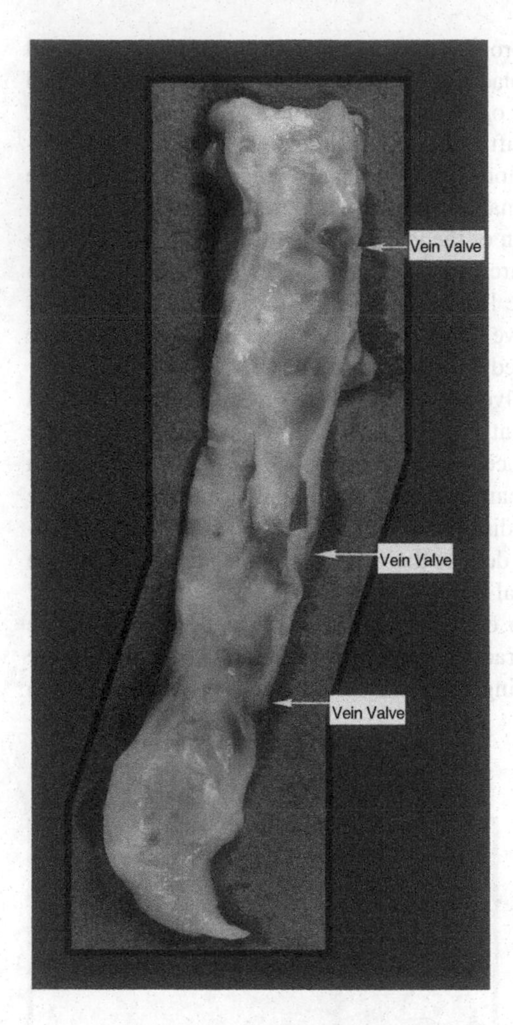

FIGURE 11.13. A vein graft removed (and opened) from a patient. All three locations where the graft developed blockages (arrows) are at the site of the vein valve. (*Please see color version on CD-ROM.*)

inverted back and cannulated. One end was attached to the aortic root and the other to the left main coronary artery, placing the vein in the reversed configuration. The markers were viewed under X-ray and after magnification, the marker movement was recorded during fluoroscopy. Figure 11.15 shows the positions of the two markers in the open and the closed valve. They recorded the flow through the coronary graft by placing an external flow probe on the graft. They did not record the pressure distal to the valve in order to avoid any perturbation in the flow channel by the pressure catheter.

The dog was on the ventilator during these experiments. To produce different coronary flow, a variety of drugs were injected in the dog, which would have an effect on both the cardiac conditions and the systemic conditions. The coronary flow was recorded simultaneously with the marker movements. Also, the flow recordings and the video recordings of the markers were synchronized. The distance between

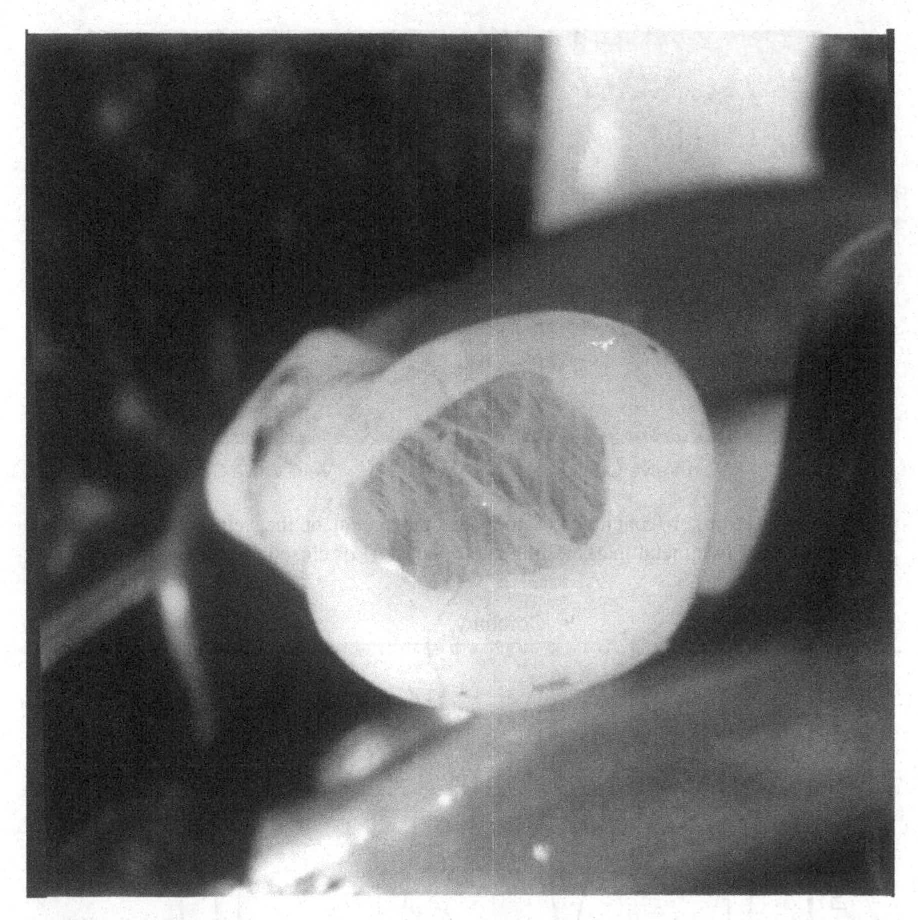

FIGURE 11.14. The closed valve in the saphenous vein. The valve leaflets are thin and transparent. A cotton sponge can be seen through the leaflets.

the two leaflet markers was measured from the video recordings and it represents the leaflet motion. Because the vein graft is small, about 4 mm in diameter, the maximum valve opening is only of the order 2–2.5 mm.

Figures 11.16 and 11.17 show the leaflet motion and the corresponding coronary flow in two different conditions. Figure 11.16 demonstrates the leaflet motion in a typical coronary flow condition. Here, the low flow reaches a negative value for a short period, and the valve can be seen to close completely for a short period of time. The periodic oscillations in the flow are similar to the periodic oscillations in the leaflet motion. A small decrease in the positive flow in the middle of the cardiac cycle appears to produce an inward motion of the leaflet. Figure 11.17 shows another example of leaflet motion responding to the flow waveform. In this case, a sustained valve closure can be noted as well as two peaks in the leaflet motion in each cardiac cycle. Points a, b, c, d, e on the flow

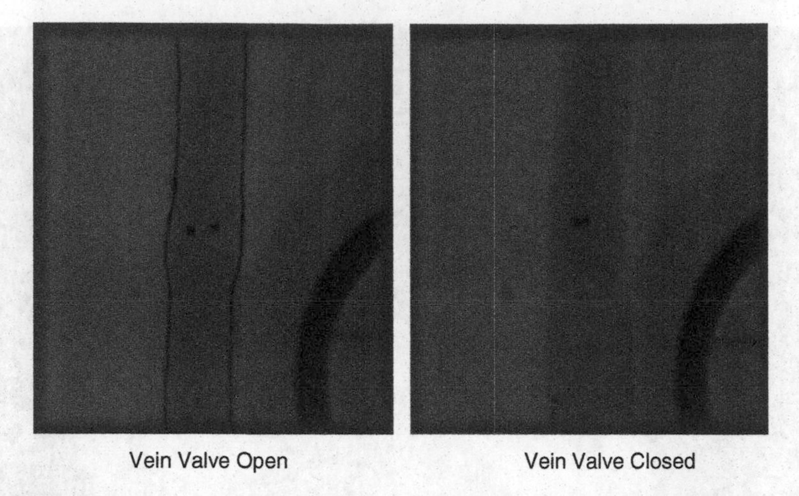

FIGURE 11.15. Marker technique to study the movement of the vein valve leaflets. The positions of the two metal markers indicate open (left) or closed (right) valve.

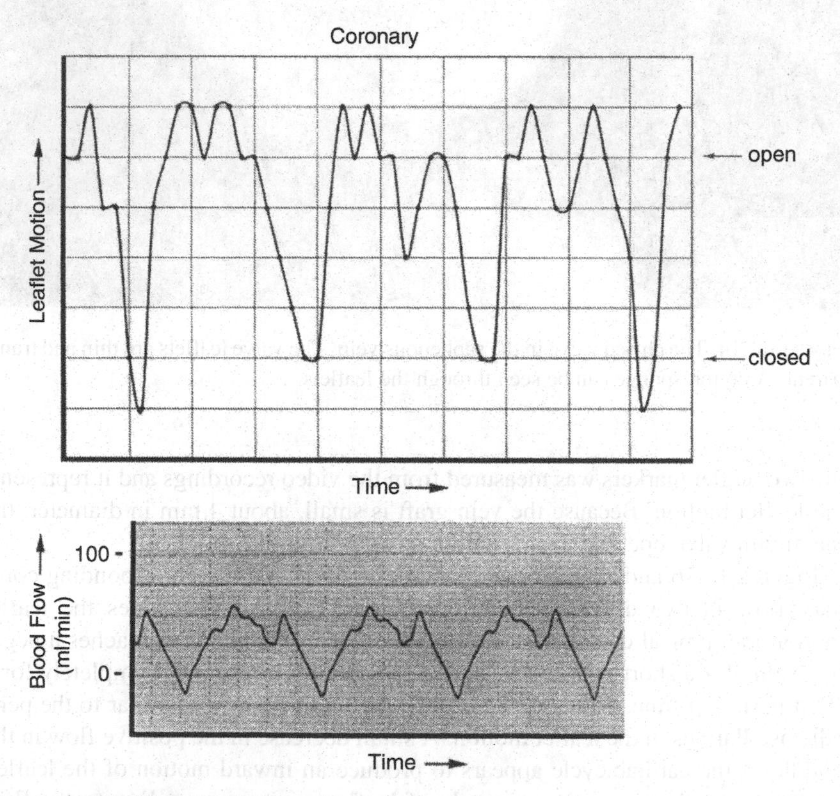

FIGURE 11.16. Human saphenous vein used as coronary artery bypass graft in a canine in the experiment. Leaflet motion was measured as the distance between the markers on the vein valve leaflets. Blood flow is the flow through the graft. The profile of the leaflet motion parallels that of the blood flow. Open, valve fully open; closed, valve fully closed.

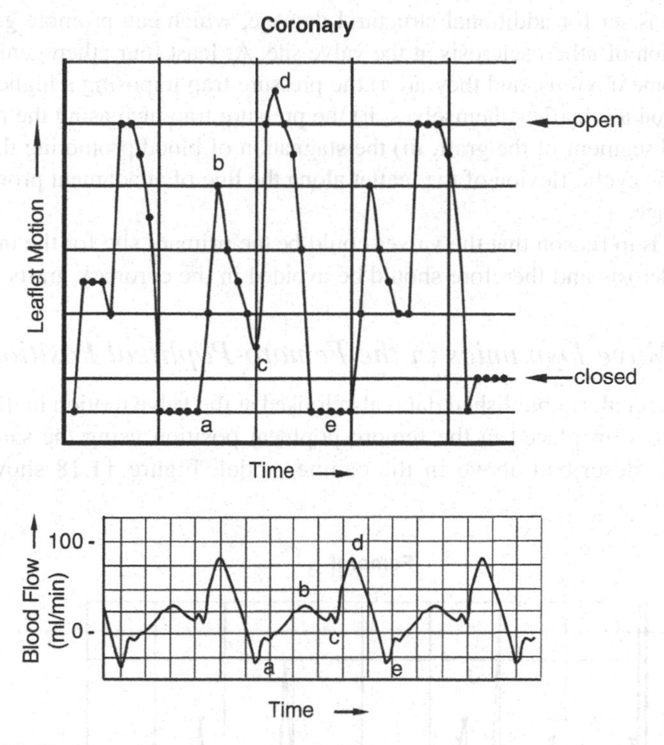

FIGURE 11.17. Similar to Figure 11.16. The points a, b, c, d, and e on the flow trace match those on the leaflet motion suggesting that the movement of the vein valve occurs in response to the blood flow. Even a small positive flow can produce a large opening in the valve (point b).

trace correspond with points a, b, c, d, e on the leaflet motion. It is important to note that a small positive flow is enough to produce a near full opening of the valve (points b and b). In conditions of abnormal cardiac contractions (such as premature ventricular contractions), there is an abnormal coronary flow. Even in such conditions, the leaflet motion responds to small changes in flow. The vein valve leaflet in the bloodstream is like an air foil in the wind. The leaflet motion occurs in response to every nuance in the blood flow. Thus, the waveform of the leaflet movement parallels that of the coronary flow.

One may debate whether the leaflets are responsible for the graft stenosis, however, there can be no debate on whether the leaflets are dynamic in the coronary graft, because they clearly are. We know from these studies then that in patients, in coronary grafts, the valve leaflets must be dynamic and the valve will close if there is a negative flow in any cardiac cycle. Of course, when the valve closes then there is a pressure trap and the stagnation of blood at the valve site. Even without a pressure trap, the leaflets are subjected to dynamic motion in every cardiac cycle. Because these leaflets are not designed to function in each heartbeat, their newly required function imposes additional demands on their structure. It is expected that the valve will go through structural changes in response to these demands. Thus, the stage is set for additional structural damage, which can promote genesis and progression of atherosclerosis at the valve site. At least four atherogenic parameters become obvious, and they are i) the pressure trap imposing a higher pressure gradient on the leaflets themselves, ii) the pressure trap increasing the pressure in the distal segment of the graft, iii) the stagnation of blood promoting thrombosis, and iv) the cyclic flexion of the leaflet along the line of attachment promoting tissue damage.

It stands to reason that the valves could be the primary site for the initiation of atherosclerosis and therefore should be avoided in the coronary grafts.

7.1. Valve Dynamics in the Femoro-Popliteal Position

Thubrikar et al. (unpublished data) also looked at the valve motion in the reversed saphenous vein placed in the femoro-popliteal position using the same marker technique described above in the canine model. Figure 11.18 shows that in

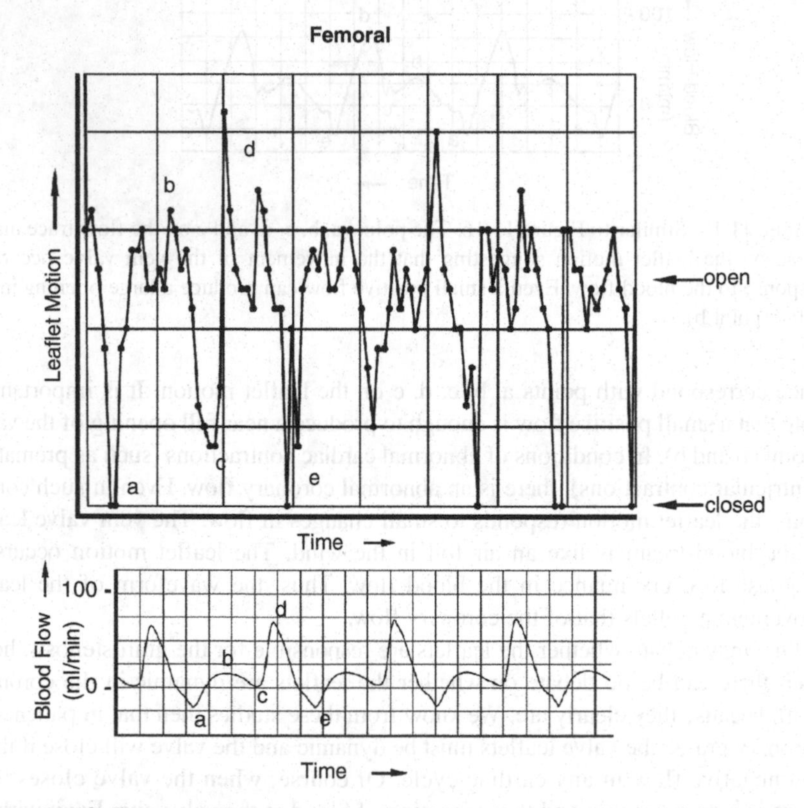

FIGURE 11.18. Human saphenous vein used as femoral artery bypass graft in a canine in the experiment. Once again, the points a, b, c, d, and e on the flow trace match those on the leaflet motion. Oscillatory flow produces oscillatory leaflet motion. The vein valve appears to close twice in a single cardiac cycle.

conditions of a typical femoral artery flow, there are cyclic oscillations in the leaflet motion, which correspond with the flow oscillations. For instance, points a, b, c, d, e on the flow trace correspond with points a, b, c, d, e, on the leaflet motion. Two humps in the flow produce two humps in the leaflet motion. In other words, leaflets go through two oscillations in each cardiac cycle, and the valve can actually close once or even twice in each cycle. This challenge of the dynamic motion imposes a significant load on the vein valve leaflets. When the forward flow is further reduced, the valve closure is enhanced and the valve can remain closed for a longer duration. In this case, the duration of flow stagnation and pressure trap is increased.

8. Mechanics of Distension of Veins

It was stated earlier that in the body, the veins are considered capacitance vessels and that they can store different amounts of blood with relative ease (i.e., without much stretching). In other words, they do not have to overcome a significant amount of elastic energy in order to increase their storage capacity. Thus, by design they are different from arteries. Because we considered their use as arterial substitutes in previous sections, it is of interest to study how their deformational mechanics work in their natural venous environment, so that we have a better appreciation of what they may be subjected to as the arterial grafts.

Veins have a very small thickness to radius ratio (≈ 0.1) and lose their circular cross section when the transmural pressure approaches zero. Their pressure-volume relationship arises from loss of a circular cross section near zero pressure. So, their cross-sectional areas change from oval to circular configuration primarily by bending their walls. Although some stretching is combined with this bending during the large volume change, the bending is much more dominant than the stretching. Once the circular cross section is reached, further increase in volume, however, is achieved primarily by stretching.

Moreno et al. (32) studied the inferior vena cava of dogs. They describe the deformation of the vein as above and illustrate it schematically in Figure 11.19. For the experiments, they removed both the intrathoracic segments and the abdominal segments of the vena cava. Also, during the experiment they used two different conditions: i) maintain the length of the vein the same as that in the animal and ii) let the length of the vein be free. The typical experiment was to distend the vein by infusing liquid into it at a constant volume flow while measuring the pressure inside it. From this experiment, they could plot a relative change in volume versus transmural pressure. Relative change in volume $v = \dfrac{V - V_o}{V_o}$, where V_o is the initial volume at transmural pressure zero and V is the final volume at a transmural pressure P_t. Transmural pressure $P_t = P_i - P_o$, where P_i is the pressure inside and P_o is the pressure outside the vein. Figure 11.20 shows the plot of v versus P_t. Curve A was obtained with the physiological longitudinal tension (constant length as in the dog) and curve B with the relaxed length of the vein. The volume

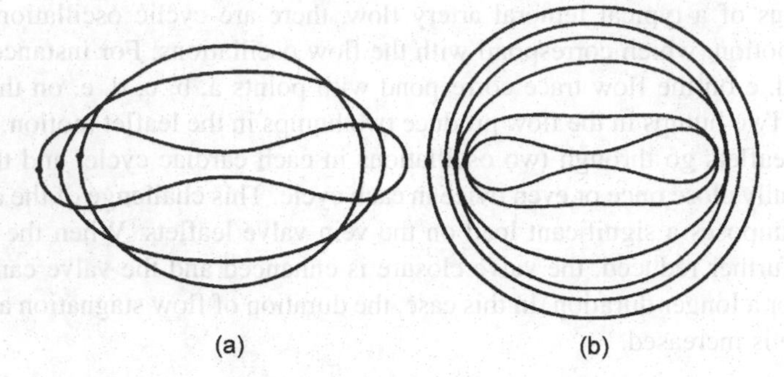

FIGURE 11.19. (a) Computer solution for the cross-sectional area of a latex tube during the bending regime of inflation. Because its modulus of flexural rigidity (EI) allows the cross section to bend at constant perimeter, the flattened configurations are wider than the ovals. (b) Computer solutions for the case of the vein with EI more than 500 times smaller than the latex tube. Solutions obtained at different perimeters and the flattened configurations are inside the oval or circular configurations. In both cases, the solutions correspond with the experimental cross sections. (Modified Schematic from Moreno AH et al., Circ Res 1970;27:1069-1080.)

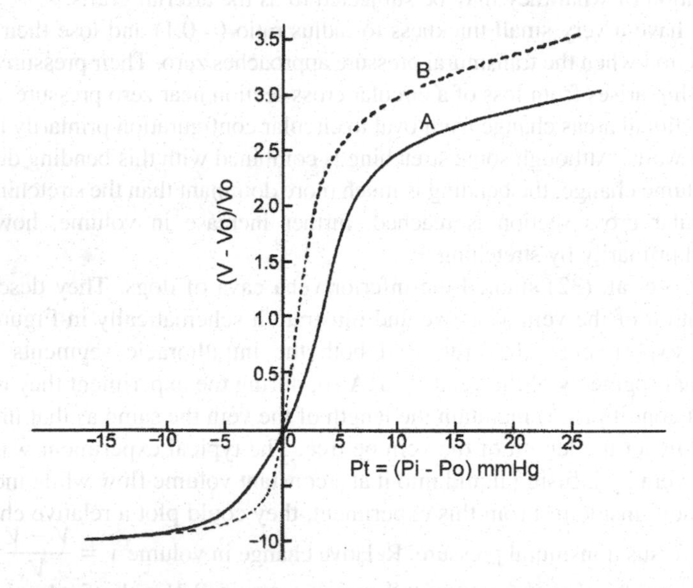

FIGURE 11.20. Pressure-volume relationships for the excised thoracic vena cava of a dog. Curve A was obtained at the physiological longitudinal tension. Curve B was obtained at the relaxed state. The plots are actual tracings generated on an x-y plotter during the experiment. (Reproduced from Moreno AH et al., Circ Res 1970;27:1069-1080, with permission from Lippincott, Williams & Wilkins.)

compliance is obtained as a ratio of the rate of change of volume over the change of pressure (i.e., the slope of the curve). It may be noted that the maximum volume compliance occurs at 4 mmHg for the vein under physiologic longitudinal tension and at 1 mmHg for the relaxed vein.

They repeated these experiments to determine the perimeter and the area of the cross section as a function of transmural pressure. They put radiopaque markers around the vein to mark a cross section and then examined these markers under X-ray. They computed the cross-sectional area from these markers at various transmural pressures. Figure 11.21 shows the shape of the cross section at various areas of the cross section as a function of transmural pressure. This figure demonstrates quantitatively what has been described earlier. Close to zero

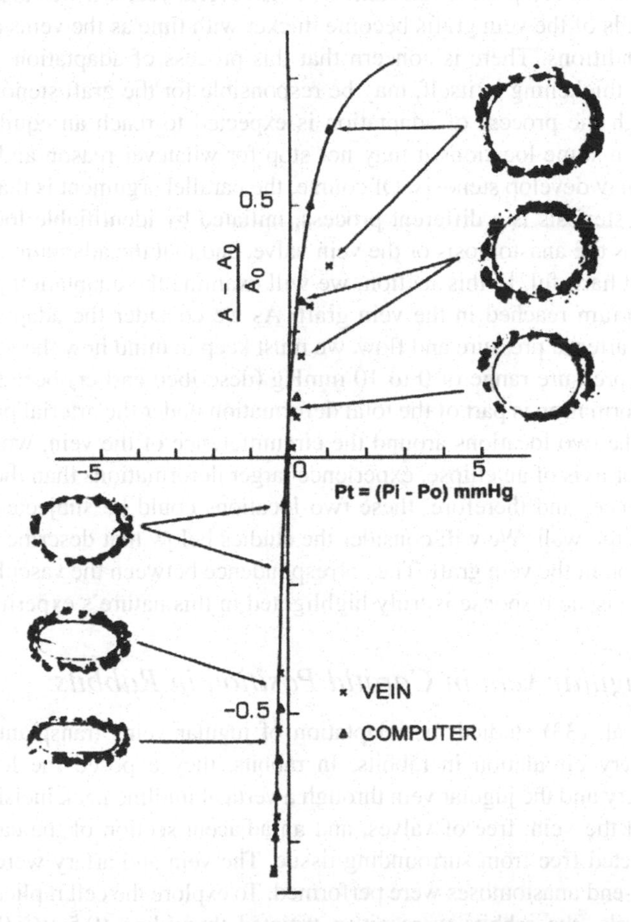

FIGURE 11.21. Comparison between pressure-area relationships and configuration of cross-sectional areas during distension of a long segment of inferior vena cava of a dog and the corresponding computer solutions. (Reproduced from Moreno AH et al., Circ Res 1970;27:1069-1080, with permission from Lippincott, Williams & Wilkins.)

pressure, there is a tremendous change in the volume, in the cross-sectional area, and in the shape of the cross section for a very little change in the pressure. There is also a small change in the perimeter as the cross-sectional area changes. This figure should be compared with Figure 11.19 for similarities in the changes in the shape of the cross section with pressure. This data demonstrates the fundamental characteristics of the vein and that is that the shape change is the primary mechanism for the volume change in the vein.

9. Adaptation of the Vein in the Arterial Position

We saw that in patients, the vein grafts work well as arterial conduits even though some of them develop localized stenosis over several years. It was also described that the walls of the vein grafts become thicker with time as the veins adapt to the arterial conditions. There is concern that this process of adaptation, which produces wall thickening in itself, may be responsible for the graft stenosis. That is, even though the process of adaptation is expected to reach an equilibrium and then stop, in some locations it may not stop for whatever reason and then those locations may develop stenosis. Of course, the parallel argument is that the development of stenosis is a different process, initiated by identifiable localized factors, such as the anastomosis or the vein valve, and that the adaptation process by itself is not harmful. In this section, we will examine the adaptation process and the equilibrium reached in the vein graft. As we consider the adaptation of the vein to the arterial pressure and flow, we must keep in mind how the vein deforms in the low pressure range of 0 to 10 mmHg (described earlier) because that history of deformation is part of the total deformation under the arterial pressure. For example, the two locations around the circumference of the vein, which are part of the major axis of an ellipse, experience larger deformations than the rest of the circumference, and therefore, these two locations could be subjected to greater changes in the wall. We will consider the studies below that describe the process of adaptation in the vein graft. The correspondence between the vascular mechanics and the tissue response is truly highlighted in this nature's experiment.

9.1. Jugular Vein in Carotid Position in Rabbits

Zwolak et al. (33) studied the adaptation of jugular veins transplanted into the carotid artery circulation in rabbits. In rabbits, they exposed the left common carotid artery and the jugular vein through a vertical midline neck incision. A 3-cm segment of the vein, free of valves, and an adjacent section of the carotid artery were dissected free from surrounding tissue. The vein and artery were transected and end-to-end anastomoses were performed. To explore the cell replication at various intervals, the rabbits were given tritiated thymidine (0.5 μCi/kg of body weight) intravenously at those intervals, and 1 h later they were sacrificed. The vein grafts were perfusion-fixed with buffered glutaraldehyde under 100 mmHg pressure. The right carotid artery and the jugular veins were also processed for controls.

For the morphometric study and for the smooth muscle cell (SMC) autoradiography, the tissue from the middle portion of the vein graft was removed. SMC proliferation was estimated from the cross-sectional autoradiography. The endothelial cell (EC) labeling index was determined from tissue segments pinned on Teflon and processed for enface autoradiography. The labeling index is defined as number of labeled cells divided by the total number of cells in the equivalent area. The authors did several additional explorations, and details of the methodology can be found in Ref. 33. Their findings are as follows:

Transplantation of a jugular vein into the carotid circulation resulted in the loss of endothelium, primarily at anastomosis and secondarily in the rest of the vein. Denuded regions were covered by a carpet of platelets, occasional microthrombi, and leukocytes. At 1 week, small patches of denudation were still present, and by 2 weeks, the endothelial layer was fully restored. The vein wall became progressively thicker. The wall contained SMCs and most of them resided in the intima; the media was composed primarily of extracellular matrix and few SMCs. At 1 week, EC labeling increased 400 times more than that in the normal (non-transplanted) vein. This activity continued for 4 weeks even though 2 weeks were sufficient for reendothelialization. At 12 weeks, there was only background activity. SMC labeling also increased massively upon transplantation of veins and reached a maximum at 1 week. This activity fell rapidly by 4 weeks and approached background activity by 12 weeks. It may be noted that SMC proliferation continued after regeneration of intact endothelium.

The thickness of the graft increased progressively and reached a maximum at 12 weeks (Table 11.5). The increased thickness represented approximately equal contributions from the intima and the media. The graft wall cross-sectional area increased in parallel with the wall thickness. Luminal radius of the grafts also increased from implant to 12 weeks (Table 11.5). In the intima, the mass of SMCs was the same at 4 weeks and at 24 weeks and thus, the increase in the intimal area after 4 weeks was due to increase in the extracellular matrix. Overall, rapid cellular replication and accumulation was almost stopped by the fourth week, and

TABLE 11.5. Dimensional analysis of vein grafts.

	Thickness (μm)			Lumen radius	Radius/thickness
	Intima	Media	Total	(mm)	ratio
Vein grafts					
1 h			19 ± 6	1.69	87 ± 19
1 wk	11 ± 3	12 ± 2	23 ± 4	1.55	70 ± 15
2 wk	23 ± 3	21 ± 5	44 ± 7	1.91	44 ± 5
4 wk	41 ± 4	35 ± 5	77 ± 7	2.36	31 ± 4
12 wk	59 ± 8	57 ± 18	116 ± 11	2.90	25 ± 4
24 wk	65 ± 19	58 ± 11	123 ± 30	2.65	23 ± 8
Right carotid	—	—	50 ± 8	0.89	18 ± 4

Mean ± standard deviation.

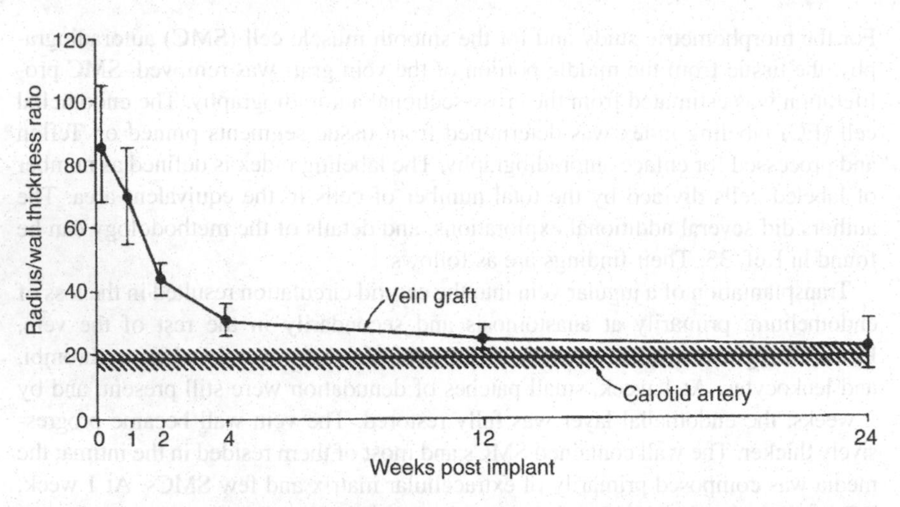

FIGURE 11.22. Graph of luminal radius/wall thickness ratio plotted against weeks after implantation. Time 0 represents vein grafts excised 1 h after implant. Values expressed as mean ± standard deviation. Carotid artery value did not change during the course of the experiment. (Reproduced from Zwolak RM et al., J Vasc Surg 1987;5:126-136, with permission from the Society of Vascular Surgery.)

the increase in wall thickness and area over the next 8 weeks was due to an increase in the extracellular matrix.

The ratio of radius to wall thickness, which reflects the tangential (circumferential) wall stress, changed as shown in Figure 11.22 (also Table 11.5). At implant, this ratio was 87, representing the thin wall, and by 1 week it fell to 70, and by 12 weeks to 25. As can be seen from this, the ratio at 12 and 24 weeks is essentially the same as that of the control carotid artery.

9.1.1. Role of Tangential Stress

The data in Figure 11.22 indicate that the vein grafts reach equilibrium by about 12 weeks, at which point the wall thickness has also reached its desired value. The authors suggest that the EC replication in the early period is in response to the denuding injury to the cells and in the later period to nondenuding injury, perhaps from exposure to the arterial flow and pressure. SMC replication is to achieve graft thickening up to 4 weeks, and beyond this, the thickness is increased by accumulation of connective tissue. The authors state that the tangential wall stress seems to be the most important factor in adaptation of the vein graft. The wall thickness increases to a value where the radius to thickness ratio for the graft equals that for the carotid artery. The average tangential stress, according to the law of Laplace, is given by $\frac{pr}{t}$, where r is the radius, t is the thickness, and p is the luminal pressure. By 12 weeks, the ratio r/t in the graft is the same as that in the carotid artery, indicating that the tangential stress is now the same in the two vessels. Thus, the adaptation of the vein occurred until it achieved the same tangential stress as that in the artery. Hence, the mean wall stress appears to determine the end point in the vein graft remodeling process.

As for the accumulation of connective tissue, it was noted earlier (see Chapter 9) that cyclic stretching of SMCs resulted in increased collagen and protein synthesis in the in vitro experiments. How the tangential stress, oscillatory pressure, and pulsatile flow may lead to connective tissue synthesis still remains to be understood. However, it seems clear from this study that the tangential stress is the main determining factor in the adaptation of the vein graft.

9.2. Adaptation of Vein Under Various Pressure and Flow Conditions

We saw in the previous section that the vein grafts adapt in response to circumferential (tangential) stress. However, the questions continue in the area of "do the vein grafts adapt to flow." Schwartz et al. (34) did complicated experiments where they exposed the veins to high pressure–low flow, low pressure–high flow, and so forth, and correlated the changes in the vein to pressure and flow to determine which parameter was the most influential. We will consider their findings here.

They studied the changes in the external jugular vein used in place of the carotid artery in the rabbit model. The four different conditions created for the vein are as follows: i) Vein used as a graft in place of the carotid artery. This is a high pressure–low flow condition for the vein. ii) Arteriovenous (A-V) fistula created by side-to-side anastomosis of the linguofacial vein and distal common carotid artery. This is a low pressure–high flow condition for the vein. iii) Placing an obstruction in the model of the A-V fistula so that the vein now has higher pressure–low flow condition, sort of like in the vein graft; and iv) in the vein graft model, create an A-V fistula so that the vein has lower pressure and higher flow, sort of like in the A-V fistula. The details of these techniques can be found in Ref. 34. They measured luminal pressure and flow through the vein in all of the preparations. The pressure and flow were measured at the time of the operation and the harvest, and the average of the two values was used in the data analysis. It should be noted that the rabbits are anesthetized during these measurements, and therefore the pressure values could be lower than those in conscious rabbits. The use of four different experimental preparations made it possible to impose various pressures and flows on the vein and to study the response of the vein to these conditions.

The external jugular veins (graft, fistula, or combination) were harvested from the rabbits after fixing them at appropriate pressures. The veins are studied histologically for various parameters including vessel area, myointimal area (media + intima area), and luminal area. Wall tension per unit length (T_w) was calculated as $T_w = P \times r_i$, where P is luminal pressure and r_i is the luminal radius.

Table 11.6 shows the pressures and flows of four different preparations of the vein. One can clearly note the high and the low pressures and the high and the low flows. Figure 11.23 shows the plot of the myointimal (media + intima) area versus wall tension. The authors state that the correlation was positive and much

TABLE 11.6. Hemodynamic measurements.

Group	Intraluminal pressure (mmHg)	Blood flow (mL/min)
Vein graft (VG) -4 wk	51 ± 4	17 ± 1
Vein graft (VG) -12 wk	62 ± 3	16 ± 4
Arteriovenous fistula (AVF) -4 wk	5 ± 1	82 ± 16
Arteriovenous fistula (AVF) -12 wk	6 ± 2	82 ± 17
AVF -Obstruction	48 ± 2	17 ± 4
VG -AVF	37 ± 2	70 ± 11

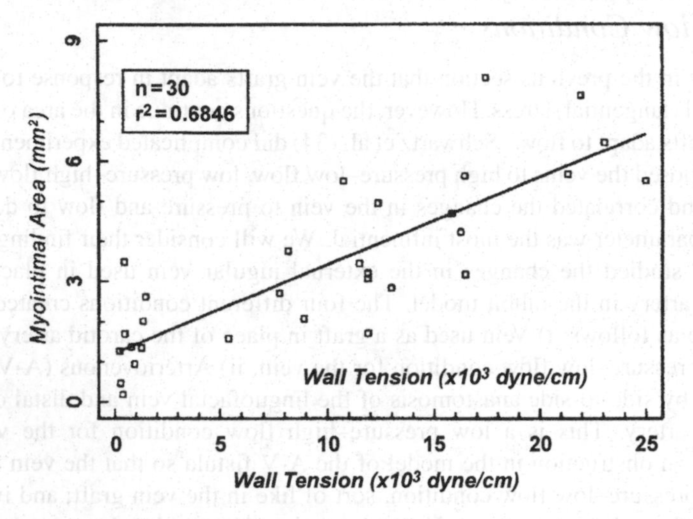

FIGURE 11.23. Least squares linear regression comparing myointimal area and wall tension. (Reproduced from Schwartz LB et al., J Vasc Surg 1992;15:176-186, with permission from the Society of Vascular Surgery.)

stronger in this case ($r^2 = 0.685$) compared with the correlation between the myointimal area versus blood flow. Thus, myointimal thickening of the vein develops in response to wall tension (or wall stress as noted in the earlier section) rather than in response to flow or shear stress (34).

9.3. Intimal and Medial Changes in Vein Grafts in Response to Pressure and Flow

In the two previous sections, we saw that the increase in the wall thickness of the vein graft was correlated with the circumferential stress or the wall tension. The wall thickness was composed of the medial thickness and the intimal thickness. It would be another step forward if we knew whether the medial thickness as well as the intimal thickness responded to the same stimulus: wall stress. To determine whether the intimal hyperplasia and the medial thickening responds to the same stimulus or to different stimuli, Dobrin et al. (35) performed the following studies in dogs. Their findings are important in providing further details and so we will consider them below.

In dogs, they implanted reversed autogenous femoral vein grafts to bypass the femoral arteries on both the right and the left side. On the right side, they tied off the femoral artery so that all of the flow would go through the vein graft. On the left side they left the femoral artery open so that the vein graft could experience the same pressure as the artery but carry less flow through it (i.e., the graft was in parallel to the artery). They harvested the grafts after 3 months and performed a histological examination. The authors made quantitative measurements of the cross-sectional areas of the intima and media. The details of the technique can be found in Ref. 35.

They report that the intimal hyperplasia was greater (1.14 mm^2) on the side where the femoral artery was left patent (low graft flow) than on the side where the artery was tied (0.87 mm^2) (high graft flow). In other words, intimal hyperplasia was greater in the graft exposed to low flow velocity. The medial thickening occurred in the graft on both sides, and it was the same on both sides (2.7–2.9 mm^2). Thus, the low or the high flow did not affect the medial thickening. Instead, the pressure in the graft, which caused the distension of the graft, was considered responsible for the medial thickening. The authors conclude that the intimal hyperplasia and medial thickening in vein grafts are responses to separate stimuli. The intimal hyperplasia is best associated with low flow velocity, whereas the medial thickening is best associated with circumferential distention of the vein graft due to pressure.

10. Adaptation of Vein Grafts Under External Support

Because the bypass vein grafts are used extensively in patients, there continues to be a great deal of interest in how to prevent stenosis of the graft. Also, as we saw earlier, the graft adapts to the arterial conditions. The process of adaptation is desirable and harmless, and the graft stenosis should be considered to occur from adverse localized conditions such as those present at the anastomosis or the vein valves. In spite of this, the increase in the wall thickness, which occurs in the adaptation process, is an integral part of the total thickness of the lesion, which is responsible for the graft stenosis. It is for this reason that attempts have been made by several investigators to reduce wall thickness of the vein graft by providing external support to the graft. The reason for using external support to reduce the wall thickness stems from the earlier data, which indicated that the wall stress due to pressure is a modulating factor for the adaptation of the graft. Obviously, the external support reduces the wall stress and consequently should reduce the wall thickness of the graft. This approach is one more excellent example of vascular mechanics at work, and here we can see how the application of the principles of vascular mechanics provides a benefit in a clinical setting. We will consider some key observations in this area.

10.1. Perivenous Mesh

Barra et al. (36) did experiments in sheep with vein grafts that were wrapped with Dacron constrictive mesh, and later they used a similar approach in some of their patients. In sheep, the native carotid artery was resected and a jugular vein interposition graft was attached by end-to-end anastomosis. In some animals, jugular veins were surrounded by a preformed 7-mm-diameter mesh tube before being anastomosed. The mesh was sutured to the artery at the anastomoses. Figure 11.24 shows the vein grafts without external support (control) and with perivenous mesh support. After 4 months, the grafts were removed and processed for

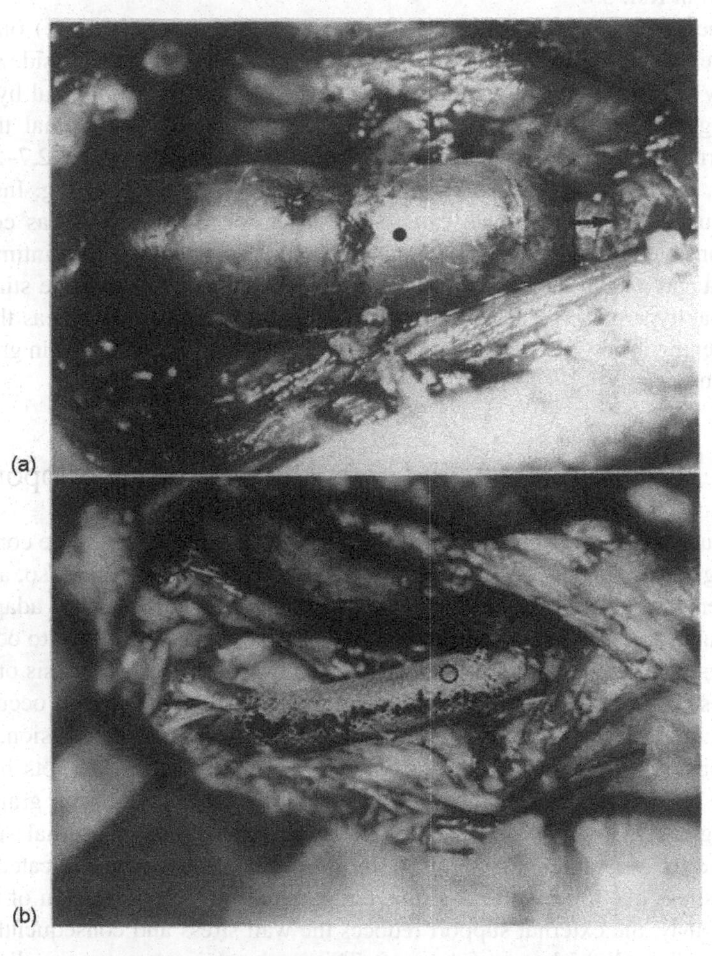

FIGURE 11.24. (a) Group A control. Free jugular vein graft at implantation after the graft has been subjected to 10 min of arterial pressure. (b) Group B "surrounded" graft in the same condition. • = free graft, ○ = mesh, → = carotid artery. (Reproduced from Barra JA et al., J Thorac Cardiovasc Surg 1986;92:330-336. with permission from Elsevier.)

histology and scanning electron microscopy. The grafts were opened longitudinally and tissue samples were taken at least 1 cm away from the anastomoses. Wall thickness was measured from the subendothelial basal lamina to the beginning of the elastic external layer. The largest and the smallest thickness were measured in both (unsupported and supported) groups. The details of the technique can be found in Ref. 36.

The average diameter of the unsupported vein grafts was 14 ± 1 mm compared with 7 ± 0.5 mm for externally supported grafts. These are external diameters measured 10 min after implantation. In the control vein grafts, the scanning electron microscopic examination revealed that endothelial cells had many abnormalities, such as cells were swollen with large nuclei, many cells were shrunken, areas of basement membrane were exposed, intercellular junctions were lost, and often the cells were separated from each other. In the externally supported vein grafts, however, the endothelial cell layer was normal (Fig. 11.25), the cells were oriented normally, and no cellular degradation or junctional breakdown was observed. Histologic examination of the control grafts showed irregular intimal hyperplasia, noninflammatory fibrosis, absence of elastic fibers in the intima, and thickening of the media. These changes in the control vein grafts are similar to those described earlier in the chapter. In the externally supported grafts also, intimal thickening was observed, however, the thickness of the intima was reduced and it was more regular. The maximum intimal thickness was 0.99 ± 0.32 mm in the control grafts versus 0.46 ± 0.08 mm in the supported grafts. The minimal

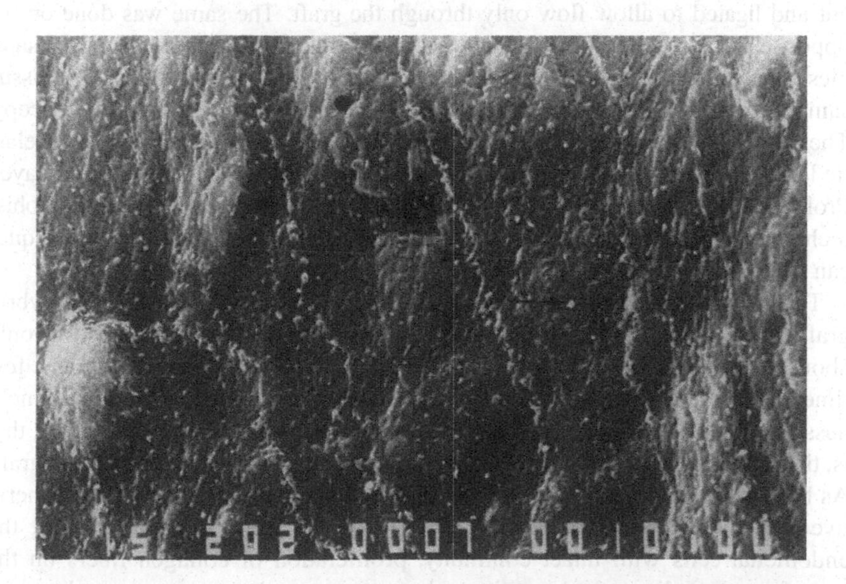

FIGURE 11.25. Group B. Scanning electron micrograph of luminal surface (original magnification, ×2000). The endothelium surface of Group B grafts is very regular. • = nucleus, > = patency of junctional system, → = microvillosity. (Reproduced from Barra JA et al., J Thorac Cardiovasc Surg 1986;92:330-336. with permission from Elsevier.)

intimal thickness was 0.45 ± 0.35 mm in the control grafts versus 0.31 ± 0.09 mm in the supported grafts.

The mesh itself was surrounded by resorptive granuloma with giant cells. There were numerous elastic fibers between the mesh and the intima; numerous capillaries were also seen in the deepest part of the intima, and this phenomenon is considered as neovascularization of the vein wall. The authors conclude that the mesh maintains the integrity of the venous wall layer and avoids endothelial and medial stretching. It results in better endothelium and thinner intima and thus in a better graft overall. They have implanted similar externally supported vein grafts in four patients and report that at 2 months the grafts were patent and had a very regular caliber.

10.2. A New Outside Stent

Krejca et al. (37) did similar studies as above in sheep where they implanted externally supported vein grafts. They constructed a 4-mm-diameter flexible mesh tubing out of polyester fiber (torlen/Dacron). They call this extravascular stent, which they used to provide external support to the vein. The mesh tubing has an ability to change its diameter depending upon the forces applied along its long axis. In sheep, the radial vein was harvested and the mesh tubing was wrapped around it and glued to the vein using tissue fibrin glue. The mesh *closely covered* the vein before glue was applied. This hybrid graft was used in place of a carotid artery in end-to-side anastomosis configuration. The carotid artery was cut and ligated to allow flow only through the graft. The same was done on the opposite side of the neck, but with the vein without the mesh. The vein graft samples were examined at various intervals between 5 days and 12 weeks. The tissue samples were used for histology, immunocytochemistry, and electron microscopy. The intima was measured from the subendothelial basal lamina to the internal elastic lamina and the media was measured from this location to the adventitial layer. Proliferation of medial and intimal cells was assessed with the use of immunohistochemical reaction with Ki-67 monoclonal antigen. The details of the techniques can be found in Ref. 37.

The wall of the vein graft thickened in both the control graft and the hybrid graft. The difference in the total wall thickness between the two groups was only about 10% in this study. However, the neointima in the control graft was a few times thicker than that in the hybrid graft. This difference in the neointimal thickness increased with time. Figure 11.26 illustrates both of these observations; that is, the intima and the media are thicker in the control graft than in the hybrid graft. As before, they also observed that in control grafts, there were several degenerative changes and vasa vasorum were seen occasionally. In hybrid grafts, the endothelial cells with intact continuity, proliferation of collagen fibers on the boundary of media and adventitia, and numerous vasa vasorum were observed. The proliferating cells were counted as percentages of stained proliferating cells' nuclei. In both types of grafts, the proliferating cell nuclei were about 8% at 4 weeks. This percentage in control grafts was 7%, 7.5%, 6.5%, and 5.5% at 6,

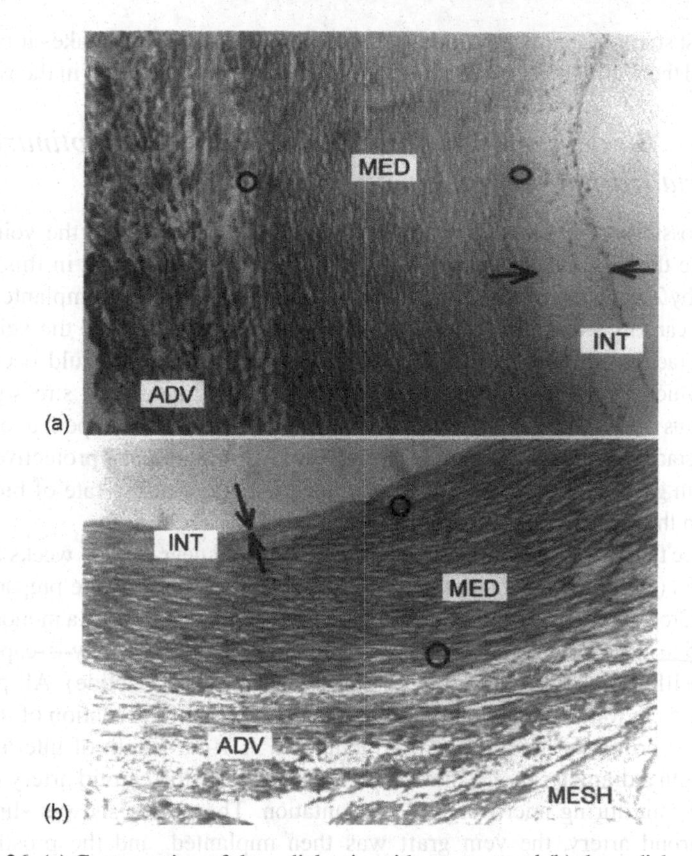

FIGURE 11.26. (a) Cross section of the radial vein with no stent and (b) the radial vein with stent (mesh) assessed after 12 weeks from grafting. Optical microscope, magnification ×150. Eosin, hematoxylin, van Gieson's staining. INT, intima between the arrows; MED, media between the circles; ADV, adventitia. (Reproduced from Krejca M et al., Eur J Cardiothoracic Surg 2002;22:898-903, with permission from Elsevier.)

8, 10, and 12 weeks, respectively. In hybrid grafts, this percentage declined more rapidly and was almost terminated at 10 weeks.

The findings indicate that the extravascular mesh prevents the hypertrophy of the graft wall, impedes the overgrowth of the intima, and decreases the proliferation rate of venous grafts' cellular elements.

It is an important point to note that the wall thickness is reduced by external support in both the studies mentioned above, however, the thickness is reduced by different amounts in the two studies. The thickness was reduced by a larger amount in the first study (10.1) than in the second study (10.2). This once again points to the important role of vascular mechanics principles. In the first study, the *tight wrap* most likely reduced the wall stress by a larger amount, whereas in the second study, the *matching wrap* most likely reduced the wall stress by a smaller amount. Consequently, the reduction in the wall thickness was more in the first study than in the second study. This reasoning in fact makes it possible to control the wall thickness reduction by "dialing in" the reduction in the wall stress.

10.3. Biodegradation of External Support to Optimize Arterialization of Vein Grafts

The possibility of "dialing in" the reduction of wall stress in the vein graft to achieve the optimum adaptation of the graft is indirectly tried in this excellent study by Zweep et al. (38). They used autologous jugular veins implanted in place of the carotid artery in rabbits where they externally supported the veins with a biodegradable prosthesis. Biodegradation of the prosthesis would occur gradually, which would expose the vein graft to increased tangential stress gradually, and thus it would regulate the arterialization. They developed a compliant, biodegradable prosthesis, which would work as a temporary protective scaffold for vein grafts. Furthermore, they examined the effect of the rate of biodegradation on the arterialization of vein grafts.

Three types of prosthesis were made, which would degrade in 3 weeks (Group I), 6 weeks (Group II), or 3 months (Group III). The prostheses were prepared as follows: Group I, copolymer of ∈-caprolactum/morpholinedione with a monomere ratio of 95.5 to 4.5; Group II, 9 to 1 (wt%) mixture of polyurethane/poly-∈-caprolactone; Group III, 9 to 1 (wt%) mixture of polyurethane/poly-(L-lactide). All prostheses were 1.5 cm long with an internal diameter of 2.5 mm. Degradation of these prostheses is primarily a matter of fragmentation. A 1-cm length of internal jugular vein with a diameter of 3.2 mm was used in place of the carotid artery of diameter 2.5 mm during microsurgical implantation. The prosthesis was slipped over the carotid artery, the vein graft was then implanted, and the prosthesis was slipped over the vein graft thus reducing the diameter of the vein graft from 3.2 mm to 2.5 mm.

In each group, the grafts were evaluated at 1, 3, and 6 weeks. The grafts were pressure-controlled perfusion-fixed and evaluated by routine light microscopy. Wall thickness (intima and media) of the vein graft, luminal radius of the graft, as well as wall thickness and luminal radius of the carotid artery were measured. The details of these techniques can be found in Ref. 38.

In Group I, the prosthesis degraded in 3 weeks. Within this time, the vein wall thickness increased and so did the graft radius (Table 11.7). After 3 weeks, the wall thickness increased but the graft radius did not. At 6 weeks, the R/t ratio, proportional to the tangential stress, did not equal that of the carotid artery. In Group II, the prosthesis degraded in 6 weeks. Within this time, the wall thickness increased gradually reaching that of the adjacent artery. The radius did not change, and it equalled that of the artery. Therefore, the R/t ratio was similar to that of the carotid artery. In Group III, the prosthesis degraded only about 50% at 6 weeks. The vein wall thickness increased in the first 3 weeks and then did not change. The radius of the graft did not change. The R/t ratio can not be compared because the full pressure load ΔP was not on the vein wall. However, the ratio is higher than that for the artery due to the thinner wall of the graft.

TABLE 11.7. Vein graft dimensions.

	Mean wall thickness (t) (μm)			Mean luminal radius (R) (mm)			(At 6 wk)
Group	1 wk	3 wk	6 wk	1 wk	3 wk	6 wk	Average ratio $\frac{R}{t}$
I	67	142	163	1.08	1.54	1.47	9.4
II	67	138	194	1.00	1.08	1.11	5.9
III	63	121	125	1.04	1.06	1.11	—
Carotid artery			206			1.05	5.0

A very important observation here is that at 6 weeks, in Group II, the vein wall thickness, graft radius, and *R/t* ratio equaled those of the adjacent artery. This is true because in this group, the degradation of the supporting structure occurred in concert with arterialization of the vein graft. Faster degradation, as in Group I, resulted in dilation of the vein graft, whereas slower degradation, as in Group III, delayed arterialization. These findings fully support the hypothesis that "the tangential stress across the vein wall seems to determine the ultimate success of arterialization." Thus, the vascular mechanics determines the final equilibrium between the stress in the graft wall, as the external support changes, and the pressure load and causes the wall thickness and the radius to equal that of the artery. Hence, if the rate of degradation of external support is properly chosen, the vein graft can achieve optimal arterialization.

References

1. FitzGibbon GM, Leach AJ, Kafka HP, Keon WJ: Bypass graft atherosclerosis: A severe long-term limitation. CL 1 Coronary Artery Disease, Cardiology Board Review 1992;9(1):83-89.
2. Davies MG, Hagen PO: Pathophysiology of vein graft failure: A review. Eur J Vasc Endovasc Surg 1995;9:7-18.
3. LiCalzi LK, Stansel HC Jr: Failure of autogenous reversed saphenous vein femoropopliteal grafting: Pathophysiology and prevention. Clin Rev Surg 1982;91(3):352-358.
4. Abbott W, Wieland S, Austen WG: Structural changes during preparation of autogenous venous graft. Surgery 1974;76:1031-1040.
5. Berguer R, Higgins RF, Reddy D: Intimal hyperplasia. Arch Surg 1980;115:332-335.
6. Blackshear WM Jr, Thiele BL, Strandness DE Jr: Natural history of above- and below-knee femoro-popliteal grafts. Am J Surg 1980;140:234-241.
7. Buxton B, Lambert RP, Pitt TTE: The significance of vein wall thickness and diameter in relation to the patency of femoro-popliteal saphenous vein bypass grafts. Surgery 1980;87:425-431.
8. Codd JE, Barner HB, Kaminski DL, Ramey A, et al.: Extremity revascularization: A decade of experience. Am J Surg 1979;138:770-776.
9. Corson JD, Johnson WC, LoGerfo FW, Bush HL Jr, et al.: Doppler ankle systolic blood pressure. Arch Surg 1978;113:932-935.
10. Darling RC, Linton RR: Durability of femoro-popliteal reconstructions: Endarterectomy vs vein bypass grafts. Am J Surg 1972;123:472-479.

11. Szilagyi DE, Elliott JP, Hageman JH, Smith RF, Dall'Olmo CA: Biologic fate of autogenous vein implants as arterial substitutes: Clinical, angiographic and histopathologic observations in femoro-popliteal operations for atherosclerosis. Ann Surg 1973;178(3):232-244.
12. Moody AP, Edwards PR, Harris PL: The aetiology of vein graft strictures: A prospective marker study. Eur J Vasc Surg 1992;6(5):509-511.
13. Mills NL, Ochsner JL: Valvulotomy of valves in the saphenous vein graft before coronary artery bypass. J Thorac Cardiovasc Surg 1976;71(6):878-879.
14. Whitney DG, Kahn EM, Estes JW: Valvular occlusion of the arterialized saphenous vein. Am Surg 1976;42(12):879-887.
15. Singh RN: Flow disturbance due to venous valves: A cause of graft failure. Catheterization and Cardiovascular Diagnosis 1986;12:35-38.
16. Chaux A, Ruan XM, Fishbein MC, Sandhu M, Matloff JM: Influence of vein valves in the development of arteriosclerosis in venoarterial grafts in the rabbit. J Thorac Cardiovasc Surg 1995;10(9):1381-1390.
17. Lajos TZ, Graham SP, Guntupalli M, Raza ST, Hasnain S: Comparison of long-term patency of "horseshoe" saphenous vein grafts with and without valves. Eur J Cardiothorac Surg 1996;10(10):846-851.
18. Weissenhofer W, Schueller EF, Schenk WG: The fate of venous valves in the arterial tree. J Surg Res 1974;17:200-203.
19. Bond MG, Hostetler JR, Karayannacos PE, Geer JC, Vasko JS: Intimal changes in arteriovenous bypass grafts: effects of varying the angle of implantation at the proximal anastomosis and of producing stenosis in the distal runoff artery. J Thorac Cardiovasc Surg 1976;71:907-916.
20. Bosher LP, Deck JD, Thubrikar MJ, Nolan SP: Role of the venous valve in late segmental occlusion of vein grafts. J Surg Res 1979;26:437-446.
21. Phillips SJ, Okies JE, Starr A: Improvement in forward coronary blood flow by using a reversed saphenous vein with a competent valve. Ann Thorac Surg 1976;21(1):12-15.
22. Phillips SJ: The augmentation of peripheral forward blood flow by prevention of flow reversal with a vein valve. J Thorac Cardiovasc Surg 1971;61(5):746-751.
23. Baba H, Djordjevic M, Kiso I, Hamada O, Moskowitz MS, von Recum A, Kantrowitz A: Hemodynamic effects of venous valves in aorto-coronary bypass grafts. J Thorac Cardiovsc Surg 1976;71(5):774-778.
24. Thubrikar MJ, Robicsek F, Fowler BL: Pressure trap created by vein valve closure and its role in graft stenosis. J Thorac Cardiovasc Surg 1994;107(3):707-716.
25. Urschel HC, Razzuk MA, Wood RE, Paulson DL: Factors influencing patency of aortocoronary artery saphenous vein grafts. Surgery 1972;72:1048-1063.
26. Stinson EB, Olinger GN, Glancy DL: Anatomical and physiological determinants of blood flow through aortocoronary vein bypass grafts. Surgery 1973;74:390-400.
27. Mundth ED, Darling RC, Moran JM, Buckley MJ, Linton RR, Austen WG: Quantitative correlation of distal arterial outflow and patency of femoro-popliteal reversed saphenous vein grafts with intraoperative flow and pressure measurements. Surgery 1969;65:197-206.
28. Barner H, Judd DR, Kaiser GC, Willman VS, Hanlon CR: Blood flow in femoropopliteal bypass vein grafts. Arch Surg 1968;96:619-627.
29. Grondin CM, Lespérance J, Bourassa MG, Pasternac A, Campeau L, Grondin P: Serial angiographic evaluation in 60 consecutive patients with aorto-coronary artery vein grafts 2 weeks, 1 year, and 3 years after operation. J Thorac Cardiovasc Surg 1975;67:1-6.

30. Bourassa MG, Lespérance J, Campeau L, Simard P: Factors influencing patency of aortocoronary vein grafts. Circulation 1972;45 and 46(Suppl):179-185.
31. Robicsek F, Thubrikar MJ, Fokin A, Tripp HF, Fowler B: Pressure traps in femoropopliteal reversed vein grafts. Are valves culprits? J Cardiovasc Surg 1999;40(5): 683-689.
32. Moreno AH, Katz AI, Gold LD, Reddy RV: Mechanics of distension of dog veins and other very thin-walled tubular structures. Circ Res 1970;27:1069-1080.
33. Zwolak RM, Adams MC, Clowes AW: Kinetics of vein graft hyperplasia: Association with tangential stress. J Vasc Surg 1987;5(1):126-136.
34. Schwartz LB, O'Donohoe MK, Purut CM, Mikat EM, Hagen PO, McCann RL: Myointimal thickening in experimental vein grafts is dependent on wall tension. J Vasc Surg 1992;15:176-186.
35. Dobrin PB, Littooy FN, Endean ED: Mechanical factors predisposing to intimal hyperplasia and medial thickening in autogenous vein grafts. Surgery 1989;105(3):393-400.
36. Barra JA, Volant A, Leroy JP, Braesco J, Airiau J, Boschat J, Blanc JJ, Penther P: Constrictive perivenous mesh prosthesis for preservation of vein integrity. Experimental results and application for coronary bypass grafting. J Thorac Cardiovasc Surg 1986;92:330-336.
37. Krejca M, Skarysz J, Szmagala P, Plewka D, Nowaczyk G, Plewka A, Bochenek A: A new outside stent – does it prevent vein graft intimal proliferation? Eur J Cardiothorac Surg 2002;22:898-903.
38. Zweep HP, Satoh S, van der Lei B, Hinrichs WLJ, Feijen J, Wildevuur CRH: Degradation of a supporting prosthesis can optimize arterialization of autologous veins. Ann Thorac Surg 1993;56:1117-1122.

12
Anastomosis

1. Introduction

Vascular anastomosis, or joining of two blood-carrying vessels, is one of the most interesting examples in the study of mechanics and pathology. Some of the commonly practiced anastomoses are artery to artery, artery to vein, and artery to synthetic tube graft. Also, the two most common forms of an anastomosis are end to end and end to side. In the science of materials, when two dissimilar materials come together, the stresses are increased at their joining. These stresses are also called *discontinuity stresses* and they produce *stress-concentration* at the joint (see Chapter 4). The discontinuity stresses are produced because two different materials expand (deform) differently upon pressure loading. In the vascular system, as we will see in this chapter, there are enhanced strains and stresses at the anastomosis. Most often, this phenomenon of enhanced strains and stresses is localized in a small region at or near the anastomosis. At the anastomosis, we also find some of the most common pathologic changes, which are of clinical significance. For instance, *subintimal hyperplasia*, which is one of the most common causes of graft failure, occurs quite frequently at the anastomosis. Another pathology is aneurysm of the vessel, which does occur at the anastomosis and not that infrequently, and in this case, it is called *anastomotic aneurysm*. So, once again, we concern ourselves with the study of stresses and strains at the anastomosis with a view toward understanding stenosis or aneurysm, which may also occur at the anastomosis.

2. Increased Compliance Near Artery-to-Artery, End-to-End, Anastomosis

Hasson et al. (1) studied the compliance near artery-to-artery, end-to-end, anastomosis in the femoral artery in dogs. The dogs weighing 20–30 kg were anesthetized and end-to-end vascular anastomosis was constructed in the mid-portion of femoral arteries with (continuous) running 6-0 Prolene sutures. Arterial diameter and wall displacement were measured over a 2-cm length centered at the anastomoses with an A-mode pulsed ultrasound. Pressure was measured via tubing inserted into a distal side branch of the femoral artery. Circumferential compliance was defined as follows:

$$C = \frac{(D_{systolic} - D_{diastolic})}{(D_{diastolic}) \times (P_{systolic} - P_{diastolic})}$$

Figure 12.1 shows a typical profile of the diameter and compliance in the femoral artery on both sides of the anastomosis. The compliance value reported in Figure 12.1 is that as defined above, but multiplied by 10,000. For instance, $C = 6$ in Figure 12.1 may come from $\frac{D_{systolic} - D_{diastolic}}{D_{diastolic}} = 2.4\% = 0.024$, and $P_{systolic} - P_{diastolic} = 40$ mmHg. Then $C = \frac{0.024}{40} \times 10{,}000 = 6$, in units of $\left(\frac{10^4}{\text{mmHg}}\right)$. In other words $C = 6$ could represent 2.4% radial change over a pulse pressure of 40 mmHg (i.e., BP 120/80).

At the suture line itself, the compliance is minimum but it is increased on both sides of the anastomosis. This increased compliance region may extend up to 5 mm on either side of the anastomosis, and the compliance may reach its maximum value at 2–4 mm from the anastomosis. The authors called this region the para-anastomotic hypercompliant zone (PHZ). This phenomenon of increased compliance was observed in 87% of the arteries. In 57% of the arteries, it was present on both sides of the suture line (1). The normal compliance of the artery can be taken as that about 10 mm away from the suture line. Using this as a reference value, the increase in compliance near the anastomosis was 46 ± 32%. The location of peak compliance was at 3.6 ± 1.2 mm from the suture line.

They also reported that the increase in compliance in the native artery was observed when a synthetic graft (PTFE; polytetrafluoroethylene) or glutaraldehyde-fixed canine carotid artery was attached to the femoral artery in dogs. They observed an increase in compliance in cephalic vein grafts connected end-to-end with the femoral artery and also observed the same in the host artery. On the venous side of the anastomosis, peak compliance was increased by 99% when compared with adjacent reference value within the vein graft. PHZ in the vein was located at 2.7 ± 1.5 mm from the anastomosis.

The authors also point out that the location of maximum subintimal hyperplasia is within 5 mm proximal and distal to PTFE grafts inserted end-to-end in

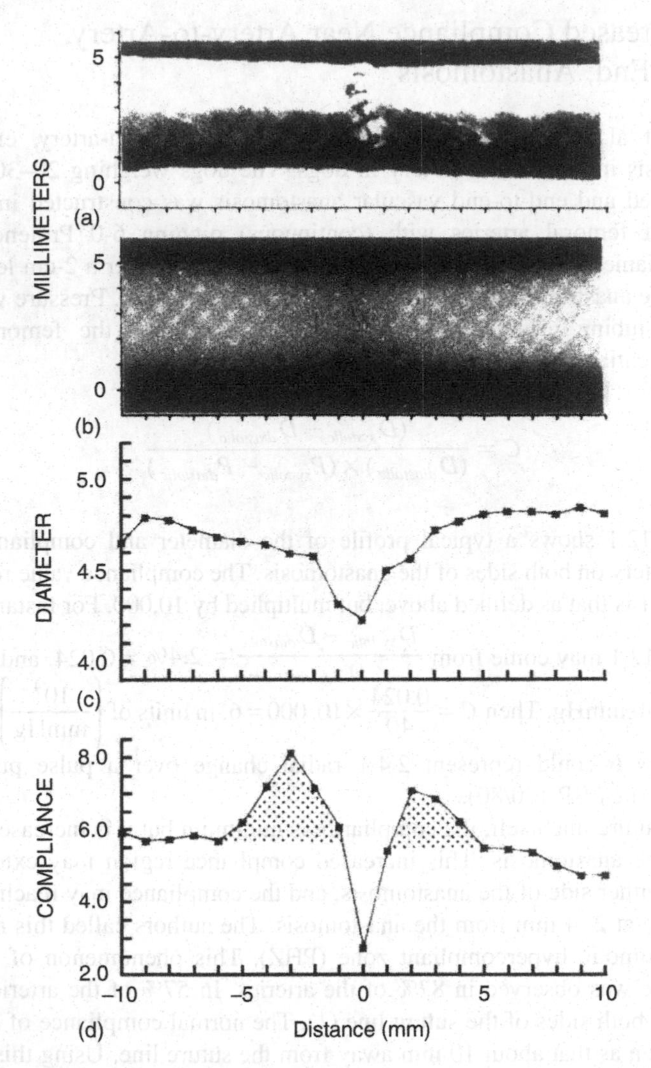

FIGURE 12.1. Diameter and compliance near the artery-to-artery end-to-end anastomosis. (a) Photograph of anastomosis constructed with continuous suture. (b) Angiogram showing contour of the lumen. (c) Ultrasonically measured diameter (in mm) versus distance from the anastomosis. (d) Compliance versus distance. Anastomosis is located at 0 mm. The regions of PHZ, in which compliance is greater than reference, are stippled. (From Hasson JE et al., J. Vasc Surg 1985;2:419-423, with permission from the Society of Vascular Surgery.)

the primate aorta (2) and just distal to Dacron grafts inserted in the canine femoral artery (3). Therefore, they conclude that the location of subintimal hyperplasia appears to be the same as the location of PHZ or increased cyclic stretching.

3. Influence of Interrupted versus Continuous Suture Technique on Compliance Near an Anastomosis

The two most common techniques used for anastomosis are continuous running sutures or interrupted sutures. In general, it would appear that continuous sutures would form a reinforcement around the circumference at the anastomosis and thereby add stiffness to the artery, thus decreasing the compliance at the suture line itself, while, as discussed earlier, increasing the compliance in the adjacent areas. The interrupted sutures, on the other hand, do not form a continuous circumferential band and therefore are expected to have less influence. Because both techniques are in common practice in surgery, we will consider their effect on the compliance of the anastomosis and explore a possible relationship with pathologic findings.

Hasson et al. (4) studied whether the suturing technique for anastomosis influences the compliance. They used the same model described above. Adult dogs were anesthetized and end-to-end anastomosis was constructed in the middle of the femoral artery using either continuous or interrupted sutures. The arterial diameter and the compliance were measured. We saw in Figure 12.1 an example of increase in compliance on both sides of the anastomosis. In this study, the authors report two other possibilities, one where the compliance is increased only on one side of the anastomosis, and the other where the compliance is unchanged near the anastomosis (Fig. 12.2). They examined whether the type of suture technique used was associated with how the compliance changed near the anastomosis. According

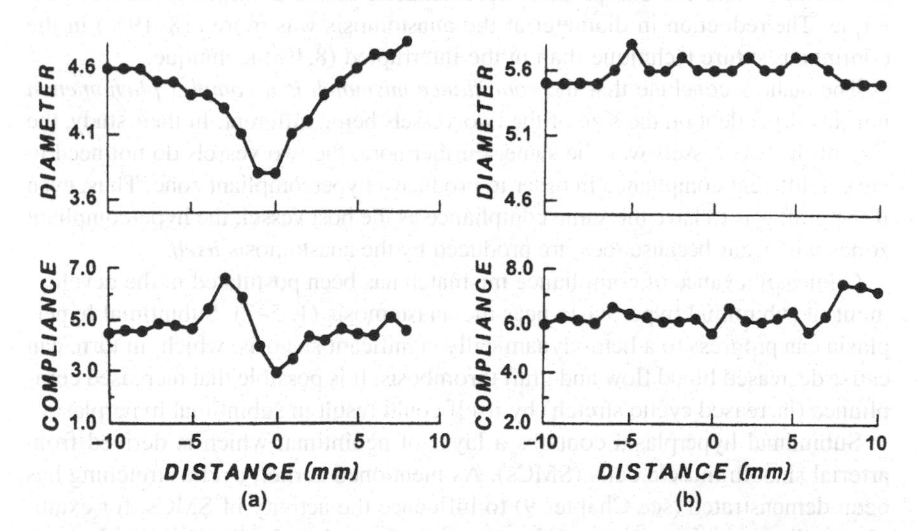

FIGURE 12.2. Diameter and compliance near the artery-to-artery anastomosis as in Figure 12.1. (a) A case (with continuous suture) where PHZ is present only proximal to anastomosis. (b) A case where interrupted sutures are used for the anastomosis and PHZ is absent. (From Hasson JE et al., J Vasc Surg 1986;3:591-598, with permission from the Society of Vascular Surgery.)

TABLE 12.1. Incidence of increased compliance for each side of each anastomosis.

Location of increased compliance	Suture technique	
	Interrupted	Continuous
Proximal only	2	5
Distal only	2	1
Proximal and distal	8	24
Total sides with increased compliance	12	30
Total sides without increased compliance	20	12

to Table 12.1, all possibilities, for example, increase in compliance only on the proximal side, only on the distal side, and on both the proximal and the distal sides, occurred in both the interrupted and the continuous suture technique. However, no increase in compliance at all seems to have occurred only with the interrupted suture technique. In Figure 12.2A we note that for a continuous suture anastomosis, the compliance first increased beyond the reference value to a maximum at about 3 mm proximal to the anastomosis, then decreased to a minimum at the suture line itself. With this suture technique, there is often a decrease in the diameter. The hypercompliant zone on both sides of the anastomosis occurred more frequently with continuous sutures (67%) than with interrupted sutures (50%). Along the same lines, an increased compliance region was present in 71% of all available sides in continuous suture techniques but in only 38% of all sides in interrupted suture technique. The peak increase in compliance occurred at 3.8 ± 1.2 mm from the anastomosis in both techniques. At the anastomosis itself, both the diameter and the compliance were reduced in the continuous suture technique. The reduction in diameter at the anastomosis was more (18–19%) in the continuous suture technique than in the interrupted (8.4%) technique.

The authors conclude that the *compliance mismatch is a complex phenomenon* not just dependent on the size of the two vessels being different. In their study, the size of the two vessels was the same. Furthermore, the two vessels do not need to have a different compliance in order to produce a hypercompliant zone. Thus, even if the graft was to have the same compliance as the host vessel, the hypercompliant zones will occur because they are produced by the anastomosis *itself*.

Clinical relevance of compliance mismatch has been postulated in the development of subintimal hyperplasia near the anastomosis (1, 5–7). Subintimal hyperplasia can progress to a hemodynamically significant stenosis, which, in turn, can cause decreased blood flow and graft thrombosis. It is possible that increased compliance (increased cyclic stretch) by itself could result in subintimal hyperplasia.

Subintimal hyperplasia contains a layer of neointima, which is derived from arterial smooth muscle cells (SMCs). As mentioned earlier, cyclic stretching has been demonstrated (see Chapter 9) to influence the activity of SMCs, for example, cyclic stretching increased the amounts of rough endoplasmic reticulum (8) and amounts of collagen and chondroitin sulfate (9) in arterial SMCs cultured in vitro. Also, protein transport in the arterial wall increases with arterial stretching (10, 11) as seen in Chapter 8. Platelet factors and other proteins have been shown to stimulate SMC proliferation (12).

It is proposed that increased compliance (cyclic stretching) near an anastomosis could lead to subintimal hyperplasia through multiple mechanisms that include stimulation of SMCs by stretch, by proteins that have been transported in increased amounts, by mitogenic factors that have been transported due to increased stretching, and by synergistic action with other concomitant processes such as blood-prosthetic material interaction. On the basis of these observations, interrupted suture technique may be preferable.

4. Arterial Anastomosis with Vein and Dacron Grafts: Stress Studies

In the earlier section, it was mentioned that in case of end-to-end anastomosis between artery and PTFE and between artery and vein studied in vivo, there was an increase in compliance in the native artery in both cases and in the vein graft in the latter case. Such experimental measurements are important in understanding the mechanical deformations at the anastomosis. Also, it is important to know how the stresses may change at the anastomosis. It may be good to note here that the general impression has been that the vascular diseases such as intimal hyperplasia or atherosclerosis are associated with the flow disturbances while aneurysm is associated with mechanical deformation and wall stress. To change these notions, we continue to establish a link between the above listed diseases and mechanical deformations and accompanying stresses. An association between mechanical deformation and intimal hyperplasia was already suggested in the earlier sections.

In this section, we will consider additional examples of vascular pathology, particularly with respect to vein and polymeric grafts. Autogenous vein bypasses as well as synthetic grafts are commonly used as arterial bypass conduits. Even though prosthetic graft replacement of the aorto-femoral arteries is satisfactory, problems with loss of patency is a significant limitation when they are used with infrainguinal bypasses. Experimentally, it has been found that 12 weeks after implantation, 80% of the vein grafts remained patent in an animal model, whereas only 30% of Dacron grafts and 15% of PTFE grafts maintained patency (13).

Gaylis (14) observed true para-anastomotic aneurysms of venous conduits adjacent to anastomoses with intact suture lines. However, with prosthetic conduits, pseudoaneurysms developed instead. These observations point to the occurrence of both kinds of vascular pathology, that is, intimal hyperplasia and aneurysm at the anastomosis, and lead to the search for a common cause.

Mechanical stress at the anastomosis was studied by Chandran et al. (15). They used a finite element modeling technique to determine stresses and deformations in end-to-end anastomosis of artery and vein graft and artery and Dacron graft.

The two geometric models created are shown in Figure 12.3. The diameter, wall thicknesses, and modulus of elasticity used in the model are listed in Table 12.2. The internal diameters for artery, vein and Dacron graft are the same (4 mm). The wall thickness for the artery and Dacron graft are the same (1 mm)

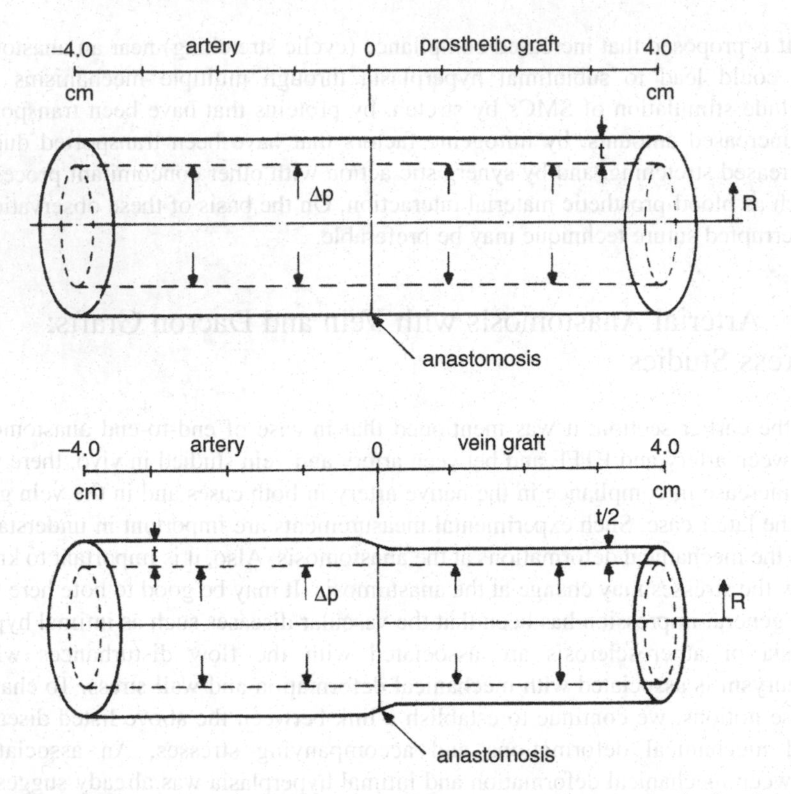

FIGURE 12.3. Finite element model of a simulated artery-graft anastomosis. R = internal radius, t = thickness of the artery, and ΔP = pressure load. (From Chandran KB et al., Med Biol Eng Comput 1992;30:413-418, with permission.)

and that of the vein graft is 0.5 mm. The wall material for all three is considered to be isotropic, homogenous, incompressible, and linearly elastic. The magnitudes of the elastic modulus for the artery, as well as the bypass grafts was specified such that the compliance distal to the anastomosis would simulate values reported from *in vivo* studies (16). The compliance of a vessel can be defined using the relationship

$$C = \frac{\Delta D}{D} \frac{1}{\Delta P}$$

where C is the compliance, D is the internal diameter, and ΔP is the pulse pressure. The elastic moduli for the artery and the various graft segments used in the analysis are included in Table 12.2. A comparison of the computed and the measured compliance distal to the anastomosis is also included. In the static analysis performed in the study, the pulse pressure was replaced by an assumed mean arterial pressure of 13.3 kPa (100 mmHg). It can be observed from Table 12.2 that the elastic modulus value assumed for the wall of the vein graft is very high compared with that of the artery. Because the wall of the vein is thinner than the

TABLE 12.2. Dimensions, elastic modulus, and the compliance of the grafts used in the model.

	Internal diameter (mm)	Wall thickness (mm)	Elastic modulus (Pa × 10^5)	Compliance, % per kPa	
				Computed from model	Measured In vivo
Artery	4.0	1.0	4.55	0.585	0.586
Vein	4.0	0.5	17.55	0.234	0.233
Dacron	4.0	1.0	19.00	0.140	0.140

arterial wall, specification of a higher elastic modulus was necessary to simulate the compliance characteristics of the vein graft measured in vivo. The general-purpose finite element program ANSYS was employed in the static analysis. The wall was divided into two layers and up to 108 eight-noded isoparametric three-dimensional elements (STIF45) with up to 355 nodes. To simulate the incompressibility of the vessel wall, a Poisson's ratio of 0.49 was specified for the artery as well as for the grafts. Taking advantage of the axisymmetry of the model, only a segment of the artery/graft anastomosis was used in the analysis by the appropriate specification of the boundary conditions at the nodes. The distance between the anastomosis and the free end of the artery and graft was specified to be 4 cm. To account for the difference in wall thickness for the artery and the vein, prismatic elements were employed in the finite-element mesh. At the free ends (distal to the anastomosis), the nodes were restricted from motion along the axial direction to simulate the tethering of the artery in vivo. More details of the technique can be found in Ref. 15.

The results of the analysis are shown in Figures 12.4 and 12.5. The distribution of compliance and the nondimensional principal stress (maximum tensile stress normalized with respect to the applied mean pressure) around the anastomosis with a vein graft subjected to a mean pressure of 13.3 kPa (100 mmHg) is shown in Figure 12.4. An increase in compliance is observed on the arterial side 4 mm from the anastomosis. A region of high tensile stress in the wall is also found 0.25 mm from the anastomosis on the vein side. A similar distribution of compliance and stress is observed with the artery-Dacron graft model (Fig. 12.5). With the Dacron graft, the increase in compliance on the arterial side is larger than the corresponding value for the artery-vein model. Moreover, the maximum nondimensional principal stress with the Dacron graft is smaller than the corresponding value for the vein graft. The location and the magnitude of the maximum compliance as well as the maximum nondimensionalized principal stresses with the vein and Dacron graft models are shown in Table 12.3.

Results of Figures 12.4 and 12.5 need elaboration. The circumferential stress in the artery away from the anastomosis is given by $\frac{PR}{t}$, where P is the pressure, R is the radius, and t is the thickness. When this stress is divided by the pressure (P), then we have the normalized stress to be R/t. With $R = 2$ mm and $t = 1$ mm (Table 12.2), the normalized stress would be 2 if there was no expansion (i.e., no increase in R). However, the normalized stress is higher, about 2.8, because there is an expansion, which is of a magnitude matching the compliance (Table 12.2).

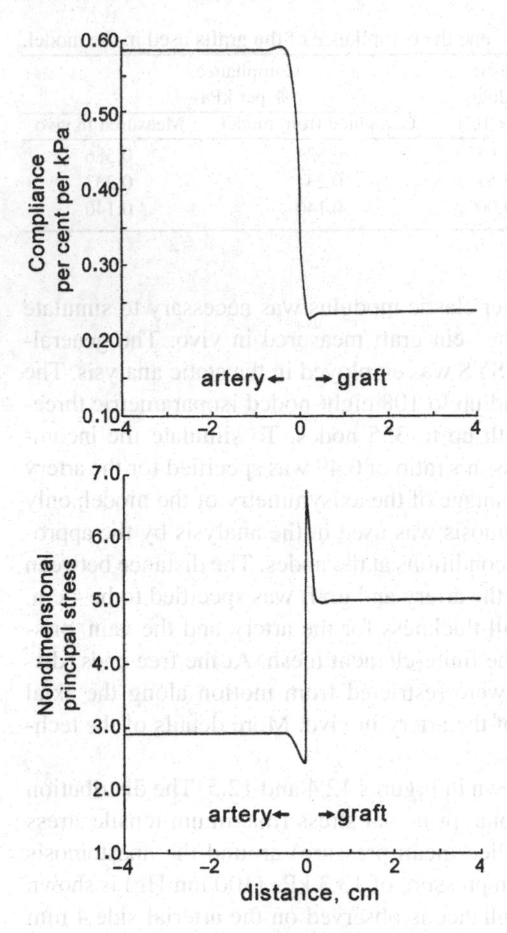

FIGURE 12.4. Distribution of compliance and nondimensional principal stress (maximum tensile stress normalized with respect to the applied mean arterial pressure) near the artery–vein graft anastomosis. (From Chandran KB et al., Med Biol Eng Comput 1992;30:413-418, with permission.)

The stress in the vein graft (Fig. 12.4) is about 5, which is almost double of that in the artery. The reason for this doubling is likely to be the wall thickness (t = 0.5 mm), which is almost half of the thickness of the artery (Table 12.2). In case of Dacron graft (Fig. 12.5), we note that the normalized stress away from the anastomosis is the same in both the artery and the graft, most likely because the radius and the thickness are the same for both.

At the anastomosis, the stress in the artery decreased and that in the graft increased (Figs. 12.4 and 12.5). The basic mechanism for this phenomenon is illustrated in Figure 12.6. If the artery was separated from the graft, it would expand more than the graft (Fig. 12.6b). But because they are joined, the radius of the artery actually decreases and that of the graft increases at the anastomosis (Fig. 12.6c). This causes a decrease in the stress in the artery and causes an increase in the stress in the graft as shown in Figures 12.4 and 12.5.

There are certain limitations to this model. This study was restricted to static loading, whereas the artery-graft anastomosis is subjected to a pulsatile load

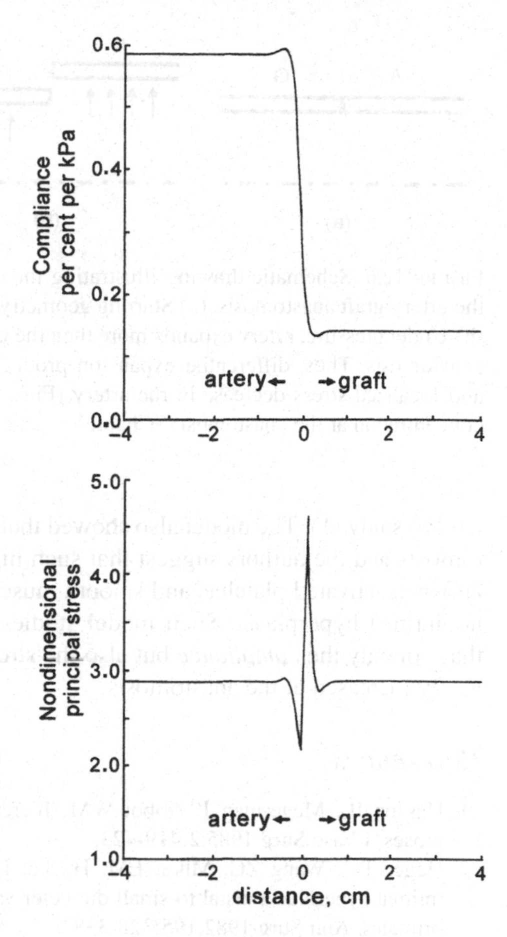

FIGURE 12.5. Distribution of compliance and nondimensional principal stress near the artery–Dacron graft anastomosis. (From Chandran KB et al., Med Biol Eng Comput 1992;30:413-418, with permission.)

TABLE 12.3. Magnitudes of maximum compliance in the hypercompliant zone and the maximum nondimensional principal (maximum tensile stress normalized with respect to the applied mean arterial pressure) stress with their locations from the anastomosis.

Graft model	Maximum compliance, % per kPa	Distance in mm from anastomosis (on arterial side)	Maximum nondimensional principal stress	Distance in mm from anastomosis (on the side of the graft)
Artery-vein	0.590	4.0	6.37	0.25
Artery-Dacron	0.595	3.5	4.61	0.5

in vivo. The effect of sutures at the junction between the artery and the graft on the mechanics of the anastomosis was not taken into account. The material property of the artery was assumed to be linear and isotropic, and hence any nonlinear effects were neglected.

In spite of these limitations, the results of the model showed that the locations of maximum increase in the compliance were similar to that observed in the

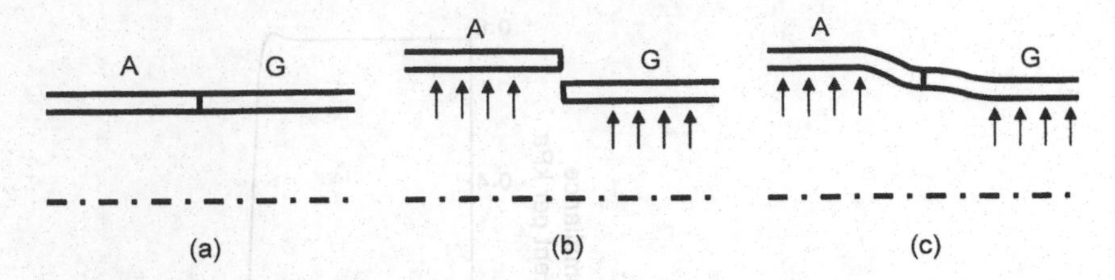

FIGURE 12.6. Schematic drawing illustrating the differential expansion under pressure at the artery-graft anastomosis. (a) Starting geometry of the artery (A, graft; G, anastomosis). (b) Under pressure, artery expands more than the graft. (c) The continuity is maintained by anastomosis. Thus, differential expansion produces localized stress increase in the graft and localized stress decrease in the artery (Figs. 12.4 and 12.5), which results in stress-concentration at the anastomosis (Chapters 4 and 5).

in vivo study (1). The model also showed that very high stresses occur at the anastomosis and the authors suggest that such high stresses could induce interaction between activated platelets and smooth muscle cells resulting in the formation of neointimal hyperplasia. Such model studies are important and they emphasize that not only the *compliance* but also the stress and the *stress-concentrations* are vastly increased at the anastomosis.

References

1. Hasson JE, Megerman J, Abbot WM: Increased compliance near vascular anastomoses. J Vasc Surg 1985;2:419-423.
2. Hagen PO, Wang ZG, Mikat EM, Hackel DB: Antiplatelet therapy reduces aortic intimal hyperplasia distal to small diameter vascular prostheses (PTFE) in nonhuman primates. Ann Surg 1982;195:328-338.
3. LoGerfo FW, Quist WC, Nowak MD: Downstream anastomotic hyperplasia: A mechanism of failure in Dacron arterial grafts. Ann Surg 1983;197:479-483.
4. Hasson JE, Megerman J, Abbott WM: Suture technique and para-anastomotic compliance. J Vasc Surg 1986;3:591-598.
5. Kinley CE, Marble AE: Compliance: A continuing problem with vascular grafts. J Cardiovasc Surg 1980;21:163-170.
6. Madras PN, Ward CA, Johnson WR, Singh PI: Anastomotic hyperplasia. Surgery 1981;90:922-923.
7. Megerman J, Abbott WM: Compliance in vascular grafts. In: C. Wright (ed), Vascular Grafting. John Wright-PSB, Boston, 1983:344-364.
8. Sottiurai VS, Kollros P, Glagov S, Zarins CK, Mathews MB: Morphologic alteration of cultured arterial smooth muscle cells by cyclic stretching. J Surg Res 1983;35: 490-497.
9. Leung OY, Glagov SG, Mathews MB: Cyclic stretching stimulates synthesis of components by arterial smooth muscle cells in vitro. Science 1975;191:475-477.
10. Chien S, Usami S, Fan FC, Skalak R, Weinbaum S, Caro CE: Effects of mechanical disturbances on uptake of macro-molecules by the arterial wall. In: Nerem RM, Cornhill JF (eds), The Role of Fluid Mechanics in Atherogenesis. Proceedings of

a Specialists Meeting sponsored by the National Science Foundation, Columbus, Ohio, 1978:16-1-16-44.
11. Fry DL, Mahley RW, Oh SY: Effect of arterial stretch on transmural albumin and Evan's blue dye transport. Am J Physiol 1981;240:H645-H649.
12. Rutherford RB, Ross R: Platelet factors stimulate fibroblast and smooth muscle cells quiescent in plasma serum to proliferate. J Cell Biol 1976;699:196-203.
13. Kidson IG: The effect of wall mechanical properties on patency of arterial grafts. Ann R Coll Surg England 1983;65:24-29.
14. Gaylis H: Pathogenesis of anastomotic aneurysms. Surgery 1981;90:509-515.
15. Chandran KB, Gao D, Han G, Baraniewski H, Corson JD: Finite-element analysis of arterial anastomoses with vein, Dacron and PTFE grafts. Med Biol Eng Comput 1992;30:413-418.
16. Abbott WM, Bouchier-Hayes J: The role of mechanical properties in graft design. In: Dardik H (ed), Graft Materials in Vascular Surgery. Year Book Medical Publishers, Chicago, 1978:59-78.

13 Anastomotic Aneurysms and Anastomotic Intimal Hyperplasia

1. Introduction

Anastomosis, or joining of two blood vessels, is one of the most important locations where vascular mechanics can play a very important role in occurrence of vascular pathology. In Chapter 12, we have covered some aspects of vascular mechanics for the configuration of end-to-end anastomosis. However, much remains to be explored for end-to-side anastomosis. The resulting geometry in the end-to-side anastomosis is similar to that of an arterial branch, and therefore significant stress increase in local regions is expected. Consequently, it does not come as a surprise that the anastomotic regions are prone to development of pathologies.

The two most common pathologies, which occur at the arterial anastomosis, are *anastomotic aneurysms* and *intimal hyperplasia*. Anastomotic aneurysms, as all aneurysms, pose a danger of rupture and bleeding, whereas anastomotic intimal hyperplasia creates stenosis and blocks the blood flow. It is to be noted that two different pathologies—aneurysm and stenosis—can occur at the same site (anastomosis), and this is of utmost significance from the vascular mechanics point of view. Perhaps a study of anastomosis could be a gateway to understanding the pathogenesis of both aneurysms and atherosclerosis (intimal hyperplasia). In broad terms, we may also consider that apoptosis of smooth muscle cells is at the core of forming aneurysms, whereas proliferation of smooth muscle cells is at the core of forming atherosclerotic lesions (intimal hyperplasia). In this context then, we may pose the question as to how might stresses in the vessel wall cause cellular apoptosis in some situations and cellular proliferation in others? Much work remains to be done to answer this question. However, in this chapter we will examine relationships between vascular mechanics and both anastomotic aneurysms and anastomotic intimal hyperplasia.

2. Anastomotic Aneurysms

An anastomotic aneurysm (junctional aneurysm, suture line aneurysm, pseudoaneurysm) in general is a *false aneurysm* that occurs at the site of any vascular repair. It occurs most commonly at the junction of *prosthesis* and *host artery*, usually several years after the original operation (Fig. 13.1). Instead of a prosthesis, if an autogenous *vein* is used, then the aneurysm can still occur at the anastomosis but this is now a *true aneurysm*, usually in the vein itself. In other words, prosthetic graft material is associated with false aneurysm while a vein graft is associated with true aneurysm. This fact also is of significant importance when we consider the relationship between vascular mechanics and vascular pathology.

Reconstructive vascular surgery has made a big impact on patient care over the past three decades and most of it was made possible by the introduction of vascular prosthesis. Aorto-femoral bypass grafting for occlusive aorto-iliac disease is one of the most common examples of the success of these procedures. In a small percentage of patients (1–4%), however, development of an anastomotic aneurysm has been well recognized as an important complication of prosthetic implantation. As stated earlier, the anastomotic aneurysm is a false aneurysm and it occurs when the prosthesis separates from the artery, partially or totally, and blood escapes into a fibrous tissue capsule, which surrounds the graft and the host artery. It is the stretching of the fibrous capsule that creates the false aneurysm (Fig. 13.2). An anastomotic aneurysm differs from a true aneurysm in that the wall consists of fibrous tissue, whereas a true aneurysm contains the anatomic layers of an artery, including intima, media, and adventitia. The fibrous capsule still maintains vascular continuity but poses the dangers of rupture, thrombosis, embolism, or pressure on adjacent structures typical of all aneurysms.

The disruption of a vascular anastomosis and formation of a false aneurysm can happen in three possible ways: 1) a defect in the fabric of the prosthesis; 2) breakage of the suture; or 3) weakness of the host artery. The suture material, especially silk, which was in common use during the 1950s, was often incriminated. More recently, the introduction of newer and more durable suture material has resulted in the frequent finding of an intact suture line. This has shifted the focus to graft dilation or to the weakness of the arterial wall as the cause.

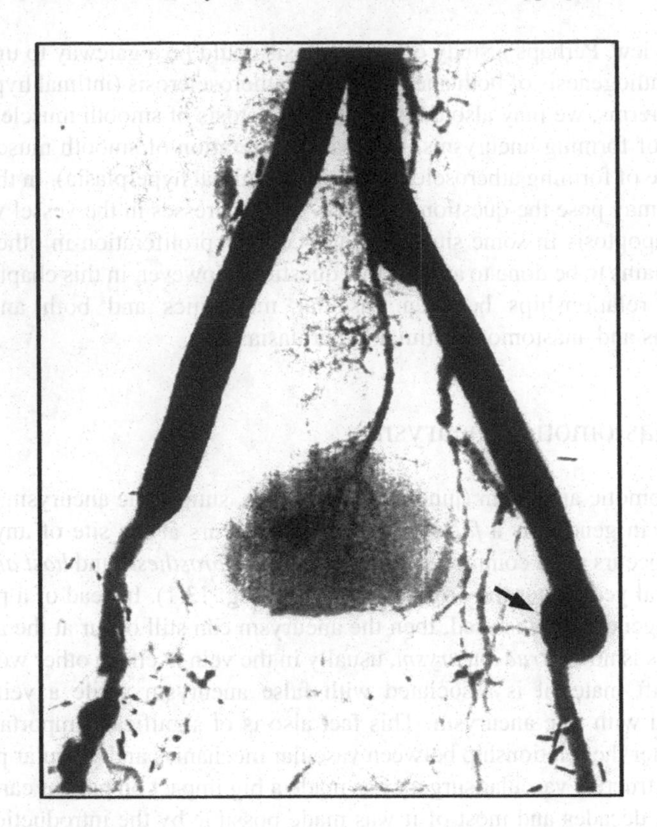

FIGURE 13.1. Anteroposterior view of an abdominal aortogram showing aortobifemoral graft with a left femoral anastomotic aneurysm (arrow). (Reproduced from Carson SN et al., Am J Surg 1983;146:774-778, with permission from Excerpta Medica Inc.)

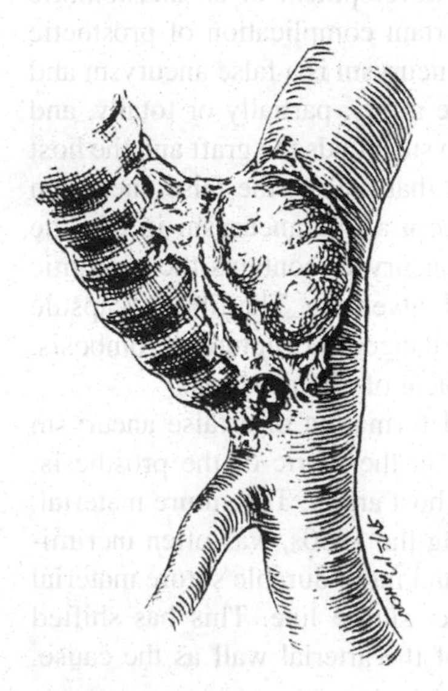

FIGURE 13.2. Anastomotic aneurysm developing because suture loops pulled out from the edge of the arterial wall. (Reproduced from Moore WS: Vascular Surgery, 4th edition,1995;1:605-611, with permission from Elsevier.)

The frequency of graft dilation is also much less and that has led to the arterial wall weakness as a more dominant cause of anastomotic aneurysm. The arterial wall weakness is not necessarily inherent to the vessel; instead, it is caused by external factors, which in turn are related to unphysiologic properties of the prosthesis. For instance, the noncompliant nature of the prosthesis and the abnormal stresses at the host-graft junction could be the main causes of anastomotic aneurysms. We must isolate the two parameters—the noncompliant nature of the prosthesis and stresses at host graft junction—and understand how each one by itself as well as in combination participate in the process. For example, the two similarly compliant arteries (the graft and the host) will still have abnormal stresses at the anastomosis. A noncompliant prosthesis and a compliant host artery will also have abnormal stresses at the anastomosis and in this case, the stresses will be even greater than those in the previous case. Furthermore, the size of the prosthesis and the size of the opening in the artery *and* how they are joined at the anastomosis can influence the stresses at the graft-host junction. In all of the above cases, the stresses at the anastomosis will always be increased and the larger the mismatch either in geometry or in compliance the higher will be the stresses. The fundamental nature of mechanics at the anastomosis, covered in Chapter 12, can be applied to studies of end-to-side anastomoses. Such studies in mechanics should have been done some 20 years ago but unfortunately must still wait for the future. With the fundamental vascular mechanics at work at the anastomosis (i.e., causing increased stresses at the graft-host junction), it is easy to see why the suture, the prosthesis, and the adjacent arterial wall are all subjected to the stress damage. Then, it is only a matter of which one will be the first to give in. We will consider more details of the mechanics a little later. For now, let us examine additional pathologic aspects of the anastomotic aneurysms.

2.1. Clinical Findings

There are a number of excellent reviews on this clinically (not mechanically) mature subject. Here we will consider some of the findings. Gaylis et al. (1) reported their experience on 101 anastomotic aneurysms occurring in 74 patients over a 25-year period. Of these, 93 were prosthesis related, 7 were associated with vein grafts, and 1 occurred after a thrombo-endarterectomy.

The seven vein graft aneurysms were *true* aneurysms. These occurred in autogenous reversed femoropopliteal vein grafts, and five were located at the femoral anastomosis and one at the popliteal anastomosis. The femoral and popliteal anastomoses were all end-to-side. These aneurysms were not at the suture line but adjacent to it in the walls of the vein. They were true aneurysms in the vein grafts at the distal anastomosis. We should keep this observation in mind when we examine, later in the chapter, intimal hyperplasia at the vein graft–artery anastomosis.

The 93 prosthesis-related anastomotic aneurysms were all (typical) *false* aneurysms. The larger percentage (80%) of these occurred at the femoral anastomoses. Of 93 prostheses, 88 were Dacron, 4 were Teflon, and 1 was PTFE. In most cases, the suture material was either braided polyester or polypropylene. The proximal anastomosis with the aorta was either end-to-end or end-to-side, whereas the distal anastomosis with the femoral or popliteal artery was always end-to-side. The number of aneurysms at various locations is listed in Table 13.1 (1).

In patients, the aneurysms presented with swelling (58%) and pain (39%) and the diagnosis of anastomotic aneurysm was confirmed by ultrasonography, CT, arteriography, MRI, or a combination of modalities (Fig. 13.1). The time of onset for the anastomotic aneurysm, after the initial operation, varied from 1 to 23 years. The most common mode of treatment was resection of a portion of the original prosthesis and bridging of the defect with a new prosthesis. The most common cause of anastomotic aneurysm in 90% of the cases was the host artery failure, in 5% it was suture failure, and in the other 5% prosthetic failure.

2.2. *Pathogenesis*

The disruption of a vascular anastomosis to form a false aneurysm can occur from suture failure, loss of holding power of the host artery, disruption of the prosthesis, or a combination of these events. The findings of the above study indicate that suture failure is not a dominant factor and neither is disruption of the prosthesis. In 90% of the cases, the sutures had pulled out of the host artery implying the loss of holding power or weakness of the host artery. Histologic studies of the arterial tissue adjacent to the anastomosis indicated that the artery had a decreased wall thickness due to loss of smooth muscle cells, disruption of elastic laminae, and proliferation of elastic fibers (2). At issue is the mechanism responsible. Some of the factors that could play a role in the process are weakening of the artery from endarterectomy, excessive mobilization of the artery, placement of sutures under excessive tension, finely placed sutures, ongoing atherosclerotic disease process, wound-healing problems that delay graft incorporation into the tissue, and so forth.

The authors noted that the incidence of the anastomotic aneurysm is very high when prostheses are used compared with the low incidence when artery-to-artery repair or autogenous vein-to-artery anastomosis is present. For instance, Mehigan et al. (3) reported that anastomotic aneurysms at the Dacron–femoral artery junction occurred in 7.9% of cases, whereas vein-artery anastomotic (true) aneurysms occurred in 1.4% of the cases. Thus, they believe that the prosthesis itself is the prime initiating factor. The unphysiologic characteristics of the prosthesis, such as a propensity to dilate over time and nondistensibility, are primarily responsible for the development of anastomotic aneurysms.

TABLE 13.1. Location of prosthesis-related anastomotic aneurysms.

Site of aneurysm	No.	%
Aortic	17	18.2
Iliac	7	7.5
Femoral	66	70.9
Popliteal	3	3.2
Total	93	100

All polyester materials exhibit initial instantaneous deformation caused by elasticity and slackness in the structure. This is followed by creep or slow deformation under stress. Thickness of the fiber and configuration of the weave plays a part in it. In some polyester grafts, after 33 months of implantation, an average dilation of 15% in normotensive patients and 21% in hypertensive patients has been seen (1). The authors also suggest that the radius of an end-to-side anastomosis is larger than the artery itself, which increases the wall tension on the suture line. Also, if the graft dilates, then according to Laplace's law, the increase in graft diameter leads to an increase in shear tension of the suture, which can lead to cutting of the arterial tissue by the suture and then pulling out of the host artery.

Even more importantly, the nondistensibility, which leads to *compliance mismatch* between the prosthesis and the host artery, is the single most important factor in the formation of anastomotic aneurysms. With each pulse wave, the artery may change the diameter by up to 10%, whereas the prosthesis does not. This compliance mismatch causes suture line shear stress. In follow-up, the prosthesis may become more rigid while the artery remains unchanged. The suture line shear stress leads to weakness of the arterial wall adjacent to the prosthesis and the eventual loss of holding power. This mechanism may be further enhanced by factors such as hypertension and traction on the anastomosis. Thus, the possible sequence of events leading to the formation of anastomotic aneurysms may be as that shown in Figure 13.3.

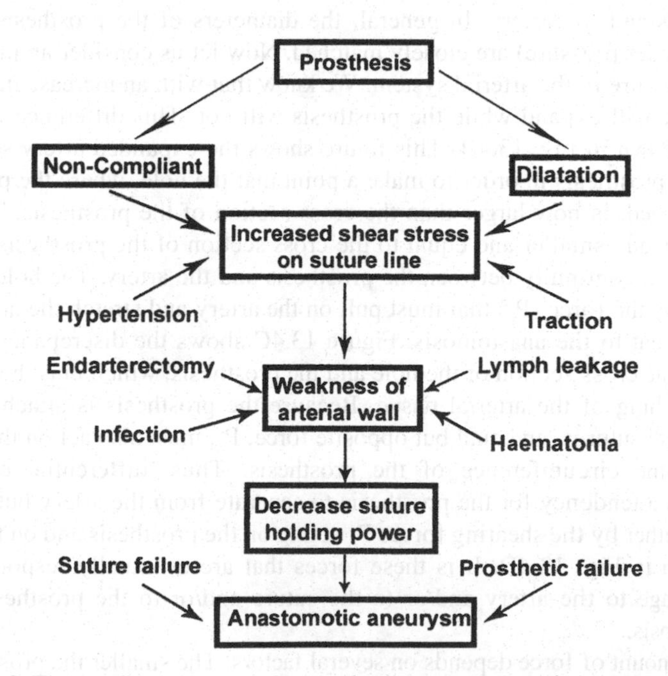

FIGURE 13.3. The possible sequence of events leading to anastomotic aneurysm. (Surgery Annual 1990;22:317-341.)

3. Mechanical Stress at Prosthesis-to-Artery, End-to-Side Anastomosis

For the pathogenesis of anastomotic aneurysms, we saw in the previous section that the most important mechanism is "weakening of the arterial wall adjacent to the suture line, leading to the arteries' loss of holding power and eventual pulling out of the sutures from the tissue causing a separation between the prosthesis and the artery." We intuitively know that the mechanical stresses are significantly enhanced in the artery wall at the anastomosis. We have examined the enhancement of stresses, strains, and compliance in the end-to-end anastomosis in Chapter 12. The end-to-side anastomosis is similar to a "T" or a "Y" junction geometry, considered in Chapters 5, 6, and 7, where we also saw that the stresses are increased (stress-concentration) at the junction. Therefore, we expect the stresses to increase (i.e., the stress-concentration to occur) at the end-to-side anastomosis. In fact, we expect even a larger stress-concentration at the prosthesis-artery anastomosis in comparison with that at the artery-artery anastomosis or the vein-artery anastomosis. Unfortunately, the detailed stress analysis at the prosthesis-artery end-to-side anastomosis has not been done. Therefore, in this section we will consider the development of mechanical stress at the anastomosis, in a conceptual fashion, so as to understand the basic principles at work in causing the stress increase.

Figure 13.4A shows a schematic of the end-to-side anastomosis between the prosthesis and the artery. In general, the diameters of the prosthesis and the artery (under pressure) are closely matched. Now let us consider an increase in blood pressure in the arterial system. We know that with an increase in pressure the artery will expand while the prosthesis will not. This difference in expansion is shown in Fig. 13.4B. This figure shows the expanded artery separately from the prosthesis in order to make a point that the hole, where the prosthesis was attached, is now larger than the cross section of the prosthesis. This hole must be made smaller and equal to the cross section of the prosthesis in order to maintain continuity between the prosthesis and the artery. The hole is made smaller by the force (P_o) that must pull on the artery and stretch the arterial tissue adjacent to the anastomosis. Figure 13.4C shows the discrepancy (shaded area) in the cross section of the hole and the prosthesis, which must be filled by the stretching of the arterial tissue. Because the prosthesis is attached to the artery with sutures, an equal but opposite force, P_o, must also act on the sutures and on the circumference of the prosthesis. Thus, differential expansion produces a tendency for the prosthesis to separate from the artery but they are kept together by the shearing forces P_o acting on the prosthesis and on the artery as shown in Fig. 13.4D. It is these forces that are eventually responsible for the damage to the artery and/or to the suture and/or to the prosthesis at the anastomosis.

The amount of force depends on several factors. The smaller the prosthesis for the same size hole, the larger will be the force P_o. Suture placement also affects

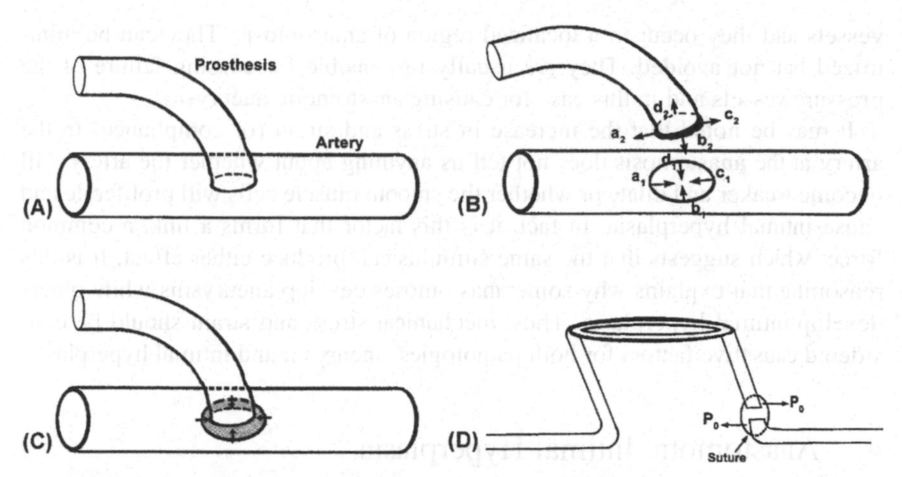

FIGURE 13.4. Schematic representation of stress-concentration at the prosthesis-artery anastomosis. (A) Prosthesis and artery after anastomosis. (B) Under pressure, hole in the artery is larger than the end of the prosthetic graft. (C) Near the anastomosis, the artery tissue is pulled inward and develops stress-concentration all along the region of anastomosis (shaded area). (D) Two opposing shear forces (P_o) act on the prosthesis and on the artery at the anastomosis. The shear force is also present in the suture.

the force. If sutures are placed farther into the artery, or if they are pulled harder, then that will increase the force. Tethering of the artery will also influence the force P_o. For instance, due to the presence of tethering, the force P_o along the axis of the artery may not change much with change in pressure, however, the circumference of the artery is free to change. So, with increased pressure, force P_o at location b_1 or d_1 (Fig. 13.4B) may increase while that at location a_1 or c_1 may not. In other words, there is no definite pattern that suggests that forces at a_1 and b_1 and c_1 are in any particular order. The force at any location around the anastomosis depends on the details of how the anastomosis has been accomplished technically. The one factor that does remain certain, however, is that the stresses are significantly enhanced all around the anastomosis and therefore the entire suture line is at risk. In other words, the arterial tissue may be damaged anywhere along the circumference of the anastomosis. It may also be pointed out that the same phenomenon is at work when the artery or vein is used in place of a prosthesis with one obvious difference. The grafted artery will expand with the host artery and the grafted vein will also expand with the host artery, although to a lesser extent. This expansion reduces the discrepancy between the graft cross-section and the hole in the host artery. Consequently, there is a lesser amount of force (P_o) produced at the anastomosis. This is perhaps the reason why the anastomotic aneurysms are more common with prosthetic grafts but are uncommon with artery or vein grafts.

The stresses at the anastomosis are also called *discontinuity stresses* (see Chapters 4 and 5) and they are significantly higher than the stresses in the main vessels and they occur in a localized region of anastomosis. They can be minimized but not avoided. They are usually responsible for causing failure of the pressure vessels and in this case for causing anastomotic aneurysms.

It may be noted that the increase in stress and strain (or compliance) in the artery at the anastomosis does not tell us anything about whether the artery will become weaker and dilate or whether the smooth muscle cells will proliferate and cause intimal hyperplasia. In fact, it is this factor that forms a link, a common force, which suggests that the same stimulus can produce either effect. It is this reasoning that explains why some anastomoses develop aneurysms while others develop intimal hyperplasia. Thus, mechanical stress and strain should be considered causative factors for both pathologies, aneurysm and intimal hyperplasia.

4. Anastomotic Intimal Hyperplasia

Intimal hyperplasia is at the core of the formation of atherosclerotic plaques. An anastomotic site is one of the well-recognized sites that has predilection for atherosclerosis. In fact, it is recognized that the most common reason for failure of prosthetic grafts, used as an arterial substitute, is the formation of intimal hyperplasia at the anastomosis. In case of vein grafts, as we saw in Chapter 11, the graft stenosis occurs at the anastomosis among other locations. In case of artery-to-artery anastomosis also the most probable site of intimal hyperplasia is at the anastomosis. However, the frequency of intimal hyperplasia is the highest when the anastomosis is between the graft made of prosthetic material and the artery. In the previous section, we have noted that the frequency of an anastomotic aneurysm also is the highest at the prosthetic graft-to-artery anastomosis. Therefore, the site of anastomosis becomes a common factor for occurrence of the two different pathologies, namely anastomotic intimal hyperplasia and anastomotic aneurysm. As we continue to explore along these lines, it becomes apparent that the compliance mismatch could be considered responsible for the anastomotic intimal hyperplasia also, as it was proposed to be the cause of anastomotic aneurysms. With this in mind, we will consider reports that address the role of compliance mismatch in anastomotic intimal hyperplasia. It needs to be pointed out that the subject of anastomotic intimal hyperplasia has been studied extensively from a clinical perspective. However, the parallel studies addressing the role of vascular mechanics in the anastomotic intimal hyperplasia still remain to be carried out.

4.1. Vein-to-Artery, End-to-Side Anastomosis

Trubel et al. (4) studied the formation of anastomotic intimal hyperplasia at the vein graft-to-artery, end-to-side, anastomosis, and they studied the effect of compliance on the intimal hyperplasia by changing the compliance of the vein graft with the application of external Dacron mesh tube. In sheep, the reversed femoral vein was implanted as a femoro-popliteal bypass graft by the end-to-side anastomosis, and the original femoral arteries were ligated. The graft preparations were divided into four groups. In groups 1 and 3, native venous grafts were implanted while in groups 2 and 4, the venous grafts were first inserted into tubes made of Dacron mesh fabric and then implanted (Fig. 13.5). The Dacron mesh tube, covering the vein, was included in the suture lines at the proximal and distal anastomosis.

Because the veins (≈8 mm) are generally larger than the arteries, in groups 1 and 2 the diameters of the grafts were approximately twice those of the host arteries. In groups 3 and 4, the bypass graft diameters were adapted to the host artery diameter (4 mm) as follows: In group 3, the natural venous graft lumen was adapted to the host arterial lumen by transversely placed single stitches. In group 4, the venous graft was inserted into a mesh tube of 4 mm inner diameter. The details of the technique can be found in Ref. 4.

The blood flow was measured in the native femoral artery before its ligation and in bypass grafts after implantation. The elastic properties of the grafts, of the distal anastomotic regions, and of the host arteries near anastomosis were evaluated using sonographic crystals. The crystals were placed on the external surface to determine the diameter change in the vessel. The arterial pressure was also recorded. Figure 13.6 shows various locations for both the diameter measurements and determination of intimal hyperplasia. Wall elasticity is determined by the following equation:

$$\text{Wall elasticity} = \frac{\Delta d}{d\,(Psys - Pdia)}$$

where d is the diastolic diameter given by the crystals.

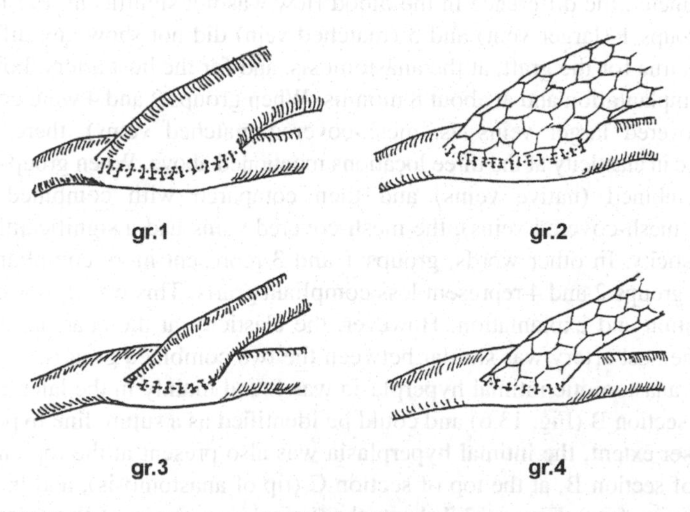

FIGURE 13.5. Reconstructions were divided into four groups consisting of native and mesh-constricted venous grafts with natural and adapted graft lumen. (Reproduced from Trubel W et al., ASAIO J 1994;40:273-278, with permission.)

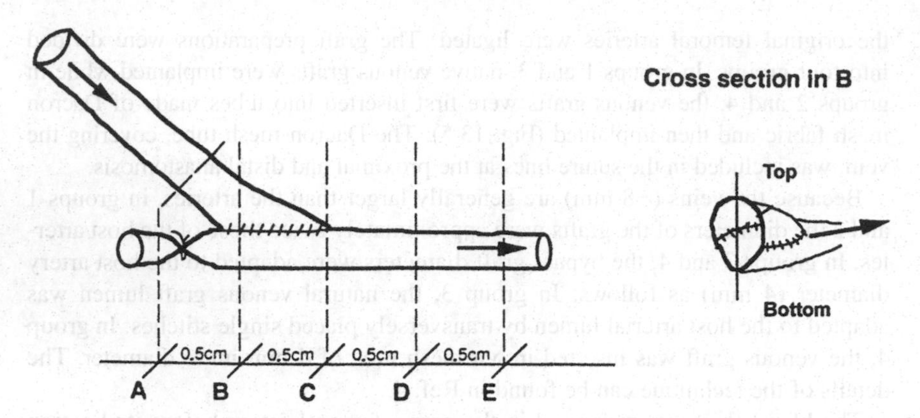

FIGURE 13.6. Cross section of the distal bypass anastomosis for histologic examination of distal anastomotic intimal hyperplasia. (Reproduced from Trubel W et al., ASAIO J 1994;40:273-278, with permission.)

After surgery, the animals recovered and the grafts were left in place for about 8 months. The measurements of diameter and flow were repeated just prior to sacrificing the animals. After sacrificing the animals, the graft and the host artery were fixed with 3% glutaraldehyde under 100 mmHg pressure and then removed. Cross sections were taken at various locations and the samples examined histologically and morphologically for intimal hyperplasia.

We will consider some of their findings below. The authors report that the blood flow was similar in different groups and that despite the difference in the graft diameter, the difference in the blood flow was not significant. For the elasticity, groups 1 (larger vein) and 3 (matched vein) did not show any difference. This was true for the graft, at the anastomosis, and for the host artery, both at the time of implantation and at about 8 months. When groups 2 and 4 were compared (mesh-covered larger veins vs. mesh-covered matched veins), there was no difference in elasticity at the three locations mentioned above. When groups 1 and 3 were combined (native veins) and then compared with combined groups 2 and 4 (mesh-covered veins), the mesh-covered veins had a significantly lower wall elasticity. In other words, groups 1 and 3 represent more compliant grafts whereas groups 2 and 4 represent less compliant grafts. This was the case at both implantation and explantation. However, the elasticity at the anastomosis itself and of the host artery was similar between the two combined groups.

Distal anastomotic intimal hyperplasia was found mainly in the lateral regions of cross section B (Fig. 13.6) and could be identified as a suture line hyperplasia. To a lesser extent, the intimal hyperplasia was also present at the top and at the bottom of section B, at the top of section C (tip of anastomosis), and in the section D (host artery). Figure 13.7 shows the typical appearance of anastomotic intimal hyperplasia. Clearly, a significant hyperplasia is seen at the suture line in the lateral portion of anastomosis. The appearance of anastomotic hyperplasia can be

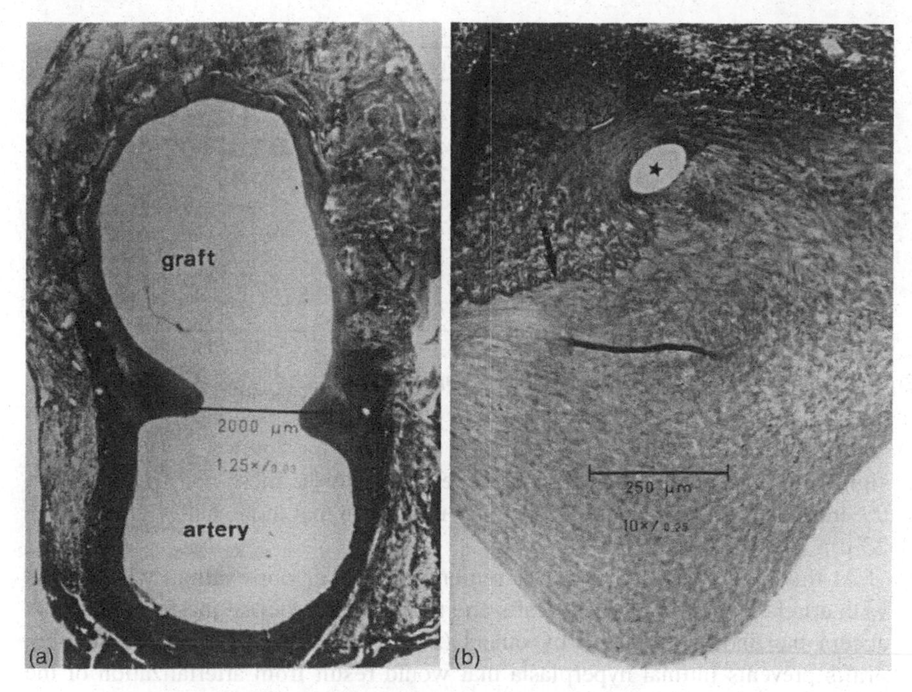

FIGURE 13.7. Cross section in the distal bypass anastomosis (in section B) of a mesh tube constricted venous graft with a lumen adapted to the host artery diameter (group 4). (a) Overview of typical narrowing suture line hyperplasia. (b) Detailed view of the border between the intima and media (lamina elastica interna [arrow]). Elastica – Van Gieson's stain; asterisk indicates suture line hole. (Reproduced from Trubel W et al., ASAIO J 1994;40:273-278, with permission.)

compared with the intimal hyperplasia in the artery at the branch. The overall thickening seen at the natural junction of the arterial branch (see Chapter 3) resembles the thickening produced by the intimal hyperplasia at the junction of the graft and the host artery. Thus, the role of vascular mechanics can be appreciated once again, for example, the wall needs to be thicker at the junction to counterbalance the stress-concentration at the junction, whether the junction is created by the natural process of arterial branching or by anastomosis. Hence, the intimal hyperplasia could be considered to occur in response to stress-concentration at the junction.

Table 13.2 (4) shows the extent and distribution of distal anastomotic intimal hyperplasia in all four groups and at various locations. In all groups, the greatest hyperplasia can be seen at the lateral wall (location B). The hyperplasia is less at location C (tip) and at location D in the host artery.

When the larger vein grafts (groups 1 and 2) were compared with the smaller grafts (groups 3 and 4), there was no difference in the intimal hyperplasia. However, when the mesh-constricted grafts (groups 2 and 4, less compliant) were compared with the native grafts (groups 1 and 3, more compliant), there was a

TABLE 13.2. Extent and localization of distal anastomotic intimal hyperplasia (μm).

Cross section	Group 1	Group 2	Group 3	Group 4
B (0.5 cm before tip)	55.9 ± 45.5	103.1 ± 41.3	105.3 ± 28.9	176.2 ± 106.3
C (anastomotic tip)	27.3 ± 12.2	80.5 ± 54.4	16.6 ± 12.6	44.2 ± 25.2
D (0.5 cm after tip)	6.7 ± 7.8	47.4 ± 34.5	<5 ± 0	30.2 ± 22.1
E (1 cm after tip)	7.6 ± 1.3	6.7 ± 4.7	<5 ± 0	6.7 ± 7.8
Top	5.8 ± 5.8	5.3 ± 5.8	<5 ± 0	22.1 ± 11.8
Right lateral	25.4 ± 11.3	75.7 ± 18.4	32.9 ± 9.4	85.1 ± 48.7
Bottom	17.8 ± 18.7	42.7 ± 17.7	24.7 ± 27.5	34.6 ± 25.8
Left lateral	23.5 ± 13.9	65.2 ± 24	39.9 ± 7.4	63.9 ± 28.2
Total (mean B-E)	18.1 ± 10.3	46.7 ± 18.5	24.4 ± 6.6	51.4 ± 28.5

Groups 1, 2, larger veins; groups 3, 4, smaller veins.
Groups 1, 3, more compliant; groups 2, 4, less compliant.

significant difference in the intimal hyperplasia. In the mesh-constricted grafts, the intimal hyperplasia was significantly more than that in the native grafts (49 ± 23 μm vs. 21 ± 9 μm).

At this point, it is important to remember another observation, which deals with arterialization of the vein grafts and the effect of external mesh. The authors accept the findings reported by others (5–8) that external support of the vein grafts prevents intimal hyperplasia that would result from arterialization of the grafts, whereas the same support increases the hyperplasia at the anastomosis. They state that the increased hyperplasia is most likely due to compliance mismatch at the anastomosis. Therefore, they suggest that the external support of the vein grafts should not be brought too close to the distal anastomosis and this way small untreated segments of the vein can act as a cushion between the mesh graft and the host artery.

We may consider briefly the mechanics involved here. The external support of the vein graft relieves stresses in the wall of the vein graft and thus reduces intimal hyperplasia. The same external support of the graft causes increased stresses in the host artery because it is the artery that has to stretch more to maintain the continuity at the junction. There is no relief of the stress in the artery at the junction because there is no support to the artery at the junction. Instead, there is an enhancement of stress on the arterial side as discussed earlier in Section 2. Thus, we may conclude that the increased stress-concentration at the junction, due to the graft being noncompliant while the host artery is compliant, is once again the most likely cause of the increased intimal hyperplasia at the anastomosis.

4.2. Effect of Graft Compliance Mismatch on Arterial Intimal Hyperplasia at the Anastomosis

To further explore the effect of graft compliance on the intimal hyperplasia at anastomosis, Wu et al. (9) performed experiments in the canine model that are quite interesting. These experiments not only deal with the compliance mismatch but they incorporate additional aspects, which offer further insight into the pathogenesis of intimal hyperplasia and the role of vascular mechanics. The experiments fall into three distinct categories:

1. Dacron prosthetic grafts attached to the artery by end-to-end anastomosis.
2. Dacron prosthetic grafts attached to the artery by end-to-side anastomosis.
3. Autogenous artery used as a graft and attached by end-to-end anastomosis.

These experiments allow us to compare the Dacron graft with the native artery graft and to compare end-to-end anastomosis with end-to-side anastomosis. Wu et al. (9) used dogs with low thrombotic potential to be able to study Dacron grafts. They used externally supported EXS Dacron grafts (Bard Vascular Systems, C.R. Bard, Inc.). For the native arterial grafts, they used an autogenous carotid artery in the femoral artery position. Compliance of both types of grafts was measured in vitro. The Dacron graft is almost noncompliant. For the carotid artery, they measured the compliance while keeping the length of the artery the same as that in vivo. They also measured the compliance of the femoral artery. They expressed the results as the compliance ratio of graft-to-host artery. In dogs (21–38 kg), the grafts were implanted as follows. A 4.5-cm length of the right carotid artery was removed. A 6-cm length of 5-mm-diameter Dacron graft was implanted in the right carotid artery position by end-to-end anastomosis. Another 6-cm length of Dacron graft was implanted in the left carotid artery position but with the end-to-side anastomosis technique. The left carotid artery was ligated to allow flow only through the graft. The left femoral artery was freed and a 4-cm length of the previously removed carotid artery was implanted in the femoral artery position by end-to-end anastomosis. Arteriograms were performed periodically to check all the grafts. The dogs were sacrificed, some at 6 months and some at 11 months. The grafts were flushed and fixed and tissue samples processed for scanning electron microscopy, transmission electron microscopy, immunohistochemistry, and routine histology to determine intimal thickness in the host artery at the anastomosis. Tissue samples were taken to include about 8-mm length of the host artery from anastomosis. Other details can be found in Ref. 9.

Their findings are as follows. The flow in the carotid artery in the end-to-end anastomosis group was 143 ± 32 mL/m and the femoral artery it was 118 ± 24 mL/m. In both the noncompliant graft (Dacron) and the compliant graft (carotid artery), there was very little intimal thickening of the host artery, and most of it was within 2 mm of the anastomosis. Table 13.3 shows the intimal thickness in the three groups (9). At the Dacron graft–host artery junction, the thickness ranged from 40 to 96 μm. The thickness decreased in a slope-like manner away from the anastomosis. Histologically, the neointima consisted of multiple layers of fibroblasts and smooth muscle cells covered by endothelial cells. Some of their data is shown in Table 13.4. The authors state that there was no significant difference in the intimal thickness between the noncompliant group and the compliant group. However, there was a significant difference between the noncompliant end-to-end versus end-to-side groups. The authors further state that the limited thickening is not caused by the compliance mismatch but may be associated with tissue response to surgical trauma and "self-smoothing" of the anasto-

TABLE 13.3. Thickness (μm) of the host arterial intima at the anastomotic interface 1 year after implant

Study group	Distance from anastomotic suture line (mm)						
	0	0.25	0.5	0.75	1	2	3
Noncompliant, end-to-end, proximal	53.2 ± 15.7	22.7 ± 15.7	0.6 ± 2.4	0	0	0	0
Noncompliant, end-to-end, distal	50.6 ± 24.8	17.1 ± 25.0	9.6 ± 22.7	8.4 ± 17.8	1.9 ± 7.2	0	0
Noncompliant, end-to-side, proximal	47.1 ± 30.0	28.5 ± 20.0	18.6 ± 17.2	13.6 ± 16.7	9.6 ± 14.6	11 ± 4.5	
Noncompliant, end-to-side, distal	71.4 ± 37.9	57.1 ± 35.2	53.4 ± 38.5	32.8 ± 32.6	21.4 ± 28.2	15 ± 4.2	0
Compliant control, proximal	51.0 ± 19.1	36.0 ± 23.7	19.3 ± 21.0	16.1 ± 20.4	5.1 ± 10.9	0	0
Compliant control, distal	87.1 ± 50.6	67.0 ± 66.7	57.0 ± 78.6	35.4 ± 61.1	12.4 ± 30.0	8.4 ± 33.7	2.8 ± 112

TABLE 13.4. Average thickness (μm) of the host arterial intima within 5 mm of anastomosis.

	Proximal	Distal
Noncompliant end-to-end	8.5 ± 18.4	9.6 ± 16.3
Noncompliant end-to-side	13.2 ± 16.1	26.4 ± 28.4
Compliant artery-to-artery, end-to-end	14.2 ± 17.4	30.1 ± 32.9

Implant period 1 year.

motic junction. They also state that the difference between the end-to-end and end-to-side noncompliant groups may be because the anastomosis in the end-to-side group is not as smooth and there may be more complex flow patterns in this group.

Let us reexamine these results from the vascular mechanics point of view. In Dacron grafts, the intimal thickening is greater in the end-to-side group than in the end-to-end group (Table 13.4). In case of end-to-end anastomosis, intimal thickening is greater for the autogenous arterial grafts than for Dacron grafts (Table 13.4). Could these results be explained on the basis of vascular mechanics?

Let us consider end-to-end versus end-to-side anastomosis first. We saw earlier in this chapter that the stress-concentration and strains are much greater in the end-to-side anastomosis than in the end-to-end anastomosis. Therefore, these results are consistent with the expectations. Now, let us consider end-to-end anastomosis in case of Dacron grafts versus arterial grafts. It is possible that Dacron grafts will produce larger stresses and strains at the anastomosis in the host artery compared with the arterial grafts. However, the neointima is likely to be formed with the cellular contributions from both the graft and the host artery, wherever possible. In case of Dacron grafts, only the host artery can contribute to neointima, whereas in case of arterial grafts both the graft and the host artery can contribute. Thus, this double contribution in case of artery-to-artery anastomosis could be important in the final outcome. Also, as stated earlier, at anastomosis there are other surgical details. For instance, a continuous suture produces an equivalent of a noncompliant "ring" at the anastomosis. Thus, for either the host artery or the graft artery, the anastomotic end is noncompliant. In fact, in this scenario, both the host artery and the graft artery may respond as if each one is attached to a noncompliant graft. In other words, the response can be twice as high as that in case of the artery-to-Dacron graft. The important points here are that the details of the mechanics at the anastomosis still remain to be established and that the final outcome must include possible contribution of the graft to the formation of neointima.

We may further speculate that when the vein or artery is used as a graft, then both the graft and the host may contribute to the neointima, and in this case, intimal hyperplasia has a greater probability of occurrence. On the other hand, when the graft is made of synthetic material, then only the host artery can contribute to the formation of neointima; and if this response is insufficient, then cellular apoptosis may ensue resulting in anastomotic aneurysm. Thus, the stress/strain concentration as a stimulus factor at the site of anastomosis seems to offer a cohesive explanation for the occurrence of either of the two pathologies—the anastomotic aneurysm or the anastomotic intimal hyperplasia.

4.3. *Suture Line Stresses at the Anastomosis: Effect of Compliance Mismatch*

In the previous section on anastomotic aneurysms, we have covered mechanical stresses at anastomosis in a purely qualitative sense because much work remains to be done in this area. In this section, we will describe the determination of mechanical stresses at anastomosis, carried out by Ballyk et al. (10) in relation to the anastomotic intimal hyperplasia. Although the work by Ballyk et al. represents a step forward, we will see that much remains to be done in the determination of mechanical stresses at the anastomosis. Particularly, analytical studies combined with experimental observations are needed. Ballyk et al. analyzed the end-to-end and end-to-side anastomoses in the analytical model in which they changed the compliance of the graft to explore how the compliance mismatch might influence the stresses. They further focused on the stress in the tissue due to the presence of sutures.

4.3.1. Material Properties and the Model Geometry

They used a large-strain thick-shell numerical model. The large strain properties of the vessels are characterized by the following strain energy density function *W*:

$$\rho_o W = \frac{1}{2} a I_E^2 - b II_E$$

where *a* and *b* are material constants with units of stress, ρ_o is the density before deformation, and I_E and II_E are the first and second Green's strain invariants defined in terms of the Green's strain tensor E_{ij}:

$$I_E = E_{ii}$$

$$II_E = \frac{1}{2}(E_{ii} E_{jj} - E_{ij} E_{ij}).$$

The components of the second Piola-Kirchhoff stress tensor S_{ij} are obtained using

$$S_{ij} = \frac{\partial(\rho_o W(E_{ij}))}{\partial E_{ij}}$$

which is related to the Cauchy stress tensor σ_{kl} through the deformation gradient tensor F_{ki}

$$\sigma_{kl} = \frac{\rho}{\rho_o} F_{ki} S_{ij} F_{ij}$$

where $\rho_o/\rho = \det[F_{kl}]$ is the density ratio between the underformed and deformed geometries. Different values for the constants *a* and *b* can be chosen to represent different stiffness of the graft material. The stress distributions at end-to-end and end-to-side graft-artery anastomoses were calculated for grafts of three different compliances: 1) artery (compliant), 2) vein (less compliant), and 3) Dacron (stiff). Case 1 represents perfect graft-artery compliance matching, while cases 2 and 3 represent different degrees of compliance mismatch. A simple discontinuous suture model was used, which assumes that the graft and the artery are attached at discrete locations around their junction. Thus, this model considers that the anastomosis is accomplished by using interrupted sutures. The sutures were placed approximately 1.0 mm apart in the end-to-end simulation and 1.5 mm apart in the end-to-side simulations.

The angle between the graft and artery in the end-to-side geometry was set to 45 degrees. Quarter symmetry was used in the end-to-end simulations while a plane of symmetry was used in the end-to-side simulations (Fig. 13.8). The same finite-element meshes and suture locations were used for each graft material in the two graft-artery geometries.

In the end-to-end cases, the free ends of the graft and artery were subjected to enough axial prestress to cause 15% axial elongation of the artery. In the end-to-side simulations, this prestress was only applied to the host artery. These prestress boundary conditions were chosen to simulate the axial tension (tethering force), which blood vessels experience in vivo. In all cases, the graft-artery configuration was inflated to 13.3 kPa (100 mmHg).

Vessels with a mean diameter of 5.0 mm and a thickness-to-diameter of ratio 1/10 were modeled to approximate the dimensions of the human superficial

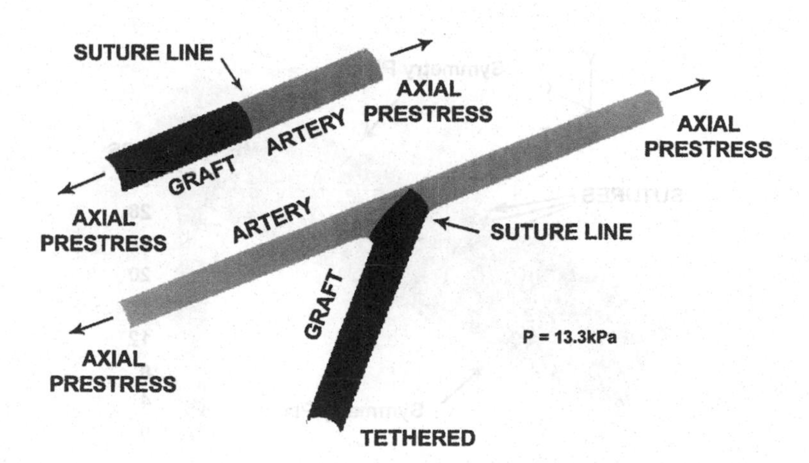

FIGURE 13.8. Schematic showing the computational domains for the end-to-end and end-to-side junctions. Quarter symmetry was used in the end-to-end simulations while a plane of symmetry was used in the end-to-side simulations. In the end-to-end cases, the free ends of the graft and artery were subjected to enough axial prestress to cause 15% axial elongation of the artery. In the end-to-side simulations, this prestress was only applied to the host artery. In all cases, the graft-artery configuration was statically inflated to 13.3 kPa (100 mmHg). (Reproduced from Ballyk PD et al., J Biomechanics 1998;31:229-237, with permission from Elsevier.)

femoral artery. The material constants *a* and *b* in the strain energy density function were chosen such that calculated compliances would match measured values for the various vessels used. The values of both *a* and *b* are higher for the less compliant vessels. More details on the model can be found in Ref. 10.

4.3.2. Model Results

In the end-to-end Dacron graft–artery anastomosis, stresses are concentrated at the suture attachment points, with slightly reduced stresses occurring between the sutures (Fig. 13.9). Away from the anastomosis, the mean stress is constant at about 10 times the inflating pressure. Each suture around the anastomosis experiences the same stress concentration, which is approximately eight times larger than stresses along the distal host artery.

In the end-to-side Dacron graft–artery anastomosis, stresses are again concentrated at the sutures, however, the stress concentration at each suture is not the same (Figs. 13.10 and 13.11). The sutures near the toe of the anastomosis experience the greatest stresses, and large stress gradients are seen around the suture line. The toe is also the location of the largest stress difference between the graft materials, with the stiffer grafts yielding greater stresses. Away from the anastomosis and along the host artery, the mean stress is constant at about 10 times the inflating pressure. Along the graft, the mean stress is constant at about 3 times the inflating pressure. The stress-concentration at the sutures is 3 to 36 times larger than the stresses along the distal host artery. Near the heel and sidewall of the

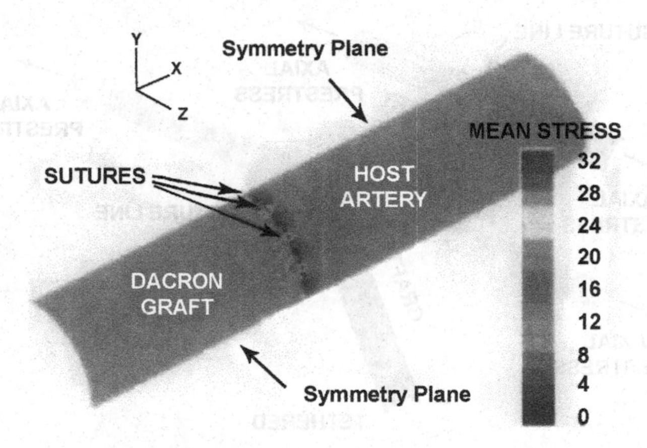

FIGURE 13.9. Mean stress distribution at the end-to-end Dacron graft–artery anastomosis projected onto a three-dimensional image of the geometry. Stress values are normalized by the inflating pressure and only one-quarter of the geometry is shown due to symmetry considerations. Stresses are concentrated equally at each suture attachment point, and the suture-induced stress concentrations are approximately eight times larger than stress values along the distal host artery. Similar results were obtained for the artery and vein grafts in this geometry. (Reproduced from Ballyk PD et al., J Biomechanics 1998;31:229-237, with permission from Elsevier.) (*Please see color version on CD-ROM.*)

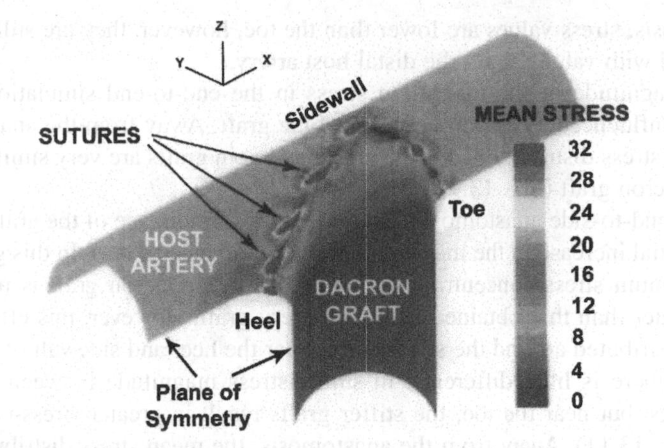

FIGURE 13.10. Mean stress distribution at the distal end-to-side Dacron graft–artery anastomosis projected onto a three-dimensional image of the geometry. Stress values are normalized by the inflating pressure and only one-half of the geometry is shown due to symmetry considerations. Stresses are concentrated at each suture attachment point, however, the stress-concentration at each suture is not the same. The suture-induced stress-concentrations range from 3 to 36 times the stress values along the distal host artery. Similar analyses were performed using artery and vein grafts in this geometry. (Reproduced from Ballyk PD et al., J Biomechanics 1998;31:229-237, with permission from Elsevier.) (*Please see color version on CD-ROM.*)

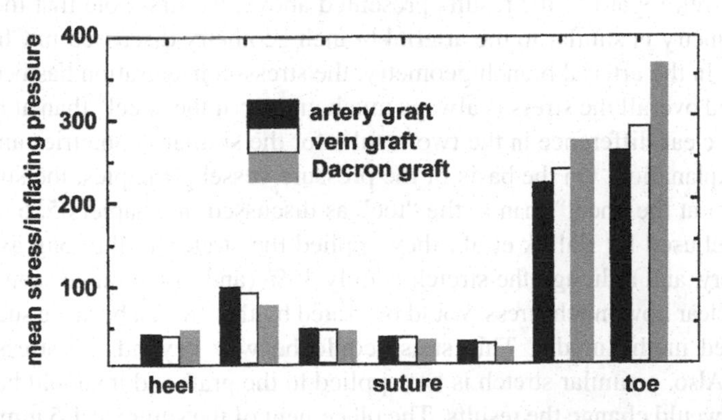

FIGURE 13.11. Plot showing the mean stress at each suture in the end-to-side simulations for each of the three graft materials simulated. Stress values are normalized by the inflating pressure. The heel and toe sutures are identified and the intermediate sutures follow sequentially from left to right (see Fig. 13.10). For all three types of graft material, the maximum stresses occur at the sutures near the toe. (Reproduced from Ballyk PD et al., J Biomechanics 1998;31:229-237, with permission from Elsevier.)

anastomosis, stress values are lower than the toe, however, they are still elevated compared with values along the distal host artery.

The magnitude of the maximum stress in the end-to-end simulations is not strongly influenced by the compliance of the graft. Away from the anastomosis, the mean stress distributions for the artery and vein grafts are very similar to that in the Dacron graft (Fig. 13.9).

In the end-to-side anastomosis, decreasing the compliance of the graft leads to a substantial increase in the maximum anastomotic mean stress. In this geometry, the maximum stress-concentration obtained using a Dacron graft is more than 40% greater than that obtained using an artery graft. However, this effect is not evenly distributed around the suture line. Near the heel and sidewall of the anastomosis, there is little difference in suture-stress magnitude between the three graft types, but near the toe, the stiffer grafts result in greater stress-concentrations (Fig. 13.11). Away from the anastomosis, the mean stress distributions for the artery and vein grafts are again very similar to the Dacron case (Fig. 13.10).

4.3.3. Comments

The above analysis represents an excellent approach to the determination of stresses in the tissue when sutures are present. However, let us consider why more work is needed in this area. First, at the anastomosis, interrupted sutures are rarely used. The continuous sutures, used normally, present a very complex scenario where not only the increase in the circumference of the suture line is largely prevented but any increase in the circumference tends to shear the tissue included in the suture line. There is a substantial increase in the thickness at the anastomosis because the walls of the graft and the host artery overlap, as they are included in the suture line.

Now, with regard to the results presented above, we first note that the end-to-side geometry is similar to the arterial branch geometry discussed in Chapters 5, 6, and 7. In the arterial branch geometry, the stress-concentration has been established and overall the stress is always much greater at the "heel" than at the "toe." This is a clear difference in the two results for the similar geometries and it calls for an explanation. On the basis of the pressure vessel principles, the stress must be greater at the "heel" than at the "toe" as discussed in Chapters 5, 6, and 7. In the model used by Ballyk et al., they applied the stretch in the long axis of the host artery, and although the stretch is only 15% (and in the range seen in vivo), it is not clear how much stress would be crated by this stretch because such details are buried in the model. This stress could be way beyond the stress present in vivo. Also, a similar stretch is not applied to the graft and it should have been, and that would change the results. The placement of the suture at 1.5 mm distance along the anastomosis reveals that, due to the angle of the graft, the suture at the "toe" has to endure longitudinal stress from a larger amount of tissue than the suture at the "heel," because the larger amount of circumference of the host artery is associated with the suture at the "toe" than with the suture at the "heel." In other words, a greater amount of longitudinal stress would be imposed on the suture at

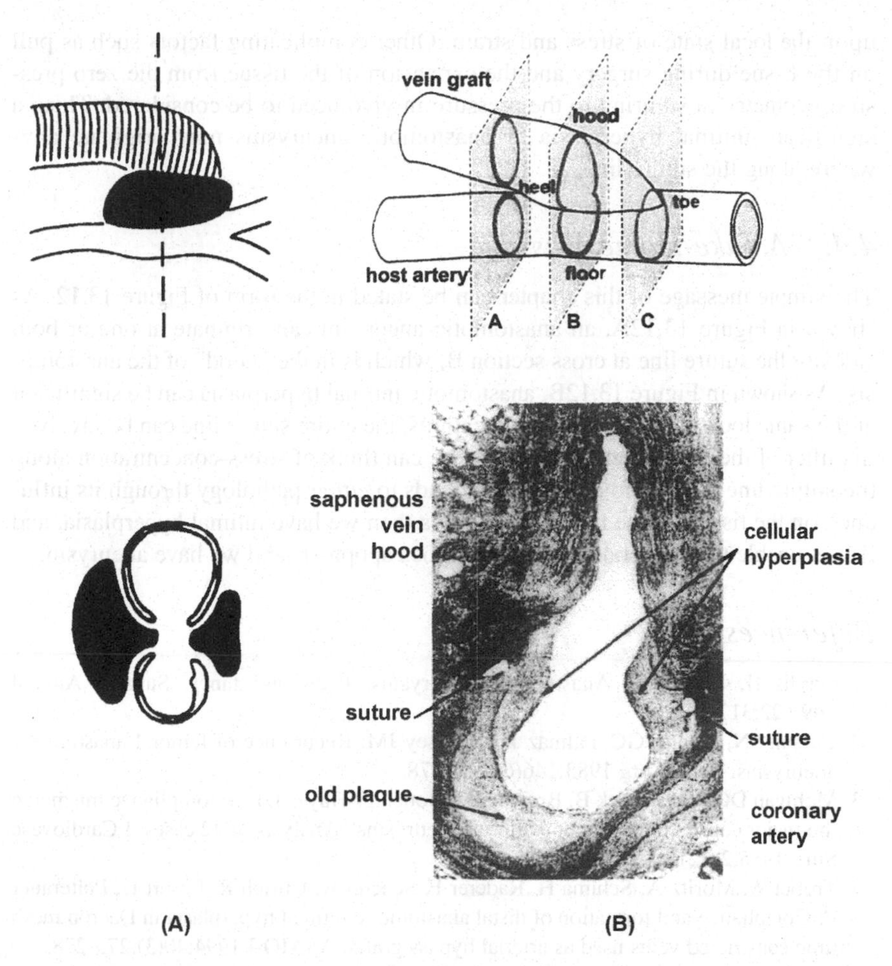

FIGURE 13.12. (A) Diagrammatic representation of one type of femoral anastomotic aneurysm. (Reproduced from Carson SN et al., Am J Surg 1983;146:774-778, modified, with permission from Excerpta Medica Inc.). (B) A 5-week-old human aortocoronary saphenous vein graft. This transverse section was taken in the midregion of the distal anastomosis (slice "B"). Preliminary studies indicate that in early grafts, cellular hyperplasia develops primarily at the suture sites. (Reproduced from Ballyk PD et al., J Biomechanics 1998;31:229-237, with permission.) Note that the sites of anastomotic aneurysm (A) and anastomotic intimal hyperplasia (B) are identical.

the "toe" than on the suture at the "heel." Overall, the suture placement and the modeling parameters also influence the difference in the results between the end-to-end anastomosis geometry and the end-to-side geometry. Given these data and the lack of detailed studies, it is prudent to conclude that the stresses are considerably enhanced all along the suture line and that a significant tissue response could occur anywhere along the suture line. The responses could vary depending upon the local state of stress and strain. Other complicating factors such as pull on the tissue during surgery and the expansion of the tissue from the zero pressure geometry at suturing to the pressure in vivo need to be considered. Thus, a significant intimal hyperplasia or anastomotic aneurysms may originate anywhere along the suture line.

4.4. *A Take-Home Message*

The simple message of this chapter can be stated in the form of Figure 13.12. As shown in Figure 13.12A, an anastomotic aneurysm can originate at one or both sides of the suture line at cross section B, which is in the "hood" of the anastomosis. As shown in Figure 13.12B, anastomotic intimal hyperplasia can be significant at the same location. Besides these locations, the entire suture line can be involved in either of the two pathologies. Thus, we can think of stress-concentration along the suture line as a common cause that leads to *either* pathology through its influence on the tissue. If the tissue proliferates then we have intimal hyperplasia, and if it is unable to proliferate then it undergoes apoptosis and we have aneurysm.

References

1. Gaylis H, Dewar G: Anastomotic aneurysms: Facts and fancy. Surgery Annual 1990;22:317-341.
2. Carson SN, Hunter GC, Palmaz J, Guernsey JM: Recurrence of femoral anastomotic aneurysms. Am J Surg 1983;146(6):774-778.
3. Mehigan DG, Fitzpatrick B, Browne HI, Bouchier-Hayes DJ: Is compliance mismatch the major cause of anastomotic arterial aneurysms? Analysis of 42 cases. J Cardiovasc Surg 1985;26(2):147-150.
4. Trubel W, Moritz A, Schima H, Raderer F, Scherer R, Ullrich R, Losert U, Polterauer P: Compliance and formation of distal anastomotic intimal hyperplasia in Dacron mesh tube constricted veins used as arterial bypass grafts. ASAIO J 1994;40(3):273-278.
5. Barra JA, Volant A, Leroy JP, et al.: Constrictive perivenous mesh prosthesis for preservation of vein integrity. J Thorac Cardiovasc Surg 1986;92:330-336.
6. Moritz A, Grabenwoeger F, Raderer F, et al.: Mesh tube-constricted varicose veins used as bypass grafts for infrainguinal arterial reconstruction. Arch Surg 1992;127: 416-420.
7. Karayannacos PE, Hostetler JR, Bond MG, et al.: Late failure in vein grafts: Mediating factors for subendothelial fibromuscular hyperplasia. Ann Surg 1978;187:183-188.
8. Kohler TR, Kirkman TR, Clowes AW: The effect of rigid external support on vein graft adaptation to the arterial circulation. J Vasc Surg 1989;9:277-285.
9. Wu MHD, Shi Q, Sauvage LR, Kaplan S, Hayashida N, Patel MD, Wechezak AR, Walker MW: The direct effect of graft compliance mismatch per se on development of host arterial intimal hyperplasia at the anastomotic interface. Ann Vasc Surg 1993;7(2):156-168.
10. Ballyk PD, Walsh C, Butany J, Ojha M: Compliance mismatch may promote graft-artery intimal hyperplasia by altering suture-line stresses. J Biomechanics 1998;31: 229-237.

14
Intracranial Aneurysms

1. Introduction

Intracranial aneurysms are also called cerebral aneurysms, saccular aneurysms, berry aneurysms, or vessel branching aneurysms (VBAs). As the names imply, these aneurysms have a sac and they look like berries due to their size and spherical shape. They almost always occur at the branches and bifurcations of the arteries. Intracranial aneurysms represent a vascular pathology where the vascular mechanics is the most applicable science and yet, ironically, it remains most understudied. As we develop the subject in this chapter, we will note that this field will remain fertile for new discoveries for quite some time, particularly in the application of vascular solid mechanics.

Intracranial aneurysms originate from junctional angles in the arteries, in and around the region of the circle of Willis. Once started, the aneurysm tends to enlarge slowly but relentlessly and it may become irregular in shape and may wrap itself around adjacent vessels. Thus, an aneurysm with a narrow neck may appear to have a broad origin from its parent vessel. The aneurysm may also become bilocular. Usually, the aneurysms may appear as bubbles erupting from the surface, and these "bubbles" are extremely thin-walled. They may produce clinical effects once they become 6–15 mm in diameter, however, some aneurysms may never produce effects. Some aneurysms will have thrombus in them and look much smaller on angiograms. Often, these aneurysms are associated with medial gaps in the vessel wall, atherosclerotic changes in the arteries, and hypertension. Because of our interest in the application of vascular mechanics to this pathology, we will examine these aneurysms for their anatomical, pathologic, and mechanical features.

2. The Anatomy

The vast majority of intracranial aneurysms are associated with the arteries that form the circle of Willis. The arterial anatomy leading up to the circle of Willis and beyond is rather complex. The blood supply to intracranial structures is through the carotid and vertebral arteries, the internal carotid artery being the dominant one. The anatomy can be best understood from the illustrations that follow. Figure 14.1 shows the common carotid artery (originating from the aortic arch), which gives rise to the internal carotid artery (ICA), the carotid siphon, and then bifurcation of the ICA into the middle cerebral artery (MCA) and the anterior cerebral artery (ACA). Figure 14.2 shows the vertebral artery (originating from the subclavian artery) and two of these joining together to form the basilar artery. Figure 14.3 shows the circle of Willis. The basilar artery bifurcates into the two posterior cerebral arteries and these are connected to the two internal carotid arteries via two posterior communicating arteries. The two anterior cerebral arteries are connected with each other by an arterial segment called the anterior communicating artery. The circle of Willis, thus complete, ensures a blood supply in all of the arteries of the circle, even if one of the internal carotid arteries or one of the vertebral arteries is completely blocked. The circle of Willis contains anterior cerebral artery, anterior communicating artery, middle cerebral artery, internal carotid artery, posterior communicating artery, posterior cerebral artery (PCA), and basilar artery (BA). The three-dimensional arrangement of arteries of the circle of Willis is quite complex as can be appreciated by examination of Figures 14.1 to 14.3. It is important to take note of this complex geometry because the geometry of the arteries dictates what type of mechanical forces can be produced by the blood flow and by the blood pressure. We will be considering such mechanical forces when we examine the pathogenesis of the diseases, such as aneurysms and atherosclerosis, in this vasculature. Figure 14.4 shows the dimensions of various arteries in the circle of Willis. Such basic information is vital for any type of engineering analysis. It should also be noted that in people, some variation in the anatomy of the circle of Willis has been observed.

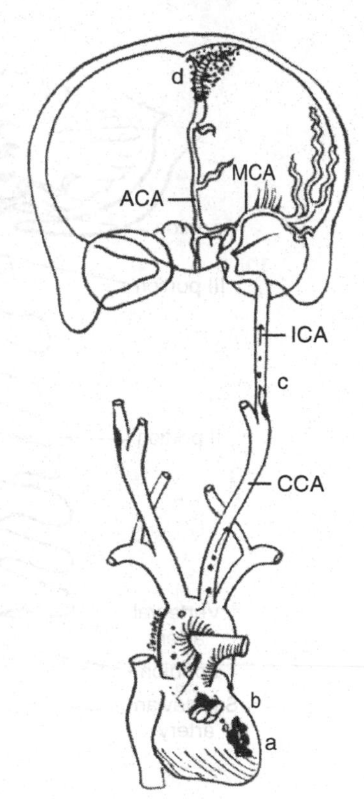

FIGURE 14.1. Anterior view depicting the aortic and cerebral circulation. CCA, common carotid artery; ICA, internal carotid artery; ACA, anterior cerebral artery; MCA, middle cerebral artery. (Figure also shows potential sources of embolism: a, cardiac thrombus; b, vegetation on heart valve; c, emboli from carotid plaque; d, infarcted cortex.) (Reproduced from Caplan LR, Stroke – A Clinical Approach, Second ed. Butterworth-Heinemann, Boston, 1993:27, with permission from Elsevier.)

3. Incidence

From various autopsy series, the incidence of occurrence of cerebral aneurysms in the general population is reported to be in the range of 0.3% to 5%. Fortunately, the incidence of rupture of these aneurysms is less, about 0.01% (1 in 10,000) in the population. Despite the low incidence of rupture, the high prevalence of aneurysms and the poor prognosis of patients with cerebral hemorrhage make the social and economic consequence of the disease tremendous. Most aneurysms are found in patients aged 40 to 70 years, with the peak incidence in ages of 50 to 60 years. Intracerebral hemorrhage due to rupture of these aneurysms is also most frequent in the age group of 40 to 60 years. In patients below the age of 40, aneurysms are more common in men, while in patients above the age of 40 aneurysms predominate in women (2:1). Multiple aneurysms occur in 25% to 30% of patients and bilateral aneurysms occur in about 17% of patients. Less than 10% of aneurysms occur below the age of 30 and about 2% occur under the age of 20. The aneurysms are rare in children during the first decade.

Most aneurysms arise at the major branches of the anterior portion of the circle of Willis (e.g., anterior communicating artery, anterior cerebral artery, middle cerebral artery, and internal carotid artery; Table 14.1). One group reported that, of

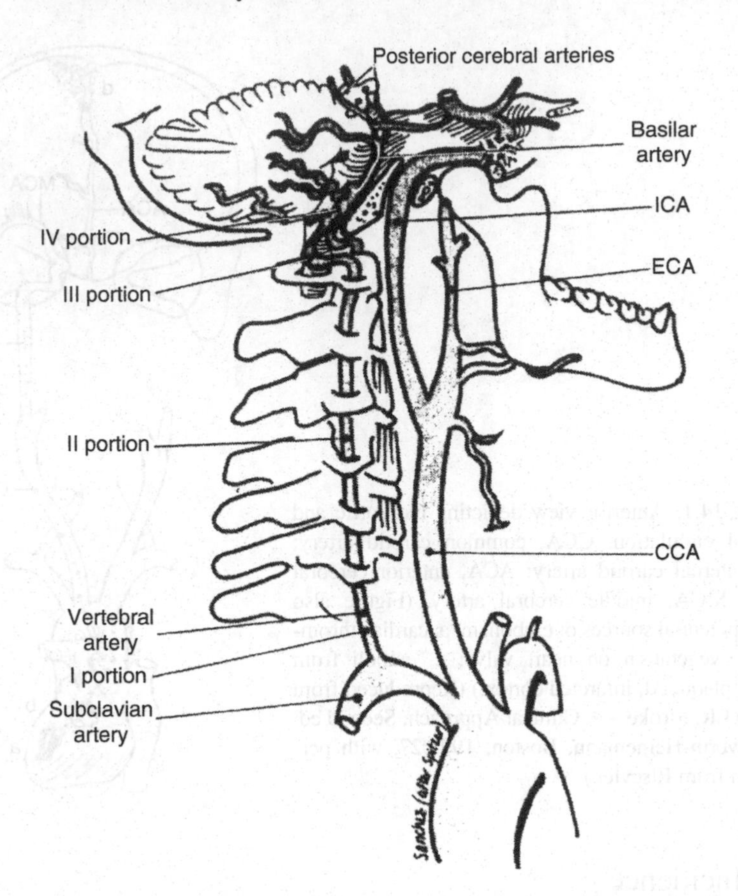

FIGURE 14.2. Lateral view depicting the aortic and cerebral circulation. ECA, external carotid artery. (Reproduced from Caplan LR, Stroke – A Clinical Approach, Second ed. Butterworth-Heinemann, Boston, 1993:35, with permission from Elsevier.)

the 1135 aneurysms examined, 87% were found in association with the anterior segment of the circle of Wills (including the internal carotid segment) while 13% were found in conjunction with the posterior or basilar-vertebral segment (1). Another group reported an even more dramatic distribution, 94% in the anterior circulation of the circle of Willis and 6% in the posterior circulation (2). Internal carotid artery aneurysms are more common in women than in men (2:1), and middle cerebral artery aneurysms are also slightly more common in women (3:2).

The distribution of aneurysms in children is considered to be different from that in adults. Some reports state that in children, the aneurysms involve the basilar artery and its branches more commonly while other reports state that the internal carotid bifurcation and the middle cerebral artery are more involved, but both reports agree on less involvement of the anterior cerebral arteries. In one series (1), the size of the ruptured aneurysm was reported to be in the range of

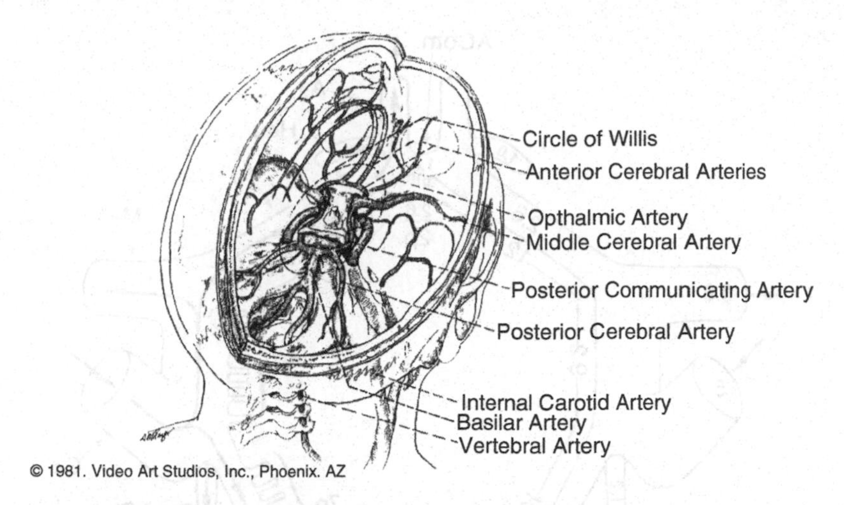

FIGURE 14.3. Intracranial cerebrovascular anatomy showing anastomotic connections of the circle of Willis. Note that the principal blood supply to intracranial structures is through the carotid arteries. (Reproduced from Zwiebel WJ, Introduction to Vascular Ultrasonography. W.B. Saunders Co., Philadelphia, 2000:106, with permission from Elsevier.)

3 to 28 mm (average 9.9 mm) while that of the unruptured aneurysm was in the range of 1 to 12 mm (average 3.9 mm).

4. Pathology

4.1. Aneurysms

Almost all of the aneurysms are located above the carotid siphon on the circle of Willis. Figure 14.5 shows the locations and the number of ruptured aneurysms in one series (3). Of the 143 ruptured aneurysms examined, 48 occurred at the junction of the anterior cerebral artery with the anterior communicating artery, 24 at the bifurcation of the middle cerebral artery, and 51 at the junction of internal carotid artery with the anterior cerebral artery, the middle cerebral artery, the posterior communicating artery, and the ophthalmic artery. In the posterior basilar-vertebral system, 14 ruptured aneurysms were noted. The sites involved include the apex of the basilar artery and the origins of the cerebral artery, the posterior communicating artery, the superior cerebellar artery, and the posterior inferior cerebellar artery. Additional sites of ruptured aneurysms are branching of the anterior cerebral artery and the origins of branches from the vertebral artery. These data establish once again that the vast majority of ruptured aneurysms involve the anterior portion of the circle of Willis, which include ACA, MCA, and ICA. The data also establish that the aneurysms occur *almost exclusively* at the origins of branches and bifurcations.

It is both impressive and intriguing to look at the images of the aneurysms. We will consider some examples. Figure 14.6 shows a large saccular aneurysm in the

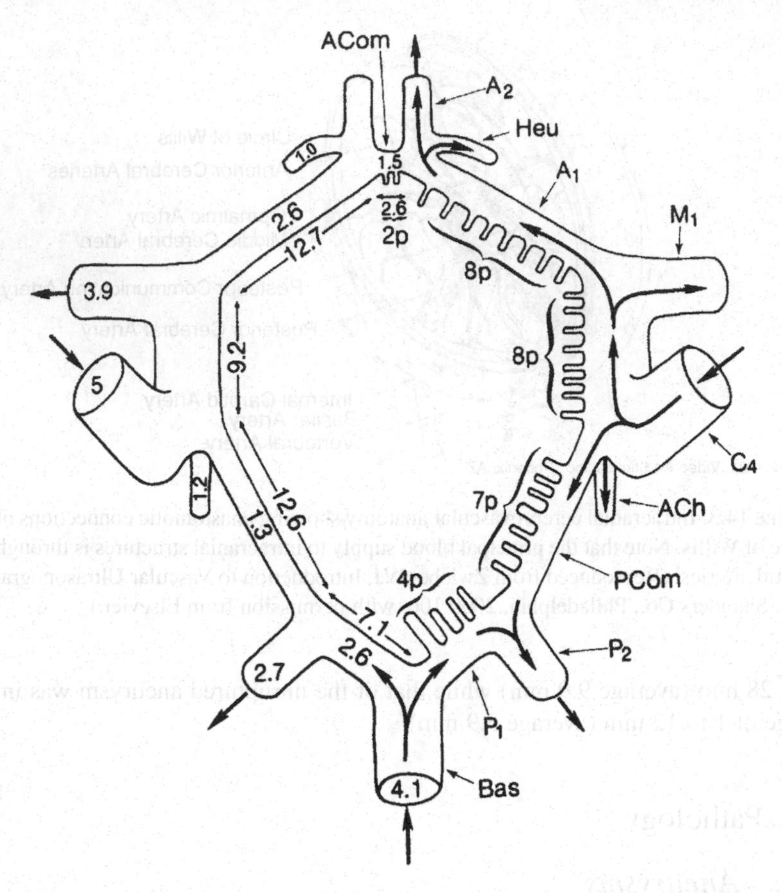

FIGURE 14.4. "Average" circle of Willis. The numbers inside the arteries indicate average diameter in millimeters. Numbers inside the circle equal mean length of segments (p, perforator). An idealized circle, combining data in various papers by A. L. Rhoton Jr. and his colleagues (viewed from below.) (Reproduced from Weir B, Aneurysms Affecting the Nervous System. Williams & Wilkins, Baltimore, 1987:311, with permission.)

TABLE 14.1. Distribution of aneurysms.

Arteries	Localized by angiogram (1) (%)	Postmortem incidence (1) (%)	Ruptured aneurysm (%) (1)	(2)
1. Anterior cerebral–anterior communicating	25	21	22	41
2. Internal carotids	45	40	48	19
3. Middle cerebral	26	23	22	34
4. Basilar-vertebral	4	16	8	6
Total	100	100	100	100

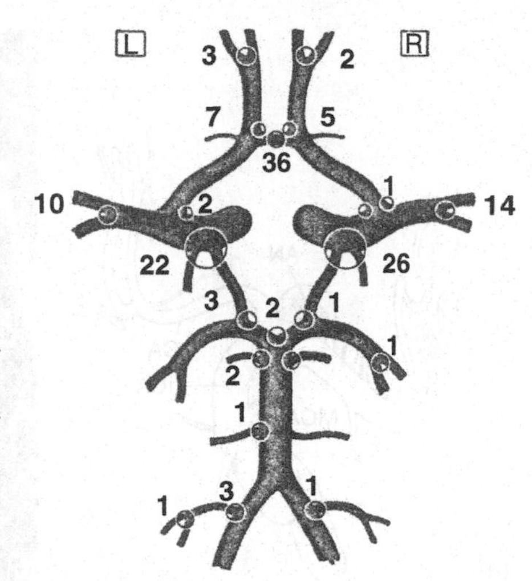

FIGURE 14.5. Locations and number of ruptured aneurysms in the circle of Willis. Note that the largest number (36) of aneurysms occurred at the anterior communicating artery. (Reproduced from Wilson G et al., J Neurosurg 1954;11: 128-134, with permission.)

circle of Willis at the junction of the ACA with the anterior communicating artery, one of the most frequent locations for aneurysm formation. In this figure, all of the arteries of the circle of Willis can be seen. Figure 14.7 is an angiogram demonstrating the aneurysm at the bifurcation of the ICA into the MCA and the ACA. One can clearly see the "berry"-like appearance of the aneurysm on the

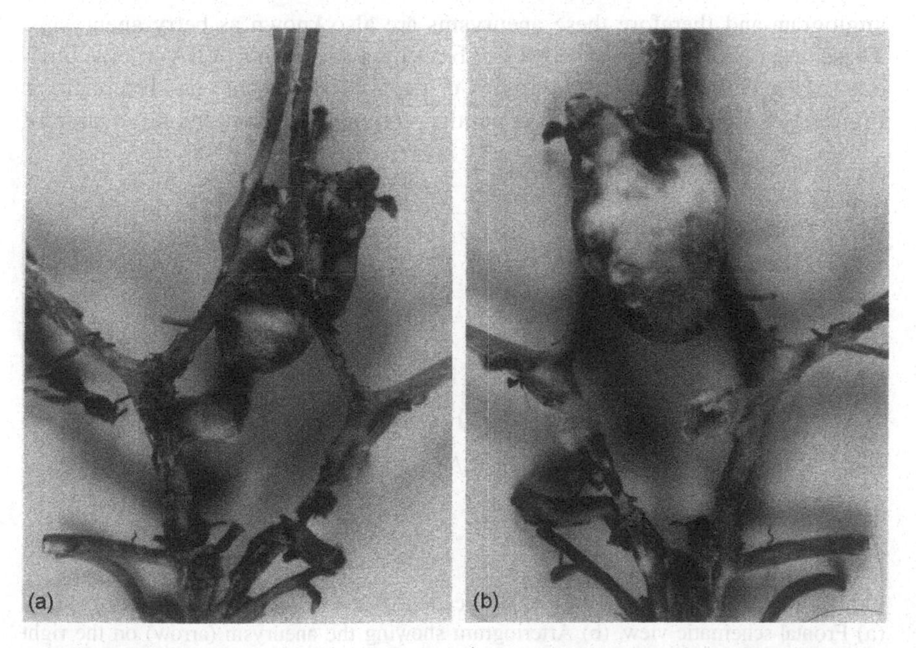

FIGURE 14.6. Photographs of the circle of Willis showing an aneurysm at the anterior communicating artery: (a) view from above, (b) view from below.

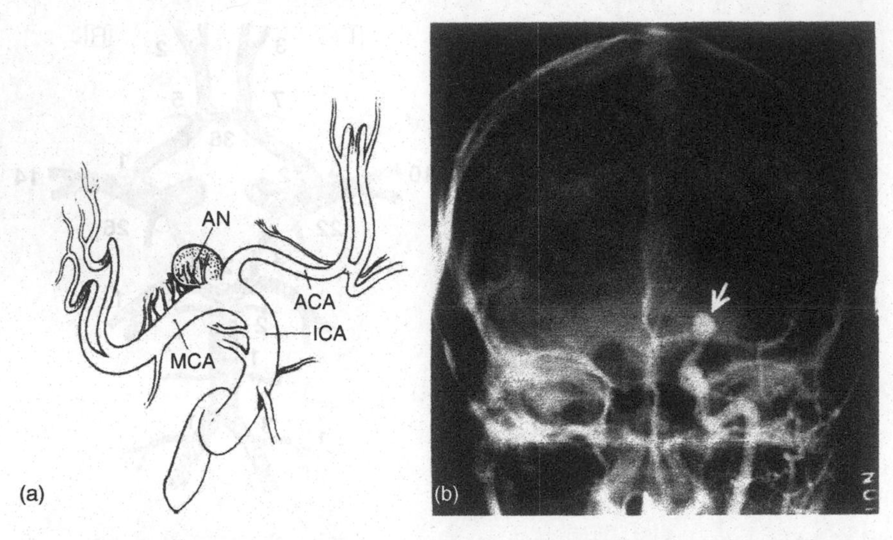

FIGURE 14.7. Aneurysm (AN) at the internal carotid artery (ICA) bifurcation. MCA, middle cerebral artery; ACA, anterior cerebral artery. (a) Schematic of the frontal view of the right side of the patient. (b) Arteriogram of the left side showing aneurysm (arrow). (Reproduced from Youmans JR, Neurological Surgery. W.B. Saunders Co., Philadelphia, 1996:1293, with permission from Elsevier.)

angiogram and therefore these aneurysms are also known as berry aneurysms. These aneurysms are also called vessel branching aneurysms (VBA) for the obvious reason as they occur at the origins of the arterial branches and bifurcations. Figure 14.8 shows an aneurysm at another common location, the bifurcation of

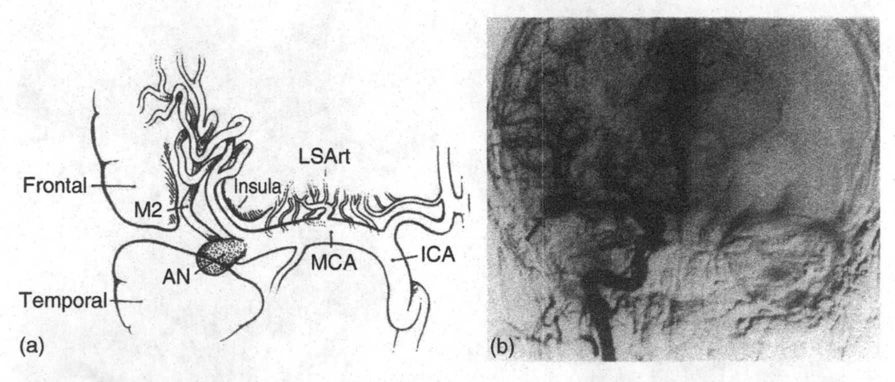

FIGURE 14.8. Aneurysm (AN) at the bifurcation of the middle cerebral artery (MCA). (a) Frontal schematic view. (b) Arteriogram showing the aneurysm (arrow) on the right side of the patient. (Reproduced from Youmans JR, Neurological Surgery. W.B. Saunders Co., Philadelphia, 1996:1295, with permission from Elsevier.)

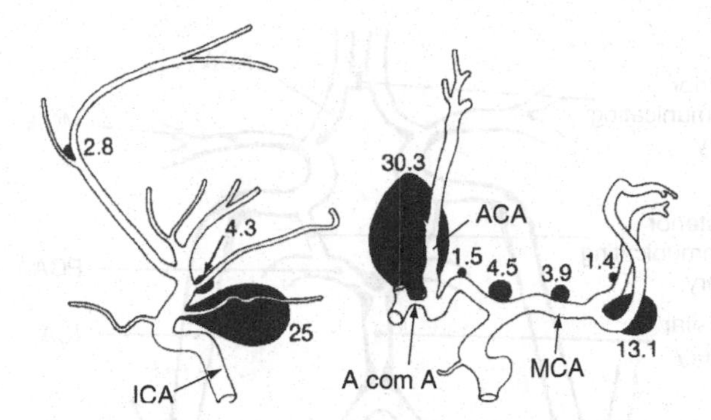

FIGURE 14.9. Relative frequencies of aneurysms at different sites. Largest frequency is at the anterior communicating artery (30.3). (Reproduced from Weir B, Aneurysms Affecting the Nervous System. Williams & Wilkins, Baltimore, 1987:345, with permission.)

the MCA. All of these *intracranial aneurysms* are also known as *saccular aneurysms* because they are in the form of a sac at the branch locations.

Figure 14.9 shows the distribution of single ruptured aneurysms producing *subarachnoid hemorrhage* (4). Once again, we note the largest percentage of aneurysms at the ACA/anterior communicating artery (30.3%), at the MCA bifurcation (13.1%), and at the posterior communicating artery/ICA (25%) (Figs. 14.3, 14.4). These three specific locations seem to be the most frequent sites for the occurrence of the aneurysms. In one of the studies, which analyzed 2630 aneurysms, the frequency and locations are reported to be ICA 41%, ACA/A Com A 34%, MCA 20%, basilar artery-vertibral artery 4%, and others 1% (compare with Table 14.1).

4.2. *Atherosclerosis*

We may remind ourselves of an interesting fact here in regard to aneurysms and atherosclerosis. In the intracranial arteries, aneurysms occur much more frequently than does atherosclerosis, while in the extracranial arteries of comparable size, atherosclerotic plaques occur quite frequently while aneurysms are rare. As we continue to discuss these two diseases, we note here that in some of the intracranial arteries, the locations of atherosclerotic plaques are similar to the locations of the aneurysms. This is an important observation and it is similar to that noted earlier in the case of anastomosis. Because aneurysms are more predominant in the intracranial vessels, we will consider this subject in greater detail a little later. For now, we examine the occurrence of atherosclerotic plaques in the intracranial arteries and pay particular attention to the locations of these plaques.

Figure 14.10 shows the locations where atherosclerotic plaques form frequently. In the internal carotid artery, the plaques are localized at the origins of the anterior cerebral artery and the middle cerebral artery. These are also the locations for aneurysm formation. The plaques are located at the apex of the basilar

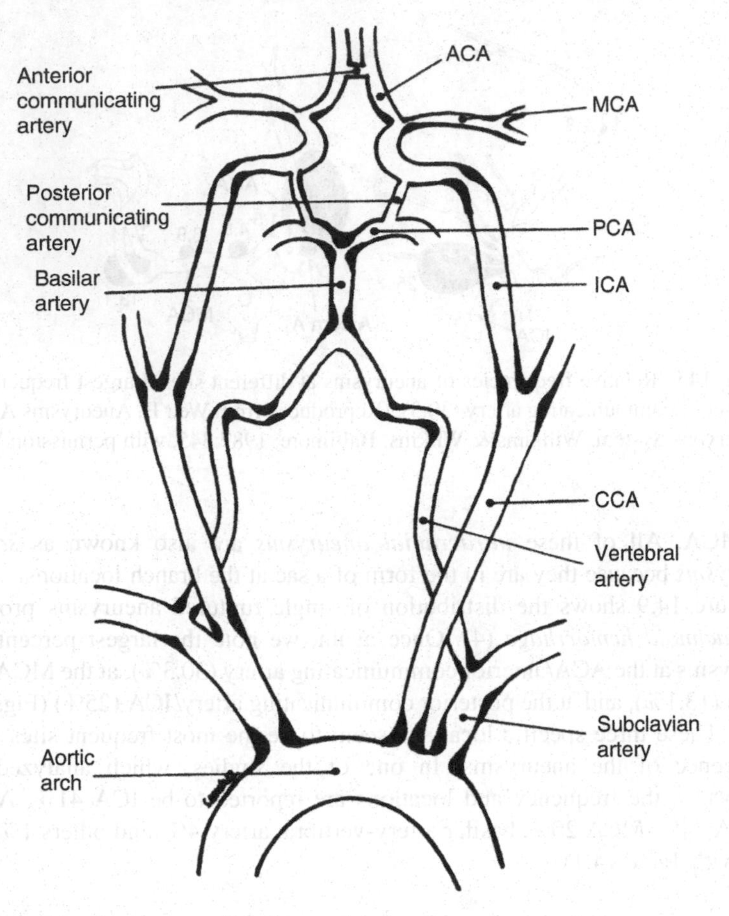

FIGURE 14.10. Sites of predilection for atherosclerotic narrowing; black areas represent plaques. (Reproduced from Caplan LR, Stroke – A Clinical Approach, Second ed. Butterworth-Heinemann, Boston, 1993:38, with permission from Elsevier.)

artery and at the origins of posterior communicating arteries. These are also the locations of aneurysms. Various other locations where plaques are seen are in the carotid siphon, at the bifurcation of common carotid artery, at the origins of vertebral arteries, and at the origins of great vessels from the aortic arch. We have also noted earlier that plaque formation occurs at the bends, bifurcations, and branches. Bifurcations and the ostia of the branches are also the sites where aneurysms occur in intracranial arteries. Figure 14.11 shows the angiogram where atherosclerotic plaques can be localized. The stenosis is seen in the middle cerebral arteries and this location seems to be at the bifurcation of MCA.

Atherosclerosis causes reduction of the blood supply to the brain tissue, which causes softening of the tissue, while rupture of the aneurysm causes flooding of

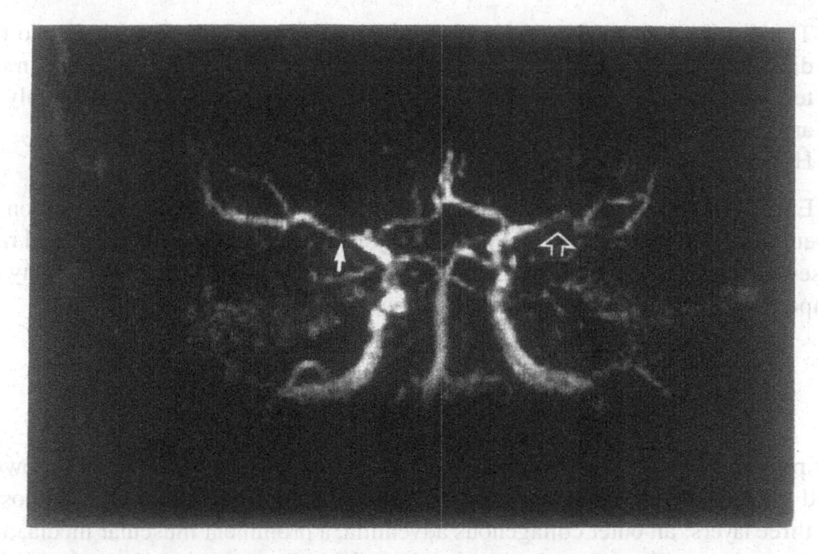

FIGURE 14.11. Magnetic resonance angiography showing the narrowing in the circle of Willis. The open arrow points to a severe stenosis of the left middle cerebral artery (MCA) while the solid arrow shows less severe stenosis of MCA. (Reproduced from Caplan LR, Stroke – A Clinical Approach, Second ed. Butterworth-Heinemann, Boston, 1993:120, with permission from Elsevier.)

the brain tissue with blood. Both qualify as strokes. The two major categories of brain damage in stroke patients are thus 1) ischemia: lack of blood flow, depriving brain tissue of needed oxygen and nutrition; and 2) hemorrhage: the release of blood into the extravascular space within the cranium. Furthermore, ischemia can be caused by thrombosis, embolism, or stenosis. Hemorrhage can be blood leaking out onto the surface of the brain tissue (subarachnoid) or bleeding directly into the brain substance (intracerebral).

4.3. *Association Between Aneurysm and Atherosclerosis*

At multiple times, it has been pointed out that in some cases the locations of aneurysms and atherosclerotic plaques are the same, for example at the anastomosis and at the origins and bifurcations of some intracranial arteries. An association between aneurysms and atherosclerosis in the intracranial arteries has also been noticed by others as far back as 1939 (5). More recently also such an association has been made (2, 6). Such an association can also be seen from the following observations:

a) All large aneurysms and many small ones show gross atherosclerotic changes. Severe cerebral atheroma is much more common in patients who die of subarachnoid and cerebral hemorrhage from ruptured aneurysms than in noncerebral deaths. Even in young patients, atherosclerosis is remarkably common when aneurysms have been demonstrated.

b) The distribution of atherosclerosis in the cerebral circulation is similar to the distribution of aneurysms, both in its predilection for a larger vessel and in the tendency for both conditions to occur most severely and most commonly at arterial branchings (7).
c) Hypertension worsens both the diseases.

Exactly how this association operates in formation, growth, or stabilization of aneurysms is not clear. The fact that aneurysms are common in the head and rare elsewhere suggests that the structure of the artery and other local factors play an important role in both the disease processes.

5. Mechanism of Aneurysm Formation

At present, there is no completely satisfactory explanation of the origin, growth, and rupture of saccular aneurysms. Cerebral arteries are normally composed of three layers: an outer collagenous adventitia, a prominent muscular media, and an inner intima lined by a layer of endothelial cells. An internal elastic lamina separates the intima from the media. There is no external elastic lamina.

Intracranial arteries are more susceptible to aneurysm formation than extracranial vessels because, intracranially, the arterial walls are thin, there is less elastin, there is no external elastic lamina, and there are "medial defects" present frequently. Around bifurcations of arteries, the intima may have focal thickenings called *intimal pads* at lateral angles, face, dorsum, and apex. Figure 14.12 shows the distribution of intimal pads in the cerebral vessels. Sometimes the muscularis media ceases abruptly at the apex, leaving a gap consisting only of intima, elastic lamina, and adventitia. These are the "medial defects." These defects are characteristically located on the distal carina at the point of branching and at the apex of bifurcation. These defects may occur at the lateral angles also. Figure 14.13 shows a typical medial defect at the apex of the cerebral artery bifurcation. Medial defects occur in the cerebral arteries in many more places than the sites

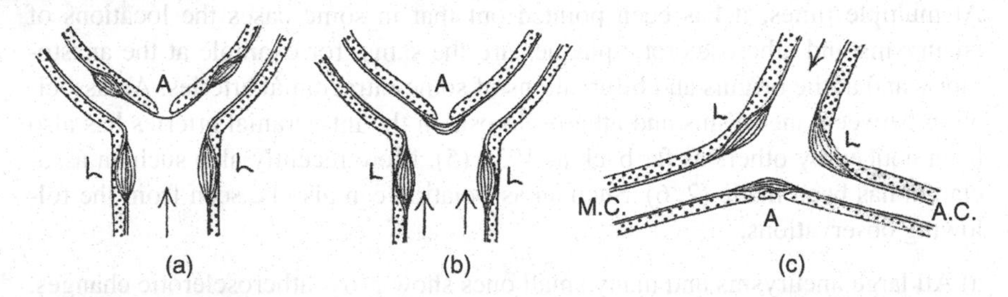

FIGURE 14.12. Distribution of intimal cushions at various sites. (a) Middle cerebral artery bifurcation, (b) basilar artery bifurcation, (c) internal carotid artery bifurcation. (Reproduced from Crompton MR, Brain 1966;89:797-814, with permission from Oxford University Press.)

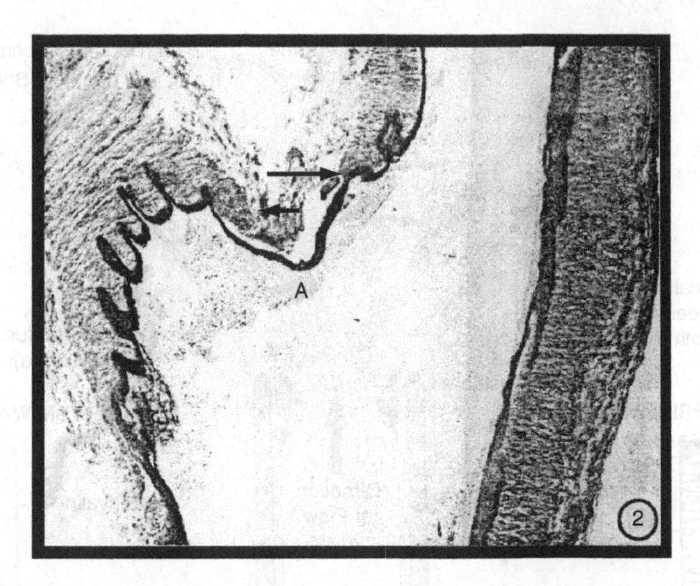

FIGURE 14.13. Bifurcation of artery of circle of Willis, showing a typical medial defect (arrows). The adventitia and intima are in loose approximation at A, the apex of the fork. The elastic layer is preserved (Weigert's elastic stain). (Reproduced from Sahs AL, J Neurosurg 1966;24:792-806, with permission.)

where the aneurysms form. Similar medial defects have been observed in the extracranial arteries, particularly in renal, mesenteric, splenic, and coronary arteries. However, in comparison with the cerebral arteries, the extracranial arteries show the defects that are smaller and less frequent, and these arteries rarely develop aneurysms.

Most saccular aneurysms occur at the apices of arterial bifurcations. The wall of the aneurysm is devoid of media, which usually ceases abruptly proximal to the neck (Figure 14.14). Elastica is either absent completely or exists as fragments. Small aneurysms are thin-walled. Their adventitia and intima are thin. Atherosclerotic changes may be seen but usually are confined to the intimal pads at the entrance to the sac. Large aneurysms have thicker walls, which mainly consist of hyaline connective tissue (Fig. 14.14). The intima often shows atherosclerotic changes consisting of smooth muscle proliferation, cholesterol clefts, lipid-laden macrophages, and ulceration.

Rupture of the aneurysms occurs most commonly in the dome region (84%), less commonly in the body of the aneurysm, and rarely at the neck (Fig. 14.14). When the aneurysm grows, there must be extreme proliferation of tissue occurring concomitantly with the great expansion of the sac wall. An increase in size thus involves true growth rather than only distention and thinning of preexisting tissue.

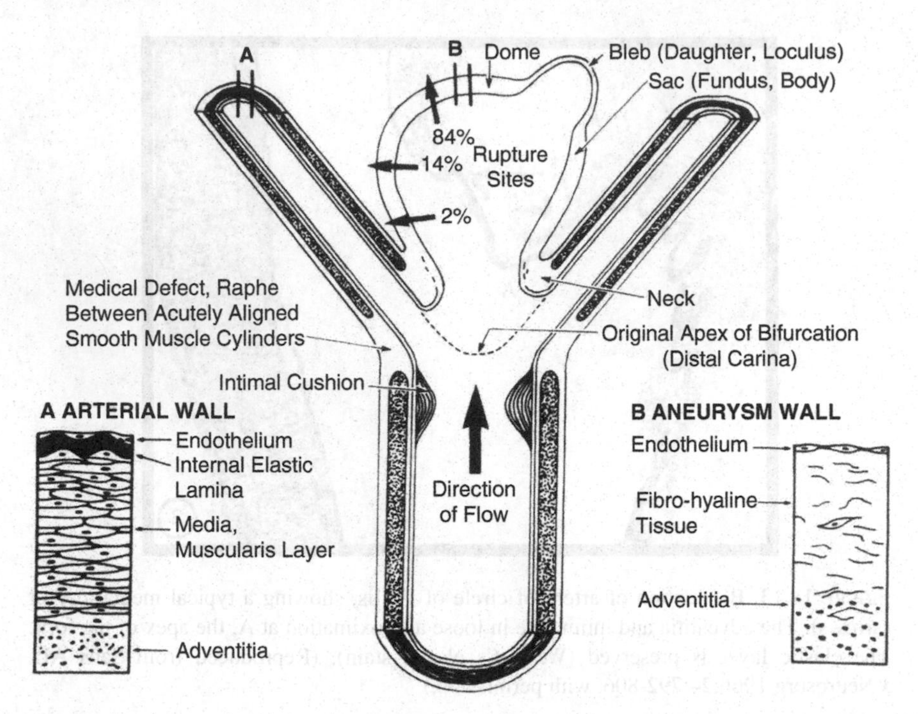

FIGURE 14.14. Aneurysm pathology. (Reproduced from Weir B, Aneurysms Affecting the Nervous System. Williams & Wilkins, Baltimore, 1987:224, with permission.)

5.1. *Pathogenesis of Aneurysms (Ref. 8)*

Eppinger (9) indicated that certain aneurysms occurred at the site of *defects* of the elastic layer of arteries and suggested that these were inborn or congenital. Forbus (10) studied arteries from 14 children and 19 adults. Medial defects (gaps in the muscle layer) were found in about two-thirds of both groups. He concluded that aneurysms might be acquired lesions arising from a *combination* of degeneration of the elastica and defects in the muscular layer. It was considered that the elastica could be damaged by constant overstretching from hemodynamic stresses occurring at the apexes of bifurcations. Glynn (11) noted that saccular aneurysms occurred more frequently in cerebral vessels than any other muscular arteries. Medial defects at bifurcation in the circle of Willis were found in 81% of patients with aneurysms and 80% of those without. Glynn subjected arteries, including bifurcations, to pressures of 400 to 600 mmHg without being able to visualize any signs of localized bulging. It was believed that the greater frequency with which aneurysms occur in the circle of Willis was probably due to the difference in their elastic tissue. It was his conclusion that "the medial defect did not constitute a locus minoris resistentiae and could play no part as such in the development of an aneurysm."

Forster and Alpers (12) studied histologically eight cases of aneurysms. They found changes in the internal elastica ranging from a complete membrane, through splitting and fragmentation, to the complete absence of the membrane with a sharp defect at the site of origin of the aneurysm. Thirteen small aneurysms were studied, and it was concluded that they all owed their origin to the combined effects of developmental deficiency of the media and arterial degeneration (6). The aneurysms were always found to arise at sites where there were substantial breaches in the muscular and elastic coats. Carmichael (13) was of the opinion that both medial and internal elastic lamina defects must be present for the formation of cerebral aneurysms. He considered that the medial defect was usually congenital and that the elastic defect resulted from atheromatosis, although it could occur by itself. In his opinion, fibrotic repair could prevent the development of aneurysms even in the presence of both medial and elastic defects.

Sahs (14) noted that the controversy surrounding the genesis of aneurysms hinged mainly on the degenerative changes in the internal elastic lamella. He concluded that medial defects were not the major etiologic factor in the formation of aneurysms but suggested that they might play a contributory role. He was particularly impressed by changes in the elastic lamellae that are present in even the smallest identifiable outpouchings. Not all such protrusions appeared to take place directly through intimal cushions. He also considered that hemodynamic factors and atherosclerosis may both have a role to play. Stehbens (15) was of the opinion that a "true" aneurysm is formed by dilatation of the components of the vessel wall, which frequently undergo extensive structural degenerative change. In their earliest stages, aneurysms are about 1 mm in diameter and appear to be a filling out of the apical angle at the bifurcation. The naturally occurring degenerative changes in the walls of arterial forks are related to atherosclerosis and hemodynamic stresses. There is a tendency for less compensatory intimal thickening at the apex and distal sides of the daughter branches than elsewhere in the fork. Focal failure of tensile strength or cohesive properties occurs at the apex of the vessel wall, permitting aneurysmal dilatation. The areas of thinning and funnel-shaped dilatations have never been observed in infants, other animals, or extracranial arteries. Electron microscopic studies of early aneurysmal changes indicate loss of elastic tissue and much cellular debris.

The entrance to most cerebral aneurysms is much wider than medial defects, but even so, the media terminates abruptly at the entrance to these aneurysms. It seems certain therefore that secondary widening of the sac entrance must develop. Underlying hemodynamic factors are almost certainly responsible for the development of cerebral aneurysms against a background of degenerative arterial wall changes. The evidence in support of this are the facts that aneurysms increase with age, probably are more common in hypertensives, and appear to be related to atherosclerosis.

Sheffield and Weller (16) studied 20 patients without aneurysms ranging in age from 1 month to 80 years. Histologic examination revealed gaps in the media at the carina and lateral walls of the bifurcation in 60% of specimens from arteries at all ages, although these were larger in older patients. The most striking change associated with age was the development of inelastic pads of intimal thickening in relation to bifurcations. These were observed at and distal to the carina of bifurcations. Sheffield and Weller suggest that hemodynamic stress stimulates the development of these inelastic intimal pads and that these pads, in turn, alter the stresses and strains at bifurcations. These pads, as well as medial gaps, may predispose to arterial aneurysm formation at bifurcations.

Ferguson (17) expressed his view that aneurysms arise preferentially at the apex, not because of congenital weakness but because this is the site of maximum hemodynamic stress. An unbalanced circle of Willis may increase the flow rate across a particular bifurcation, thereby subjecting it to increased stress and the likelihood of the development of an aneurysm. This is the only discernible congenital aspect to the pathogenesis of aneurysms. Once the internal elastic membrane has been damaged, aneurysmal outpouching can begin in response to the pulsatile pressure head transmitted to the weakened apex by the impinging axial streams. The controversy surrounding the pathogenesis of saccular aneurysms is not ended. The future is likely to see focus shift from histology to chemistry of the arterial wall and its defects.

5.2. *Mechanisms of Growth and Rupture*

Ferguson (18) considered that turbulence itself could damage vascular tissue. It can excite an artery to vibrate at its resonant frequency so that even relatively low forces result in relatively high strains, thereby damaging the wall. The normal role of elastin is to maintain tension against the normal blood pressure, whereas collagen has a protective supporting role. Because the aneurysm wall is composed of collagen alone, it is therefore less able to resist pulsatile stress. Once aneurysms are formed, turbulence readily develops. The hemodynamically generated forces resulting from the direct impingement of the central streams at the apex of bifurcations are probably the most important factor contributing to the focal degeneration of the internal elastic membrane and the early origin of aneurysms. These forces could enlarge medial defects already present. The impingement of central axial streams results in a much greater velocity gradient and sheer stress at the apex than is experienced in the main stem or branches of bifurcations. The impact and sudden deflection of the central stream at the apex results in the transmission of a pulsatile impulse to that region of bifurcation. The peak force will be great, because there is a brief impact time. At the moment of impact, the kinetic energy of the moving blood is changed to pressure. The total pressure acting at the apex equals the sum of the transmural pressure and the kinetic energy equivalent pressure.

The incidence of systemic aneurysms increases linearly with age. The media gives the wall dynamic recoil, and the adventitia provides dynamic strength. Maintenance of structure depends on a balance between intraluminal pressure and the wall's ability to resist stress, which in turn depends on optimal proportions and concentration of elastica, muscle, and fibrous content. Old arteries become stiff. Therefore, systemic aneurysms appear to occur with greater frequency in arteries that have undergone structural changes with age.

Macfarlane and co-workers (19) studied shape changes at the apex of isolated human cerebral bifurcations with changes in transmural pressure. The outline of the internal apical curve was measured. As pressure increased, the central region flattened, the shoulders broadened, and the arterial wall thinned. These changes tend to concentrate stress on the apex, which would be even greater if there were medial gaps. Scott and Ballantine (20) obtained static pressure-volume curves of human saccular aneurysms and cerebral arteries obtained at autopsy. The walls of aneurysms were found to be quite brittle in contrast with the walls of arteries, presumably because aneurysm walls consist only of collagen, whereas normal arteries contain collagen and elastin. Increasing intra-arterial pressure to 200 mmHg on two or three runs dramatically reduced the distensibility, presumably by breaking the internal elastic layer. The wall stress in aneurysms is also much greater than in arteries, as the aneurysm wall is relatively thinner. At a given pressure, an aneurysm is more likely to rupture with increasing radius and decreasing wall thickness.

5.3. *Aneurysm Wall Histology*

The histology of aneurysms is summarized in Figure 14.14. Walker and Allegre (21) frequently noted intimal and elastic lesions in aneurysms, suggestive of atheromatous change. In 45% of the aneurysms, they found fragments of elastica in the sac. They illustrated examples of endothelial cushions in arteries near aneurysms and atheromatous plaque within aneurysms. Subendothelial proliferation was documented in arteries bearing aneurysms. These intimal changes were attributed to atheromatous degeneration.

Sahs (14) carried out histologic examinations of 50 saccular aneurysms in 130 (mainly from adult) bifurcations and made the following observations. The earliest precursors of aneurysms were small outpouchings through areas of the media. At these points, the elastic layer had undergone conspicuous fragmentation. Hyperplastic changes were always noted in the zone beneath defects in the elastica. In larger outpouchings, the media usually broke sharply and elastica progressively fragmented as it entered the sac of the outpouching. Only rarely could strands of elastica be followed through the entire aneurysm. Hence, for the most part the wall consisted of collagen interposed between adventitia and intima, with abundant fibroblasts and occasional macrophages. In larger aneurysms, abrupt cessation of the media at the neck and absence of the elastic layer within the wall were noted.

6. Elasticity of the Aneurysm

When a saccular aneurysm is present, one is greatly concerned with the possibility of its rupture. Any information that would allow us to determine the probability of its rupture is very important. The first step in this direction is understanding the elasticity of the aneurysm. The elasticity of the intact aneurysm was experimentally studied by Scott et al. (22) and they also studied the elasticity of the normal cerebral arteries for comparison. The experiments were performed on fresh specimens of intracranial arteries and aneurysms obtained at autopsy. In the experiments, a pressure volume apparatus was used where a known volume of fluid was introduced into the specimen and the developed pressure was recorded. They used the following approach for presentation of their results.

The aneurysm is considered to have a spherical shape, which is most often the case. The artery has a cylindrical shape. For the spherical object, according to the law of Laplace, $p = \frac{2T}{R}$, and for the cylindrical object $p = \frac{T}{R}$, where p is the luminal pressure, T is the wall tension (in any direction for the spherical object and in the circumferential direction for the cylindrical object), and R is the radius of curvature. The pressure is converted to wall tension for expressing the results. The volume is converted to strain as follows. Strain was defined as the ratio of change in the surface area to the initial surface area. Figure 14.15 shows their results. The tension-strain curve for the aneurysm is shifted to the left, indicating that the aneurysm wall is much stiffer than the artery wall. This change in stiffness is expected because there is a change in the tissue structure of the aneurysm wall. The aneurysm wall consisted mainly of collagen, whereas the artery wall consisted of collagen and elastin. They also found that distensibility curves for cerebral aneurysms changed abruptly after being subjected to pressures of 200 mmHg three times. The aneurysm wall became brittle, presumably due to the disruption

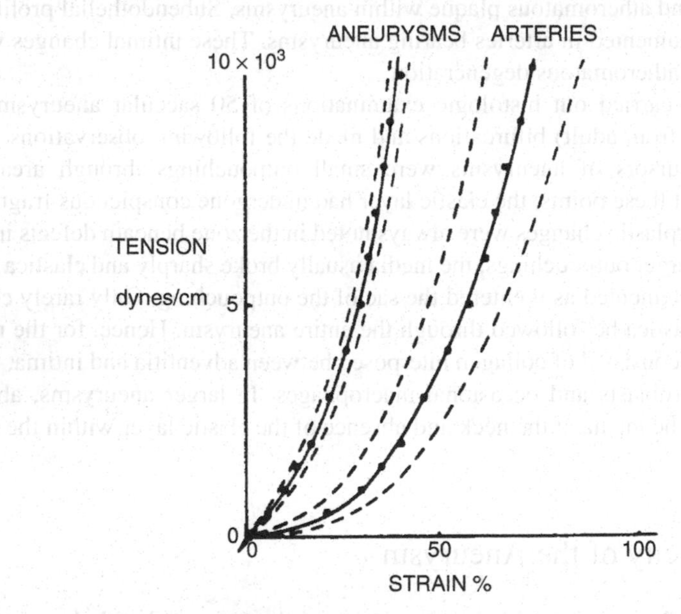

FIGURE 14.15. Mean tension-strain curves from 7 aneurysms and 16 arteries. Dashed lines show 1 standard error of the mean. Note the two slopes in the arterial curve, presumably due to the elastin and collagen of the wall. The aneurysmal curve is shifted to the left, indicating that the aneurysm wall is stiffer than the arterial wall. (Reproduced from Sekhar LN et al., Neurosurgery 1981;8:248-260, with permission.)

of elastic components of the wall. They calculated the stress in the wall, which is given by wall tension/wall thickness. Because the aneurysm wall is much thinner in comparison with the artery wall, the stress in the aneurysm is much greater than that in the artery. Obviously, any local thinning of the aneurysm wall poses an even greater threat of aneurysm rupture, due to further increase in the wall stress. Also, increase in the radius of the aneurysm can have a devastating effect by causing the stress to increase not only due to the increase in the radius but also due to a likely decrease in the wall thickness. Conversely, large aneurysms with thick walls may not rupture. Another obvious consequence of the above considerations is the effect of pressure. The larger the pressure, the greater the wall stress. In patients, hypertension has been associated with increase in the number of aneurysms and the number of ruptures (23, 24) while there are also reports that this effect of hypertension was not always present (23, 25).

7. Mechanical Forces in Aneurysm Formation

7.1. *Fluid Dynamics–related Forces*

In the earlier section we described that according to Ferguson (17, 18), forces generated from direct impingement of the central stream on the apex of bifurcation contribute to the focal degeneration of the internal elastic membrane and thus to the origin of aneurysms. Here we will consider these studies in more detail.

More than 30 years ago, Ferguson (18) carried out studies based on the premise that "turbulence causes vibration in the vessel wall and this vibration produces and accelerates degenerative changes in vascular tissue." He made glass models of bifurcation aneurysms and experimentally studied the turbulence and the flow pattern in the aneurysm sac (Fig. 14.16). He observed that in the aneurysm model, there was a significant turbulence even at remarkably low flow rates. The turbulence is superimposed upon a circular, whirlpool-like fluid motion within the sac. On the basis of Reynolds number (Re), he concluded that the turbulence within

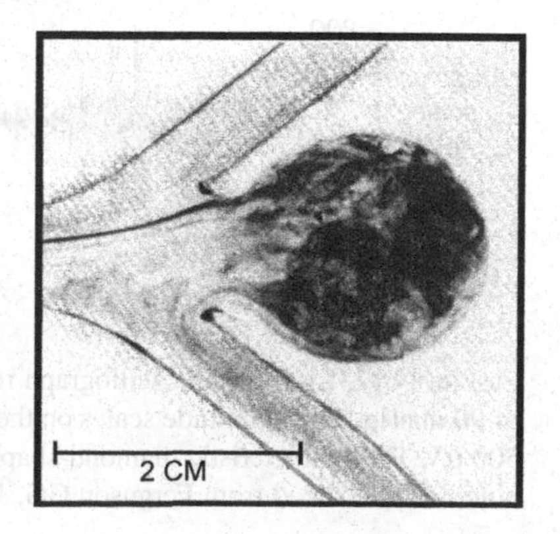

FIGURE 14.16. Turbulence in the sac of the large spherical glass model aneurysm. The Reynolds number (Re 500) corresponds with the flow rate at which the photograph was taken. The direction of flow is from left to right. Evans blue dye has been injected to demonstrate the flow pattern. Flow in the stem and both branches is streamlined. (Reproduced from Ferguson GG, J Neurosurg 1970;33:485-497, with permission.)

the sac can occur at a flow rate less than one-half of that, which may be necessary for the turbulence to occur at a normal bifurcation.

He continued the exploration of turbulence in the aneurysms in patients. To record *sound* from intracranial aneurysms exposed during surgery, an intracardiac phonocatheter with a tip diameter of 2 mm was used. The phonocatheter was applied directly to the external surface of the aneurysms, or close to it (18). The bruits were monitored audibly and also recorded on an oscilloscope. He also did a series of control recordings from major intracranial vessels and bifurcations (without aneurysm) at normal systemic pressure. In all control cases, there was no bruit present. In casc of aneurysms, there was bruit present in 10 out of 17 cases. In these patients, the pressure was low because of administration of the hypotensive drug Arfonad, which was a practice during these surgical procedures. He obtained the data in only those patients where the pressure was 50 mmHg or above. All aneurysmal bruits had similar features; they were musical in quality and quite high-pitched in tone. The average predominant frequency was 460 ± 130 Hz. The bruits had a diamond-shaped oscillographic profile (Fig. 14.17). He stated that the bruits recorded were the result of turbulent blood flow within the aneurysm. In normotensive individuals, it should be noted that the total cerebral blood flow remains near normal, as a result of autoregulation, even if the blood pressure falls, until the mean pressure of 50 mmHg is reached.

From these studies, he concluded that there is no turbulence at the normal bifurcation and therefore turbulence does not play a role in the initiation of aneurysm. However, once the outpouching begins, then, according to his postulate, the turbulence within the sac contributes to the enlargement of the sac. This brings us to his next set of experiments, which again explore the flow-related cause of initiation of the aneurysm.

Ferguson (17) prepared glass models of cerebral artery bifurcations (without aneurysms) using various angles for bifurcation. In these models, he injected Evans blue dye to examine the flow patterns. A reproducible sequence of flow patterns was observed with both steady and pulsatile flow, as the flow rates were increased in each model. At low flow rates, the axial and periaxial streams

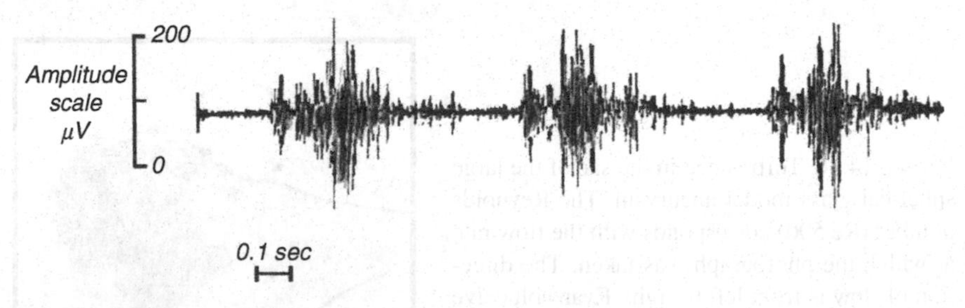

FIGURE 14.17. Light-beam oscillograph record of bruits at mean systemic arterial pressure of 90 mmHg. The amplitude scales on the left are related to a standard calibrating signal of 500 μV. The characteristic diamond-shaped profile of the bruits from the aneurysm can be noted. (Reproduced from Ferguson GG, J Neurosurg 1970;33:485-497, with permission.)

impinged directly upon the apex of the bifurcations, corresponding with the site where saccular aneurysms normally arise. With an increase in the flow rate, turbulence arose in the proximal portions of the branches but not in the region of the apex. The flow rates in the circle of Willis are in the range where axial stream impingement will occur. He stated that the impingement results in a much greater velocity gradient and shear stress at the apex than in the main stem or branches. The momentum per unit volume of flow is greatest in the central streams due to high velocity. The impact and sudden deflection of central streams at the apex results in the transmission of a pulsatile impulse to the apex region of bifurcation. At the impact, the kinetic energy of moving blood is converted to pressure energy (stagnation pressure) at the apex. This extra pressure is the force responsible for focal degeneration of the internal elastic membrane and thus the cause of initiation of aneurysms at the apex. Thus, the aneurysms arise at the apex because this is the site of maximum impingement force. Apparently, no other theories existed, which could explain why the aneurysms occur at the site of bifurcation.

7.2. Solid Mechanics–related Forces

The theories proposed by Ferguson, considered above, which are based solely on the forces generated by blood flow have been the only theories that have existed, over at least a 30-year period, for the explanation of aneurysm formation. There had been a consideration, as proposed by Macfarlane et al. (19), that the change in shape of the apex at the bifurcation as a function of pressure would tend to increase the stress at the apex and that could contribute to the aneurysm formation. However, the fundamental principles of solid mechanics, which reveal that there is a stress-concentration at the apex of bifurcation and at the ostia of branches, remain unmentioned and unexplored. From the perspective of solid mechanics, it would appear that the most obvious cause of aneurysm formation is greatly increased wall stress—due to stress concentration—which could cause tissue degeneration at the location of stress concentration and lead to aneurysm formation at the apex of bifurcation and at the ostia of branches. We only need to refer to previous chapters (e.g., Chapters 5, 6, and 7) to see how and why the stress-concentration exists and how it is made more harmful by an increase in pressure. In this context, we must also ask what could be the reason why some medial gaps develop aneurysms while others do not. The medial gaps will further increase the stress concentration at that location. Also, we must ask why the anterior region of the circle of Willis has the lion's share of aneurysms while the posterior region has only a small number of aneurysms. The detailed geometry of the intracranial arteries is extremely complex and fascinating and, as we know, plays a very important role in the development of wall stress due to pressure. I believe that the answer to many of these questions will come from detailed exploration of solid mechanics–related mechanical forces (i.e., pressure-induced wall stress and stress-concentration). Thus, the pathogenesis of cerebral aneurysms remains unknown at this point and must wait until more research is undertaken in vascular mechanics, perhaps in combination with fluid dynamics.

References

1. Alpers BJ: Aneurysms of the circle of Willis. Morphological and clinical considerations. In: Intracranial Aneurysms and Subarachnoid Hemorrhage.
2. Crawford T: Some observations on the pathogenesis and natural history of intracranial aneurysms. J Neurol Neurosurg Psychiatry 1959;22:259-266.
3. Wilson G, Riggs HE, Rupp C: The pathologic anatomy of ruptured cerebral aneurysms. J Neurosurg 1954;19:128-134.
4. Weir B: Aneurysms affecting the nervous system. Williams & Wilkins, 1987: 344-345.
5. McDonald CA, Korb M: Arch Neurol Psychiatry 1939;42:298.
6. Carmichael R: J Pathol Bacteriol 1950;62:1.
7. du Boulay GH: Some observations on the natural history of intracranial aneurysms. Br J Radiol 1965;38:721-757.
8. Weir B: Aneurysms affecting the nervous system. Williams & Wilkins, Baltimore, 1987:209-226.
9. Eppinger H: Pathogenesis (Histogenesis und Aetiologie) der Aneurysmen einschliesslich des Aneurysma equi verminosum. Pathologisch-anatomische studien. Arch Klin Chir 1887;35(Suppl 1):1-563.
10. Forbus WD: On the origin of miliary aneurysms of the superficial cerebral arteries. Bull Johns Hopkins Hosp 1930;47:239-284.
11. Glynn LE: Medial defects in the circle of Willis and their relation to aneurysm formation. J Pathol 1940;51:213-222.
12. Forster FM, Alpers BJ: Anatomical defects and pathological changes in congenital cerebral aneurysms. J Neuropathol Exp Neurol 1945;4:146-154.
13. Carmichael R: Gross defects in the muscular and elastic coats of the larger cerebral arteries. J Pathol Bacteriol 1945;57:345-351.
14. Sahs AL: Observations on the pathology of saccular aneurysms. In: Sahs AL, Perret GE, Locksley HB, Nishioka H (eds), Intracranial Aneurysms and Subarachnoid Hemorrhage – A Cooperative Study. Lippincott, Philadelphia, 1969:22-36.
15. Stehbens WE: Intracranial arterial aneurysms. In: Pathology of the Cerebral Blood Vessels. CV Mosby, St. Louis, 1972:351-470.
16. Sheffield EA, Weller RO: Age changes at cerebral artery bifurcations and the pathogenesis of berry aneurysms. J Neurol Sci 1980;46:341-352.
17. Ferguson GG: Physical factors in the initiation, growth, and rupture of human intracranial saccular aneurysms. J Neurosurg 1972;37(6):666-677.
18. Ferguson GG: Turbulence in human intracranial saccular aneurysms. J Neurosurg 1970;33(5):485-497.
19. Macfarlane TWR, Canham PB, Roach MR: Shape changes at the apex of isolated human cerebral bifurcations with changes in transmural pressure. Stroke 1983;14:70-76.
20. Scott RM, Ballantine HT Jr: Spontaneous thrombosis in a giant middle cerebral artery aneurysm: Case report. J Neurosurg 1972;37:361-363.
21. Walker AE, Allegre GE: Histopathologie et pathogenei des anevrysmes arteriels cerebraux. Rev Neurol 1953;89:477-490.
22. Scott S, Ferguson GG, Roach MR: Comparison of elastic properties of human intracranial arteries and aneurysms. Can J Physiol Pharmacol 1972;50:329-332.
23. Andrews RJ, Spiegel PK: Intracranial aneurysms: Age, sex, blood pressure and multiplicity in an unselected series of patients. J Neurosurg 1979;51:27-32.

24. Kwak R, Mizoi K, Katakura R, Suzuki J: The correlation between hypertension in past history and the incidence of cerebral aneurysms. In: Suzuki J (ed), Cerebral Aneurysms: Experience with 1000 Directly Operated Cases. Neuron Publishing Co., Tokyo, 1979:20-24.
25. McCormick WF, Schmalstieg EJ: The relationship of arterial hypertension to intracranial aneurysms. Arch Neurol 1977;34:285-287.

15
Aortic Aneurysms

1. Introduction

Aortic aneurysm is a bulbous enlargement of the aorta (Fig. 15.1). The most common aortic aneurysm is the abdominal aortic aneurysm (AAA). An aneurysm also may occur in the ascending aorta, descending thoracic aorta, and sometimes it may involve more than one segment of the aorta. There is another pathology called dissecting aneurysm where the aortic wall dissects and blood enters the wall causing the enlargement. This condition is called aortic dissection. We will consider aortic dissection in Chapter 16; here we will focus on the abdominal aortic aneurysm. It may be emphasized that in case of AAA, the application of vascular mechanics is of paramount importance because it can have a direct bearing on the clinical outcome for a patient. The mechanics of AAA continues to be developed, and we are sure to reach a point when it will be applied frequently for making a clinical decision. AAA provides yet another example of vascular mechanics and pathology being inseparable.

2. Incidence of AAA

Abdominal aortic aneurysms are present in 8% of men over 60 years of age (1). The diagnosis of AAA in many patients is made only after the aneurysm ruptures, after which mortality is high. Depending on the size of the AAA, oftentimes the only treatment available after its discovery is "wait and watch." Unfortunately, in this process of waiting, some aneurysms do rupture and have fatal consequences. Thus, a better assessment of risk of rupture, as well as treatment to slow the growth of AAA, would have an important therapeutic value. AAAs are among the most common and most lethal conditions that a vascular surgeon is called upon to treat. Despite the success in characterizing the pathologic characteristics of established AAAs, relatively little is known about the initiating etiologic factors. Most studies investigating aneurysm pathogenesis use tissues from established aneurysms. Because this tissue represents the end-stage of a pathologic process, it is difficult to determine the causative factors from the observation of degenerative changes.

Reed et al. (2) conducted a 20-year follow-up study of more than 8000 men of Japanese ancestry who were living in Hawaii in 1965. Among 7682 men, who were free of clinical aortic aneurysms at the initial examination, 151 developed aortic aneurysms and 23 developed aortic dissections. Of the aneurysms, 138 were abdominal (AAA), and 13 were thoracic. Of the 151 men with aneurysms, 85 had died and 24 (28%) of these had "aneurysm" noted as a cause of death. Of the 23 men with aortic dissection, 16 (70%) had died, and in all cases "aneurysm" was noted as the cause of death, but not always as dissecting aneurysm.

Figure 15.2 shows the incidence rates per 100,000 person-years for aortic aneurysms and dissections by age at diagnosis. Aortic aneurysms increased steadily with age after age 60. Aortic dissections, on the other hand, peaked between ages 70 and 75 and decreased after that. Reed et al. also studied certain

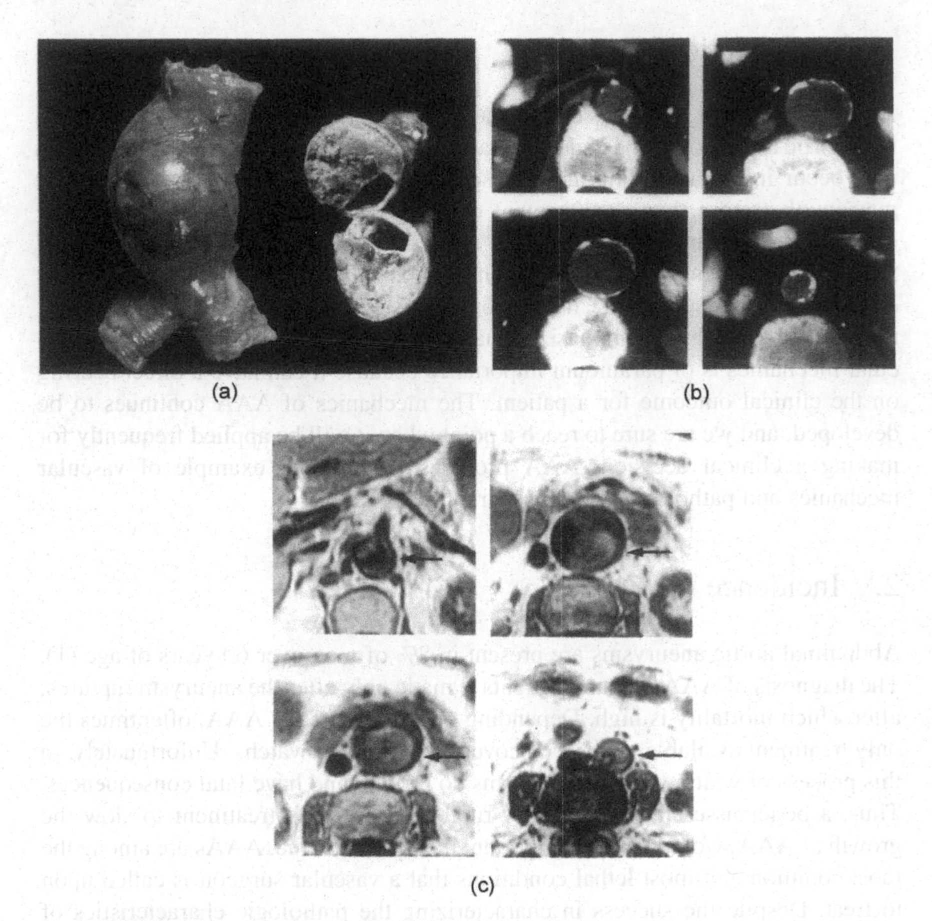

FIGURE 15.1. (a) Photograph of a typical AAA extending up to the aortic bifurcation. The size and the contour of the aneurysm can be appreciated. This aneurysm, cut in the middle circumferentially, shows the presence of a huge thrombus (clot) within it. Only a small opening is available for the blood flow. (b) CT scans of an abdominal aortic aneurysm (AAA 2) of a 70-year-old male showing the cross sections of the aorta at four different levels beginning proximal to the aneurysm (top left) and continuing distally through it. Top right, bottom left: Within the aneurysm bulb. Bottom right: Just after the aneurysm. In each scan, the aorta/aneurysm can be seen in the center next to the vertebra. The presence of the thrombus in the aneurysm can be seen as a darker shade of gray although its boundaries and character are difficult to determine. (c) MRI scans of an abdominal aortic aneurysm (AAA 1) of another 70-year-old male showing cross sections of the aorta at four different levels. Top left: Just proximal to the aneurysm. Top right, bottom left: Within the aneurysm bulb continuing distally. Bottom right: Just after the aneurysm. In each scan, the aorta/aneurysm can be seen in the center next to the vertebra. The presence of the thrombus in the aneurysm can be seen clearly, however, its boundaries and character are difficult to determine. The aneurysm is 11.0 cm long.

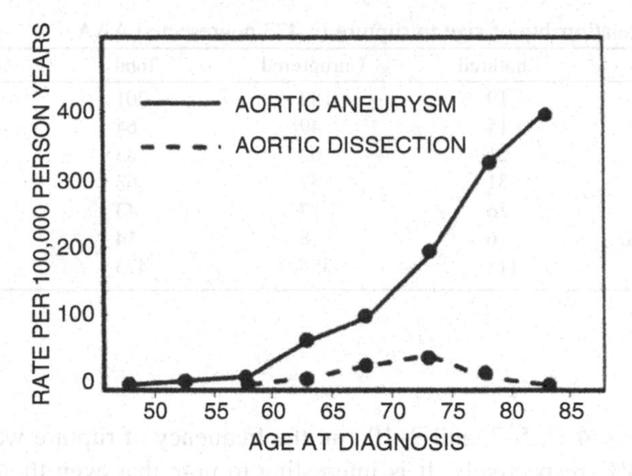

FIGURE 15.2. Plots of incidence rates of aortic aneurysm and dissection by age at diagnosis. (Reproduced from Reed D et al., Circulation 1992;85:205-211, with permission.)

risk factors and found the following correlations. There was a clear association of aneurysms with systolic blood pressure, serum cholesterol, cigarette pack-years, and a person's height. They therefore proposed that atherosclerosis may be involved in the pathogenesis of aneurysms. From the mechanics point of view, we recall two important observations: i) increase in systolic pressure is associated with increase in wall stress, and ii) increase in the age is associated with increase in wall stiffness (modulus of elasticity). Both of these parameters are expected to synergistically participate in causing tissue degeneration from stress and thus could correlate with increase in the aneurysm formation. Aortic dissections, interestingly enough, were associated only with systolic blood pressure and serum triglyceride. Lastly, we must note that AAAs also occur in women.

3. Association Between Aneurysm Size and Rupture

Darling et al. (3) reported their findings on 24,000 autopsies performed at the Massachusetts General Hospital over a 23-year period. There were 473 patients who had AAA. Of these nonresected AAAs, 118 (24.9%) were ruptured. In the latter group there were 52 patients with known AAA who were monitored and who had died either from ruptured aneurysm (about 50%) or from other causes. In this group also the size of the aneurysm at autopsy was tabulated. In this study, the size refers to the widest diameter measured. Table 15.1 shows the relationship between size of the aneurysm and percentage that were ruptured in 473 patients. The percentage of ruptured aneurysms increased as the size of the aneurysms increased. One additional observation of note is that even small aneurysms (4 cm or less) rupture. Also, in the 52 patients with known AAA that were monitored, the rate of rupture increased as the size of the aneurysm increased. For the AAAs

TABLE 15.1. Relationship of size to rupture in 473 nonresected AAA.

Size (cm)	Ruptured	Unruptured	Total	% Ruptured
4 or under	19	182	201	9.5
4.1–5.0	15	49	64	23.4
5.1–7.0	21	62	83	25.3
7.1–10.0	31	37	68	45.6
10.1 and over	26	17	43	60.5
No size recorded	6	8	14	
Total	118	355	473	24.9

of diameter <4, 4–5, 5–7, and 7–10 cm, the frequency of rupture was 8%, 25%, 50%, and 64%, respectively. It is interesting to note that even though these 52 patients were not considered suitable candidates for aneurysm resection, due to their other illnesses, almost one-half of them died of aneurysm rupture. The authors state that they did not find factors other than the size that correlated with the likelihood of aneurysm rupture.

Juvonen et al. (4) studied descending thoracic or thoraco-abdominal aneurysms in a total of 88 patients, where 24 patients had ruptured aneurysms and 64 patients were still being monitored. They reported that for a maximum diameter of 2–3, 3–4, 4–5, 5–6, 6–7, 7–8, and 8–9 cm, the rate of rupture was 0%, 6%, 12%, 36%, 50%, 100%, and 50%, respectively. Thus, for thoracic and thoraco-abdominal aneurysms also, the rate of rupture increases with the size of the aneurysm.

Scott et al. (5) studied 218 AAAs in patients in the age group of 65 to 80 years. The patients were offered surgery if their aneurysm expanded more than 1 cm/year or if the diameter reached 6 cm. They report the *potential rupture rate*, which is the combination of actual rupture plus elective surgery, as follows. The potential rupture rate was 2.1% per year for small AAAs (3 to 4.4 cm diameter) and 10.2% per year for AAAs of 4.5 to 5.9 cm diameter.

Noel et al. (6) studied 418 patients that had ruptured AAAs. There were 344 men and 74 women between the ages of 49 and 97 years. Forty (9.6%) of these patients, in whom the aneurysms ruptured, had AAAs of less than 5.5 cm diameter. They concluded that rupture of a *smaller aneurysm* is as lethal as rupture of larger aneurysms and that this group of patients would have been appropriate candidates for elective repair. The important message here is that at this time, there is no good way to predict which smaller aneurysms are likely to rupture. Consequently, elective repair is not performed, and when these aneurysms do rupture, major damage is done to the patient.

Coady et al. (7) studied 230 patients with thoracic aortic aneurysms. They reported that median size at the time of rupture or dissection was 6.0 cm for *ascending aortic aneurysms* and 7.2 cm for *descending aortic aneurysms*. If the median size is used as criteria for intervention, then half of the patients would suffer devastating complications before the operation. Therefore, they recommend that elective resection may be done at 5.5 cm for ascending aortic aneurysms and at 6.5 cm for descending aortic aneurysms. From the vascular mechanics point of view, which we will consider a little later, we should note here that ascending aortic aneurysms rupture at a smaller size (6 cm) than do descending aortic aneurysms (7.2 cm).

For AAAs, Juvonen (8) suggested that elective repair should be carried out for an aneurysm of 7 cm or greater while Cronenwett (9) suggested that elective repair should be carried out when the diameter reaches 5–6 cm.

4. Pathogenesis of Aortic Aneurysm

In spite of advances in the treatment of AAA, the pathogenesis of aneurysmal disease remains obscure. There are several reviews on the subject, and we will consider some of them (10–14) in brief, while trying to put them in the context of vascular mechanics.

4.1. Histology/Morphology

4.1.1. Histology of Normal Aorta and AAA

The normal artery has three layers: The intima, the media, and the adventitia. The innermost component of the intima, lining the lumen of the artery, is a continuous layer of thin endothelial cells, which offer little resistance to outward pressure but are resistant to the shearing force of the flowing blood. The media, which is responsible for most of the strength and elasticity of the arterial wall, is composed primarily of elastin and smooth muscle cells and some fibrous material. The elastic walls stretch to accommodate the volume of blood ejected from the heart and then they recoil, acting as subsidiary pump. The adventitia is a fibrous sheath composed largely of collagen. Because the adventitia has little elasticity, it lacks the dynamic recoil of the media, but it accounts for much of the static strength of the arterial wall.

The basic ingredients of the aortic media are smooth muscle cells, elastin fibers, collagen fibers, and some amorphous ground substance. These components are arranged in an orderly fashion. The smooth muscle cells are arranged circumferentially. Interposed among them are variable amounts of elastin, collagen, and ground substance. The media contains a series of concentrically arranged perforated elastic membranes, or laminas, and between these membranes are smooth muscle cells, thin collagen fibrils, and an amorphous ground substance. The basic structured unit of aortic media is two parallel elastic laminas with smooth muscle cells, collagen fibers, and ground substances sandwiched between them. The number of elastic laminas (lamellar units) generally increases with age (40 in the newborn; 70 in the adult). Elastin is easily stretched and can double its length and spring back to its original dimensions. Collagen has very different properties. Fibrillar collagen has a tensile strength more than 20 times greater than that of elastin and is very difficult to stretch. It cannot extend beyond a small percentage of its length before structural damage occurs. Aortic collagen is coiled up in such a way that, initially, the load of the aorta is borne by elastin, and as the load increases, collagen fibers uncoil and are progressively recruited as load-bearing elements. This allows the aorta to stretch easily initially and to become less and less distensible progressively with load.

The difference in the growth of the media after birth between the thoracic and the abdominal aorta is quite interesting. In the thoracic segment, the increase in the thickness of the media is accomplished mainly by an increase in the number of lamellar units (from 35 to 56), while the thickness of each unit is relatively unchanged. In contrast, in the abdominal segment the number of lamellar units increases minimally (from 25 to 28), but the thickness of each unit increases significantly (from 0.012 to 0.026 mm). This increase is mostly due to proliferation of smooth muscle cells. Thus, the supporting tissue for the thoracic aorta are mainly elastin and collagen, whereas those for the abdominal aorta are smooth muscle cells. Also, the thoracic aorta contains more elastin and the abdominal aorta more collagen in comparison (14).

The destruction of elastin is considered a key factor in the pathogenesis of an aneurysm. Aneurysms occur with greater frequency in the abdominal aorta where the number of elastic lamellae (and therefore elastin) is markedly decreased in comparison with the thoracic aorta (11). Elastin destruction is a striking histologic feature of an aneurysm wall. Mature elastin is not synthesized in the adult abdominal aorta and appears to have a half-life of 70 years. Disruption of elastin by elastase produces aneurysms in animal models (11). Elasticity of the aneurysm wall has been observed to be reduced, and it is correlated with reduced elastin content.

Collagen must also fail for significant dilation and rupture to occur. However, unlike elastin, collagen is continually synthesized throughout life. A deficiency in collagen type III, the most extensive form in the aortic wall, has been identified as an underlying abnormality in Ehlers-Danlos syndrome, a rare inherited disorder that includes aneurysms among its manifestations. A deficiency of type III collagen has been associated with ruptured intracerebral aneurysms, and these intracerebral aneurysms are also more common in patients with AAA (12). Also, some studies have reported that relative collagen concentration and absolute collagen content of the aneurysmal aorta are elevated compared with control aortic tissue (13).

4.1.2. Morphology of Ascending Aortic Aneurysms

It should be noted that the vascular mechanics of AAA, discussed in several sections that follow, apply equally well to the aneurysm of the ascending aorta. In particular, determination of wall stress and change in wall stress during growth of the aneurysm could be applicable to the ascending aortic aneurysm. We also expect some similarities between the AAA and the aneurysm of the ascending aorta. Thus, it is of interest to examine the morphology of the aneurysm of the ascending aorta.

Klima et al. (15) studied the morphology of the ascending aortic aneurysm, and we will consider some of their findings. Tissue samples were obtained from 339 patients who underwent resection of an ascending aortic aneurysm. Of these samples, 53 had clinical signs of Marfan syndrome and their observations will not be considered. Of the total of 286 non-Marfan's patients, 203 were men and 83 were women. The age of the patients ranged from 19 years to 87 years. Multiple histologic sections were prepared for microscopic examinations and stained with hematoxylin-eosin or other special stains.

Histologically, *cystic medial change* was found in 67% of the non-Marfan's patients. The lesion decreased with age and fewer older patients had cystic medial changes. *Elastic fragmentation* was found in 93% of non-Marfan's patients. Elastic fragmentation of grade 3 was more common than that of grade 2 or 1 (Fig. 15.3). It may be recalled that elastic fragmentation is also a striking feature of AAA. *Medial fibrosis* occurred in 59% of non-Marfan's patients. Grade 3 medial fibrosis was more common than grade 2 or 1 and with age grade 3 medial fibrosis increased (Fig. 15.4). When intimal atherosclerotic plaques were present,

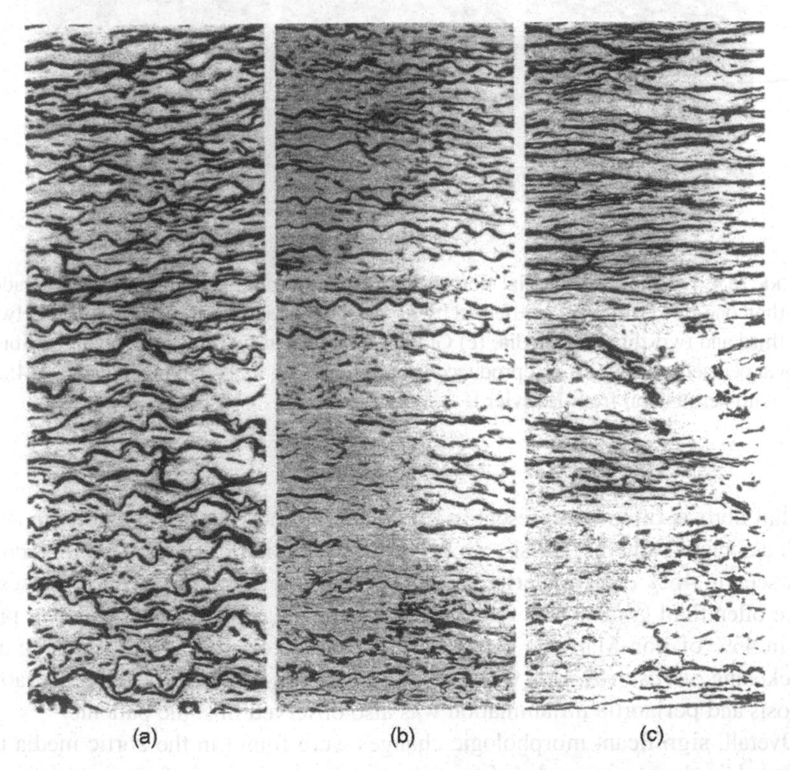

FIGURE 15.3. Examples of elastic fragmentation grades shown at the same magnification. (a–c) Grades 1 through 3, respectively, show increased fragmentation of the elastic lamellae (Movat pentachrome stain). (Reproduced from Klima T et al., Human Pathol 1983;14:810-817, with permission from Elsevier.)

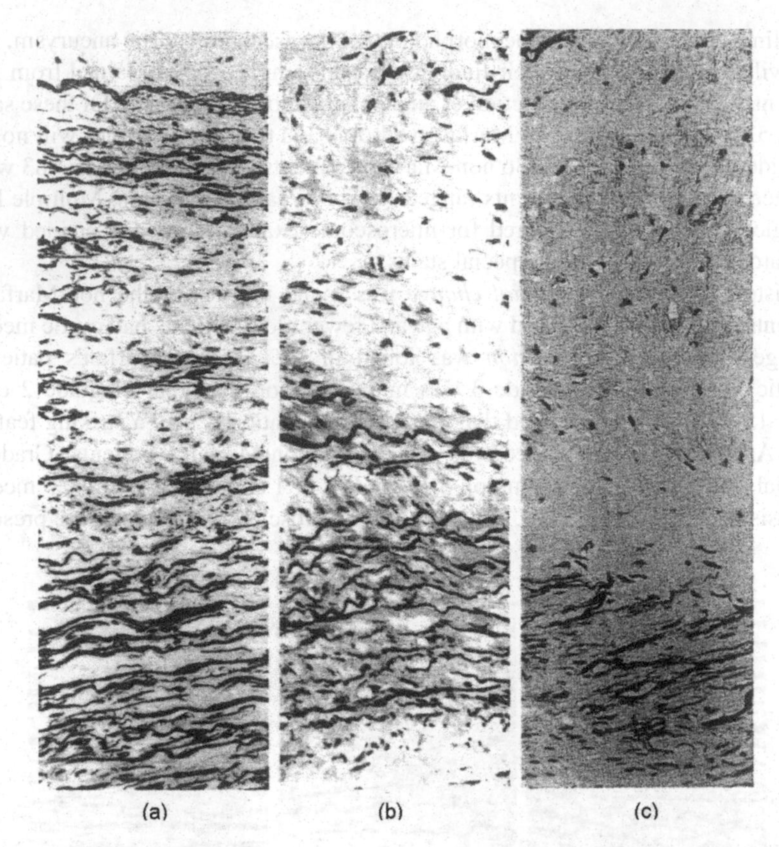

FIGURE 15.4. Examples of medial fibrosis shown at the same magnification. (a) Grade 1: less than one-third of media is affected by fibrosis. (b) Grade 2: fibrosis occupies between one-third and two-thirds of media. (c) Grade 3: more than two-thirds of media is fibrotic (Movat pentachrome stain). (Reproduced from Klima T et al., Human Pathol 1983;14:810-817, with permission from Elsevier.)

medial fibrosis often was present in the subintimal portion of the media. In AAA with associated atherosclerosis, we expect to see similar changes. *Medial necrosis* was seen in 46% of non-Marfan's patients. Grade 3 medial necrosis was present more often than grade 2 or 1 and it increased with age. *Atherosclerosis* was present in 45% of non-Marfan's patients. It was more common with increasing age. Thickening of *vasa vasorum* was seen in 21% of non-Marfan's patients. Periaortic fibrosis and periaortic inflammation was also observed in some patients.

Overall, significant morphologic changes were found in the aortic media that included, in decreasing order of frequency and extent, elastic fragmentation, cystic medial change, medial fibrosis, and medial necrosis. Both cystic medial change and elastic fragmentation were inversely correlated with the severity of atherosclerosis while fibrosis and medial necrosis were directly correlated with increasing age. The finding that many younger non-Marfan's patients have high-grade lesions of the media could suggest a "tissue insufficiency" in early life leading to aortic dilatation.

4.2. *Other Contributing Factors*

4.2.1. Genetic Factors in AAA

A high incidence of AAA among members of the same family suggests the presence of a genetic marker for this condition. Some of the possible mechanisms by which genetic variables could play a role in the development of AAA could be as follows: Inherited defects in elastin and collagen could result in a weaker aortic wall. Increased enzymatic destruction of structural components could be brought about by genetic variables. The precise genetic bases of aneurysm formation remain unresolved.

4.2.2. Atherosclerosis and AAA

The traditional view for AAA has been that the most common cause of aortic aneurysms is atherosclerosis (12). Aneurysms generally develop in older people and therefore they are invariably associated with atherosclerosis. They are even described as "atherosclerotic AAA." It seems logical that atherosclerosis can lead to progressive weakening of the arterial wall and eventually to aneurysm formation. The correctness of this epithet, however, has been challenged.

The most common site of AAA is below the renal arteries and above the aortic bifurcation. The same aortic segment is also the site of occlusive atherosclerosis (10). Hypertension and smoking enhances both atherosclerotic disease and aneurysm formation. However, there are distinct differences in the two conditions and their connection may be coincidental. Many individuals with aorto-iliac occlusive disease do not develop an aneurysm. The incidence of atherosclerotic occlusive disease peaks at the age of about 55, whereas the peak incidence of AAA occurs at age 70 or more. Atherosclerosis is a systemic disease that affects many arteries while aneurysms occur in far fewer arteries. Individuals with aneurysms do not show evidence of increased atherosclerosis in an age-matched group of people without aneurysm. Although clear epidemiological differences exist between patients with dilating arterial disease and those with stenosing disease, it is recognized that peripheral arterial disease carries a high risk of AAA (11).

4.2.3. Smoking and AAA

The strong association between cigarette smoking and AAA has been recognized for more than 20 years (11). One study showed that death from ruptured aneurysm was four times more common in smokers than in nonsmokers and 14 times more common in those who smoked hand-rolled cigarettes (16). It has also been reported that aneurysm expansion rates are greater in patients who continue to smoke (17).

4.2.4. Inflammatory Response and AAA

In 4–10% of patients undergoing surgery for AAA, the aneurysm is found to have a thick white wall that is usually adherent to surrounding structures. This so-called inflammatory aneurysm is characterized by an intense inflammatory cell infiltrate that often extends beyond the aortic wall into surrounding tissue (11). It is thought that this aneurysmal variant may be caused by an autoimmune response to components of the aortic wall leaking into surrounding tissue.

4.2.5. Load on the Aortic Wall and AAA

Hypertension is considered a risk factor for AAA for both prevalence and rupture. However, results of screening hypertensive subjects for aneurysms have been inconclusive (11).

The localization of aortic aneurysms to the infrarenal abdominal aorta is a striking feature of the condition, which calls for consideration of relevant hemodynamic factors. The pulse pressure in the aorta increases as the wave passes from the aortic root to the aortic bifurcation. Also, branching of the aorta at the bifurcation gives rise to reflection of pressure waves. The reflected pressure wave is met with a sudden expansion of the geometry from below the renal aorta to above the renal aorta. With age there is a significant reduction in the elasticity of the aorta. The combination of loss of elasticity and the "water hammer" effect, when the reflected pressure wave collides with the forward moving pressure wave of the next heartbeat, results in a greater lateral pressure on the aortic wall and greater stress. Furthermore, the increase in diameter of the aorta with progression of the aneurysm increases the stress in the wall on the basis of a larger diameter. The increase in pressure load and consequently in stress in the infrarenal aorta could cause degenerative changes in the aortic wall.

Overall, the pathogenesis of AAA seems to involve complex interaction of a variety of factors, acting over many years, which weaken the aortic wall.

5. Expansion Rate of AAA

For assessing the risk of rupture of AAA, the size of the aneurysm continues to be used as the most important parameter, as described earlier. Some groups have studied other factors, such as the rate of expansion of AAA, to explore additional parameters for determining the risk of rupture. The rate of aneurysm expansion can be used when an aneurysm is modeled in mechanics, and therefore consideration of this factor has multiple advantages.

Wolf et al. (18) examined CT scans of AAA in 80 patients who underwent scanning two times at least 6 months apart. They measured the dimensions of the luminal thrombus and the arc of the aneurysm wall covered by thrombus (TARC). They reported that the mean aneurysm expansion rate was 0.26 ± 0.25 cm/year. There was a good correlation between the rate of AAA expansion and mean TARC ($r = 0.43$), thrombus volume fraction, and thrombus area fraction. In 19%

of the patients, a rapid expansion (>0.5 cm/year) was observed. The parameters that correlated with *rapid expansion* were mean TARC and the presence of carotid artery disease. They concluded that increased thrombus load on AAA is associated with a rapid expansion rate. In other words, the presence of thrombus may be considered favorable to aneurysm growth and rupture. This should point toward early surgical repair.

Cronenwett et al. (9) also studied the rate of expansion of AAA of less than 6-cm diameter in 73 patients. This group included 54 men and 19 women, 51 to 89 years of age, and their AAAs had an initial mean size of 4.1 cm anteroposterior and 4.3 cm lateral. After a mean of 37 months of follow-ups with ultrasound technique, they found that the mean rate of aneurysm expansion was 0.4 to 0.5 cm/year. Elective operations were performed at a rate of 10% per year when the mean AAA size increased to 5.1 cm. Patients who underwent surgery had more rapid aneurysm expansion. The final aneurysm size was predictable by initial size, duration of follow-up, and by both systolic and diastolic pressure (pulse pressure). They state that small aneurysms may be safely monitored and elective repair performed when the size reaches 5–6 cm. It is important to note that, in their study, individual AAA expansion rates could not be predicted by any of the parameters tested.

6. Thrombus and AAA

A thrombus is almost always present in the aneurysm and particularly in AAA. The presence of a thrombus has many implications in terms of both its effect on the rupture of AAA and its role in the etiology of the disease. We will be examining several aspects of these relationships throughout this chapter.

Satta and Juvonen (19) studied the effect of thrombus thickness on the aneurysm rupture. In a group of 51 patients with ruptured AAAs, they measured the diameter of the aneurysm in all of them and thrombus thickness in 29 of them. They also made the same measurements on 26 patients with nonruptured, expanding, symptomatic AAAs. They used B-mode ultrasonography for the measurements. Hypertension and obstructive pulmonary disease were more common in ruptured AAA (RAAA) patients than in expanding AAA (EAAA) patients. For instance, hypertension was present in 70% of RAAAs compared with 31% of EAAAs. The distribution of maximum diameters of RAAAs and EAAAs is shown in Figure 15.5 and there seems to be no significant difference between the two groups. The median diameter for RAAA was 7 cm versus 6 cm for EAAA.

The same figure also shows the distribution of maximum hyperechoid endoluminal thrombus in the two groups. The average thrombus thickness was significantly greater in the RAAA group than in the EAAA group (3.5 cm vs. 2.0 cm). The number of small aneurysms was 14% in the RAAA group and 19% in the EAAA group. They state that because the percentage of small aneurysms was similar in the two groups, the size of the AAA is not sufficiently sensitive to be the only indicator of aneurysm rapture. Because thrombus thickness is associated

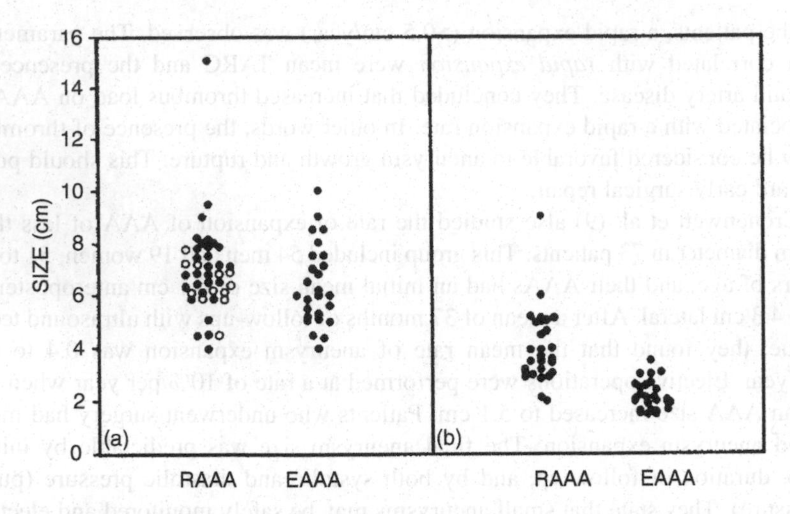

FIGURE 15.5. Distribution of the largest diameter (a) and thickness (b) of the maximum hyperechoid endoluminal thrombus in patients with RAAA (n = 29) and EAAA (n = 26). Open dots denote diameters in RAAA patients with no data on thrombus thickness (n = 22). RAAA, ruptured aneurysm; EAAA, expanding aneurysm. (Reproduced from Satta J et al., J Vasc Surg 1996;23:737-739, with permission from the Society for Vascular Surgery.)

with the risk of rupture, they suggest that this should be used as a parameter in addition to the size. In other words, the larger the thrombus thickness, the greater the probability of aneurysm rupture. Thus, as in the previous section where we considered the use of AAA expansion rate as one more risk factor, here we see a proposal that thrombus thickness should also be considered as a risk factor. The case for the thrombus thickness as a risk factor, however, is not supported by the observations we are about to discuss in the next section.

Pillari et al. (20) studied 55 patients with AAAs using computed tomography (CT) scans. The entire aneurysm was scanned at an interval of 20 mm and the scan thickness was 10 mm. The patients' age was in the range of 60 to 86 years and there were 43 men and 12 women. The aneurysm sac, lumen, and thrombus were measured at the level of the greatest aneurysm diameter. They divided the aneurysm into three groups: Group 1, less than 5 cm in size; Group 2, 5–7 cm; and Group 3, greater than 7 cm (Fig. 15.6). Table 15.2 shows the distribution of aneurysm size, thrombus area, lumen area, and thrombus arc. As the AAA size increases, the thrombus area increases and the lumen area also increases. However, the thrombus arc increases from Group 1 to Group 2 and then decreases for Group 3. The two notable observations here are that in Group 2 (5–7 cm), as the aneurysm increases in size, the thrombus volume increases with it while the lumen remains mainly unchanged. In other words, in Group 2 the aneurysm growth is associated primarily with the growth of the thrombus. In Group 3 (>7 cm), the increase in aneurysm size is associated with the increase in the lumen area and no

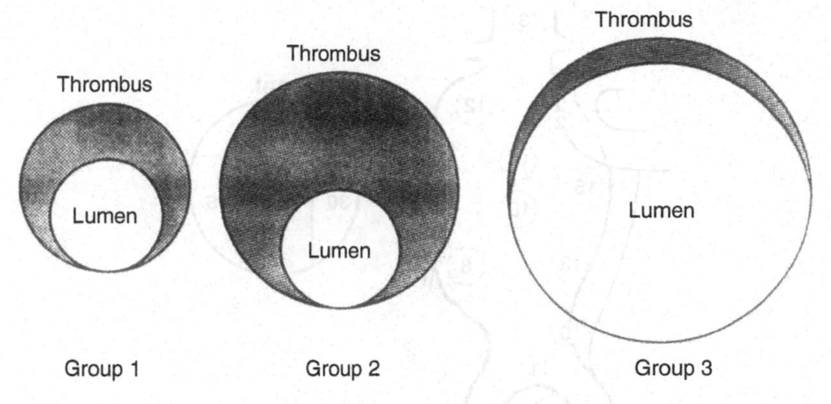

FIGURE 15.6. Illustration of internal composition of aneurysm at different sizes. Group 2 aneurysms (5 to 7 cm) show marked increase in thrombus volume associated with overall increase in sac diameter. Group 3 aneurysms (>7 cm) are characterized by increasing size of patent lumen associated with overall increase in sac diameter and no appreciable increase in thrombus volume. (Reproduced from Pillari G et al., Arch Surg 1988;123:727-732, with permission from American Medical Association.)

TABLE 15.2. Parameters for description of internal composition of aneurysms.

Group	Aneurysm size (cm)	Thrombus area (cm^2)	Lumen area (cm^2)	Thrombus arc (cm)
1	<5	5.18	9.40	168.9
2	5–7	13.37	13.56	261.5
3	>7	26.95	41.05	225.6

appreciable increase in thrombus volume. Overall, the aneurysm growth up to 7 cm is associated with a disproportionate increase in thrombus volume, whereas further growth is associated with increase in the patent lumen and no increase in thrombus volume. According to this data, the authors raise questions regarding whether the presence of thrombus could be a protection against aneurysm rupture, while absence of thrombus could promote growth and rupture. We note here that the thrombus thickness may not increase with the size of AAA, particularly from Group 2 to Group 3, and therefore greater thrombus thickness may not be used as an indicator of increased risk of aneurysm rupture.

7. Location of Rupture in AAA

The rupture of AAA occurs when stress in the aneurysm wall exceeds the strength of the wall. Therefore, it is important not only to know the stress in the wall at different locations and in different directions but also to know the strength of the wall at different locations and in different directions. Naturally then, it is of interest to

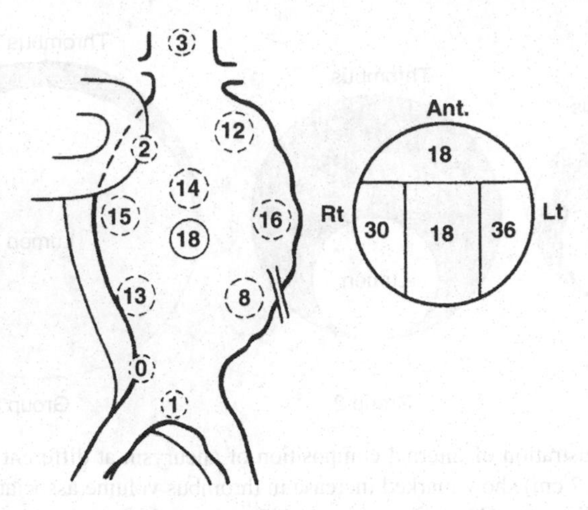

FIGURE 15.7. Sites of rupture in 118 nonresected AAAs as noted at autopsy. There was no predilection for site of perforation although the majority was into the retroperitoneum. (Reproduced from Darling RC et al., Circulation 1977;56:II-161-II-164, with permission from Lippincott, Williams & Wilkins.)

know whether the aneurysms rupture more frequently at certain locations and in certain directions. Darling et al. (3) recorded the details in 102 ruptured AAAs at autopsy. Figure 15.7 shows the site of aortic rupture. In 18 cases, the rupture occurred anteriorly and directly into the peritoneal cavity. In the vast majority (84 cases), the rupture occurred retroperitoneally. They state that there was no specific predilection for the site of the rupture. However, when we examine Figure 15.7 closely, we note that a vast majority (63 out of 102) of ruptures occurred in the bulbous region (midregion) of the aneurysm. Thus, it is possible to conclude that the rupture is more likely to occur in the bulb of the aneurysm. In this study, they did not report on the orientation of the tear. Thus, we do not know if the rupture was caused by the circumferential stress, which would produce a longitudinal tear, or by the longitudinal stress, which would produce a circumferential tear. The tear could also be oriented obliquely in which case the causative stress is also oriented obliquely.

8. Effect of Thrombus on AAA Dilation and Stress

It is obvious that an AAA rupture is caused by stress in the wall when it exceeds the strength of the wall. It is also understood that a certain number of small aneurysms rupture because we are not able to predict the risk of rupture. Clinically, their risk of rupture is small and therefore they go untreated. The prediction of risk of rupture depends on the accurate knowledge of the wall stress and on the knowledge of the wall strength. Because the wall strength may not be known in patients, the decision to intervene may have to be based primarily on the knowledge of the wall stress. Thus, the more accurately we can determine the wall stress, the more precisely we will be able to predict the risk of rupture.

In this and the following sections, we will see that the determination of wall stress in AAA is complicated. We will also note that there are several aspects to the behavior of the intact aneurysm, which must be considered because they have a strong bearing on wall stress. We will describe some of these aspects individually to gain an insight into them. However, we know fully well that only comprehensive knowledge can lead to the accurate determination of stress and to the accurate determination of risk of rupture. Because the majority of AAAs have a thrombus in them, we begin with an inquiry of how does the thrombus affect the pressure on the aortic wall and how does it affect the dilation of the aneurysm. Thubrikar et al. (21) studied intact aneurysms, which contained thrombus, and we will describe some of their findings.

8.1. Pressure Transmission Through the Mural Thrombus

Two abdominal aortic aneurysms with mural thrombus were removed intact from two patients, both age 70. The diameters of the aneurysms after removal were measured to be 5.3 cm (AAA 1) and 4.5 cm (AAA 2), respectively. The aneurysms were cannulated and filled with normal saline solution under 100 mmHg pressure, keeping the body of the AAA horizontal and free to expand in response to static pressurization. The pressure transmission from the lumen through the thrombus was measured using a Millar microtip needle catheter pressure transducer inserted through the wall from the outside inward and advanced through the thrombus in 2-mm steps. At each depth, the needle was fully rotated in 90-degree increments and pressure readings taken at each rotation. Once the transducer tip had reached the lumen, it was withdrawn in 2-mm decrements and the pressure was measured again. The pressure transmission measurements were repeated in the anterior, lateral, and posterior regions of the AAA. Figure 15.8 shows the typical distribution of pressure through the thrombus at one site of aneurysm 1. At this location, the thrombus was more than 8 mm thick. The pressure drop through the thrombus was at no time larger than 13%, and it was progressively less as the catheter approached the lumen. The results obtained from multiple sites in each of the anterior, lateral, and posterior regions of the aneurysm were averaged. The maximum pressure reduction through the thrombus was 9 ± 8% at 2 mm from the aortic wall, and it decreased to 4 ± 3 % at 6 mm. The pressure acting on the aneurysm wall was 92 ± 6 % of the luminal pressure.

In vivo, in one patient, the measurement of the pressure through the thrombus was done during surgery, just after exposure of the AAA. The patient was a 67-year-old male and the aneurysm was 5.7 cm in diameter. A 16-G angiocatheter, modified with a side port and connected to a fluid pressure transducer, was inserted through the aortic wall and advanced through the thickness of the thrombus in 2-mm increments. The pressures inside the thrombus were averaged and compared with radial artery pressures. In vivo, the pressure measurements showed a close match between the pressure inside the thrombus and that in the

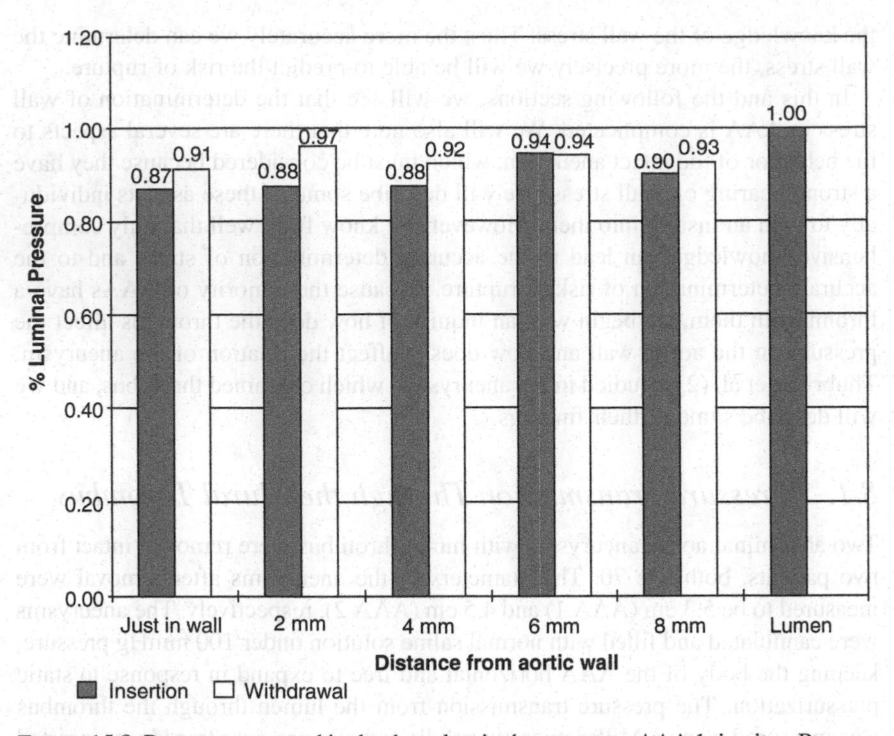

FIGURE 15.8. Pressure measured in the thrombus in the aneurysm AAA 1, in vitro. Pressure during insertion and withdrawal of the catheter through the thrombus is shown. Luminal pressure is taken as a reference pressure of 1 (or 100%).

lumen. This was the case regardless of the depth. When the patient had a systemic pressure of 124/68 mmHg, the pressure through the thrombus was measured to range from 125/55 to 127/55 mmHg. Hence, in a patient the complete transmission of pressure through the thrombus was in agreement with the results of in vitro experiments.

8.2. AAA Dilation Under Intraluminal Pressure

The two complete AAAs were used in the dilation experiments. To determine the dilation of the aneurysm, the marker technique was used (21). In the anterior, lateral, and posterior regions of the aneurysm, a finely ground potassium permanganate powder was sprinkled to create randomly distributed dots. The proximal end of the aneurysm was cannulated and the distal end was tied. A container of normal saline solution was connected to the cannula and was used to produce the desired hydrostatic pressure within the aneurysm. The pressure was increased from 0 to 120 mmHg in 10–20 mmHg increments and then decreased in a similar manner. A video camera was used to record the expansion of the aneurysm as a function of pressure.

By using still frames of the videotape, the distance between two dots oriented circumferentially and two dots oriented longitudinally was measured. Then, the thrombus was carefully separated from the aneurysm wall, removed, and the dilation experiments repeated.

Figure 15.9 shows a typical result for the distance between two dots, oriented longitudinally, in the lateral region of aneurysm 1. With the thrombus present, the distance between the two markers increased slightly up to a pressure of 40 mmHg and then remained almost constant. After removal of the thrombus, the distance increased almost linearly up to a pressure of 60 mmHg and then remained roughly unchanged.

Comparison of strains with the thrombus present and removed was carried out for the two ranges of pressures; 0 to 40 mmHg and 40 to 100 mmHg (Table 15.3). This separated the range of the large strains (0–40 mmHg) from that of the small strains (40–100 mmHg), which not only allowed better comparison but also predicted what may occur under systemic pressure.

For aneurysm 1, in the circumferential direction, the strains ranged from 4% to 7.8% for 0 to 40 mmHg pressure and only from 1.0% to 1.8% for 40 to 100 mmHg pressure in the presence of the thrombus. These strains ranged from 4.4% to 8% and 1.1% to 1.8%, respectively, when the thrombus was removed. In the longitudinal direction, for the pressure range of 0 to 40 mmHg the strains were in the range of 2.8% to 8% when the thrombus was present and higher (3.6–11.4%), when the thrombus was removed. For the higher pressure range of 40 to 100 mmHg, the strains were again lower (0.7–1.8%) when the thrombus was present and did not change much (0.9–2.5%) when the thrombus was removed.

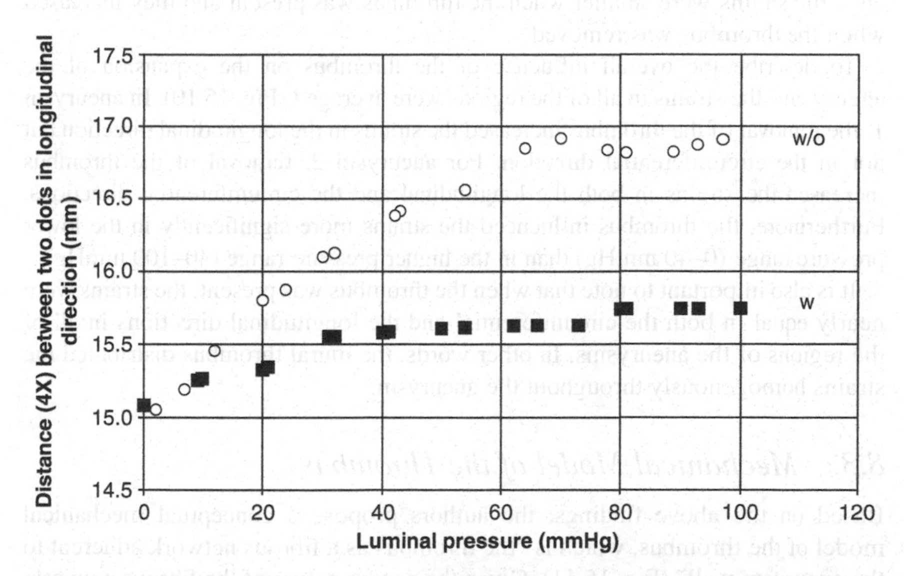

FIGURE 15.9. A typical plot of distance between two markers in the longitudinal direction versus luminal pressure in AAA 1. The measurements were done in the aneurysm with thrombus (w) and also in the same aneurysm after the thrombus was removed (w/o). This figure is representative of the effect of the thrombus on dilation in various regions.

TABLE 15.3. Dilation (%) of the aneurysm under pressure.

Aneurysm		With thrombus 0–40 mmHg	With thrombus 40–100 mmHg	Without thrombus 0–40 mmHg	Without thrombus 40–100 mmHg
AAA1					
Circumferential	Anterior	4.3	1.0	4.7	1.1
	Lateral	7.8	1.8	4.4	1.1
	Posterior	4.0	1.0	8.0	1.8
Longitudinal	Anterior	2.8	0.7	11.4	2.5
	Lateral	8.0	1.8	10.7	2.4
	Posterior	5.3	1.3	3.6	0.9
AAA2					
Circumferential	Anterior	5.6	1.3	15.1	3.3
	Lateral	2.4	0.6	5.8	1.4
	Posterior	3.5	0.8	7.2	1.7
Longitudinal	Anterior	3.0	0.7	7.0	1.6
	Lateral	1.6	0.4	7.8	1.8
	Posterior	2.5	0.6	4.5	1.1

For aneurysm 2, in the circumferential direction, the strains ranged from 2.4% to 5.6% for 0 to 40 mmHg pressure and only from 0.6% to 1.3% for 40 to 100 mmHg pressure when the thrombus was present. These strains increased considerably when the thrombus was removed; 5.8% to 15.1% for 0 to 40 mmHg pressure and 1.4% to 3.3% for 40 to 100 mmHg pressure. In the longitudinal direction also, the strains were smaller when the thrombus was present and they increased when the thrombus was removed.

To describe the overall influence of the thrombus on the expansion of the aneurysms, the strains in all of the regions were averaged (Fig. 15.10). In aneurysm 1, the removal of the thrombus increased the strains in the longitudinal direction but not in the circumferential direction. For aneurysm 2, removal of the thrombus increased the strains in both the longitudinal and the circumferential directions. Furthermore, the thrombus influenced the strains more significantly in the lower pressure range (0–40 mmHg) than in the higher pressure range (40–100 mmHg).

It is also important to note that when the thrombus was present, the strains were nearly equal in both the circumferential and the longitudinal directions in all of the regions of the aneurysms. In other words, the mural thrombus distributed the strains homogenously throughout the aneurysm.

8.3. *Mechanical Model of the Thrombus*

Based on the above findings, the authors propose a conceptual mechanical model of the thrombus, which is "the thrombus as a fibrous network adherent to the aneurysm wall" (Fig. 15.11). Given the porous nature of the fibrous network, while the thrombus transmits most of the luminal pressure to the aneurysm wall, it reduces the dilation of the aneurysm under pressure, as the fibrous network has to stretch with the aneurysm. Because the thrombus may not be uniformly

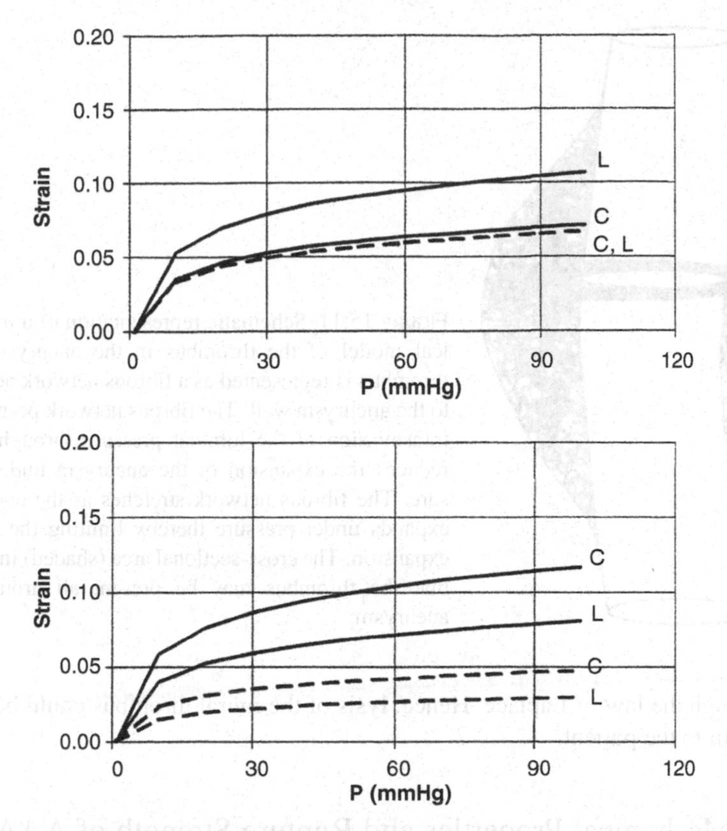

FIGURE 15.10. Average curves of the engineering strain versus the luminal pressure P for AAA 1 (Top) and AAA 2 (Bottom). These averages are after combining anterior, lateral, and posterior region data. Dashed line, aneurysm with thrombus; solid line, aneurysm without thrombus; C, circumferential direction; L, longitudinal direction.

distributed inside the aneurysm, removing the thrombus may not increase the dilation uniformly.

Obviously, the benefit of less dilation (strain) on the aortic wall is to reduce the wall stress. To serve as an effective reinforcement, the thrombus has to be firmly adherent to the aneurysm wall, and this was the case in both aneurysms.

8.4. Clinical Significance of Thrombus in Aneurysm

Because the strains are reduced by the presence of the thrombus, the stress in the aneurysm wall is also reduced. Thus, for two aneurysms of identical size, the one containing a thrombus has a lower wall stress and consequently is less likely to rupture compared with the other without the mural thrombus. If for some reason the thrombus undergoes lysis, then such an aneurysm will have two factors increasing the stress on its wall; it will have to bear the stress previously borne by the thrombus and it will also dilate immediately, thereby increasing the stress on

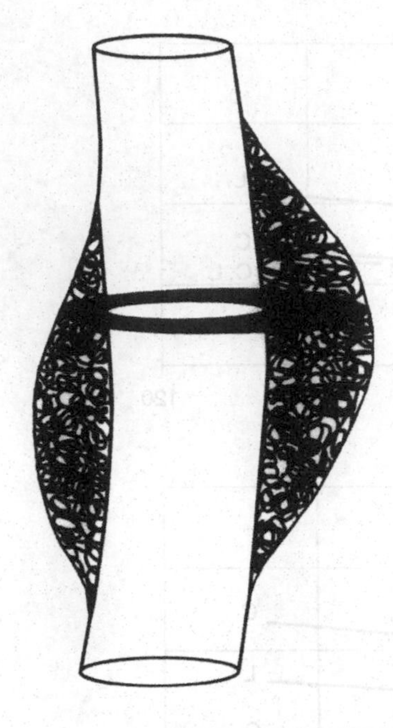

FIGURE 15.11. Schematic representation of a mechanical model of the thrombus in the aneurysm. The thrombus is represented as a fibrous network adherent to the aneurysm wall. The fibrous network permits the transmission of the luminal pressure through it but reduces the expansion of the aneurysm under pressure. The fibrous network stretches as the aneurysm expands under pressure thereby limiting the overall expansion. The cross-sectional area (shaded) indicates that the thrombus may be present all around the aneurysm.

it through the law of Laplace. Hence, lysis of the mural thrombus could be quite harmful to the patient.

9. Mechanical Properties and Rupture Strength of AAA

In the previous section, we noted that thrombus in the AAA allows luminal pressure to be transmitted to the aneurysm wall but reduces the amount of dilation under pressure. In this section, we will examine the mechanical properties of the aneurysm wall, which would be required for modeling the aneurysm for determination of wall stress. We will also examine the rupture strength of AAA in different regions of the aneurysm and in different directions. As stated earlier, the knowledge of both the wall stress and the rupture strength is necessary in order to understand when, where, and why the AAA may rupture.

Thubrikar et al. (22) studied the mechanical properties and rupture strength of AAA, and we will consider some of their findings. Five aneurysms, 5 cm or greater in diameter, were obtained from patients (60 to 87 years of age) undergoing elective surgical repair. Forty-seven test specimens were prepared by cutting the aneurysms into small rectangular pieces. Twenty-one specimens were longitudinally oriented, and 26 were circumferentially oriented. The typical sample size was 16 mm × 4.3 mm. The thickness of the specimens was measured.

The uniaxial tensile tests were run on an Instron Mini 44 load frame. The specimens were held between the load cell and the crosshead using specially designed grips made from surgical clamps. The initial length of the samples under zero load was recorded (typically 10 mm). In order to make sure that no slippage was present, as well as to obtain reproducible data, each specimen was preconditioned by three successive loadings to 5% strain and unloadings at a constant strain rate of 10% min^{-1}. Then, the sample was stretched at the same strain rate until rupture.

The force and displacement data were transformed into true stress σ and engineering strain ε. The true stress (or Cauchy stress) is the ratio of the force f applied to the current cross-sectional area a of the sample. Using the assumption that the volume of the specimen was conserved, the current cross-sectional area was

$$a = \frac{l_0 wt}{l_0 + \Delta l}$$

where l_0 is the initial length, w is the initial width, t is the initial thickness, and Δl is the elongation of the specimen. The true stress was calculated as

$$\sigma = \frac{f}{a} = \frac{f}{wt}(1 + \varepsilon)$$

where the engineering strain ε is defined as

$$\varepsilon = \frac{\Delta l}{l_0}.$$

The curve of true stress σ versus engineering strain ε was plotted for each specimen. This type of curve typically exhibits a strong nonlinearity. Different forms of curve fit can be used. They used a power equation with two parameters a and b shown below:

$$\sigma = a\varepsilon^b$$

with $a > 0$, $b > 1$ and $0 \leq \varepsilon \leq 1$, and that provided an excellent fit. The best-fit parameters a and b were determined for each curve. When several samples from different aneurysms, regions, or directions were considered for averaging, the data from these samples were pooled into a single data set. The least squares regression was then used to obtain the best-fit parameters a and b describing the average curve.

The incremental modulus E at any point of the stress-strain curve is equal to the value of the slope. Thus, we get:

$$E = \frac{d\sigma}{d\varepsilon} = ab\varepsilon^{b-1} = b\frac{\sigma}{\varepsilon}.$$

The parameters a and b for each specimen were determined from the stress-strain curve between zero load and the yield point. The yield point is defined here as the point of the stress-strain curve where the slope ($d\sigma/d\varepsilon$) begins to decrease with increasing strain. Because permanent damage could occur beyond the yield point, the yield stress σ_Y and the yield strain ε_Y of the specimens are considered important and are reported. The mean yield stress and yield strain were determined as the arithmetic average.

The aneurysmal aorta specimens were separated into six groups according to i) the region (anterior, posterior, or lateral) and ii) the direction (circumferential or longitudinal).

First, by examining the data on a single aneurysm, one can appreciate the variations in the material properties in different regions of the aneurysm (22). The experimental data obtained from multiple samples taken from the anterior, lateral, and posterior regions of one aneurysm show that in any given region there is an appreciable variation in the stress-strain properties. Figure 15.12 presents the average best-fit curves for the anterior, lateral, and posterior regions of one aneurysm. In both circumferential and longitudinal directions, the stress-strain curves of the lateral region are above those of the anterior region, which are above those of the posterior region, suggesting that the wall stiffness decreases from lateral to anterior to posterior regions. Furthermore, for each of the regions, the relative position of the curves suggests that the tissue is stiffer in the circumferential than in the longitudinal direction.

The thickness of the specimens was also measured. In the five aneurysms, the thickness gradually decreased from posterior (2.73 ± 0.46 mm) to lateral (2.52 ± 0.67 mm) to anterior (2.09 ± 0.51 mm) region.

For the five aneurysms, the averages of the parameters *a*, *b*, yield stress and yield strain, for each of the six groups are shown in Table 15.4.

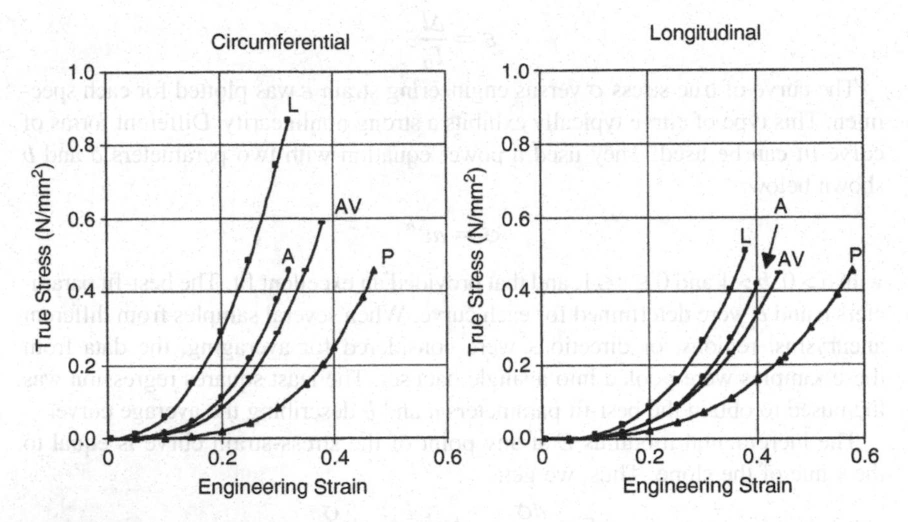

FIGURE 15.12. Average best-fit stress-strain curves in each of the three regions of the aneurysm. Also, the stress-strain curve representing the grand average of the three regions is shown. L, A, P: lateral, anterior, and posterior region, respectively. AV: grand average. Wall stiffness increases from posterior to anterior to lateral regions of the aneurysm. Also, the stiffness is greater in the circumferential direction than in the longitudinal direction.

TABLE 15.4. Material properties of the aneurysm wall, $\sigma = a\varepsilon^b$ (*five aneurysms*).

Region	Direction	a (N/mm²)	b	σ_Y (N/mm²)	ε_Y
Anterior	Circ.	17.06 ± 1.1	3.01 ± 0.1	0.52 ± 0.2	0.33 ± 0.1
Lateral	Circ.	12.39 ± 1.2	2.21 ± 0.1	0.73 ± 0.2	0.28 ± 0.1
Posterior	Circ.	10.14 ± 1.6	3.54 ± 0.3	0.45 ± 0.1	0.39 ± 0.1
Anterior	Long.	7.84 ± 1.3	2.97 ± 0.1	0.38 ± 0.2	0.32 ± 0.1
Lateral	Long.	7.43 ± 1.1	2.77 ± 0.1	0.51 ± 0.1	0.38 ± 0.0
Posterior	Long.	2.08 ± 1.5	2.80 ± 0.3	0.47 ± 0.3	0.58 ± 0.0

σ_Y = yield stress; ε_Y = yield strain.

9.1. Comparing Different Regions

In the *circumferential direction*, the yield stress of the lateral region was greater than that of the anterior or posterior region (0.73 ± 0.22 N/mm^2 vs. 0.52 ± 0.20 N/mm^2 or 0.45 ± 0.14 N/mm^2). For a given strain, stress in the lateral region was greater than that in the anterior region, which was greater than that in the posterior region (Fig. 15.12).

In the *longitudinal direction*, the yield strain in the posterior region was larger than that in the anterior or lateral region (0.58 ± 0.04 vs. 0.32 ± 0.12 or 0.38 ± 0.02). For a given strain, stress in the lateral or anterior regions was greater than that in the posterior region (Fig. 15.12).

9.2. Comparing Different Directions

For any given region, there was no marked difference between the yield stress in the circumferential direction and that in the longitudinal direction (Table 15.4). However, in the posterior region, the yield strain was higher in the longitudinal direction than in the circumferential direction (0.58 ± 0.04 vs. 0.39 ± 0.13). Also, in any region, the stress-strain curve in the circumferential direction was above its counterpart in the longitudinal direction.

The yield stress in different regions and in different directions is shown also in Figure 15.13. From these results, the anterior region appears to be the weakest part of the aneurysm, especially in the longitudinal direction (σ_Y = 0.38 ± 0.18 N/mm^2, ε_Y = 0.32 ± 0.12).

The incremental modulus as a function of strain was also determined. The modulus varies from <1 N/mm^2 to almost 6 N/mm^2 in the aneurysm. It is of interest to know which value of the modulus should be used when modeling the aneurysm for determination of stress under systemic pressure. For a 5-cm-diameter aneurysm, modeled as a closed-end cylinder with a 2-mm-thick wall, the law of Laplace yields about 0.33 N/mm^2 and 0.16 N/mm^2 for the circumferential and the longitudinal stresses, respectively, under 100 mmHg pressure. By reading the corresponding strains for the different regions and directions, one can determine that the incremental modulus at 100 mmHg is about 4.0 N/mm^2 in the circumferential direction and about 1.5 N/mm^2 in the longitudinal direction. These values may be used as a first-order approximation when the aneurysm is modeled as having linear orthotropic material properties at systemic pressure.

It is apparent that in the aneurysm, the stress-strain properties are different in different regions and in different directions. However, for the stress analysis, it may not be possible to consider this degree of complexity given the available analytical software. Also, for practical reasons it may be necessary to carry out the stress analysis with different levels of complexity. For these reasons, the material properties of the aneurysm may be represented in a couple of different ways. The material properties can be consolidated to represent orthotropic behavior as well as isotropic behavior for the entire aneurysm. The results are as follows (Table 15.5; Fig. 15.14). When all of the regions are combined to

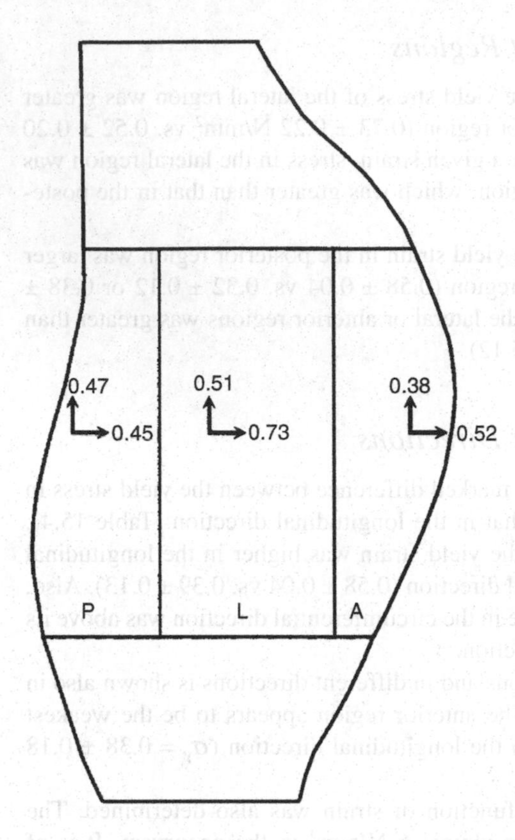

FIGURE 15.13. Schematic drawing of an aneurysm showing the approximate areas of anterior (A), lateral (L), and posterior (P) regions from which the samples for uniaxial tests were taken. Also, the value of the yield stress (N/mm^2) in each region in the circumferential (horizontal) and longitudinal (vertical) directions are shown. The silhouette of AAA is drawn from an actual angiogram of a patient. Upper end represents the proximal end of the aneurysm.

represent the orthotropic model of the aneurysm, the stress-strain curve for the circumferential direction lies above that for the longitudinal direction indicating that the wall is stiffer circumferentially. The material properties of the wall may be described by equations $\sigma = 12.89\varepsilon^{2.92}$ in the circumferential direction and $\sigma = 4.95\varepsilon^{2.84}$ in the longitudinal direction. Also, the wall is stronger in the circumferential direction than in the longitudinal direction ($\sigma_Y = 0.55 \pm 0.21$ N/mm^2 vs. $\sigma_Y = 0.42 \pm 0.19$ N/mm^2), whereas the yield strains are not significantly different. For an isotropic model, the material properties may be described by $\sigma = 7.89\varepsilon^{2.88}$ and the yield stress and strain are respectively $\sigma_Y = 0.50 \pm 0.21$ N/mm^2 and $\varepsilon_Y = 0.35 \pm 0.11$.

TABLE 15.5. Average material properties, $\sigma = a\varepsilon^b$.

Model	Direction	a (N/mm^2)	b	σ_Y (N/mm^2)	ε_Y
Orthotropic	Circ.	12.89±1.4	2.92±0.2	0.55±0.2	0.33±0.1
Orthotropic	Long.	4.95±1.2	2.84±0.1	0.42±0.2	0.38±0.1
Isotropic	All	7.89±1.2	2.88±0.1	0.50±0.2	0.35±0.1

σ_Y = yield stress; ε_Y = yield strain.

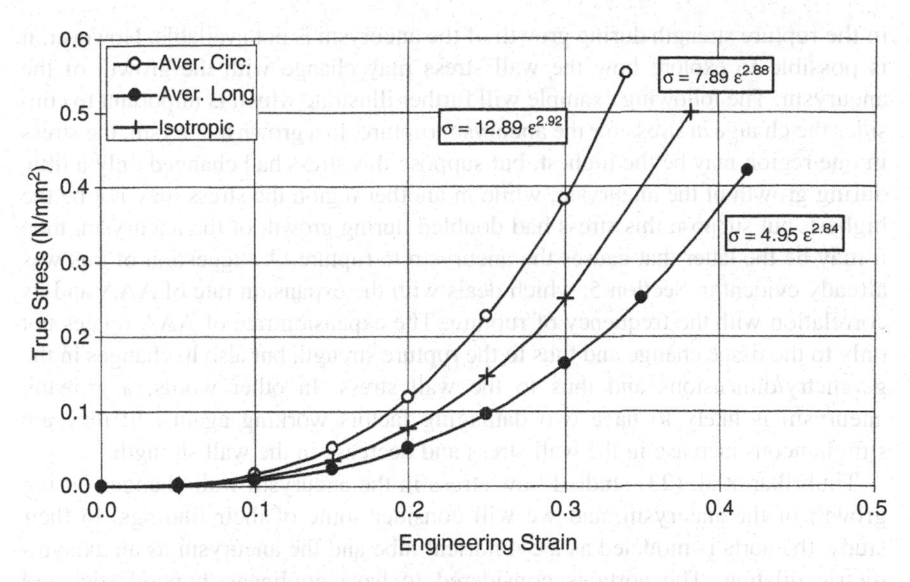

FIGURE 15.14. Average stress-strain curves for all regions and all aneurysms. Orthotropic properties are represented by the curves for circumferential and longitudinal directions, while isotropic properties are determined by further averaging these two curves.

In summary, the stress-strain properties of the aortic wall are different in anterior, lateral, and posterior regions of the aneurysm. The wall thickness is also different in different regions of the aneurysm. The yield stresses and strains are also different in different regions of the aneurysm as well as in different directions (circumferential vs. longitudinal). In general, the tissue is stiffer and stronger in the circumferential direction than in the longitudinal direction, and the material properties are highly nonlinear in both directions. The anterior region seems to be thinnest and weakest, particularly in the longitudinal direction, which suggests that if the aneurysm ruptures in this region, the tear is likely to be circumferential. As mentioned earlier, this detail knowledge is necessary for understanding where and why the aneurysms may rupture.

10. Changes in Wall Stress During Growth of the Aneurysm

The previous section points out clearly that in the aneurysm, both the material properties and the rupture strength of the wall are different in different regions of the aneurysm. Furthermore, these properties are changing as the aneurysm is growing. Therefore, it is not enough to simply know the stress in the aneurysm wall and the rupture strength but it is important to know how both of these change during the growth of the aneurysm. Unfortunately, the information on changes in the rupture strength during growth of the aneurysm is not available. However, it is possible to explore how the wall stress may change with the growth of the aneurysm. The following example will further illustrate why it is important to consider the change in stress for the aneurysm rupture. In a grown aneurysm, the stress in one region may be the highest, but suppose this stress had changed only a little during growth of the aneurysm, while in another region the stress may not be the highest, but suppose this stress had doubled during growth of the aneurysm, then it may be the latter that causes the aneurysm to rupture. A suggestion of this was already evident in Section 5, which deals with the expansion rate of AAA and its correlation with the frequency of rupture. The expansion rate of AAA relates not only to the tissue change and thus to the rupture strength but also to changes in the geometry/dimensions and thus to the wall stress. In other words, a growing aneurysm is likely to have two damaging factors working against it; they are simultaneous increase in the wall stress and decrease in the wall strength.

Thubrikar et al. (23) studied how stress in the aneurysm wall changes during growth of the aneurysm, and we will consider some of their findings. In their study, the aorta is modeled as a cylindrical tube and the aneurysm as an axisymmetric dilation. The aorta is considered to have nonlinear, hyperelastic, and isotropic material properties. The "growth" of the aneurysm (which in patients occurs over several months or years) is simulated by creating multiple models of the aneurysm with increasing bulb radius. Finite element analysis is used to determine the stresses, which develop under internal pressure load.

10.1. Material Properties of the Normal Aorta

Although the material properties of the aneurysmal aorta are reported (24) to be different from those of the normal aorta, the authors chose to use the properties of the normal aorta, which they determined during the study. This may be justified on the basis that the study deals with the "growing" aneurysm where the material properties are likely to be changing between those of the normal aorta and those of the aneurysmal aorta. Furthermore, the main results of the study are not influenced significantly by the material properties.

In the study, the aorta is considered to be homogeneous, isotropic, and nonlinearly hyperelastic. Its nonlinear hyperelastic material properties are described by a polynomial form of strain energy density function used by ANSYS (a finite element analysis software). The polynomial form of strain energy density function used has five coefficients. These coefficients were determined from the experiments performed on a human ascending aorta. In brief, a human ascending aorta with aortic arch was cannulated and subjected to a static internal pressure. In the lateral portion of the aorta (lateral to the curvature of the arch), small marks were made in the circumferential and the longitudinal directions using potassium permanganate powder. The positions of the marks were then recorded under an internal pressure of 0, 20, 40, 60, 80, 100, and 120 mmHg. From the distance between the marks, both the circumferential and the longitudinal strains were determined at each pressure. Also, the circumferential and the longitudinal stresses were determined considering the aorta as a thin cylindrical vessel. From these stresses and strains, the five coefficients of the strain energy density function were determined. The experimental details and calculations of coefficients have been described previously in Chapter 7 for the case of the aortic arch.

One set of coefficients was determined for the circumferential direction and another set for the longitudinal direction. From these two sets, a new set of coefficients was computed by taking the averages, which represented the aorta as an isotropic material. The average values of the five coefficients were A = 2.839, B = 0, C = 24.189, D = −24.364, and E = 7.227.

10.2. Geometry of the Aneurysm Model

The aneurysm was modeled as an axisymmetric dilation in a cylindrical aorta of 1-inch diameter (Fig. 15.15). The length of the straight segment was 2 inches on either side of the aneurysm. To simulate the growth of the aneurysm, multiple such models were generated by increasing the radius of the bulb in steps of 10% up to 100% while keeping the length of the bulb constant at 2 inches. Thus, the largest bulb in the model had a radius that was double the original value. This geometry is close to that seen in adults. For instance, the diameter of the ascending aorta is approximately 1 inch or slightly greater in some individuals while that of the abdominal aorta may be slightly less. The length of the aneurysm bulb used is 2 inches. The aneurysm in the ascending aorta could be 2–3 inches long (Fig. 15.1), while that in the abdominal aorta could be close to 3 inches long. In all of the models, an initial wall thickness of 0.19 inches was used for the entire length of the aorta. This gives a radius to wall thickness ratio (R/t) of 2.6 for the undilated aortic segment under zero pressure. Others, in analyzing the artery, have used this ratio to be 2 or 3 (25, 26). In vivo for the normal undilated aorta under pressure, this ratio is in the range of 8 to 10. In the fully developed aneurysm, this ratio may be slightly greater. In their model, when the bulb radius is twice that of the straight segment, under pressure this ratio at the bulb is approximately 10.

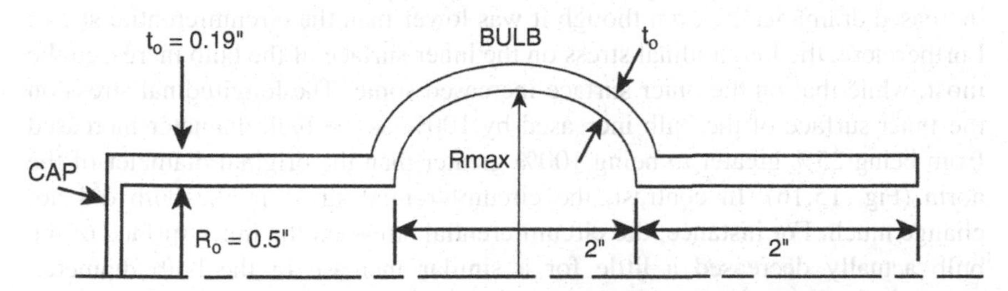

FIGURE 15.15. Schematic representation of the axisymmetric aneurysm of the aorta at zero pressure. R_{max} represents the radius of the bulb at its maximum dilation. t_o represents thickness of the aortic wall. R_0 represents radius of the undilated aorta.

10.3. *Finite Element Analysis of the Model*

The material properties were kept the same for the entire aorta including the bulb. One end of the model was constrained in both the radial and the axial directions while the other end was capped and permitted to move along the axis (i.e., the aorta was allowed to lengthen under pressure). Although in the body the aorta is prestretched and tethered and thus does not change length, its prestretched length is close to and perhaps slightly greater than that which it acquires under a pressure of 120 mmHg (27). Thus, by capping the end and allowing the aorta to lengthen, the geometry of the aorta is allowed to become closer to that present in vivo. The internal pressure of 120 mmHg was applied to each model of the aneurysm. The stress analysis was carried out using commercially available finite element analysis (FEA) software ANSYS 5.0. Quadrilateral elements were used. There was a minimum of four elements through the thickness in the bulb region. 2D element-Hyper 56 was used.

After the aorta was loaded with the internal pressure, various parameters such as axial and circumferential stress and strain, thickness, and radial displacements were determined for both the straight segment and the bulb at the following locations: in the middle of the straight segment toward the fixed end and in the bulb at the maximum radius R_{max} in Figure 15.15.

Figure 15.16 shows the plot of various stresses in the aneurysm bulb and in the straight segment as a function of initial bulb diameter. The stresses in the bulb were always greater than the corresponding stresses in the straight segment. In the bulb, both the circumferential and the longitudinal stresses were greater than those in the straight segment. This was true for the stresses on the inner surface as well as on the outer surface of the bulb. In all of the models, the circumferential stresses were greater than the longitudinal stresses both in the bulb and in the straight segment. In the bulb, both the circumferential stresses and the longitudinal stresses were greater on the inner surface than on the outer surface. In the straight segment, the circumferential stresses were greater on the inner surface than on the outer surface, however, the longitudinal stresses were slightly greater on the outer surface than on the inner surface.

As the diameter of the bulb increased, the longitudinal stress in the bulb increased dramatically, even though it was lower than the circumferential stress. Furthermore, the longitudinal stress on the inner surface of the bulb increased the most, while that on the outer surface increased some. The longitudinal stress on the inner surface of the bulb increased by 100% as the bulb diameter increased from being 25% greater to being 100% greater than the original diameter of the aorta (Fig. 15.16). In contrast, the circumferential stress in the bulb did not change much. For instance, the circumferential stress on the inner surface of the bulb actually decreased a little for a similar increase in the bulb diameter, although it was larger than the longitudinal stress. In the straight segment of the aorta, the growth of the aneurysm did little to change the circumferential stress or the longitudinal stress. Furthermore, the larger the bulb diameter to begin with, the greater was its radial distention. In other words, the larger the aneurysm, the more it distends under pressure.

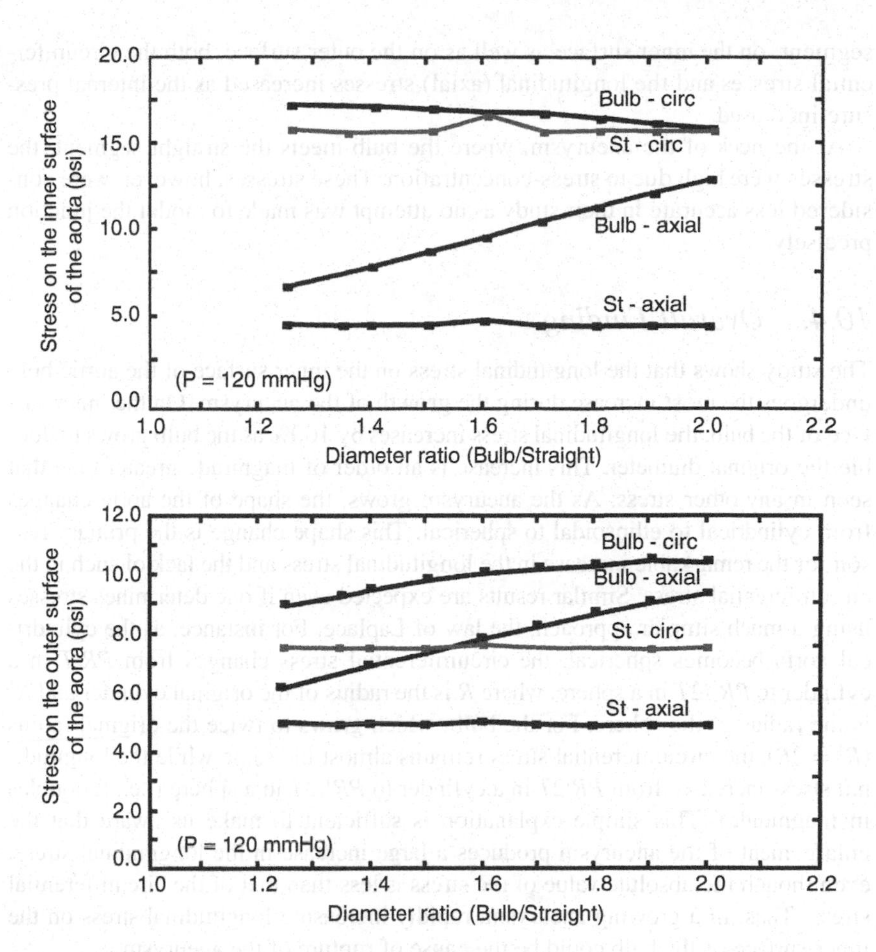

FIGURE 15.16. Stresses in the aortic wall as a function of diameter of the bulb for a pressure load of 120 mmHg. Both the circumferential (transverse) and the longitudinal (axial) stresses in the bulb and in the straight segment are shown. As the aneurysm becomes larger (as the diameter ratio increases), the longitudinal stress in the bulb increases more rapidly than any other stress. Bulb-circ, circumferential stress in the bulb; Bulb-axial, longitudinal stress in the bulb; St-circ, circumferential stress in the straight segment; St-axial, longitudinal stress in the straight segment.

They explored how the thickness of the bulb and that of the straight segment changes under pressure as a function of bulb diameter. Under the pressure load of 120 mmHg, the thickness of the aorta decreased, both in the bulb and in the straight segment. Furthermore, under the same pressure load, the bulb got considerably thinner as the bulb size increased, whereas the thickness of the straight segment changed only slightly.

They also explored how the stresses in the aorta changed as a function of the internal pressure load. All of the stresses in the bulb and in the straight segment increased as the luminal pressure increased. In the bulb as well as in the straight segment, on the inner surface as well as on the outer surface, both the circumferential stresses and the longitudinal (axial) stresses increased as the internal pressure increased.

At the neck of the aneurysm, where the bulb meets the straight segment, the stresses were high due to stress-concentration. These stresses, however, were considered less accurate in their study as no attempt was made to model the junction precisely.

10.4. Overall Finding

The study shows that the longitudinal stress on the inner surface of the aortic bulb undergoes the most increase during the growth of the aneurysm. On the inner surface of the bulb, the longitudinal stress increases by 100% as the bulb grows to double the original diameter. This increase is an order of magnitude greater than that seen in any other stress. As the aneurysm grows, the shape of the aorta changes from cylindrical to ellipsoidal to spherical. This shape change is the primary reason for the remarkable increase in the longitudinal stress and the lack of such in the circumferential stress. Similar results are expected even if one determines stresses using a much simpler approach, the law of Laplace. For instance, as the cylindrical aorta becomes spherical, the circumferential stress changes from PR/T in a cylinder to $PR^1/2T$ in a sphere, where R is the radius of the original cylinder and R^1 is the radius of the sphere. For the bulb, which grows to twice the original radius ($R^1 = 2R$), the circumferential stress remains almost the same while the longitudinal stress increases from $PR/2T$ in a cylinder to $PR^1/2T$ in a sphere (i.e., it doubles in magnitude). This simple explanation is sufficient to make us aware that the enlargement of the aneurysm produces a large increase in the longitudinal stress, even though the absolute value of the stress is less than that of the circumferential stress. Thus, in a growing aneurysm, rapidly increasing longitudinal stress on the inner surface of the bulb could be the cause of rupture of the aneurysm.

For the range of bulb radius shown in Figure 15.16, the longitudinal stress is lower than the circumferential stress. However, if the bulb continues to grow and the bulb radius becomes more than twice that of the aorta, then the longitudinal stress will become greater than the circumferential stress. In Figure 15.16, this occurs at a radius ratio of 2.2. In an eccentric aneurysm, this is often the case, and therefore, the probability of aneurysm rupture from the longitudinal stress becomes even greater.

10.5. Justification of the Model

The model of the growing aneurysm used here simulates the real situation only partially because it considers only the increase in the bulb radius to represent growth of the aneurysm. In the real situation, there are changes in the material properties and often there is even the presence of atherosclerotic plaque and/or thrombus in the aneurysm cavity. In their study, changes in the material properties had some influence on the magnitude of the stress, however, it did not affect the trend that the longitudinal stress increased rapidly with the size of the bulb, while the circumferential stress did not.

10.6. Orientation of Tear

The wall stress and the breaking strength of the tissue must be considered together for understanding the aneurysm rupture. For the aneurysmal abdominal aorta in humans, Raghavan (28) reported that the yield stress was slightly lower in the longitudinal direction than in the circumferential direction (65 ± 9 N/cm^2 vs. 71 ± 12 N/cm^2). Recent data from Thubrikar et al. on a similar aorta also suggests that the yield stress is slightly lower in the longitudinal direction than in the circumferential direction (53 ± 7 N/cm^2 vs. 61 ± 15 N/cm^2). Because the current study demonstrated that the longitudinal stress increased most rapidly in the growing aneurysm, it appears most likely that the longitudinal stress will be the first to reach the yield stress, particularly if the yield stress is lower in that direction. Rupture caused by the longitudinal stress will produce circumferential tear. In concurrence with this, the circumferential stress may be greater than the longitudinal stress and yet not cause rupture because the yield stress is also greater in that direction.

11. Wall Stress in AAA

In the previous section, we considered how not only the absolute value of the stress but also the change in the stress over time is important in determining the likelihood of aneurysm rupture. In this section, we will examine how the stress is determined in AAA. Because in a patient the rupture strength of the aneurysm wall and other pertinent details remain unknown, the best information available—on which to base the decision on whether to treat the aneurysm or not—is the wall stress. The wall stress takes into account the wall thickness, pressure, aneurysm geometry, and material properties, and it allows some insight into how other parameters such as different material properties, tethering force, and so forth, may influence the wall stress. As stated earlier, it is the wall stress that eventually causes the rupture of the aneurysm.

Recent work by Fillinger et al. (29) showed that ruptured or symptomatic AAAs had a significantly higher wall stress than those that underwent elective repair, even when their diameters were identical. The smallest ruptured AAA of 4.8 cm diameter had a wall stress that was equivalent to the average electively repaired AAA of 6.3 cm diameter. More accurately determined wall stress had a better correlation with AAA rupture than did the wall stress determined using the law of Laplace, which considers only diameter and pressure.

Thubrikar et al. (30) studied wall stress in AAA in a clinical case, and we will consider their findings. Stresses in the aneurysm wall produced by internal pressure were determined using finite element analysis (FEA). This required the following information: geometry of the aneurysm, material properties of the aneurysm wall, internal pressure, and boundary conditions.

11.1. The Approach

Geometry: Several CT scans of an AAA of one clinical case were obtained. The patient was an 80-year-old male. The aneurysm occupied most of the infrarenal region and extended close to the aortic bifurcation. From these scans, 11 sections, each 1 cm apart, were selected. These sections were used for the measurement of the diameter, wall thickness, and the distance of the center of each cross section from the examining table. To define the geometry more accurately, 11 additional cross sections were created and placed alternatively with the measured sections so that in the composite model, each section was only 5 mm apart. These 22 cross sections were connected to create the surface of the aneurysm (Fig. 15.17). The values of the wall thickness and radius of various cross sections are shown in Figure 15.17. The presence of the clot in the aneurysm was ignored.

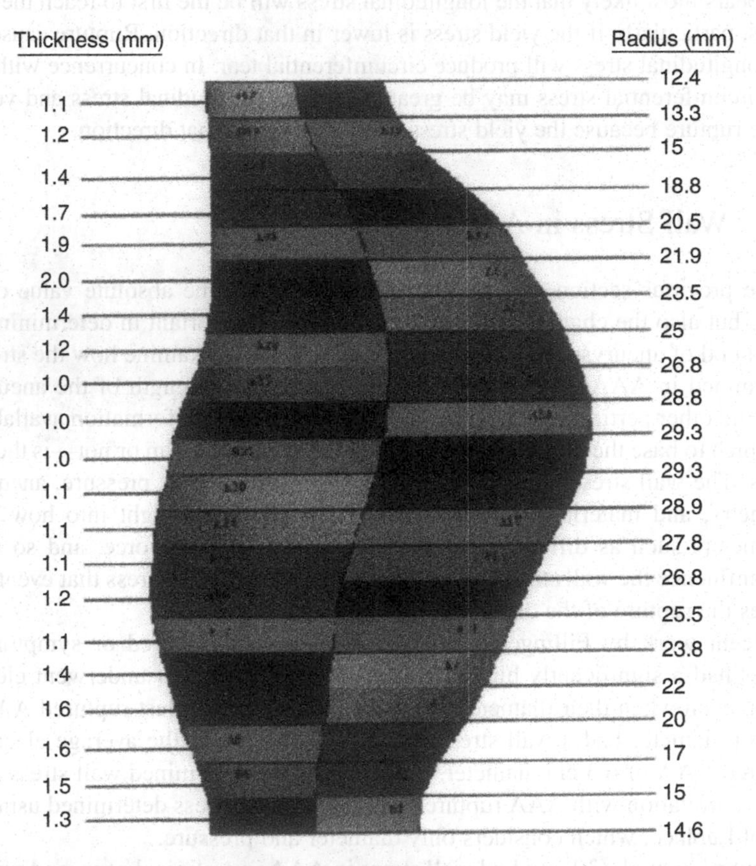

FIGURE 15.17. Model of the aortic aneurysm showing values of wall thickness and radius of cross sections at various levels. The wall thickness is kept constant between two adjacent cross sections. Eccentricity of the aneurysm is obvious. The bottom represents the distal end near the aortic bifurcation.

Material properties: The aneurysmal aorta was assumed to be homogeneous and isotropic with linear elastic material properties. The modulus of elasticity used was 4.66 N/mm^2 with a Poisson's ratio of 0.49. Although human arterial tissue acts like nonlinear material, above a pressure load of 80 mmHg the aorta behaves more like a linearly elastic material.

Internal pressure: An internal pressure of 120 mmHg (0.016 N/mm^2, systolic) was applied to the aneurysm. To account for the tethering force, which allows the aneurysm to increase in length, the aneurysm was considered to have a closed-end geometry. A force equivalent to the pressure acting on the closed end was applied to the lower end. This force mimics the tethering force on the aneurysm in an approximate sense as the exact value is unknown.

Boundary conditions: The model was constrained at the upper end, while the bottom end was allowed to move in the longitudinal (Z) direction (i.e., the aneurysm could increase in length). Due to symmetry, only half of the model was analyzed. The aneurysm was allowed to increase also in diameter.

Mesh generation and analysis: For the FEA, they used the ANSYS 5.3 program. The geometric parameters shown in Figure 15.17 were used to create a model of half of the aneurysm. Mesh size was varied and a finer mesh was used in the bulb region. 2D Shell element type 63, of both quadrilateral and triangular shapes, was used. Each belt was assigned a thickness according to the measured value. An internal pressure of 120 mmHg was applied. Also, a force at each of the nodes on the bottom edge was applied to simulate pressure at the closed end of the aneurysm. Nonlinear analysis, which allows for geometric nonlinearity and step-wise loading, was carried out. Results were obtained in terms of wall stress on the inner and middle surface of the aneurysm, equivalent stress, direction of stress, deformations, and displacements.

Parametric studies: Several parameters, including modulus of elasticity, local variation in the wall thickness, and tethering force, still remain unknown in vivo. It is necessary to determine how they might influence the results. Therefore, the following parametric studies were carried out: 1) To determine how the wall thickness may affect the results, they created two other models in which wall thicknesses of 1.31 mm and 1.58 mm were used for the entire aneurysm. 2) Because in a patient blood pressure does vary, they applied a higher internal pressure of 160 mmHg and determined the stresses. 3) Because the aneurysmal aorta is "stiffer," they determined the effect of a higher modulus by creating another model in which the modulus was five times that used in the original model. 4) The effect of the tethering force was determined by lengthening the aneurysm by an amount that was almost twice that seen in the analysis of the original model.

11.2. Observations

The aneurysm was fairly typical; it was eccentric, 10.5 cm long, about 6 cm in diameter, and had a wall thickness of 1.0–2.0 mm.

11.2.1. Model I (the Original Model): Measured Wall Thickness

Figure 15.18 shows the contours of the first principal stress (maximum stress) on the inner surface and on the middle surface of the aortic wall. The maximum stress on the inner surface is located along two circumferential belts, one at and the other below the bulb. Maximum stress on the posterior side (location A) was 0.4 N/mm^2, on the anterior side just below the bulb (location B) it was 0.3 N/mm^2, and on the anterior side at the bulb (location C) it was 0.4 N/mm^2 (Figs. 15.18 and 15.19; Table 15.6). The maximum stress on the middle surface of the wall occurred at two locations, one just above the bulb on the posterior wall and the other just below the bulb on the anterior wall. The maximum stress on the posterior wall (location A) was 0.37 N/mm^2, on the anterior wall just below the bulb (location B) it was 0.37 N/mm^2, and on the anterior wall at the bulb (location C) it was 0.24 N/mm^2. The directions of these stresses are shown in Figure 15.19. On the inner surface, the direction of maximum stress was longitudinal at the anterior region of the bulb (location C) and circumferential at other locations (A and B). On the middle surface, the direction of the maximum stress was similar, that is, longitudinal at the bulb anteriorly (location C) and circumferential at other locations (A and B). It is notable that even though the stress anteriorly at the bulb (location C) was lower than that at the other two locations, it should be

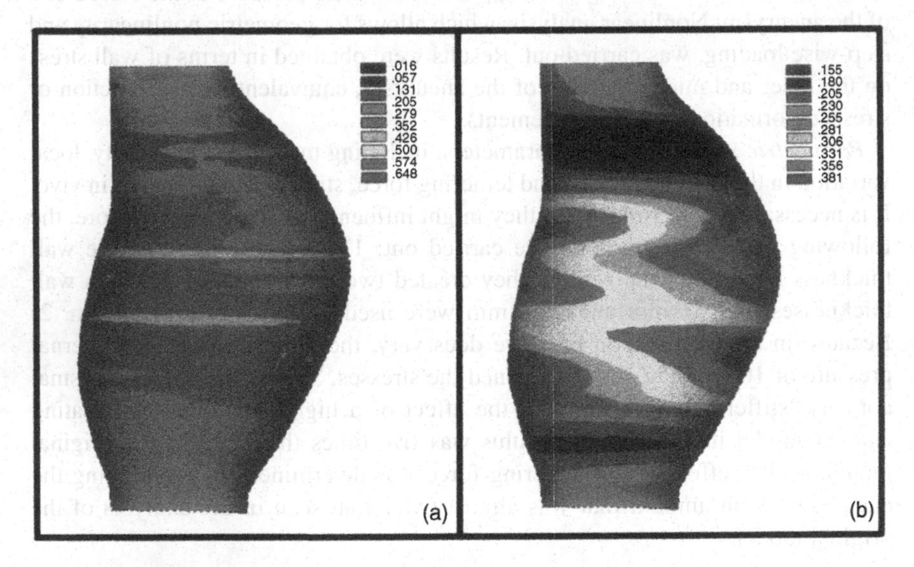

FIGURE 15.18. First principal stress (maximum stress) on inner surface (a) and middle surface (b) of aneurysmal aortic wall. Stress values are color-coded and shown in the inset in N/mm^2. (a) On the inner surface, maximum stress occurs along two circumferentially oriented belts, one at the level of the bulb and the other just below the bulb. (b) Maximum stress occurs posteriorly at the bulb and anteriorly just below the bulb. Stress distribution is for luminal pressure of 120 mmHg. (*Please see color version on CD-ROM.*)

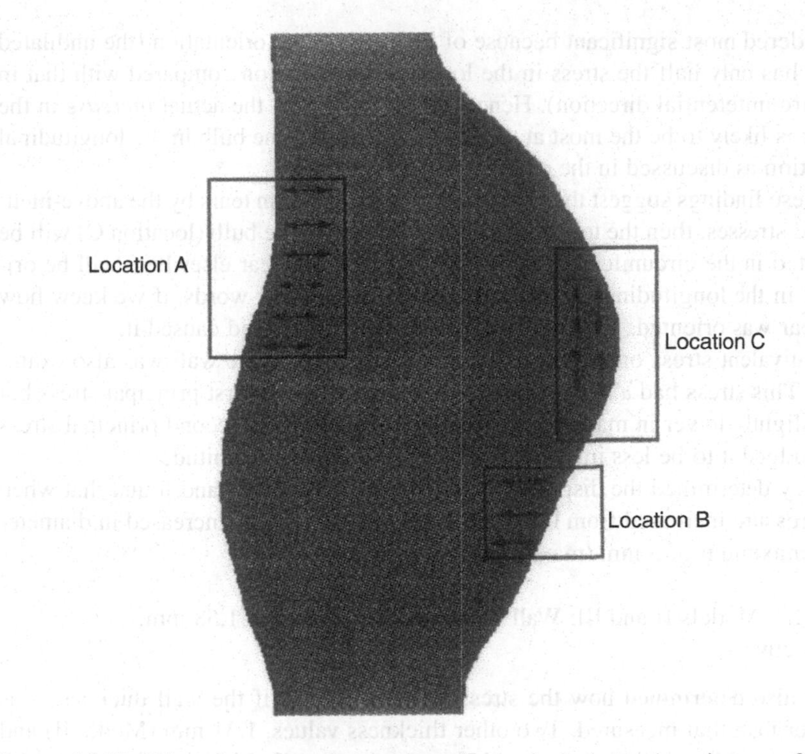

FIGURE 15.19. Orientation of maximum stress (first principal stress) on inner surface of aorta in three regions of interest. At locations A and B, the maximum stress is oriented in the circumferential direction, while at location C it is oriented in the longitudinal direction.

TABLE 15.6. First principal stress in aneurysm wall (N/mm²).

		Posterior wall at bulb[a] (location A)[a]		Anterior wall just below bulb[a] (location B)[a]		Anterior wall at bulb[a] (location C)[a]	
Model no.	Model parameter	Inner surface	Middle surface	Inner surface	Middle surface	Inner surface	Middle surface
I	Measured thickness (t)	0.4 ↔	0.37 ↔	0.3 ↔	0.37 ↔	0.4 ↕	0.24 ↕
II	Constant thickness ($t = 1.31$ mm)	0.34	0.34	0.26	0.32	0.34	0.18
III	Constant thickness ($t = 1.58$ mm)	0.27	0.28	0.23	0.26	0.27	0.15
IV	Luminal pressure ($P = 160$ mmHg)	0.49	0.48	0.39	0.48	0.49	0.31
V	Elastic modulus ($E = 5\ E_{initial}$)	0.44	0.41	0.36	0.41	0.44	0.24
VI	Displacement ($U_z = 5$ mm)	0.65 ↕[b]	0.43 ↕[b]	0.4	0.36	0.5	0.33

[a] Arrow indicates the direction of stress.
[b] Indicates the only place where the direction has changed.

considered most significant because of its longitudinal orientation (the undilated aorta has only half the stress in the longitudinal direction compared with that in the circumferential direction). Hence, in the aneurysm, the actual *increase* in the stress is likely to be the most at the anterior region of the bulb in the longitudinal direction as discussed in the previous section.

These findings suggest that if the wall of the aneurysm tears by the above-mentioned stresses, then the tear at the anterior region of the bulb (location C) will be oriented in the circumferential direction, whereas the tear elsewhere will be oriented in the longitudinal direction (Table 15.7). In other words, if we knew how the tear was oriented, we could determine which stress had caused it.

Equivalent stress on the inner and middle surface of the wall was also examined. This stress had a similar distribution pattern as the first principal stress but was slightly lower in magnitude. They also examined the second principal stress and judged it to be less important because of its lower magnitude.

They determined the displacement of the aneurysm wall and found that when the pressure increased from 0 to 120 mmHg, the aneurysm increased in diameter by a maximum of 3 mm (maximum displacement).

11.2.2. Models II and III: Wall Thickness 1.31 mm and 1.58 mm, Respectively

They also determined how the stresses might change if the wall thickness was greater than that measured. Two other thickness values, 1.31 mm (Model II) and 1.58 mm (Model III), were used for constant wall thickness for the entire aneurysm. The 1.31-mm thickness was chosen because it is close to the thickness of the undilated aorta at the ends of the aneurysm. The 1.58-mm thickness was selected because it gives a ratio *t/R* of 1/8 for the undilated segment, a ratio often

TABLE 15.7. Relationship between cause, location, and orientation of the tear.

			Orientation of tear		
Model no.	Model parameter	Stress causing the tear	Posterior wall at bulb ruptures	Anterior wall below bulb ruptures	Anterior wall at bulb ruptures
I	Measured thickness	Inner surface	L	L	C
	(t)	Middle surface	L	L	C
II	Constant thickness	Inner surface	L	L	C
	(t = 1.31. mm)	Middle surface	L	L	C
III	Constant thickness	Inner surface	L	L	C
	(t = 1.58 mm)	Middle surface	L	L	C
IV	Luminal pressure	Inner surface	L	L	C
	(P = 160 mmHg)	Middle surface	L	L	C
V	Elastic modulus	Inner surface	L	L	C
	($E = 5\ E_{initial}$)	Middle surface	L	L	C
VI	Displacement	Inner surface	C	L	C
	(U_z = 5 mm)	Middle surface	C	L	C

L, longitudinal; C, circumferential.

used for the normal aorta. It may be noted that a constant thickness implies that as the aneurysm "grows," there is a build-up in the wall thickness to counterbalance the thinning that must accompany the increase in the diameter.

With the constant wall thickness of 1.31 mm (Model II) the first principal stress on the inner surface at locations A, B, and C was 0.34, 0.26, and 0.34 N/mm^2, respectively (Table 15.6). On the middle surface, the respective stresses were 0.34, 0.32, and 0.18 N/mm^2. The pattern of stress distribution and the orientation of the stress were reasonably similar to those in the original Model I (Figs. 15.18 and 15.19). The equivalent stress and the second principal stress were also determined but considered less important because of their lower magnitude. The maximum displacement was 2.82 mm.

With the wall thickness of 1.58 mm (Model III), the first principal stress on the inner surface at locations A, B, and C was 0.27, 0.23, and 0.27 N/mm^2, respectively, and on the middle surface it was 0.28, 0.26, and 0.15 N/mm^2, respectively (Table 15.6). The pattern of stress distribution and the orientation of the stress were similar to those in the original Model I. The maximum displacement was 2.38 mm.

11.2.3. Model IV: Intraluminal Pressure of 160 mmHg

Because AAA patients are often hypertensive, stresses were also determined at an internal pressure load of 160 mmHg (conservative estimate) in the initial Model I. The first principal stress (maximum stress) on the inner surface at locations A, B, and C was 0.49, 0.39, and 0.49 N/mm^2, respectively (Table 15.6). The stress on the middle surface was 0.48, 0.48, and 0.31 N/mm^2, respectively. As expected, the rise in pressure causes all of the stresses to increase. The pattern of stress distribution and the orientation of the stress were similar to those in the original Model I. The maximum displacement was 3.83 mm.

11.2.4. Model V: Increased Wall Stiffness

The wall of the aneurysmal aorta is stiffer and when it is expressed in terms of pressure-diameter relationships, the stiffness may be greater by a factor of 2 to 4. They studied the effect of stiffness by increasing the modulus five times ($E = 5E_o$). The first principal stress on the inner surface at locations A, B, and C was 0.44, 0.36, and 0.44 N/mm^2, respectively, and on the middle surface 0.41, 0.41, and 0.24 N/mm^2, respectively (Table 15. 6) (i.e., the stresses were slightly increased). Once again, the pattern of stress distribution and the orientation of the stress were similar to those in the original Model I. The maximum displacement was 0.89 mm. The increased stiffness had more influence on the displacement than on the stresses.

11.2.5. Model VI: Effect of Pull (Tethering Force)

The aorta may be under a greater pull along the length due to tethering than what may occur by free expansion under internal pressure. Therefore, they determined stresses under conditions in which the aorta is pulled down by 5 mm along its length. This displacement is approximately twice the amount observed in the aneurysm, which lengthens freely under pressure (Model I). In the original model, they applied both the internal pressure of 120 mmHg and an axial displacement of 5 mm to the bottom edge.

The results were as follows: The areas of high stresses shifted to the posterior wall of the aneurysm (Fig. 15.20). The first principal stress on the inner surface at locations A, B, and C was 0.65, 0.4, and 0.5 N/mm^2, respectively, and on the middle surface it was 0.43, 0.36, and 0.33 N/mm^2, respectively (Table 15.6). All of the stresses were increased considerably and both the pattern of stress distribution and the orientation of stress were altered. The orientation of the first principal stress on both the inner and the middle surface on the posterior wall, at location A, changed and became longitudinal. For instance in Figure 15.19, at location A, the three bottom arrows appear vertically oriented. The orientation of the stresses just below the bulb (location B) and at the anterior region of the bulb (location C), however, remain the same as that in the original model (Model I). The maximum displacement increased to 5.27 mm and was along the long axis. These results show that if the tear on the posterior wall was oriented in the circumferential direction, then it would have been caused by the dominance of the tethering (or longitudinal) force (Table 15.7).

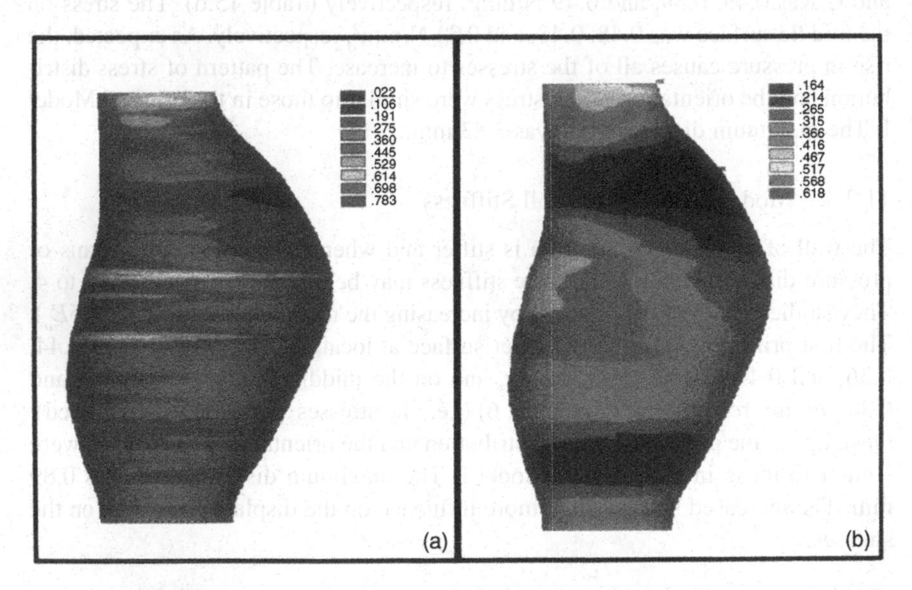

FIGURE 15.20. Maximum stress (first principal stress) on inner surface (a) and middle surface (b) of aorta for model in which luminal pressure and tethering force, which causes 5-mm increase in length, are applied. The stress distribution and orientation are altered substantially compared with that in the original model. (*Please see color version on CD-ROM.*)

11.3. The Implications

The model uses the geometry of an actual AAA of one clinical case, accounts for eccentricity, evaluates the effect of various parameters, and maps the details of stress distribution. The aneurysmal aorta was considered isotropic, homogeneous, and incompressible. Other studies support the postulate that the aneurysmal aortic tissue is isotropic (28). Although the material properties of the AAA are nonlinear (24, 28), at a pressure load of 60 mmHg or greater, they are fairly linear. Because the geometry of the aorta used in the current study is that which existed in vivo (i.e., at a systemic pressure of the patient), at this geometry the AAA wall does behave almost linearly.

11.3.1. The Wall Stress

The maximum wall stress should be considered important in three regions: on the posterior and anterior wall of the bulb and on the anterior wall just below the bulb (Figs. 15.18 and 15.19; Table 15.6). As Models II and III show, the increase in the wall thickness reduces the stress but changes neither the distribution pattern nor the orientation of the stress. An increase in the intraluminal pressure raises the wall stress, while an increase in the wall stiffness raises the wall stress minimally. Hence, the stress depends more on the wall thickness and the luminal pressure than on the material properties. The tethering force does affect the stress distribution and the orientation. If a substantial tethering force is acting on the AAA, the tear in the posterior wall should be oriented circumferentially, otherwise it will be oriented longitudinally.

11.3.2. The Significance of Wall Stress

The study explains which stresses may be responsible for the AAA rupture. First, the stress on the inner wall is greater than that on the middle wall, and therefore the tear has a predilection to the inner surface. Second, the highest stresses occur at the bulb, and thus the rupture is likely to begin here. Third, the rupture in the posterior wall region (location A) will produce a longitudinal tear, whereas that in the anterior wall will produce a longitudinal tear below (location B) and a circumferential tear at (location C) the bulb (Fig. 15.19; Table 15.7). This location and orientation of the tear is not likely to be altered by change in the wall thickness, luminal pressure, or wall stiffness. Also, the orientation of the tear is the same, whether it is caused by the stress on the inner surface or by the stress on the middle surface. If the tear on the posterior wall is circumferential, then it must have occurred due to a significant tethering force (Tables 15.6 and 15.7).

11.3.3. The Yield Stress and Location of Rupture Sites

The results of the current study predict the likely location and orientation of the tear in the aneurysm. The aortic wall starts to yield when the stress reaches the yield stress. Raghavan et al. (28) reported the yield stress to be 0.65 ± 0.09 N/mm^2 in the longitudinal direction and 0.707 ± 0.12 N/mm^2 in the circumferential direction for the anterior wall of the AAA. Thubrikar et al. (22) reported slightly lower values. The stress obtained in the current study is lower than the yield stress and that is why the aneurysm has not ruptured in the patient. However, if we consider additional parameters such as localized thinning, calcification, buckling, or weakening in the aneurysm, then the stress could equal or even exceed the yield stress. Thus, the current results can be used to determine the probability of rupture of an AAA on the basis of wall stress.

Darling et al. (3) reported that 62% of the ruptures occur near the maximum diameter (bulb) region, 34% at the midpoint, and only 4% at the junction of the dilated and undilated aorta (Section 7). The current observations support that, in the absence of any localized abnormality, almost all of the ruptures should occur in the bulb region. Hence, these observations agree with the rupture locations in a vast majority of cases. They also predict the orientation of the tear, however, there are no published reports on the tear orientation for comparison.

11.4. Comments

In summary, we noted many aspects that are part of a growing aneurysm. To be able to predict the probability of rupture of an aneurysm, more work needs to be done. The best approach seems to be to determine the wall stress at various stages during growth of an aneurysm while the patient is being observed. This approach can easily be integrated in a clinical practice.

References

1. Krohn C, Kullmann G, Rosen L, Kroese A: Ultrasonographic screening for abdominal aortic aneurysm. Eur J Surg 1992;158:527-530.
2. Reed D, Reed C, Stemmermann G, Hayashi T: Are aortic aneurysms caused by atherosclerosis? Circulation 1992;85(1):205-211.
3. Darling RC, Messina CR, Brewster DC, Ottinger LW: Autopsy study of unoperated abdominal aortic aneurysms. The case for early resection. Circulation 1977;56(3): II-161-II-164.
4. Juvonen T, Ergin MA, Galla JD, Lansman SL, Khanh N, Levy D, deAsla RA, Bodian CA, Griepp RB: J. Maxwell Chamberlain Memorial Paper: A prospective study of the natural history of thoracic aortic aneurysms. Program for the Thirty-third Annual Meeting of the Society of Thoracic Surgeons in San Diego, California, February 3-5, 1997:36-37.
5. Scott RAP, Tisi PV, Ashton HA, Allen DR: Abdominal aortic aneurysm rupture rates: A 7-year follow-up of the entire abdominal aortic aneurysm population detected by screening. J Vasc Surg 1998;28:124-128.
6. Noel AA, Gloviczki P, Cherry KJ Jr, Bower TC, Panneton JM, Hallett JW Jr: The implications of rupture of small abdominal aortic aneurysms. Abstract presented at the 1999 Joint Annual Meeting of the North American Chapter, International Society for Cardiovascular Surgery, Washington, DC, June 6-9, 1999.
7. Coady MA, Rizzo JA, Hammond GL, Mandapati D, Darr U, Kopf GS, Elefteriades JA: What is the appropriate size criterion for resection of thoracic aortic aneurysms? J Thorac Cardiovasc Surg 1997;113(3):476-491.

8. Juvonen T, Ergin MA, Galla JD, Lansman SL, Nguyen KH, McCullough JN, Levy D, deAsla RA, Bodian CA, Griepp RB: A prospective study of the natural history of thoracic aortic aneurysms. Ann Thorac Surg 1997;63:1533-1545.
9. Cronenwett JL, Sargent SK, Wall MH, Hawkes ML, Freeman DH, Dain BJ, Curé JK, Walsh DB, Zwolak RM, McDaniel MD, Schneider JR: Variables that affect the expansion rate and outcome of small abdominal aortic aneurysms. J Vasc Surg 1990;11: 260-269.
10. Johansen K: Aneurysms. Scientific American 1982;247(1):110-125.
11. MacSweeney STR, Powell JT, Greenhalgh RM: Pathogenesis of abdominal aortic aneurysm. Br J Surg 1994;81:935-941.
12. Patel MI, Hardman DTA, Fisher CM, Appleberg M: Current views on the pathogenesis of abdominal aortic aneurysms. J Am Coll Surg 1995;181:371-382.
13. Wills A, Thompson MM, Crowther M, Sayers RD, Bell PRF: Pathogenesis of abdominal aortic aneurysms – Cellular and biochemical mechanisms. Eur J Vasc Endovasc Surg 1996;12:391-400.
14. Schlatmann TJM, Becker AE: Histologic changes in the normal aging aorta: Implications for dissecting aortic aneurysm. Am J Cardiol 1977;39(1):13-20.
15. Klima T, Spjut HJ, Coelho A, Gray AG, Wukasch DC, Reul GJ Jr, Cooley DA: The morphology of ascending aortic aneurysms. Hum Pathol 1983;14:810-817.
16. Strachan DP: Predictors of death from aortic aneurysm among middle-aged men: The Whitehall study. Br J Surg 1991;78:401-404.
17. MacSweeney STR, Powell JT, Greenhalgh RM: Abdominal aortic aneurysms grow more quickly in smokers. Br J Surg 1994;81:614 (Abstract).
18. Wolf YG, Thomas WS, Brennan FJ, Goff WG, Sise MJ, Bernstein EF: Computed tomography scanning findings associated with rapid expansion of abdominal aortic aneurysms. J Vasc Surg 1994;20:529-538.
19. Satta J, Läärä E, Juvonen T: Letter to the Editor: Intraluminal thrombus predicts rupture of an abdominal aortic aneurysm. J Vasc Surg 1996;23(4):737-739.
20. Pillari G, Chang JB, Zito J, Cohen JR, Gersten K, Rizzo A, Bach AM: Computed tomography of abdominal aortic aneurysm. An in vivo pathological report with a note on dynamic predictors. Arch Surg 1988;123:727-732.
21. Thubrikar MJ, Robicsek F, Labrosse M, Chervenkoff V, Fowler BL: Effect of thrombus on abdominal aortic aneurysm wall dilation and stress. J Cardiovasc Surg 2003;44:67-77.
22. Thubrikar MJ, Labrosse M, Robicsek F, Al-Soudi J, Fowler B: Mechanical properties of abdominal aortic aneurysm wall. J Med Eng Technol 2001;25(4):133-142.
23. Thubrikar MJ, Agali P, Robicsek F: Wall stress as a possible mechanism for the development of transverse intimal tears in aortic dissections. J Med Eng Technol 1999;23(4): 127-134.
24. He CM, Roach MR: The composition and mechanical properties of abdominal aortic aneurysms. J Vasc Surg 1994;20:6-13.
25. Aoki T, Ku DN: Collapse of diseased arteries with eccentric cross section. J Biomechanics 1993;26:133-142.
26. Chandran KB, Gao D, Han G, Baraniewski H, Corson JD: Finite-element analysis of arterial anastomoses with vein, dacron and PTFE grafts. Med Biol Eng Comput 1992;30:413-418.
27. Dobrin PB: Pathophysiology and pathogenesis of aortic aneurysms: Current concepts. Surg Clin North Am 1989;69:687-703.

28. Raghavan ML, Webster MW, Vorp DA: *Ex vivo* biomechanical behavior of abdominal aortic aneurysm: Assessment using a new mathematical model. Ann Biomed Eng 1996;24:573-582.
29. Fillinger MF, Raghavan ML, Marra SP, Cronenwett JL, Kennedy FE: In vivo analysis of mechanical wall stress and AAA rupture risk. J Vasc Surg 2002;36(3):589-597.
30. Thubrikar MJ, Al-Soudi J, Robicsek F: Wall stress studies of abdominal aortic aneurysm in a clinical model. Ann Vasc Surg 2001;15(3):355-366.

16
Aortic Dissection

The history of aortic dissection dates back several centuries. There are many reviews published on the subject (1–7). The etiology of the disease has been well described, however, the role of vascular mechanics has not been fully considered in the process. As we will see, vascular mechanics plays a very important role in the pathogenesis of aortic dissection. We will consider aortic dissection, once again, with the point of view that this is one more disease of the aorta, which calls our attention to the mechanics of the aorta.

1. Morphologic Features

By definition, the dissection of the aorta is present when the aortic media has split and there is extraluminal blood in the aortic wall. The amount of splitting, and thus blood, in the aortic wall may vary greatly, from less than a millimeter in thickness to total occlusion of the true lumen. The initiating event in the aortic dissection is a tear in the intima through which blood surges into the media, thereby splitting the media and separating the intima from the adventitia (Figs. 16.1 and 16.2). The dissection usually propagates from the intimal tear to the distal aorta, although it may also extend proximally. Blood in the false channel

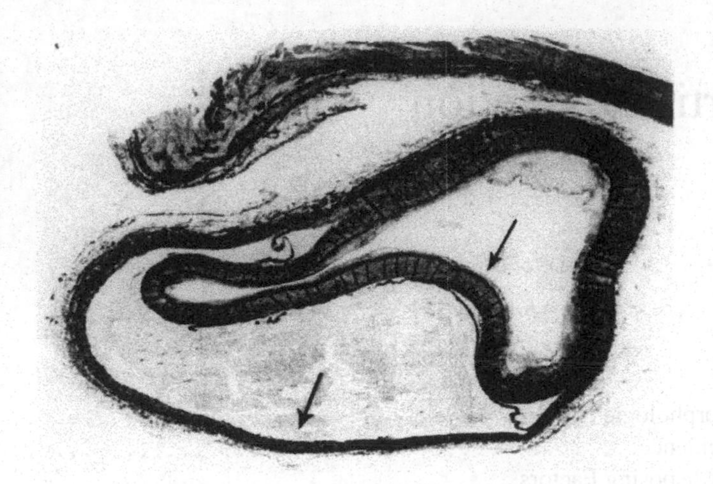

FIGURE 16.1. Dissection (arrows) within the outer third of the aortic wall (Weigert's Van Gieson stain). (Reproduced from Anagnostopoulos CE, Acute Aortic Dissections. University Park Press, 1975:47.)

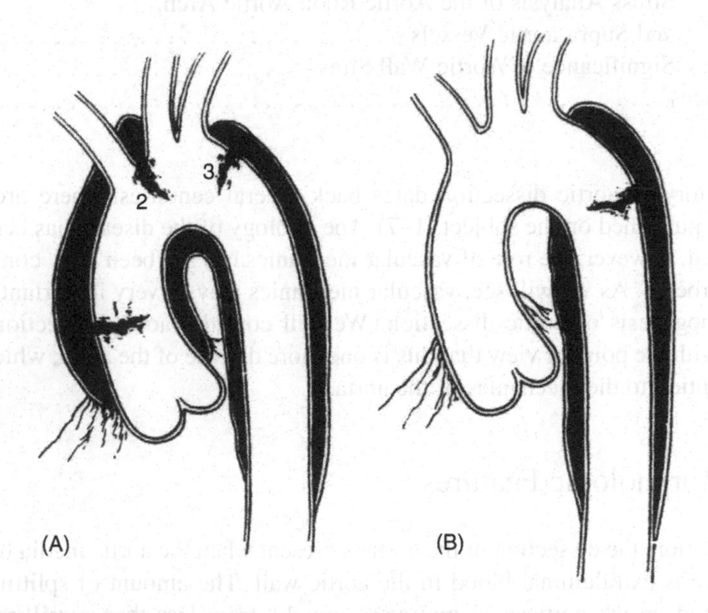

FIGURE 16.2. Classification of aortic dissections (Stanford). In type A, the ascending aorta is dissected (A). The intimal tear has always been at position 1, but it can occur at position 2 or 3. In type B dissection, the dissection is limited to the descending aorta (B), and the intimal tear is usually within 2 to 5 cm of the left subclavian artery. (Reproduced from Sabiston DC Jr, Spencer FC, Surgery of the Chest, Fifth ed., Vol. II. W.B. Saunders, 1990:1201, with permission from Elsevier.)

can reenter the true lumen anywhere along the course of dissection. Rupture of the outer wall of the aorta from dissection, the most common cause of death, occurs most frequently in the pericardial space and the left pleural cavity.

Once blood enters the media via the intimal and medial tear, the time required thereafter to dissect the media of the entire length of the aorta appears to be only a few seconds. In the vast majority of patients (nearly 70%), the tear, which marks the beginning of the dissection, is located in the *ascending aorta*, usually about 2 cm distal to the sino-tubular junction. The entrance tear is usually *transverse* and involves about one-half of the circumference of the aorta, occasionally much less, and rarely the entire circumference. The location of the dissecting channel (false channel) is in the outer half of the aortic media. Consequently, the outer wall of the false channel is very thin, usually only about one-fourth as thick as the original medial wall. The inner wall is about 3 times thicker than the outer wall. The thinness of the outer wall is the anatomic feature, which explains its high frequency of rupture outside the aorta. The rupture of the inner wall to connect the false channel with the true lumen is much less frequent. The thinner the outer wall of the false lumen, the greater is the likelihood of fatal rupture.

The portion of the aortic circumference involved in the dissection is fairly characteristic. The entrance tear usually involves the *right lateral aortic wall* in the ascending aorta and the dissection thereafter courses downstream along the greater curvature of the ascending, transverse, and descending thoracic portion of the aorta (Fig. 16.2). Because the arch arteries arise on the greater curvature of the aorta, extension of the dissection into the innominate, left common carotid, and left subclavian arteries is common.

The second most common location of the primary tear is in the descending thoracic aorta just distal to the origin of the left subclavian artery in the region of the insertion of the ligamentum arteriosum (Fig. 16.2B). Because of this and because the dissection, which originates in the ascending aorta, usually extends into the descending aorta (even the entire aorta), the descending aorta becomes the most common segment to be involved in the dissection. Quite often the dissection stops within the aorta without re-rupture. The most common cause of interruption of the longitudinal dissection appears to be atherosclerotic plaquing, which has caused underlying media to scar or atrophy or both. Another cause of interruption of a dissection is aortic isthmic coarctation. The isthmus of the aorta is a site of attachment of the obliterated ductus (ligamentum arteriosum).

2. Incidence

The exact incidence of aortic dissection each year in the United States is unknown, although it has been estimated that 2000 cases are diagnosed per year. The incidence of aortic dissection, either diagnosed or missed, is probably about 10/100,000/year. Aortic dissection is being diagnosed now more often because of newer imaging techniques such as CT, MRI, and trans-esophageal echocardiography (TEE).

3. Predisposing Factors

Hypertension is the most important predisposing factor for aortic dissection. It exists in 70–90% of these patients. Hypertension can enhance atherosclerotic plaques and thus enhance cardiovascular disease and/or it can weaken the media and promote dissection. However, hypertension is very common, whereas dissecting aneurysm is very rare. It is also known that experimentally it takes 8 times above normal pressure to cause rupture of the aorta. These observations emphasize that we still lack the understanding of the basic cause of the dissecting process.

Other factors predisposing to aortic dissection are congenital disorders of the connective tissue, especially Marfan syndrome and, to a lesser extent, Ehlers-Danlos syndrome. Also, congenitally bicuspid and unicommissural aortic valves and aortic coarctation are associated with an increased risk of aortic dissection. Dissection predominates in males, with a male:female ratio of 3:1, but there is an association with pregnancy. Half of all dissections in women under the age of 40 occur during pregnancy, usually in the third trimester.

4. Classification, Location, Frequency

Although the intimal tear occurs more commonly in the ascending aorta and at the proximal descending thoracic aorta, it may also occur in the aortic arch and other sites of the aorta (Fig. 16.2). Also, it is not always possible to determine the original site of the dissection. The aortic dissections have been classified by DeBakey (8) on the basis of the anatomic location of the dissection process.

Type I: The type I dissection involves the ascending aorta, as well as the distal portions of the aorta. Although in most patients the intimal tear arises as a transverse opening in the anterior wall of the proximal portion of the ascending aorta, it may also begin in the aortic arch or even in the descending thoracic aorta, with retrograde dissection to the aortic root and distal dissection throughout the remainder of the aorta.

Type II: In type II, the dissecting process is limited to the ascending aorta and is usually characterized by a transverse tear in the intima anteriorly just above the aortic valve with the dissection terminating just proximal to the origin of the innominate artery.

Type III: In type III, the dissection usually begins just distal to the origin of the left subclavian artery and extends distally for a varying distance.

DeBakey determined the distribution of the three types of dissections in their 527 patients. The mean age of patients was 50.2, 47.3, and 57.9 years, respectively, for DeBakey's type I, type II, and type III dissections.

There is also another type of classification used for aortic dissection, the Stanford type A and type B (Fig. 16.2). In type A, the tear begins in the ascending aorta and in type B it begins in the descending aorta.

Table 16.1 shows the frequency and location of the tear as reported by various groups. Primary tears are located in the ascending aorta in about 62% of the

TABLE 16.1. Frequency and location of the intimal tear.

Frequency/location	Reference
66% type A, 33% type B	(4,5)
63% Ascending Aorta, 10% arch, 27% proximal descending Aorta	(6)
65% Ascending Aorta, 10% arch, 20% descending Ao, <5% abdominal Aorta	(1)
61% Ascending Aorta, 9% arch, 16% isthmus, 10% descending Aorta, 3% abdominal Aorta	(7)

cases. The tears occur with decreasing frequency as the distance from the aortic root increases. Primary tears are about 5 times more likely to be transverse (circumferential) in orientation than longitudinal. Tears have been observed in a variety of shapes in the ascending aorta, including elliptical, oblique, zig-zag, or round. The preponderance of transverse intimal tears may reflect the predominately circular arrangement of the supporting elements of the media but have also been attributed to an elongating type of stress (7).

Next to the ascending aorta, the isthmus of the aorta, which is the site of attachment of the obliterated ductus (ligamentum arteriosum), is the most frequent site of entry tears. The relative fixation of the aorta at this site has been considered as the most likely explanation for the higher frequency of dissections there.

Another important parameter for classification of dissections is the duration of the dissection at the time of its first presentation. Dissections are categorized as "acute" if its presentation occurs within 2 weeks of its onset and as "chronic" if more than 2 weeks have elapsed. The importance of this classification lies in the fact that approximately 65–75% of the patients with untreated dissections die in the first 2 weeks after onset (2).

5. Pathogenesis

Schlatmann and Becker (9) studied the aortic media in patients with dilated ascending aorta and compared their findings with the data obtained from the aortas that had dissected. A particular emphasis was placed on the following features: 1) cystic medial necrosis, 2) elastin fragmentation, 3) fibrosis, and 4) medionecrosis. Each of these features showed a pattern of distribution in the diseased aorta, which was similar to that in the normal aging aorta. There were no qualitative differences between the *normal* aging aorta and the *dilated* or *dissected* aorta. Only quantitative differences were present in these histopathologic features. The authors suggest that the process of injury and repair represents part of the aging process in the aortic media and leads to wall weakening and concomitant dilation. Local hemodynamic forces could then influence these events further and could produce either incomplete or complete dissection.

Thus, we should note that the changes in the aorta due to aging and the changes seen in the aneurysmal aortic wall are more quantitative rather than qualitative. The description of histopathologic changes in the ascending aortic aneurysm can be found in Chapter 15.

The pathogenesis of aortic dissection may be summarized as follows (7):

1. Medial degeneration in the wall of the thoracic aorta sets the stage by decreasing the cohesiveness of the medial layers of the aortic wall.
2. Repeated motion of the aorta related to the beating of the heart results in flexion stresses, most marked in the ascending aorta and proximal portion of the descending thoracic aorta, 60–100 times a minute, 37 million times a year.
3. Hydrodynamic forces in the bloodstream, related to the pulse wave propagated by each myocardial contraction, as well as the level of systolic blood pressure, act upon the wall of the aorta—most markedly the proximal ascending aorta.
4. A combination of these factors eventually results in an intimal tear, which leads to a hematoma dissecting into the media of the aortic wall. Hydrodynamic forces, primarily related to the steepness of the pulse wave dp/dt_{max}, as well as the blood pressure, continue to propagate the dissection until rupture occurs.

6. The Role of Aortic Root Motion in the Pathogenesis of Aortic Dissection

The specific characteristics of aortic dissection (such as orientation of the tear most frequently being circumferential and the frequency of the dissection being the most near the aortic root and declining as the distance from the heart increases) clearly suggests that the downward movement of the aortic root during the cardiac cycle must play an important role in this pathology. In fact, this factor was alluded to in the pathogenesis described above. However, how important is this downward movement of the aortic root in aortic dissection cannot be known unless this aspect of the aortic root mechanics is explored in detail. Once again, such studies have not been done in the past, and only recently our group (10) has explored how the aortic root motion affects the stress in the aorta and how that may participate in the dissection process.

It is common knowledge that in systole, when the left ventricle contracts, it pulls the aortic root downward and causes the root to twist, due to the contractile twisting motion of the myocardium. Several angiographic studies have shown that the aortic root moves downward during systole and upward during diastole. The force causing the aortic root to move downward is the ventricular traction accompanying every heartbeat. Naturally, this force is transmitted to the aortic root, the ascending aorta, the transverse aortic arch, and the supra-aortic vessels. Thus, the aortic root motion has a direct influence on the deformation of the aorta and on the mechanical stresses developed in the aortic wall. We will consider this subject below.

6.1. Measurement of Aortic Root Motion in Patients

We analyzed aortic root contrast injections on cine films in 40 patients (45 to 87 years old, mean age 66) with coronary artery disease. Some patients had valvular disease such as aortic stenosis or aortic insufficiency. Seventeen patients had undergone coronary artery bypass grafting (CABG) and one patient had a heart transplant. Other conditions noted were myocardial hypertrophy, left-ventricular hypokinesis, and a history of hypertension. None of the patients had aortic dissection, or an obviously dilated aorta (Table 16.2). Most aortograms were done in left anterior oblique projection, some in the right anterior oblique view. The aortograms were projected frame by frame and the outlines of the aortic root in the most upward and downward positions traced on a transparency (Fig. 16.3). The base of two sinuses and the sino-tubular junction (STJ) were marked. The distances between the marked points were measured. The downward motion (axial displacement) of the aortic root, perpendicular to the plane of the sino-tubular junction, was measured and expressed also as a percentage of the STJ diameter.

The downward axial displacement of the aortic root during the cardiac cycle ranged between 0% and 49% of the STJ diameter (Table 16.2) with an average of 15 ± 11%. We also measured the absolute values of the STJ diameter and the axial displacement in a subgroup of 14 patients using the 6F (2 mm) angiocatheter for calibration. We found the STJ diameters to be normal (25 ± 4 mm) and the aortic root displacements to range between 1 and 14 mm (Fig. 16.4).

6.1.1. Effect of Cardiac Pathology on Aortic Root Movement

Because patients with aortic insufficiency (AI) have increased stroke volume, we expected them to show increased aortic root motion. Indeed, their aortic root motion was 22 ± 13% of STJ, as opposed to 12 ± 9% in patients without AI. Patients with left ventricular hypokinesis have reduced ventricular traction and therefore were expected to show reduced aortic root motion. Their aortic root motion was significantly reduced to 10 ± 9% STJ, versus 17 ± 12% in patients without hypokinesis.

The effect of myocardial hypertrophy was uncertain, and it showed a trend toward increasing aortic root movement, compared with patients without hypertrophy (24 ± 10% of STJ vs. 14 ± 11%). The fibrotic adhesions due to previous cardiac operations could be responsible for reduced aortic root motion to 12 ± 11% of STJ, compared with 17 ± 11% in patients without previous operations. The root movement did not correlate with patients' age or ejection fraction. A history of hypertension or aortic stenosis also did not influence the aortic root motion.

6.2. *Stress Analysis of the Aortic Root, Aortic Arch, and Supra-aortic Vessels*

A finite element model of the human aortic root, aortic arch, and supra-aortic vessels was built for stress analysis using ANSYS software.

Geometry: A multipronged approach was adopted to create a model representing a general, rather than patient-specific geometry. In the model, the distance between the base of the root and the brachiocephalic trunk was 70 mm, and the angle of the aortic root to the horizontal plane in the frontal view was 30 degrees

TABLE 16.2. The measured aortic root movement in patients and other clinical data.

ID	Axial displ. (% of STJ)	Age	History of hyper-tension	Ejection fraction (%)	Aortic stenosis	Aortic insuffi-ciency	Hypo-kinesis	Hyper-trophy
1	0	52	No	40			mod-sev inferior	
R2	1	70	Yes	50			mild apical	
3	1	75	No	35			mod-sev anterior	
4	2	77	No	60	mod			
R5	3	47	Yes	35			mod-sev	
R6	3	51	Yes	55		trivial		
R7	3	72	Yes	38	mild-mod			
R8	3	59	Yes	25			mod inferior	
R9	4	50	Yes	30			mod-sev	
R10	7	64	Yes	30			sev inferior	
11	7	54	Yes	60	mild	mild		
12	7	65	Yes	50				
R13	7	75	Yes	60				
14	8	51	Yes	67				
15	8	77	Yes	50		mild		
16	10	70	No		mod-sev			
17	10	58	Yes	67				
R18	11	57	No	55			mild anterior	
R19	11	71	Yes	30			mod-sev	
R20	14	75	Yes	60				
R21	14	64	Yes	67				
R22	14	62	Yes	50			mod	mod
R23	14	82	No	67		mild		
R24	15	80	No	60	mod-sev			
R25	16	64	Yes	25			mod-sev inferior	
26	17	63	No	30			mod-sev	
27	17	72	Yes	75		mild		mod
28	19	66	Yes	67				
29	20	45	Yes	20			sev inferior	
30	22	65	No			trivial		
31	22	68	Yes	60		mild		
32	23	66	Yes	65				
33	25	68	Yes	55		mild		
R34	26	87	Yes	65				
35	26	70	Yes	55	sev	trivial		mod
36	30	68	No	70	sev			
37	30	68	Yes	35	mild	mild	mod-sev anterolateral	
38	33	52	No	20				
39	37	78	Yes	75		mild		mod-sev
R40	49	82	No	15	mod	mild		

R, patient with previous coronary artery bypass grafting, except R21 with previous heart transplant; mod, moderate; sev, severe; STJ, sinotubular junction diameter in diastole.

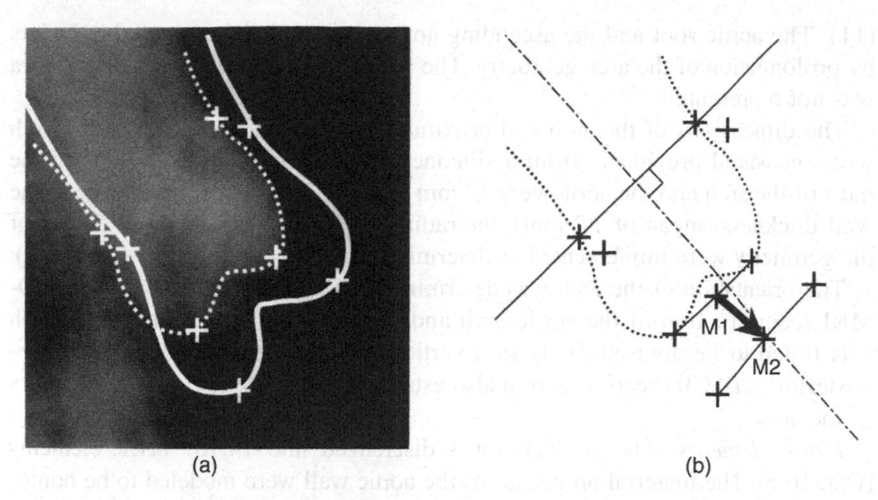

FIGURE 16.3. (a) Overlaid angiograms with outlines of the most upward and downward positions of the aortic root in a cardiac cycle. (+) Reference points for measurement at the base of the sinuses of Valsalva and at the STJ. In schematic (b), the arrow represents the movement of the base of the aortic root from M1 to M2. The projection of this displacement in the direction perpendicular to the plane of the STJ was defined as the axial displacement of the aortic root.

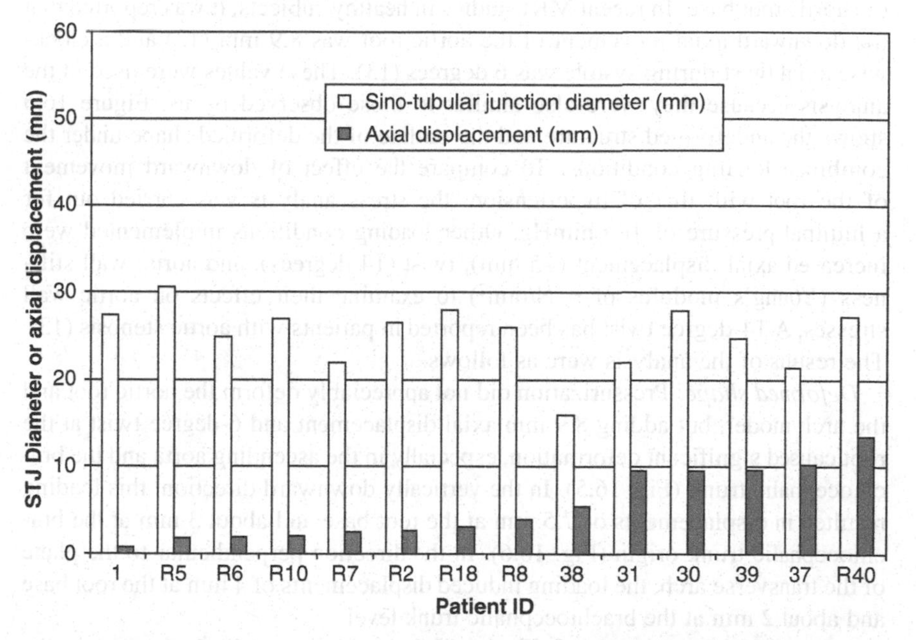

FIGURE 16.4. Sino-tubular junction (STJ) diameter and axial displacement in patients. For a similar STJ diameter, some patients have much larger axial displacements than others.

(11). The aortic root and the ascending aorta were modeled as curved cylinders by prolongation of the arch geometry. The actual shape of the sinuses of Valsalva was not represented.

The dimensions of the arch and proximal regions of the branches in the arch were measured previously from a silicone mold of a normal human aorta. The radii of the arch and the aorta were 37 mm and 12.2 mm, respectively. The aortic wall thickness (mean of 1.2 mm), the radii of the branches, and other details of the geometry were implemented as determined in our laboratory (see Chapter 7).

The orientation of the arch was determined in a healthy volunteer from a 3D-MRI reconstruction of the aortic arch and supra-aortic vessels. The aortic arch was found to lie approximately in a vertical plane oriented 20 degrees antero-posteriorly. The 3D reconstruction also established the geometry of the branches of the arch.

Finite elements: The geometry was discretized into 14,707 brick elements (Fig. 16.5). The material properties of the aortic wall were modeled to be homogeneous, incompressible, linear elastic, and isotropic, with a Young's modulus of 3 N/mm^2 and a Poisson's ratio of 0.49. In order for the aorta to deform in a physiological way, the distal ends of the supra-aortic vessels and of the aorta were fixed, and stiffer material properties (Young's modulus of 12 N/mm^2) were used to increase the influence of the tethering at the distal ends of the branches.

Loading: A luminal pressure of 120 mmHg was the load in the control model. Then, additionally, 8.9 mm axial displacement and 6-degree twist were applied to the aortic root base. In recent MRI studies in healthy subjects, it was reported that the downward axial movement of the aortic root was 8.9 mm (12) and a clockwise axial twist during systole was 6 degrees (13). These values were used in the analysis because they were also similar to those observed by us. Figure 16.5 shows the undeformed structure and the outline of the deformed shape under the combined loading conditions. To compare the effect of downward movement of the root with that of hypertension, the stress analysis was carried out for a luminal pressure of 180 mmHg. Other loading conditions implemented were increased axial displacement (15 mm), twist (14 degrees), and aortic wall stiffness (Young's modulus of 6 N/mm^2) to examine their effects on aortic wall stresses. A 14-degree twist has been reported in patients with aortic stenosis (13). The results of the analysis were as follows.

Deformed shape: Pressurization did not appreciably deform the aortic root and the arch model, but adding 8.9 mm axial displacement and 6-degree twist at the root caused significant deformation, especially in the ascending aorta and the brachiocephalic trunk (Fig. 16.5). In the vertically downward direction, this loading resulted in displacements of 7.5 mm at the root base and about 3 mm at the brachiocephalic trunk origin (Fig. 16.6). In the direction perpendicular to the plane of the transverse arch, the loading induced displacements of 4 mm at the root base and about 2 mm at the brachiocephalic trunk level.

Stresses with pressure load: Figure 16.7 shows the mechanical stress in the control model subjected to 120 mmHg pressure only. The stresses are averaged across the vessel wall thickness. The longitudinal and circumferential directions

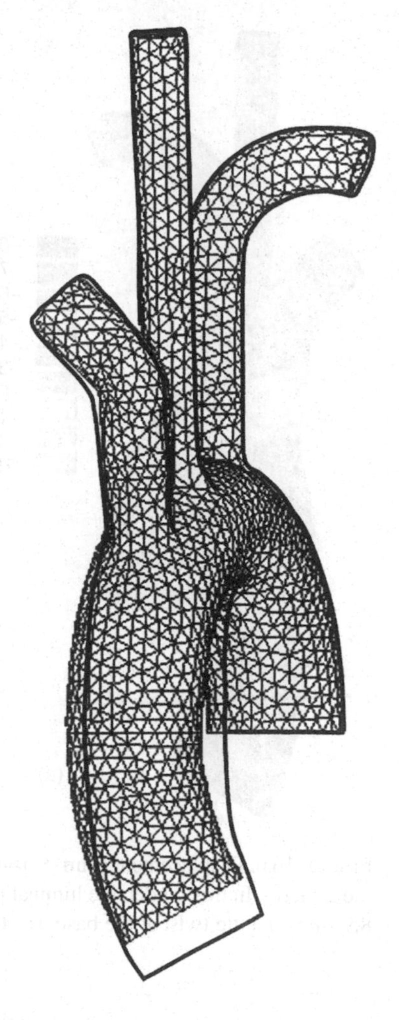

FIGURE 16.5. Front view of the finite element model of the aortic root, arch, and branches. The undeformed mesh is shown along with the deformed shape outline when 8.9 mm axial displacement and 6-degree twist were applied to the aortic root base. One may appreciate how the model deforms in space.

are defined with respect to the local orientation of the aorta. As expected, stress-concentrations were present at the ostia of the supra-aortic vessels. Between the brachiocephalic trunk and the left common carotid artery (LCCA), the circumferential stress was about 0.40 N/mm^2 and the longitudinal stress about 0.25 N/mm^2. Above the STJ, the circumferential and longitudinal stresses in the aortic wall were 0.32 and 0.21 N/mm^2, respectively (Table 16.3).

At a pressure of 180 mmHg, the stress between the origin of the brachiocephalic trunk and the LCCA was about 0.68 N/mm^2 in the circumferential direction and about 0.27 N/mm^2 in the longitudinal direction. Above the STJ, the circumferential and longitudinal stresses were 0.49 and 0.34 N/mm^2, respectively.

Stresses with pressure load and downward aortic root movement: At 120 mmHg luminal pressure, the circumferential and longitudinal stresses did not change markedly between the brachiocephalic trunk and the LCCA when aortic

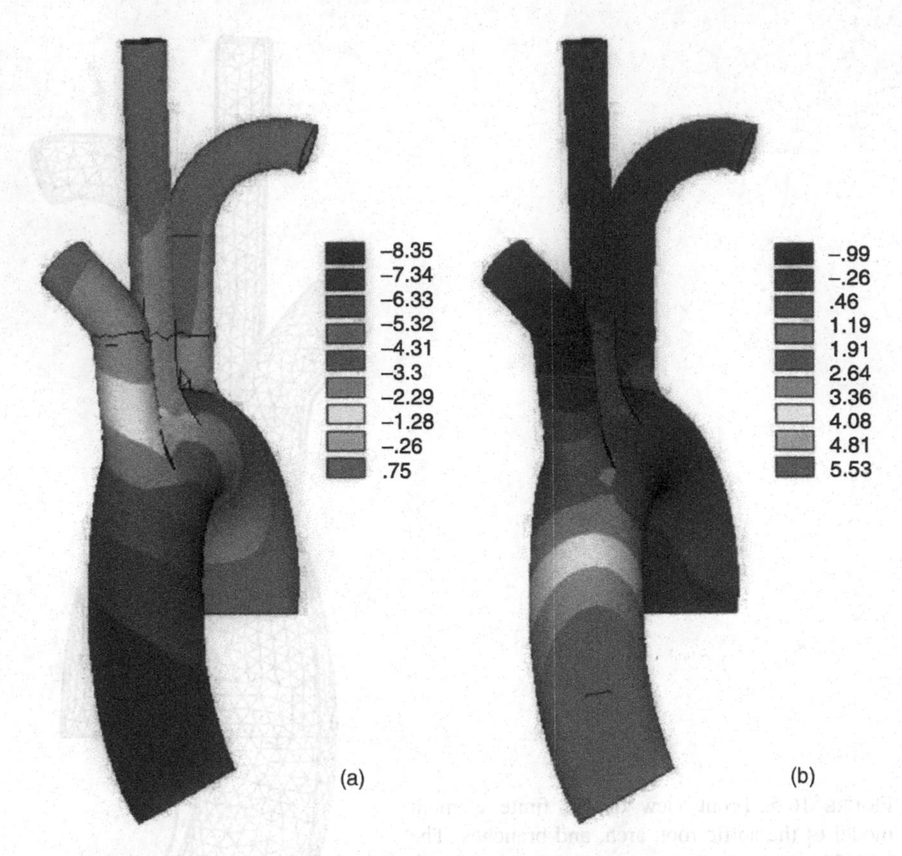

FIGURE 16.6. Displacement (mm) vertically downward (a) and perpendicular to the plane of the arch (b). In this model, the luminal pressure was 120 mmHg, the axial displacement was 8.9 mm, and the twist at the base was 6 degrees. (*Please see color version on CD-ROM.*)

root movement was applied. The area where the most significant changes occurred was about 2 cm above the STJ: while the circumferential stress was unchanged, the longitudinal stress increased by 50%, up to 0.32 N/mm^2 with 8.9 mm axial displacement. At the largest value of axial displacement (15 mm), the longitudinal stress was further increased to 0.47 N/mm^2. It ultimately reached 0.64 N/mm^2 when the stiffness of the aorta was doubled to simulate the rigidity one may encounter in older subjects (Table 16.3).

Similar results were observed with a pressure of 180 mmHg. Adding 8.9 mm axial displacement increased the longitudinal stress above the STJ from 0.34 N/mm^2 to 0.41 N/mm^2. It rose further to 0.56 N/mm^2 with 15 mm axial displacement, or to 0.57 N/mm^2 when the stiffness of the aorta was doubled with the same displacement.

Stiffening the aorta alone to simulate the aging process did not markedly increase the stresses in the aortic wall, except when aortic root motion was introduced,

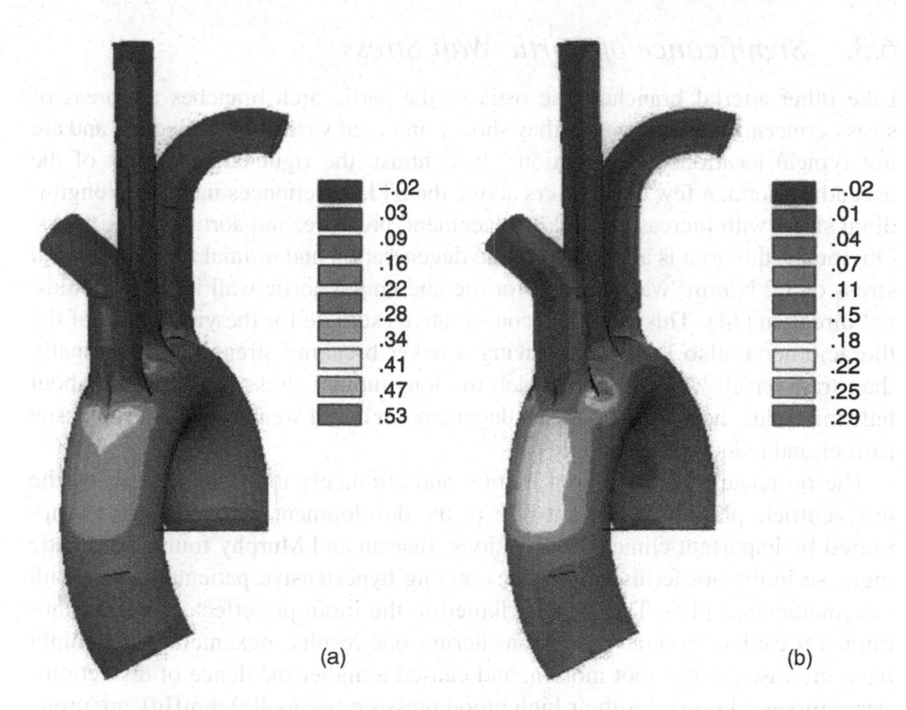

FIGURE 16.7. Distribution of circumferential (a) and longitudinal (b) stresses (N/mm^2) in the aortic arch. In this control model, the only load was 120 mmHg luminal pressure. Expected stress-concentrations around the ostia of the supra-aortic vessels were observed. (*Please see color version on CD-ROM.*)

which increased the longitudinal stress in the ascending aorta. The twist was found to exert only a small effect on the deformation of the aorta and, even when the value of 14 degrees was implemented, it did not alter the circumferential and longitudinal stresses significantly (Table 16.3).

TABLE 16.3. Wall stress in the ascending aorta approximately 2 cm above the sinotubular junction.

Pressure (mmHg)	Axial displacement (mm)	Twist (°)	Circumferential stress (N/mm^2)	Longitudinal stress (N/mm^2)
120	0	0	0.32 ± 0.03	0.21 ± 0.02
120	8.9	0	0.34 ± 0.05	0.32 ± 0.04
120	8.9	6	0.35 ± 0.04	0.32 ± 0.04
120	8.9	14	0.34 ± 0.04	0.31 ± 0.04
120	15	0	0.31 ± 0.06	0.47 ± 0.06
Stiff 120	15	0	0.33 ± 0.08	0.64 ± 0.10
180	0	0	0.49 ± 0.05	0.34 ± 0.03
180	8.9	0	0.47 ± 0.06	0.41 ± 0.05
180	15	0	0.47 ± 0.07	0.56 ± 0.07
Stiff 180	8.9	0	0.42 ± 0.06	0.57 ±0.07

In "Stiff" models, the elastic modulus of the aorta was 6 N/mm^2 instead of 3 N/mm^2 in the other models.

6.3. *Significance of Aortic Wall Stress*

Like other arterial branches, the ostia of the aortic arch branches are areas of stress-concentration. However, they show reinforced vascular architecture and are not typical locations of dissections. In contrast, the right lateral aspect of the ascending aorta, a few centimeters above the STJ, experiences increased longitudinal stress with increased axial displacement, pressure, and aortic wall stiffness. Obviously, this area is at a risk of tissue degeneration and intimal rupture. A yield stress of 1.2 N/mm^2 was reported for the abdominal aortic wall in the longitudinal direction (14). This may be a conservative estimate for the yield stress of the thoracic aorta, also known for having a lower breaking strength longitudinally than transversely (15). In our model, the longitudinal stress reached only about half this value, however, the tissue degeneration could weaken the arterial tissue further and reduce the yield stress.

The postulate that aortic root motion and ultimately the force applied by the left ventricle play an important role in the development of dissections is supported by important clinical observations. Beavan and Murphy found a dramatic increase in the aortic dissection rates among hypertensive patients treated with hexamethonium (16). This was attributed to the inotropic effect of hexamethonium on cardiac contractions. Considering our results, hexamethonium might have increased aortic root motion, and caused a higher incidence of dissections. Also, turkeys, known for their high blood pressure (up to 400 mmHg), are prone to aortic rupture. In susceptible flocks, aortic ruptures can be prevented by administration of reserpine in a dosage with little effect on the blood pressure but effectively depleting cardiac catecholamines and reducing cardiac contractility (17).

It is important to note that root motion alone is only an indicator of the force that the heart exerts upon the aorta. Thus, a large aortic root displacement may be well tolerated in a compliant aorta or may cause rupture in a stiffer aortic tissue. The aortic root motion has direct influence on the mechanical stresses acting upon the aorta. The longitudinal stress increased critically in the ascending aorta above the STJ. This may explain why circumferential intimal tears and aortic dissections occur more often at this location. Downward natural displacement of the aortic root appeared to be as much of a risk factor for dissection as hypertension. Increased stiffness was found to enhance the effects of root motion on aortic wall stress. For all the above reasons, in patients possibly at risk of dissection, aortic root movement should be monitored and thought of as an additional risk factor.

The model described above has several limitations, but it is the first model that considered the aortic root, arch, and supra-aortic vessels and also considered the effect of root movement, hypertension, and wall stiffness on the wall stress. Furthermore, it considered the implication of the increase in the wall stress due to root motion for aortic dissection. Patients who have conditions such as hypertension and/or Aortic Insufficiency would further be associated with increased wall stress from pressure load and increased root motion and thus may have higher risk of dissection. It is therefore obvious that more studies in vascular mechanics, of the kind considered above, are needed if we are to understand the pathogenesis of aortic dissection.

References

1. Svensson LG, Crawford FS: Aortic dissection and aortic aneurysm surgery: Clinical observations, experimental investigations, and statistical analyses part II. In: Wells SA Jr (ed), Current Problems in Surgery. Mosby Yearbook, Chicago, 1992;XXIX(12): 917-1011.
2. DeSanctis RW, Doroghazi RM, Austen WG, Buckley MJ: Aortic dissection. N Engl J Med 1987;317(17):1060-1067.
3. Roberts WC: Aortic dissection: Anatomy, consequences, and causes. Am Heart J 1981;101(2):195-214.
4. Wheat MW Jr: Acute dissection of the aorta. Cardiovasc Clin 1987;17(3):241-262.
5. Sabiston DC, Spencer FC: Surgery of the Chest, Fifth ed., Vol. II. W.B. Saunders, 1990:1182-1205.
6. Haverich A, Miller DC, Scott WC, Mitchell RS, Oyer PE, Stinson EB, Shumway NE: Acute and chronic aortic dissections-determinants of long-term outcome for operative survivors. Circulation 1985;72(suppl II):II-22-II-34.
7. Doroghazi RM, Slater EE: Aortic Dissection. McGraw-Hill, New York, 1983.
8. DeBakey ME, McCollum CH, Crawford ES, Morris GC Jr, Howell J, Noon GP, Lawrie G: Dissection and dissecting aneurysms of the aorta: Twenty-year follow-up of five hundred twenty-seven patients treated surgically. Surgery 1982;92(6):1118-1134.
9. Schlatmann TJM, Becker AE: Pathogenesis of dissecting aneurysm of aorta. Comparative histopathologic study of significance of medial changes. Am J Cardiol 1977;39:21-26.
10. Beller CJ, Labrosse MR, Thubrikar MJ, Robicsek F: Role of aortic root motion in the pathogenesis of aortic dissection. Circulation 2004;109:763-769.
11. Liotta D, Del Rio M, Cooley DA, et al.: Diseases of the Aorta. Domingo Liotta Foundation Medical, Argentina, 2001:1-22.
12. Kozerke S, Scheidegger MB, Pedersen EM, et al.: Heart motion adapted cine phase-contrast flow measurements through the aortic valve. Magn Reson Med 1999;42: 970-978.
13. Stuber M, Scheidegger MB, Fischer SE, et al.: Alterations in the local myocardial motion pattern in patients suffering from pressure overload due to aortic stenosis. Circulation 1999;100:361-368.
14. Raghavan ML, Webster MW, Vorp DA: Ex-vivo biomechanical behavior of abdominal aortic aneurysm: Assessment using a new mathematical model. Ann Biomed Eng 1996;24:573-582.
15. Mohan D, Melvin JW: Failure properties of passive human aortic tissue, II: biaxial tension test. J Biomechanics 1983;16:31-44.
16. Beaven DW, Murphy EA: Dissecting aneurysm during methonium therapy: A report on nine cases treated for hypertension. Br Med J 1956;14:77-80.
17. Carlson CW: Further studies with reserpine for growing turkeys and laying hens. In: Second Conference on Use of Reserpine in Poultry Production. St. Paul, MN: 1960:25.

Index